FUNDAMENTALS OF DIMENSIONAL METROLOGY

WILKIE BROTHERS FOUNDATION

TED BUSCH

DELMAR PUBLISHERS INC.
2 COMPUTER DRIVE, WEST — BOX 15-015
ALBANY, NEW YORK 12212

WILKIE BROTHERS FOUNDATION
Copyright © 1964
Third Edition 1966

LIBRARY OF CONGRESS CATALOG NUMBER: 64-12593
ISBN 0-8273-0193-6

Printed in the United States of America
Published simultaneously in Canada
by Nelson Canada,
A Division of International Thomson Limited

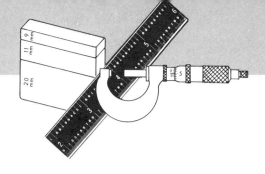

Preface

This is a text about technique, not about hardware. Its objective is to develop dimensional measurement ability for skilled workers, inspectors, technicians and for those going on into engineering and science. Its premise is that these skills can be acquired best when knowledge accompanies practice.

There are many books that treat dimensional measurement as an aside to other disciplines from machine operation to quality control. Here the primacy of measurement is recognized as the common element without which there could be no crafts, no manufacturing, and no science. This has long been recognized in England, Europe and the Soviet Union. The author has borrowed freely from their experience.

Communicative, as well as manipulative, aspects are stressed because the semantic differences separating the production worker, skilled worker, inspector, gagemaker, engineer, quality-control person, metrologist, and research worker are now a recognized deterrent to continued progress. Although these subjects are not taught as such in this text, the student will become familiar with the principles, terminology, and instrumentation common to them all. Recognizing that this book will be used in terminal courses, and that its language will carry into employment, the recently adopted decimal-inch terminology has been used wherever possible. The science student need only make a few transpositions, such as "part" to "object", "inspector" to "observer", to adapt the material for his needs. Metric information is included for the immediate needs of the science student and to prepare the industrially oriented student for the possible extension of the metric system in the United States. The relationship between systems and commercial standards is shown.

The book begins with the reasons for communication by measurement. The basic considerations are developed first with the lowest-precision instruments. Each following chapter then advances these basic considerations to successively higher levels. The importance of the standard of length, cleanliness, alignment, and recognition of uncertainty is reinforced at each level. Practical applications are included throughout to give relevance to the theoretical aspects.

Two devices have been used to adapt the book to the diverse curriculums to which it is directed. The chapters are subdivided into smaller units which are nearly complete within themselves. Although this necessitates some duplication, it enables the educator to select as much or as little as suits his needs. Similarly, one instrument is chosen in each area to best explain the basic considerations. The laboratory experiment book, however, permits the hardware aspects to be treated as extensively as each curriculum requires.

Controversial subjects have neither been avoided nor dispensed with by doctrinaire pronouncements. The student is told of the conflicting views and invited to join in the arguments.

A casual style is attempted, including free use of vernacular measurement terms and anecdotes where appropriate. No attempt, however, has been made to write down to the least interested student who will probably make the least use of the material anyway. Previous knowledge of geometry, algebra and trigonometry is very desirable, but not essential.

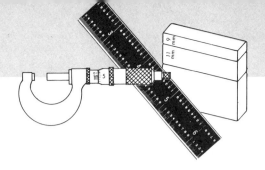

Measurement is a highly systematized discipline relying heavily on laws both informal and statutory. Recognition of these laws is vital even when the understanding of them cannot be fully expected. Therefore, attention is called to applicable federal and trade standards. To the best of the author's knowledge, nothing in the book is contrary to these standards, although they are now so complex that there are, undoubtedly, oversights.

Invaluable help was obtained from the manufacturers of measurement instruments. They have been credited along with their contributions.

It would be impossible to acknowledge all who have helped produce this text. The unique contribution of the Wilkie Brothers Foundation deserves a special mention. The interest of the Wilkies, Robert, James and particularly Leighton A. Wilkie, must have been contagious, judging by the support received by many who generously gave time and advice. Their farsighted recognition of measurement problems and determination to help solve them has made this book possible.

Particular mention must be made of Fred Colvin, whose books kindled my original interest in technical writing; of James Anderson, whose texts served as my standards; and of John Harrington, the distinguished metrologist who inspired my full-time activity in this field.

Many educators provided valuable help in formulating the text. Among these are David Allen, University of Calif.; Wm. C. Belvel, Sierra College; H. M. Black and John Morrison Green, Iowa State University; Dean John A. Jarvis, Stout State College; Walter B. Jones, University of Pennsylvania; Prof. H. H. London, University of Missouri; A. E. Mills, Hamilton Institute of Technology; Ronald F. Moon, Moline Senior High School; John R. Plenke, Wisconsin State Board of Vocational and Adult Education; Arthur J. Plourde, Metrology Training Director, Bureau of Naval Weapons; and Dennis H. Price, University of Cincinnati. Those concerned with training in industry were equally helpful. Although president of Professional Instruments Co., T. J. Arneson devotes much time to apprentice training activities. His views were essential. Earl D. Black of General Motors Institute devoted many hours of his time.

Beyond the basic objectives, the text is eclectic with contributions from many sources. For example, the late Clifford W. Kennedy's treatment of references has been borrowed from his Inspection and Gaging. From John W. Greve, Editor of Tool & Manufacturing Engineer, came the attention to language. On his staff is Dr. Felix Giordano whose fusing of the humanities with science I have flagrantly plagiarized wherever possible. The target analogy was borrowed from J. M. Juran whose name is almost synonymous to quality control.

Fundamentals of Dimensional Metrology is characterized by its use of decimal-inch notation. This resulted from the work of the B-87 joint committee of the American Society of Tool and Manufacturing Engineers and American Standards Association headed by Roy P. Towbridge, Director of Engineering Standards of General Motors Corporation. Col. Leslie S. Fletcher, Standards Director of the

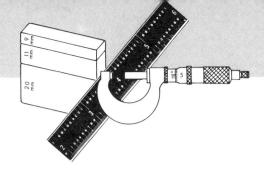

ASTME, steered me through the committee's findings. A committee member, Russell Hastings of Clark Equipment Co., has left an unmistakable imprint on <u>Fundamentals of Dimensional Metrology</u>. This is true also of Al Huff of Bell and Howell Co. Being nearby, he bore the brunt of my inquiry.

Not connected with the B-87 Committee and on the opposite side of the world was Dr. Robert C. Gilles whose measurement erudition was tapped freely. The metric issue brought strong response from many quarters. For example, L. R. Booker of Clemson Agricultural College added encouragement to face this issue.

Many other experts will find their philosophy included, correctly I hope. Joe Moody of Sandia Corp. influenced the emphasis on positional considerations. Everyone in metrology will expect that he also influenced reference surface considerations. Professor T. G. Beckwith of the University of Pittsburgh must be credited for his help with the important study of errors. Albert M. Dexter of Bausch and Lomb, Inc. made available his papers and helped resolve some of the questionable points about the use of gage blocks. Peter Flauter of Chicago Dial Indicator Co. remained on call throughout the writing of the chapters on comparison measurement. Much material throughout the text was furnished by Wm. Wingstedt, Editor of <u>Quality Assurance</u>.

Heartfelt gratitude is due H. Richard Fick, President, and Otto J. Dundr, Vice President, of Dundick Corporation. Both were extremely patient and placed their long experience in metrology freely at my disposal. Lastly, but by far the most patient of all, was my wife, Sené. For two years she has given up social life in order to stand by and fan me and feed me so that this work could become a reality.

It is recognized that this text will not fill all teaching needs and, alas, that there may be errors in it. Feedback from both students and teachers is earnestly solicited. It will assure improvements in future revisions.

Chicago, Illinois   Ted Busch

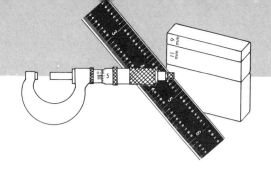

Contents

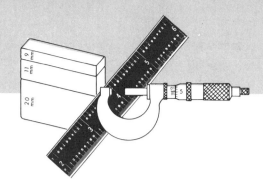

Although often dwarfed by the task, man controls his works through measurement.
This is symbolized by this inspector checking the bulkhead of a giant tanker.
(Merrill Lynch, Pierce, Fenner & Smith)

The Reasons for Measurement
The Universal Language

Metrology is the science of measurement, and measurement is the language of science. It is the language we use to communicate about size, quantity, position, condition and time, as shown in Fig. 1-1.

A language consists of grammar and composition. Grammar is a science; composition, an art. This book unfolds the grammar of measurement, but only experience will develop the art. You will find the language of measurement much easier than French or Russian, because you already have considerable skill in it. Properly applied, this skill can decrease your effort and improve your results in work, hobbies or continued studies.

There are three reasons that we all need to communicate in measurements. First of all, we need measurements to make things, whether the things we make are our own brain children or somebody else's. This applies to all skilled workers and artisans.

Second, we need measurements to control the way other people make things. This applies to ordering an engagement ring, fencing a yard, or producing a million spark plugs. Third, we need it for scientific description. It would be impossible for you to give definite information to someone else about hull designs, electron mobility or the plans for Saturday night without measurements.

MEASUREMENT TO MAKE THINGS

Less than 200 years ago, James Watt was jubilant that a certain John Wilkinson had perfected a boring mill with such great precision that it could bore the cylinders for steam engines to a tolerance of "one thin shilling." This unheard-of accuracy made Watt's dream of the steam engine come true. This same accuracy in the cylinders of your car today would make it an intolerable oil burner, if it ran at all.

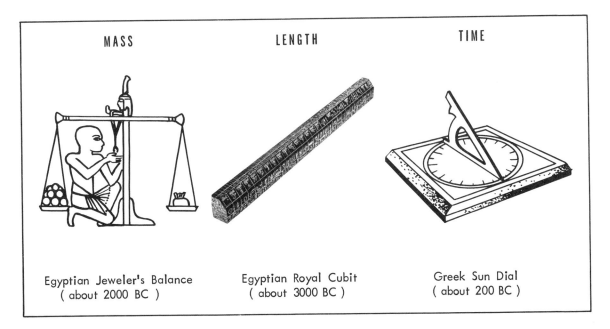

MASS	LENGTH	TIME
Egyptian Jeweler's Balance (about 2000 BC)	Egyptian Royal Cubit (about 3000 BC)	Greek Sun Dial (about 200 BC)

Fig. 1-1   If "science is measurement," then without metrology there is no science. This has been recognized since early in man's history as these early examples of fundamental quantities show. Each era produced its equivalents.

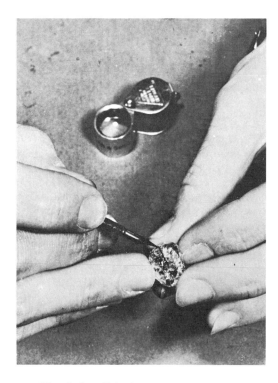 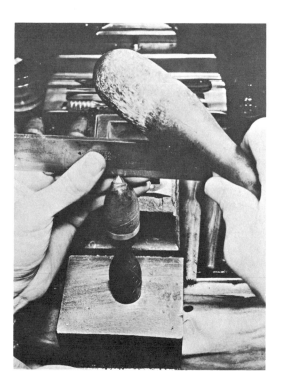

Fig. 1-2   This diamond cutter represents one of the highest of all skills. Millions of dollars depend upon his practiced hands and eyes . . . and his measurements. *(Harry Winston, Inc.)*

Apart from the role of measurement in mass production, measurement is necessary whenever we make anything. The Indian making a flint arrow had an approximate idea of the size needed to do the job. This is true in all the crafts and skilled trades.

Unless numerical values are established, things that must fit together can be made only by trial and error. Even today, some fitting is expected on one-of-a-kind jobs such as making a complex die, outfitting an ocean liner, or rebuilding an automobile engine. Measurement skill reduces hand fitting. The better your ability to measure, the faster you can turn out skilled jobs. Of course, the measurements required in one field may be very different from those required in another, Fig. 1-2; but *no craftsman can be higher skilled than his ability to measure.*

Strictly speaking, metrology is the measurement of mass, length and time. From these, all of the quantities involved in mechanics, electronics, chemistry, and hydraulics are derived. This book is restricted to *mensuration.* The dictionary defines mensuration as that branch of applied geometry concerned with finding the length of lines, areas of surfaces, and volumes of solids from certain simple data of lines and angles. This includes those measurements which are required to use tools and instruments for designing, building, operating and maintaining material objects, whether they are refrigerators or cyclotrons. This is the reason for the term *dimensional metrology.*

MEASUREMENT TO CONTROL MANUFACTURE

An extension of making things yourself is the control of manufacture by others. This is the role of inspectors and quality control personnel.

There always has been an inspection department in industry, even though the old-timers will speak of the good old days when there was none. In those days, the machinist would build an article the same way that you build storage shelves into a closet — cut and try until the parts fit. The old-timer did not recognize it, but he, himself, was the inspection department.

Because of his pride as a workman, ambition to get ahead, or fear of getting fired, he set a pace and quality in keeping with the end use. That worked for one article, but fell hopelessly behind when many were required.

In order to mass produce, specialists developed who did only a portion of the job, but who did it well and fast. This led to the "piece-part" system of manufacturing in which the worker's pay was determined by the number of parts completed. Naturally, he was paid only for the satisfactory parts. For this, he required gages and the services of an arbiter who would pass on what was acceptable and what was not. Thus, the role of the inspector was created.

Interchangeable manufacture did not stop at the factory walls. Parts might be made in widely scattered plants and even in different countries. The result was a refinement of the language of measurement. Not only was the ability to measure to very small dimensions required, but the measurements had to be based upon an accepted standard or the parts would not interchange, Fig. 1-3. This *standard of length* became as important as the means of measuring.

Now, at the threshold of the space age, "good enough" has been pushed beyond the borders of belief. Yesterday's standards and instruments can fulfill only their new functions through the imagination and ingenuity of the new breed of industrial specialists, the quality control people. Working with statistics and electronic computers, they regulate every phase of modern life which requires mass-produced products.

However, no matter how fast their computers can multiply ten-digit figures or how specifically their calculations can predict production, everything the quality control

Fig. 1-3   The brains of machines such as this are the measurements built into its components and circuits. Only accurate measurement detects malfunction – and corrects it. *(Cincinnati Milling Machine Co.)*

Fig. 1-4 This computer is limited by the accuracy of the data fed into it, most of which are measurements. Its enormous speed requires fast data collection. *(Control Data Corp.)*

people do depends upon the information they receive. These data are the measurements provided by the inspector. He is crucial to the entire arrangement. He is a technician of the highest type. His teammate, the quality control man or woman, is a recognized professional. Each depends upon the other, and we all depend upon both of them.

These are the new aristocrats. Their responsible employment is needed at every level from production inspection to comparing master standards to the wave lengths of light. Those who specialize in measurement lead adventurous lives. They are present at both the births and deaths of new products, new scientific achievements, and new campaigns. Few decisions are made without their valued counsel.

MEASUREMENT FOR PROGRESS

Man's progress has been paced by his ability to measure. This is even truer today than in antiquity. Measurement is truly a universal language. Where every other communication must be translated, all industrial people today recognize the same standards of length, and convert in and out of each other's systems of measurement. The time is near when even the systems may be the same.

This has been due largely to industrial progress, but it is needed as much in pure science as in applied science. There is no way one research worker can repeat the work of another without specific measurements, Fig. 1-4. This is quite true throughout all branches of science, from astronomy to biology. (In fact, the micrometer was invented for use in astronomy, not shop work.) For many years, progress was slowed and goods made more expensive by the "zero of ignorance." This refers to the extra decimal place added to the tolerance of a part because the designer was not sure how accurate the dimension really had to be. In few fields is that old luxury still possible. In most products today the quality control instruments are already at their limits--there is no "zero of ignorance" left.

As never before, the ability of an engineer, a chemist, a biologist, or a physicist to test his ideas hinges upon his understanding of measurement.

A LOOK AHEAD

As shown, there are three roles for measurement. Although each is different, they all use the same system of measurement and the same instruments. Moreover, they overlap.

The skilled diemaker might use his micrometer to measure a given part to one-half thousandth of an inch. His experience and skill may enable him to do so with a high degree of *reliability;* but if the production of that same part is to be *controlled*, Fig. 1-5, to the same reliability, a much more *sensitive* instrument must be chosen to replace the skill of the diemaker. Furthermore, the physicist might be able to use the same micrometer, but may also have to determine

the atmospheric pressure, temperature or some other variable before the measurement will be suitable for his purpose.

The following chapters will, first of all, assure that each of these three groups means the same thing when they refer to their measurements. The standard measuring instruments will be discussed in the order in which you probably will encounter them, from the least precise to the most precise. Far more important, some rules will be developed to guide you in the selection of the best instrument for any given dimensional measurement.

Methods will be emphasized, rather than products. Most manufacturers of gaging products have ample literature about the use of their specific instruments; and many thorough books which discuss the use and care of all imaginable measuring instruments and a few unimaginable ones are available. Frequent reference will be made to these books for specific problems.

Methods, rather than products, will be emphasized for still another reason. Industry and science finally are aware of their dependency upon precision measurement. Improvements in measuring instruments are rapidly being made. It is expected that many more will come from you, the users of this book. Tomorrow's instruments will not look much like today's. The principles will remain as true tomorrow as when they were evolved long ago. Understand the principles and get some actual experience. The rest will take care of itself.

TERMINOLOGY

● Reliability is the likelihood that actual performance will be the same as planned performance.

● Control in quality control means that the variables that affect reliability are known and performing according to plan.

Fig. 1-5 These two terms are very useful in describing measurement applications. They will be expanded as the text moves to more complex considerations.

QUESTIONS

1. What are the three general reasons that measurement is needed?
2. Are one-of-a-kind jobs outmoded? Explain.
3. What are the fundamental quantities with which metrology is involved?
4. Why is metrology restricted to these fundamental quantities?
5. Which of the fundamental quantities defines hydraulic pressure?
6. What measurements are included in dimensional metrology?
7. What is an example of a measurement not included in dimensional metrology?
8. When was the inspection function begun?
9. When did the inspector as a separate industrial skill emerge?
10. Upon what information does quality control depend?
11. Upon what information does scientific research depend?
12. Why can a skilled diemaker measure closer with a micrometer than an unskilled worker?
13. Why, in the example, was it mentioned that the physicist may have to determine atmospheric pressure and other variables before he can use the micrometer reading?
14. Why does this text emphasize methods rather than specific measurement devices?

DISCUSSION TOPICS

1. Measurement has been compared to language. There are many languages used throughout the world. Why, then, is it insisted that there must be universal agreement on measurement?
2. Was the tolerance of "one thin shilling" a fine tolerance or a coarse tolerance?
3. Explain the reason why a craftsman can be no higher skilled than his ability to measure.
4. Explain why temperature is not included as a fundamental quantity.
5. Explain the reason that mass production focussed attention on the standard of length.

> *That when you can measure what you are speaking about, and express it in numbers, you know something about it: but when you cannot measure it, when you cannot express it in numbers, your knowledge is of a meager and unsatisfactory kind; it may be the beginning of knowledge, but you have scarcely, in your thoughts, advanced to the stage of science, whatever the matter may be.*
>
> Lord Kelvin

The Language of Measurement
Let's Understand Each Other

A child is supposed to have asked President Lincoln how long a certain man's legs were. He is reported to have answered, "Long enough to reach from his body to the ground". While language in this case circumvented a measurement problem, it all too often causes them. This you experience daily.

You have your automobile engine bored for larger pistons. After burning some rubber to show that you cut the 0 to 60 mph time by 2 seconds, someone asks you how accurately the new pistons must fit for that kind of flashing performance. How do you answer?

You might answer: "Oh, just for a *good fit* with the cylinders". Or, "The maximum clearance cannot be more than .002 in. or less than .001 in". Or, "It's got to be one of Bob Smith's jobs. The Speed Shop's work isn't good enough". Or, more commonly "Dead Nuts!" Which of these answers is funny and which contains the desired information depends upon what was meant by the word *accurate* in the question.

Consider another everyday word, *size.* We all know what is meant by "2-in. sq.," as in Fig. 2-1. Specifying a 2-x 3-in. rectangle is also clear. But what is the size of C or D in the Fig. 2-1? Furthermore, in A and B, is knowing the length of the sides and the general shape enough to specify the sizes or do

we also need to know *how square?* and again, *how rectangular? ***

Unfortunately, millions of dollars are wasted yearly because the machine operator, machinist, supervisor, inspector, quality-control person, design engineer and physicist may all mean something different for such everday terms as accuracy and size.

Not long ago, outside of the metrology laboratory, the terms needed were few, and all inclusive. *Accurate* and *precise* were the same and often synonymous with *reliable* and *repeatable. Measure, inspect* and *gage* all meant the same thing. *Standardize* meant to make a private agreement with your boss that all parts hereafter would be thus and so. *The inch* meant *the inch*--you had better believe it, and forget all about the article some odd-ball wrote that there were over 300 standard inches in use in the United States alone.

The situation is complicated today in that dimensional measurement now requires the use of electronic, optical instruments and interferometry in addition to the time-honored mechanical methods.

---

*Wolowicz, C.S., Rindone, C.P., Wilcox, R.J., "Axioms of Drafting Technology - Part V", American Machinist / Metalworking Manufacturing, August 6, 1962, p. 81.

WHAT IS THE SIZE?

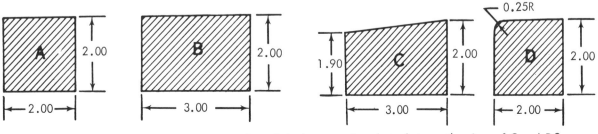

Fig. 2-1 A is a 2-inch square, B a 2- x 3-inch rectangle - but what are the sizes of C and D? Only by dimensions can we define size accurately.

There is not yet universal agreement about the meaning of the various terms. Although there is always a justifiable reluctance to introduce new terms, the only way for the confusion about measurement to be settled is for the inspector to adopt some of the metrologist's and quality-control man's terminology. The terms used in this book are consistent with all of these related fields. Moreover, the recent adoption of the decimal inch standard gives us some much needed guidelines.

HOW BIG?

It will be easier to understand the terms if you have an idea about the sizes involved.

Comparisons are shown in Fig. 2-2. A graduation on the average machinist's rule is about 0.003-in. wide. This is *three thousandths* of an inch and is read *three mil* in decimal inch terminology. The eye can see light through a crack as small as 0.0001 in. This is *one tenth* of a thousandth of an inch. In ordinary usage, it is called "one tenth". One-tenth *inch* (0.10), of course, would be enormous in comparison. This confusion is eliminated by the decimal inch terminology. With it 0.0001 inch is called *point one mil* or *one hundred mike*. There is little chance for misunderstanding. Because of its importance, the decimal inch is discussed in the next chapter.

LENGTH COMPARISON

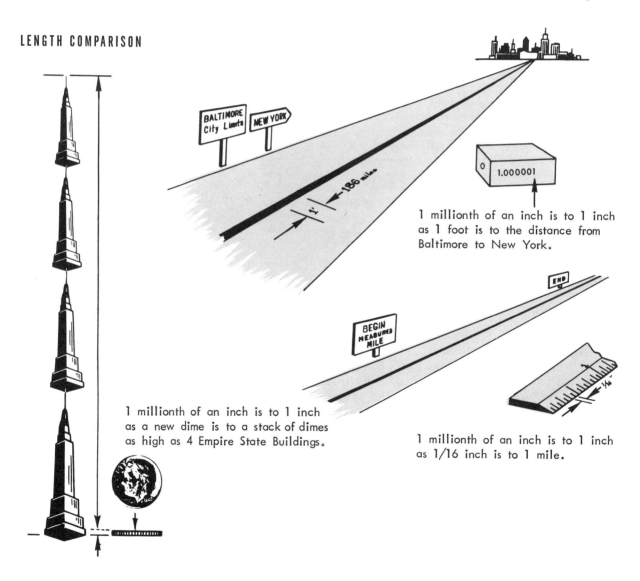

1 millionth of an inch is to 1 inch as a new dime is to a stack of dimes as high as 4 Empire State Buildings.

1 millionth of an inch is to 1 inch as 1 foot is to the distance from Baltimore to New York.

1 millionth of an inch is to 1 inch as 1/16 inch is to 1 mile.

Fig. 2-2   Comparison of relative lengths shows that in measurement we must consider widely different sizes.

HOW FAR APART?

This book is about measurement. Not all measurements, because these include the flow through pipes and the temperatures of stars, but dimensional measurements. Dimensional measurements are the ones that we use daily in designing, building and operating the objects that surround us, as well as for communicating about past, present and future objects.

The principal dimensional measurement is length. Closely related is the measurement of angles and curvature. An expression of any of these is referred to as a *dimension*. Using these we can describe surface finish, flatness and angular relationships between features. Hence, we can describe *size*. Excluding bulk products that are sized by volume or weight, you can describe shape without size but you cannot describe size without shape. In few situations would "She is a 110-lb. girl" be an adequate description.

A linear measurement is a means for expressing the distance separating two points. Note that it is *not the distance* but simply a way to *talk or write about the distance*. Unless it permits the distance to be reconstructed later, it has no use. A *measurement enables a distance to be reproduced*. A measurement as long as "the left feet of the first sixteen men to leave church on a certain Sunday" sounds absurd. Yet it fulfills our definition and was actually used in the 16th Century. One millionth of an inch is totally beyond our imagination, yet it, too, fulfills the definition.

The latter two measurements have two things in common. Each has a *unit of length* and a multiplier. The magnitude of the unit of length does not matter if everyone using it is in agreement. They seldom have been. These are the physical forms we use to preserve our units of length.

The second part of the statement of length is the multiplier. This may be cardinal, fractional, decimal or exponential, Fig. 2-3. The cardinal form simply states the number of times that the unit of length is multiplied. The fractional and decimal multipliers show the size of the parts to which the unit of length has been divided. The exponential form is a convenience when distances are very large or very small, Fig. 2-4. It is used extensively in scientific work because of the ease with which exponential forms may be used in involved calculations.

MEASUREMENT NOTATION

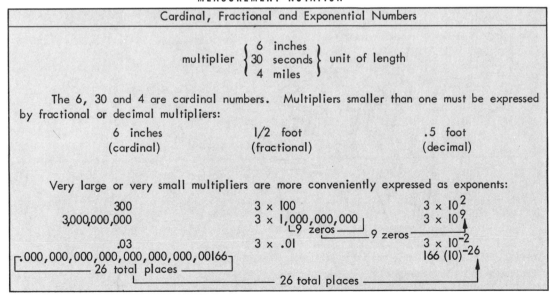

Fig. 2-3 Exponential multipliers are not only more compact, they are much easier to use in calculations. These are the reasons that they are widely used in engineering and science.

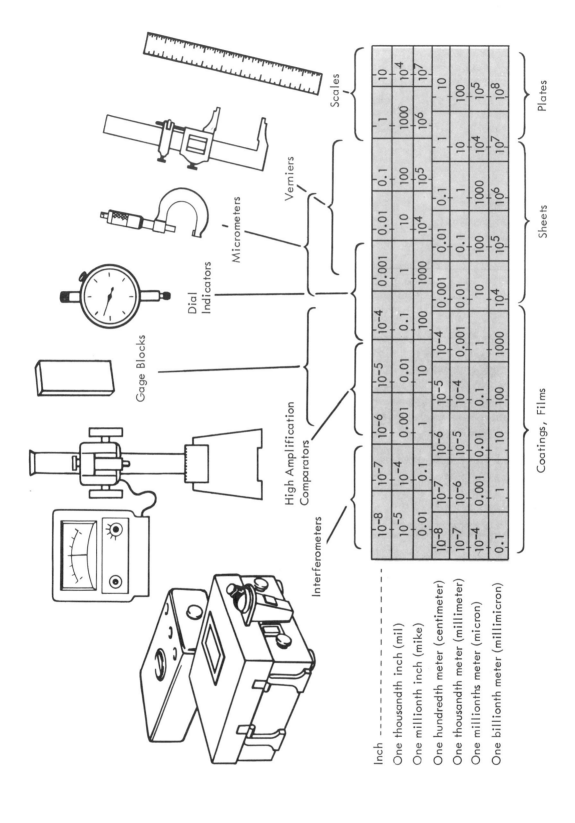

Fig. 2-4 Table comparing cardinal, fractional, decimal and exponential multipliers, showing this relationship to measurement.

FROM END TO END

Every measurement begins at the *reference point* (or reference end) and terminates at the *measured point* (jokingly called the "thinking" end for steel tapes). These are usually reversible, and lie along a *line of measurement*. The line of measurement has direction, and unless its direction is known in relation to the feature being measured, the measurement is of little value.

For a simple part such as shown in Fig. 2-5, the decision, which is the reference point and which is the measured point, need not be made. Together they are referred to as *references*. When you begin to measure it, convenience will dictate which is which. The left- and right-hand inspectors will disagree. When dimensions are interrelated the decision may be more complex as you will probably realize soon.

Every dimension can be traced back to a designer. Although he seldom thinks of himself in that role, often a skilled craftsman is the designer. A dimension is the perfect separation between two definable points,* called *features*, either on one part or between parts. The only problems are that in the real world there is no such thing as perfect, and points do not exist. The medieval dilemma about how many angels could stand on the point of a needle (scholars actually spent hours on this) was more theological than metrological. Real features on real parts are bounded by lines and areas, Fig. 2-6 -- and they are never perfect.

---

*Wolowicz, C.S., Rindone, C.P., Wilcox, R.J., "Axioms of Drafting Technology - Part VII", American Machinist / Metalworking Manufacturing, September 3, 1962, p. 101.

DIMENSIONS, FEATURES AND MEASUREMENTS

1. The designer's concept for a perfect part determined the DIMENSION.

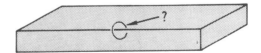

2. The toolmaker's machining resulted in the FEATURE of the part.

3. The inspector's MEASUREMENT verified the toolmaker's work to designer's concept.

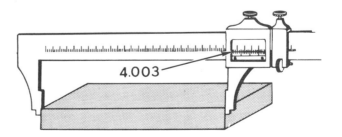

Fig. 2-5 Measurement verifies the designer's dimension to the feature of the actual part. This happens even when the designer, the machinist, and the inspector are all the same person, as is often the case.

DIMENSIONS AND PART FEATURES

DIMENSIONS MUST LIE ALONG THE LINE OF MEASUREMENT. . . . .

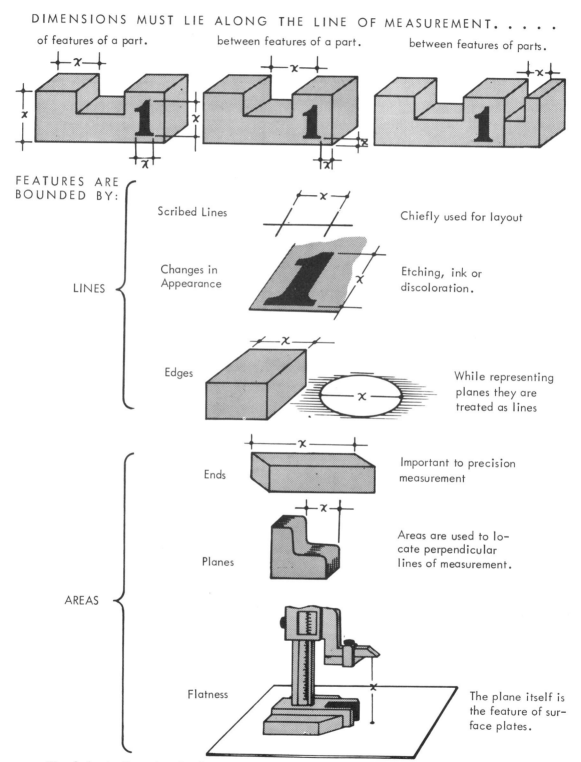

of features of a part.   between features of a part.   between features of parts.

FEATURES ARE
BOUNDED BY:

LINES

Scribed Lines   Chiefly used for layout

Changes in   Etching, ink or
Appearance   discoloration.

Edges   While representing
planes they are
treated as lines

Ends   Important to precision
measurement

Planes   Areas are used to locate perpendicular
lines of measurement.

AREAS

Flatness   The plane itself is
the feature of sur-
face plates.

Fig. 2-6 A dimension is the statement of intended size of a particular feature. A measurement is the statement of the actual size. Unlike many other terms, these can be interchanged without causing serious errors.

Most lines involved in measurement are actually edges formed by the intersection of two planes. When we discuss measurement instruments you will find that the various types of edges present very different measurement problems. To be certain that we are understood in discussing these, the familiar male and female terminology is applied, Fig. 2-7. For complete description, the *material condition* of the edge must be specified. An outside edge is less than 180° of material; an inside edge, more than 180°.

Of course, there are also angular dimensions, surface finish dimensions and others that we will get to later. Features always represent physical characteristics.

When a feature is known, it has a measurement. Thus, a feature has length and direction. Strictly it also has time and temperature. If a part is subject to wear, an old measurement of a feature may be larger or smaller than the dimension. If hot it will be larger than if cold. For most work the time and temperature can be overlooked, but their importance should never be forgotten when doing precise work. Much of this text is devoted to these points. First, however, let us deal with basic considerations.

EDGES AS MEASUREMENT LINES

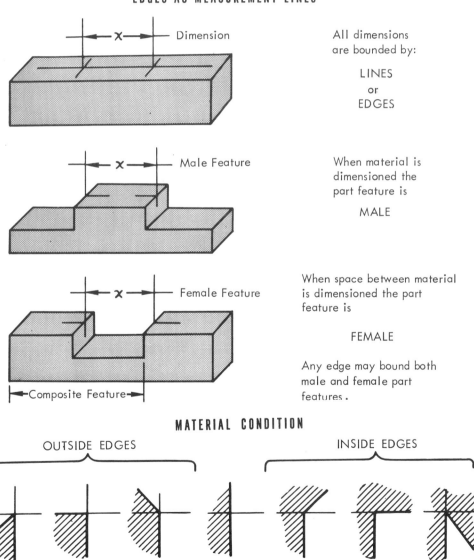

All dimensions are bounded by:

LINES
or
EDGES

When material is dimensioned the part feature is

MALE

When space between material is dimensioned the part feature is

FEMALE

Any edge may bound both male and female part features.

MATERIAL CONDITION

Fig. 2-7 Designating the dimension and the edges removes confusion about part features.

TARGET ANALOGY

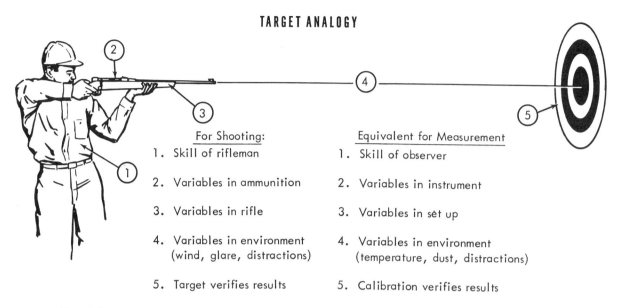

For Shooting:

1. Skill of rifleman

2. Variables in ammunition

3. Variables in rifle

4. Variables in environment (wind, glare, distractions)

5. Target verifies results

Equivalent for Measurement

1. Skill of observer

2. Variables in instrument

3. Variables in set up

4. Variables in environment (temperature, dust, distractions)

5. Calibration verifies results

Fig. 2-8 Measurement is influenced by variables similar to those which affect the score when target shooting. This comparison shows the difference in meaning of familiar terms.

ACCURACY, PRECISION AND RELIABILITY

If you were discussing architecture or freight routing, among many possible subjects, these words could be used interchangeably. In precision measurement their meanings differ enough to affect your choice of measuring instruments. And in quality control, to confuse these terms could be a serious source of problems. Fortunately, if you know their specific meanings you can use them in any circumstance and be sure of expressing yourself correctly.

Compare five men in a shooting match.* Whether or not each shot hits the target depends upon four sets of variables: the rifleman's skill, the rifle, the cartridges and the environment, Fig. 2-8. Each of these can be subdivided into further variables. For example the environment could be subdivided into wind, temperature, glare condition, distractions and many more.

It would not do to base our appraisal on just one shot for each man because, whether good or bad, it could be explained as luck. Let us use ten shots for each. Then to sim-

---

*Juran, J. M., Quality Control Handbook, New York, McGraw-Hill Book Co., 1962, 2nd edition, p. 9-4.

plify the matter let us consider the target to be a circle. Shots in it are good; outside, bad. The results are shown in Fig. 2-9.

The results of Mr. A are five good, five bad. Was this accurate? Precise? Reliable?

Mr. B produces a tight group - but all bad. This man is more *precise*, but not as *accurate* as Mr. A. Mr. C produces a group similar to Mr. A (thanks to my drawing) but all are good. Even though he is not as precise as B, he is as precise as A and thus is more accurate than either of them.

Mr. D produces his tight group, all good. He is more precise than Mr. C yet his score is the same. Then Mr. E takes the rifle and places an equally tight group squarely in the center of the target. Quite good shooting, yet his score is no better than either Mr. C or Mr. D.

There is no justice to this and we immediately try to find something that can be said for Mr. E that will show him to be the best shot. That something is *reliability*. At the county fair, horse sense would win out. If we were placing any bets on the results of the next series of shots from these five men, we would obviously give better odds, far better, for Mr. E than for any of the others. But why?

TARGET ANALOGY

ACCURACY, PRECISION AND RELIABILITY COMPARED

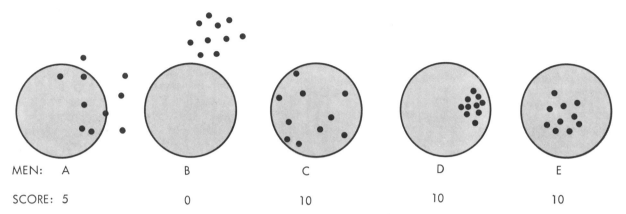

MEN: A   B   C   D   E

SCORE: 5   0   10   10   10

Fig. 2-9 Which of these targets represents accurate shooting? Precise shooting?
Reliable shooting? Can you tell why?

The marksmanship of the men is only one of the variables. Assume that the 10 rounds each fired is sufficient to represent his skill. Then assume that one of the other variables changes, a wind arises from the left to the right in front of the riflemen. The next match might look like Fig. 2-10.

Mr. A drops to a score of three. If the wind had been from right to left he would have gained. Not a very reliable situation. Mr. B, precise as ever, is beyond the help or hindrance of wind in this example. Mr. C mercifully lost only two. If the wind had been from the right to left he would have lost more.

He is in a less favorable situation than Mr. A because almost any change causes him to lose.

Mr. D's very precision caused him to lose miserably and fall from 10 to 4. If he had been less precise, probably more of his group would have remained good. On the other hand, this same precision protects him against a reverse wind. Hardly the reliable way to win rifle matches. Mr. E can tolerate interferences beyond his control because he has a comfortable margin all around his group. *That is reliability*. There is greater probability that in this match Mr. E *will perform as planned* than any of the others.

TARGET ANALOGY
CHANGE OF ONE VARIABLE

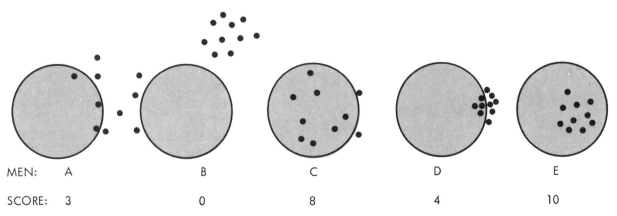

MEN: A   B   C   D   E

SCORE: 3   0   8   4   10

Fig. 2-10 A change in one variable, wind, alters the results as shown. Does this show which shooting was the most reliable?

In using any of these terms it must be clear to what they apply. In the examples given they applied to the relative skill of five men. If only one man did all the shooting but a different rifle was used for each series of shots, then the accuracy, precision and reliability of the rifles could be determined. Similarly the ammunition or the other variables could be evaluated.

Let us total our score thus far. *Accuracy is a relative matter.* It is the comparison of desired results with undesired results. Hence, it is frequently called "the quality of conformity." *Precision is a positive matter.* It is the measure of the dispersement of the results. Therefore, it is called "the quality of refinement." *Reliability is relative.* It is the probability that the results will be as predicted. They are distinctly different.

These definitions apply to the results of several shots (or measurements, or dividing amoebae or any other observable phenomena). Do they apply to single observations? For example, can we make statements that will apply to the first shot fired by each of the five men? No. We can make educated guesses, but that is all. But, after we have observed a group of shots, we can then make more reliable guesses about the next shot.

These guesses will be probabilities until the shot is fired and will be facts after the firing. In the first case they must have a range. In the second case they will be specific information. Adding the information about the new shot to the previous shots may alter the total picture and cause us to upgrade or downgrade our guess about the next shot.

Accuracy, precision and reliability must be modified to have specific meaning. When we say that a certain steel cylinder is a precision part or an accurate part we are speaking of its *history*. When we say it is a reliable part we are speaking of its expected *future*, as well as its history.

An *accurate cylinder* is one designed within a small range of permissible sizes. Inspectors examined large numbers of cylinders during production. And the measuring instruments they used were of a relatively high order of accuracy, precision and reliability.

The same remarks could be made for a *precision cylinder*. When we say *reliable cylinder* we may be referring to the reliability with which it was manufactured and inspected, or the reliability with which it is expected to perform its intended role. Unless specified there is no way to know which is meant.

When speaking about the cylinder it would be far better to specify which property is being described: an accurately inspected cylinder, a reliably designed cylinder, a cylinder with a precision tolerance, etc.

Precision refers to the fineness of the range of sizes that is allowable for a part or to which an instrument can read. It answers the question, how much? The answer may be vague as *high precision* or specific as *within one mike.* It applies to one part or to many parts.

Precision does not answer the question, *good enough?* Accuracy does. The term accuracy conveys an indication whether or not the subject complies with a standard, is too big or too small, is in or out of tolerance.

Precision never designates accuracy but accuracy may designate precision. In Fig. 2-11 the target size (specified standard to which performance must conform) has been reduced to one-half of the previous size. It is now seen that in order to have ten accurate shots the precision of the shooting can be no less than that of B, D and E. Mr. C's shooting has not changed, but because the target calls for greater accuracy his precision is not adequate, and his score drops from ten to three.

Mr. D's precision is sufficient to give him a score of ten but his accuracy reduces the score to four. The less said about Mr. B the better. Only our sharpshooter Mr. E has both the accuracy and precision required to maintain his 100% score. Note, however, that he has no margin of safety left. If the wind came up again it could only do one thing regardless of direction, reduce his score. The lesson here is that *whenever conditions are made more severe the reliability problem increases.*

TARGET ANALOGY

CHANGE OF SECOND VARIABLE

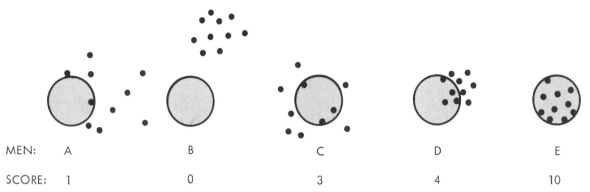

MEN:	A	B	C	D	E
SCORE:	1	0	3	4	10

Fig. 2-11 Reducing the target size by one-half shows that the accuracy requirement may dictate the precision requirement.

We have now made a full circle and can justify the popular use of precision. Without precision we cannot have accuracy or reliability. If we are speaking in generalities, such as the chart in Fig. 2-12, we need not and cannot know about specific accuracies. We speak in terms of precision.

The Bureau of Standards is attempting to find ways to split one millionth of an inch into ten parts. When that is accomplished we will have risen to a higher level of precision. It will then be up to the users of measurement in science and industry to make practical applications. It will be in the resulting practical applications that the considerations of improved accuracy and higher reliability will be involved. Without question it will alter our knowledge of the universe and result in new products.

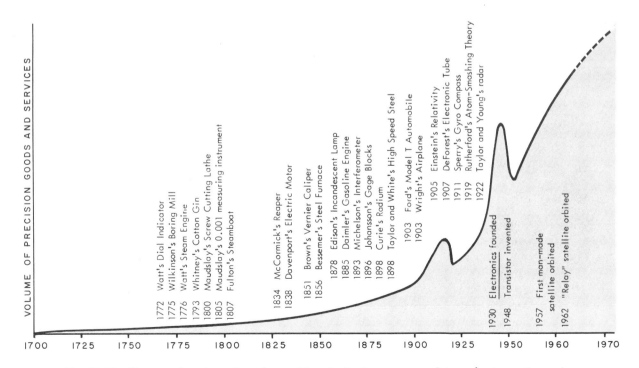

Fig. 2-12 The accelerating rise of precision indicates a severe future shortage of people skilled in measurement. (Historical events have been added to the chart used on page 1-7 of Quality Control Handbook by J. M. Juran)

ACCURACY VERSUS PRECISION

As the examples chosen have demonstrated, the word accuracy may mean many different things. Even though some of the meanings are subtle, they are all clearly related, Fig. 2-13. The word accurate is used most commonly to mean precise; therefore, it is suggested that, when in doubt, you substitute *precise* for accurate. Then when you do use accurate it will be clear that you mean something much more specific. The best part is, no one will quibble about your grammar. Precision may be used either as an adjective (a precision instrument) or a noun (working to great precision).

The important terms in this chapter are your tools and they help you to do precision work just as much as a set of gage blocks or a microscope. These definitions are not complete and will be supplemented throughout the book with additional meaning that gives them greater precision. A little thought will eliminate most errors.

The chapters that follow first introduce the systems for measurement, then show you how to take measurements from the least precise to the most precise. Ability to measure can come only from your practice of these principles.

TERMINOLOGY

	PRECISION	ACCURACY	RELIABILITY
General meanings:	Exactness Degree of exactitude	Desirability	Probability of achieving desired results
Measures:	Fineness of readings	Ratio of correct to incorrect readings	Probability of correct readings
Method for stating:	Within a 3-inch circle Plus or minus one thousandths inch	5 out of 10 50 percent of full scale	90 percent reliable
Specific meaning:	The lower the standard deviation of measurement, the higher the precision. *	The number of measurements within a specified standard as compared to those outside.	The probability of performing without failure a specific function under given conditions for a specified period of time. *

*Juran, J. M., Quality Control Handbook, New York, McGraw-Hill Book Co., 1962, 2nd edition, pp. 9-4 and 13-5.

Fig. 2-13   These definitions fit most measurements but many exceptions can be found. A good rule is to use the most precise term that the listener can understand easily.

Summary

● Because it is necessary to understand each other some terms must have specific meaning. Measurement expresses distance between points. It permits distances to be reproduced. A measurement consists of a unit of length and a multiplier. Each measurement begins at a reference point and terminates at a measured point. It lies along a line of measurement which must have a known relationship to the feature being measured.

● A feature is a measurable characteristic. It is bounded by edges, usually but not always, formed by the intersection of planes. Features are spoken of as being male, female or some combination. The dimension of the feature is the designer's concept of perfection. Features of actual parts are not perfect.

● Measurement shows the deviation from perfection. The measured conformity to the dimension is the accuracy. The refinement with which this can be known is the precision. The effect of accuracy and precision on attaining the desired results is reliability. Precision is essential for reliability but alone cannot produce it. Increased reliability requires increased accuracy and that requires increased precision. Therefore, the general term used to denote progress in measurement is precision.

QUESTIONS

1. If all we know about a figure is that it has four sides 2 1/2 inches long, what is the name that describes the figure?
2. What is the principle dimensional measurement?
3. What are the next important dimensional measurements?
4. Knowing the above, what other dimensions can be stated?
5. What do the dimensional measurements tell us about a part?
6. Name a situation in which dimensional measurements are not directly required for ordinary work.
7. What is the reference point?
8. What is the measured point?
9. When are they the same?
10. What are they both called when the same?
11. What is a dimension?
12. What is a part feature?
13. What are most features?
14. How are edges formed geometrically?
15. What are male and female features?
16. Are accuracy, precision and reliability the same?
17. Why cannot a rifleman's skill be determined by one shot?
18. Why do the scores in Fig. 2-9 fail to give a true comparison of the skill of the five shooters?
19. What one word best describes accuracy?
20. What one word best describes precision?
21. Is reliability the same as accuracy or precision?
22. Can we predict the results of the first shot?
23. What term answers the question "How good?"
24. Can precision be increased without improving accuracy?
25. Can accuracy be increased without increasing precision?
26. What general term can be used to describe the advances in our ability to measure?

DISCUSSION TOPICS

1. Explain the practical significance of "how square?" and "how rectangular?"
2. Explain relationship between size and shape.
3. Explain importance of the terms: reference point, measured point, and line of measurement.
4. Explain meaning of "A design is conceived in perfection."
5. Of what importance is time in defining size?
6. Why was the target analogy chosen in a book on measurement?
7. Discuss the reasons why accuracy and precision were said to apply to history, while reliability applies to the future.

Weights and measures may be ranked among the necessaries of life, to every individual of human society. They enter into the economical arrangements and daily concerns of every family. They are necessary to every occupation of human industry; to the distribution and security of every species of property; to every transaction of trade and commerce; to the labors of the husbandman; to the ingenuity of the artificer; to the studies of the philosopher; to the researches of the antiquarian; to the navigation of the mariner, and the marches of the soldier; to all the exchanges of peace, and all the operations of war. The knowledge of them, as in established use, is among the first elements of education, and is often learnt by those who learn nothing else, not even to read and write. This knowledge is riveted in the memory by the habitual application of it to the employments of men throughout life.

John Quincy Adams
Title Page — "Report Upon Weights and Measures"
1821

chapter 3

Systems of Measurement

Inch vs. Meter vs. Decimal Inch

Much has been said recently about adoption of the metric system in place of the inch-pound system, Fig. 3-1. This has been well publicized in the United States and throughout the British Commonwealth. The resulting fiery oratory has produced two opposing factions. One side admits that the metric system is great but claims that it would cost a fantastic amount to change to it. The other faction claims that the metric system is so much better than the inch-pound system that any cost is not too much. Which side is right? Both border on the absurd, in spite of the prominent persons they each have enlisted.*

Consider the cost argument first. If the change is made in one fell swoop the cost will be astronomical. The change from carbon steel cutting tools to high-speed-steel ones, and then on to carbide cutting tools, was made without the all-or-nothing-at-all approach. These changes involved the replacement of virtually all production machine tools. Certainly that was more drastic than the changes of some graduated dials, lead screws, and assorted gages.

The other faction states that the metric system is best *unequivocally*. They are just as far off base. In spite of its undramatic beginning, the inch-pound system provides units that in many instances are much better suited to actual requirements. One sixty-fourth of an inch naturally conforms with the resolving power of the human eye when used as the smallest steel rule division. That is far more than can be said of the smallest metric steel rule division. One millimeter is a real "fatty" in comparison.

Like almost everything else in life, neither system can be said to be best unless you define your point of view. Each has its place – at least at the present. Fortunately, apart from the sound and fury more moderate minds recognize this and are producing valuable results. One of the major advances is the *decimal inch standard being formulated* by the American Standards Association.** We will

* There is a rapidly growing third faction that firmly believes the metric system is one of history's costly mistakes and should be replaced with either the inch-pound system or an entirely new system.

** It appears that this standard will be adopted before this book is "in covers" but at the time of writing it had not been.

THE INCH-POUND AND METRIC SYSTEM

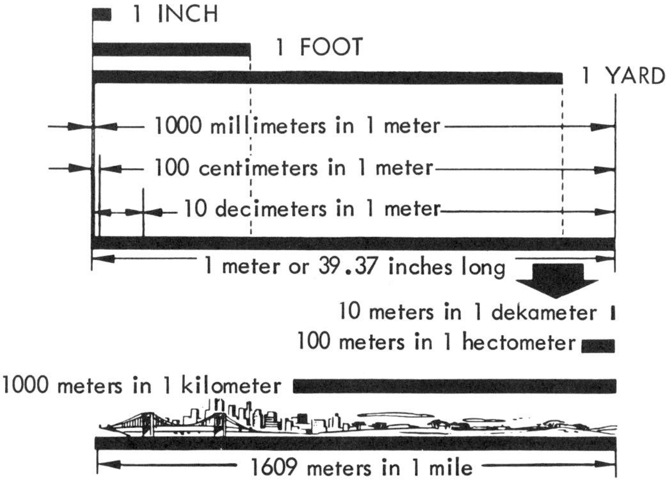

Fig. 3-1 These two systems are convertible but many authorities believe only one should be used. (Precision .... A Measure of Progress, General Motors Corp., 1952, p.20)

now examine the background of the systems of measurement so that you can make up your own mind which system best suits the metrological problems you encounter.

NEW STANDARDS FOR OLD*

Man's ability to measure has reflected his way of life since our first record of his struggle to harness natural resources. He followed instinct and ego. Parts of his body were his length standards. When man grew in number and organized into cities and then states, he had to refine these standards for bartering and crafts. The many variations in body size resulted in conflicts. These were resolved only by the monarch choosing his own body as the standard. This established standards as the responsibility of the state along with taxes and the rights of ownership.

The earliest recorded standard is the Egyptian cubit, Fig. 3-2. This standard recognized the fundamental principles of metrology. Their enforcement of these principles was somewhat more severe than ours. Failure to calibrate the working cubit to the royal cubit at each full moon was punishable by death. Many feel that this would be helpful today. There is little question that the pyramids and other structures of the Eqyptians could not have been built without their refined system of measurement.

---

* Moody, J.C., "New Standards for Old," Ordnance, May–June, 1962.

The cubit spread throughout the ancient world. Each culture that adopted it modified it somewhat. This progress halted, as did much else, with the barbarian invasions.

Romanticists would like to show a chain of continuous development linking our measurements to those of the ancients. In their efforts to do this, they overlook a far more important point that can be made only if there is not such a continuous chain. *Whenever man has developed civilization he has invented a measuring system, and these systems have always been very similar.** This appears to be as natural as taxes or fermentation.

So without trying to show King Henry I of England hand in hand with Rameses II, we can introduce his great invention, the English yard. He defined this unit as the distance from his nose to the tip of his thumb when his arm was extended. This *unit of length* was physically made into a *standard*, the iron ulna. Although the Egyptians and countless others had embodied their unit of length in material form, this was a major advance in the Dark Ages. At this point the inch (one thumb-breadth) and the foot were already in use, Fig. 3-3. Hundreds of years later in the 16th century, the English rod was defined as the combined lengths of the left feet of the first 16 men to leave church on a certain Sunday — not as funny as it now sounds.

** A striking example is that the ancient Roman ounce and modern English ounce are nearly identical.

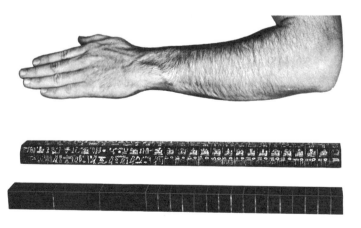

Fig. 3-2 The Egyptian Cubit shows recognition of metrology fundamentals that still apply today. The unit of length was the Pharaoh's forearm. The standard was the Royal Cubit and to this standard, the Working Cubits were added. (The Story of Measurement, DoALL Company)

MILESTONES IN MEASUREMENT

YEAR	PLACE	UNIT	CONTRIBUTION
before 4000 B.C.	Chaldea	circle	First recorded standard of measurement.
about 4000 B.C.	Egypt	cubit	Length of Pharaoh's forearm.
4000 to 2000 B.C.	Egypt	span palm digit meridian mile fathom	Outstretched hand. Middle of middle finger. Equal to 4000 cubits or 1000 fathoms. Length of outstretched arms, about 6 feet.
500 B.C.	Greece	stadia mile	1/10 of meridian mile borrowed from Egypt. 1000 paces, similar to today's mile.
	Rome	thumb-breadth	Divided foot into 12 parts. First inch.
890	England	foot	King Alfred established foot as the measure of a cubical vessel containing 1000 Roman ounces of water.
1130	England	yard	King Henry I established yard and made an iron standard.
1150	Scotland	inch	King David made inch average measure of three men's thumbs.
1324	England	inch	King Edward II made inch length of three barley-corns.
1776	U.S.		Articles of Confederation gave Congress power to establish weights and measures.
1790	France	meter	Metric system imposed by law of Republican Convention.
1795	Holland and Belgium		Forced to metric system by defeat.
1800	France		Napoleon relaxed metric system.
1828	U.S.		First effective weights and standards act passed congress.
1837	France		Louis Philippe put metric system back in force.
1859	Italy		Forced to metric system by defeat.
1859	Austria		Adopted metric system to thank France for aid against Germany.
1866	U.S.		Metric system legalized but not compulsory.
1868	Germany		Bismarck adopted metric system for Germany.
1878	U.S.		Metric Convention ratified.
1890	U.S.		Metric standards received in U.S.
1893	U.S.		Mendenhall Order links U.S. standards to metric.
1927	France	light wave lengths	7th International Conference stated the meter was to 1,553,164.13 wave lengths of cadmium-red light.
1959			English speaking countries agree on an International inch.

Fig. 3-3   Measurement has had a checkered past, often enforced with the sword. The people involved have seldom been consulted about the changes.

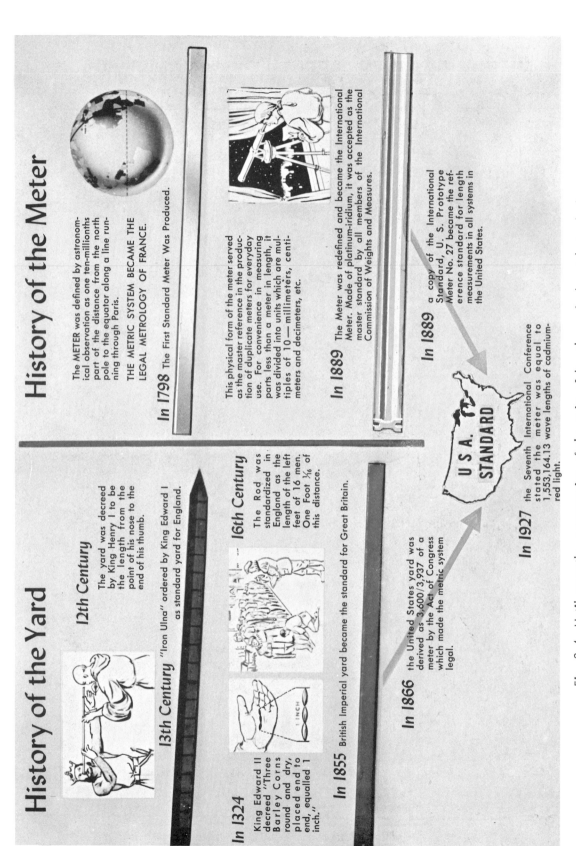

History of the Meter

In 1798 The First Standard Meter Was Produced.

The METER was defined by astronomical observation as one ten-millionths part of the distance from the north pole to the equator along a line running through Paris.

THE METRIC SYSTEM BECAME THE LEGAL METROLOGY OF FRANCE.

This physical form of the meter served as the master reference in the production of duplicate meters for everyday use. For convenience in measuring parts less than a meter in length, it was divided into units which are multiples of 10 — millimeters, centimeters and decimeters, etc.

In 1889 The Meter was redefined and became the International Meter. Made of platinum-iridium, it was accepted as the master standard by all members of the International Commission of Weights and Measures.

In 1889 a copy of the International Standard, U. S. Prototype Meter No. 27 became the reference standard for length measurements in all systems in the United States.

In 1927 the Seventh International Conference stated the meter was equal to 1,553,164.13 wave lengths of cadmium-red light.

U.S.A. STANDARD

History of the Yard

12th Century
The yard was decreed by King Henry I to be the length from the point of his nose to the end of his thumb.

13th Century "Iron Ulna" ordered by King Edward I as standard yard for England.

In 1324
King Edward II decreed "Three Barley Corns round and dry, placed end to end, equalled 1 inch."

16th Century
The Rod was standardized in England as the length of the left feet of 16 men. One Foot ⅟₁₆ of this distance.

1 INCH

In 1855 British Imperial yard became the standard for Great Britain.

In 1866 the United States yard was derived as 3,600/3,937 of a meter by the Act of Congress which made the metric system legal.

Fig. 3-4 Until recently, preservation of the units of length required metal standards. (The Story of Measurement, DoALL Company)

These standards were not clearly convertible. This invited the "sharp operators" of those days to take unfair advantage in their trading and commerce. In addition, the rulers could use overgenerous measures for the collection of taxes. This condition continued to smolder for many generations but exploded in the French Revolution in 1789.

The inch-pound system at this time was based on English practice. But they, too, had their problems. During the latter part of the eighteenth century they had at least three different miles and two units for subdivisions. To clear this up the Imperial Standard Yard became the basis for all linear measurements. Unfortunately, the one standard was lost when fire destroyed the House of Parliament in 1834. A new one was constructed; but this, of course, was slightly different from the original. The first axiom of metrology is that in order to measure there must be a standard. The world then learned the second axiom: *the standard must be reproducible.* The new standard was adopted in 1855 and since 1866 has been in the custody of the British Board of Trade.

THE METRIC SYSTEM BORN OF REVOLUTION

Because the abuses of measurement were among the causes of the French Revolution, they received immediate attention by the Republican Convention. As usual in human affairs, they went too far. Their unit of length, adopted in 1790, was the distance between the North Pole and the Equator passing through Paris. Divided into 10-million parts, this was the meter. They also set up new units for mass and time.

The people did not take well to 10-day weeks and days divided into 10 hours of 100 minutes each. It was even harder to get used to buying produce by the 10's and 20's instead of dozens. So Lavoisier, the principal member of the Metric Committee, lost his head; but the system was enforced. Napoleon won popular support by relaxing it in 1799. By 1812 he permitted the old standards to be used again, although the metric system legally remained.

Painstaking investigations showed the difficulty in using the earth as a standard of length. Therefore a metal standard was made, known as the Meter of the Archives. It was adopted as the official standard in 1799.

In 1870 the first of a series of international conferences was held for the purpose of establishing the metric system throughout the world. Forty-eight delegates represented 25 countries including France. After interruptions, a general conference in 1889 at Paris, approved the work of the committee.

Thirty-prototype meters and 40-prototype kilograms were constructed of a platinum-iridium alloy, Fig. 3-4. The prototypes were calibrated with each other. The one most nearly equal to the Meter of the Archives was selected as the International Prototype Meter. It is now located at the International Bureau of Weights and Measures, near Paris. The remaining 29 prototypes were distributed to participating countries to serve as national standards. They are returned periodically for rechecking.

LEGAL STATUS OF METRIC SYSTEM IN THE UNITED STATES

The metric system is legal in the United States. This surprises many people. Here is the way it came about. The United States received two of the prototypes. Number 27 is designated as our national prototype meter. It arrived in 1890. Number 21 followed later that year. These standards were in the custody of the Office of Weights and Measures. This was then a part of the Coast and Geodetic Survey, which in turn was part of the Department of Treasury.

In 1866, by an act of Congress, the use of metric system was legalized but not made obligatory. This act also established ratios between the corresponding units of the United States and metric systems. The yard was defined as equal to 3600/3937 meter. *

At that time the United States standard was an 82-in. brass bar prepared in London and brought over in 1813. The yard was defined as the distance between its 28th- and 64th-inch graduations. In 1893 the International Prototype Meter was declared to be the fundamental standard of length, and the yard to be determined from it. When the National Bureau of Standards was established in

---

*Smith, Ralph W., "The Federal Basis for Weights and Measures", National Bureau of Standards, Circular 593, June 5, 1958.

1901, the Office of Weights and Measures became a division.

The Mendenhall Act of 1893 established the metric International Prototype Meter as the legal standard and established its relationship to the yard. It failed, however, to specifically legalize the inch-pound system. It is not unusual for politicians to create paradoxes but this one ranks among the outstanding "bloopers". It gave the United States *a legal system that was not in popular use and a widely used system that had never been legalized.*

THE INTERNATIONAL INCH

The Mendenhall Act defined the United States inch as 25.4000508 mm, Fig. 3-5. Strangely enough, the British Imperial Standard Yard at that time was defined as 25.399978 mm. In 1922 the British standard was revised to 29.399956 mm. That increased the discrepancy between the United States and Great

*Gardner, Irvine C., "Light Waves and Length Standards", Journal of the Optical Society of America, Vol. 45, No. 9, September, 1955, p. 688.

Britain. Then in 1951 Canada defined its inch as 25.4 mm exactly. That resulted in a third - and more sensible - inch. It probably made the English speaking countries aware that they had to agree soon.

The reconciliation came in 1959 when all agreed to a 25.4-mm conversion factor. This represents 35 millionths of an inch difference in a length of 18 in. - nothing as far as most products are concerned, but an intolerable difference for a nation's master standard. The correction factor of two millionths of an inch per inch of length is used to correct for the change when calibrating gage blocks. Metrologists had apparently learned their lesson about politicians. The International inch was accepted in 1959 by general agreement without specific legislation of Congress.

One other change of great importance soon followed. Light waves replaced metal bars as the international standard. Work in that direction was begun back in 1892, but it was not until 1960 that the new standard was adopted. Light waves permit much greater precision and accuracy. They are above the partisan struggle of inch versus meter; they respond equally well for either system.

LINEAR MEASUREMENT UNITS IN THE INCH-POUND AND METRIC SYSTEMS COMPARED

LINEAR MEASURE	MICROINCH (.0000001 y) MUin.	TEN THOUSANDTHS (.0001y)	THOUSANDTHS (.001y)	INCH (y) or (z/12) in.	FOOT (12y) or (z) ft.	YARD (3z) yd.	MILE (5280z) mi.
MILLIMICRON **mµ** (.000000001x)	0.0394 **25.4**						
MICRON **µ** (.000001x)	39.37 **0.0254**	0.3937 **2.54**	25.4				
MILLIMETER **mm** (.001x)		393.7	39.37 **0.0254**	0.0394 **25.4**	304.8		
CENTIMETER **cm** (.01x)		3937.	393.7	0.3937 **2.54**	0.0328 **30.48**	91.44	
METER **m** (x)				39.37 **0.0254**	3.2808 **0.3048**	1.0936 **0.9144**	1609.3
KILOMETER **km** (1000x)					3280.8	1093.6	0.6214 **1.6093**

Fig. 3-5 To use the U.S./Metric Conversion Tables: Enter a horizontal or vertical column; combine either the light face or bold face units and numbers; and read its corresponding value in terms of one unit of the opposing scale. For example: 1.0936 yards = 1 meter (reprinted from Tools of Metrology, supplement to Quality Assurance, January, 1963).

CLEARING AWAY THE OBSTACLES

We cannot evaluate the possible measurement systems on their real merits until we clear away the smoke screen of bias that has been thrown up around this issue. Five definite types of propaganda are used to becloud the issue: (1) naturalness of the systems, (2) economic considerations, (3) syllogistic reasoning, (4) overemphasis, and (5) extension of the problem. We will consider them in that order.

Naturalness of the systems: It should now be apparent that both the inch-pound and metric systems are completely arbitrary inventions of man. Neither has any natural basis for being. It is amazing how close the metric system came to natural standards. A pendulum 0.9932 meters long at the latitude of New York and at 760 mm barometric pressure swings once a second. At the equator it would swing slightly less but even at the poles it would not reach unity. The acceleration of gravity similarly narrowly misses ten and averages 9.801 meters per second. The only real constant (according to Einstein) is 299,792,000 meters per second. The fact that the metric system came so close to these natural phenomena is moot. The consideration is now a practical one, neither psychic or philosophical. The metric system did have a natural basis at the beginning, but that was lopped off along with the head of its creator.

Similarly there is nothing "natural" about the decimal system *(whether divisions of the inch or of the meter)* except that man has four fingers and a thumb on each hand. It is likely that if man had evolved with eight fingers and a thumb on each hand, we would have a system based on 18. Whether that would be better or worse than a a system based on ten I do not know, but certainly ten is a poor base. It is divisible evenly by only two and five. If 12 had been chosen it would have been far more convenient because 12 is divisible evenly by two, three, four, and six. One of the best indications that the base ten is not a natural one is its limited use in those activities where complete freedom of choice is possible without regard to metrology or government standards. The dozen and the gross with their multiples and even divisions are far more important in commerce and trade. Interestingly enough, it is only advocates of the metric system who resort to moral, ethical and "logical" arguments. We are reminded that most of the rash

acts of history have been foisted on the world for allegedly the same reasons. Only an objective mind can make reliable measurement decisions and this is what is needed here.

Economic considerations: If any economy has thrived on obsolescence, it is the one that has flourished in the inch-pound stronghold. One of the arguments against changing to the metric system thus seems a little odd. It is the argument that it would require wholesale modification and replacement of machines, tools and gages. It is stated that this would cost $26,000,000,000 in the United States alone. That is not only $26,000,000,000 *spent*, it is also $26,000,000,000 *earned*. Much of it could be accomplished by normal attrition. A large amount of productive capacity is replaced each year even without conversion. Furthermore, it is foolish to sell short human ingenuity. A host of gadgets would appear to permit existing equipment to function in either system.

Syllogistic reasoning: A much worse bias has confused this issue. Either-or reasoning might have been fine for ancient philosophers, but gets complex, modern man into most of his problems. The issue does not have to be all or nothing at all. There is no reason why we cannot investigate combinations of the two systems as well as complete adoption of one or the other.

Overemphasis: When comparing systems, some emphasize fractions as the main reason to abandon the inch system in favor of metric. This is serious overemphasis. It implies that fractions are an exclusive problem of the inch system. The use of fractions is natural and universal. They only create a problem in one of the three factors that we will use to compare systems, the computation factor. They are a nightmare for computation, and they invite error in rounding off.

For communication fractions are invaluable. They find extensive use in the metric system, a fact often concealed by those who advocate complete decimalization. It is hard to imagine a customer in an Italian restaurant ordering 500 cc of wine. He would order a half liter. Furthermore, France uses a wine "gallon" of 4 liters. The metric-using European has no hesitancy, in fact, to specify "halves" of any metric unit, or duodecimal divisions.*

*Gilles, Dr. Robert C., "New Light on a Weighty Problem, F.B.I. Review, London, Federation of British Industries, September, 1963, p. 22.

There is a natural basis for the selection of fractions or decimals. We naturally swing to decimals when computation is the chief objective, and to fractions when communication is of greatest importance. And it is equally natural to switch from one to the other as required.* For example we regularly divide the gallon into quarts and pints, yet all gasoline is sold in tenths of gallons because of the computing-pump dispensers. We ordinarily divide the mile into yards and feet, yet the odometer of automobiles divides it into tenths. An even better example is United States currency. Although the coinage is fully decimalized we habitually refer to "halves" and "quarters". Fractions are thoroughly ingrained in human activity - regardless of the measurement and system.

Extension of the problem: There is general failure to recognize the difference between *units of measure* and *sets of sizes*.** Consider a shaft running in a bearing. The desired fit depends upon the proper tolerance of each part. The fact that the workman turning the shaft spoke French and the one who bored the hole spoke English has nothing to do with the fit. The finished parts have no way of reflecting the systems of measurement used to machine them. Whether a screw is designated 5/16 - 24UNF or M8 x 1,125 is unimportant. After all, most automobiles use metric spark plugs in inch-dimensioned engines. The important consideration is that the parts fit properly. For this, accurate convertability is the answer - not necessarily universal use of one system.

WHICH SYSTEM IS BEST?

Having removed the five obstacles to objective consideration, we can now seriously ask the question, "which system is best?" There is no one answer, as you already suspected.

By far the strongest point to be made for the metric system is its use in science. It

has been long predominant in that area. For many years science and practical work were separated by a huge chasm spanned thinly by engineering. On the practical end, engineering used the inch system to get ideas off of the drawing board and into our garages and homes. On the theoretical end it held securely to the metric system. Chemistry, for example, is firmly bonded to this system.

Automated production controlled by electronic and hydraulic circuitry and regulated by computers has moved industry far away from the man with the hammer. It depends largely on the various branches of science for smooth functioning. Now the space age has made the beards of the scientists almost as familiar in the manufacturing plants as the pipes of the engineers. Add to this, the fact that many precision products ranging from bearings, photographic film,* and spark plugs to surgical instruments are already produced to metric measurements. These are good reasons for adopting the metric system.

The most frequently heard reason, however, is certainly a poor one when analyzed. It states "all the rest of the world is on the metric system, why should we be different?" This is called a *non sequitur*. While the fact stated may be correct, the conclusion does not necessarily follow.**The correct concern is not how many countries, or how many people, or how many square miles are in the metric system. The concern is how much measurement is involved. It is impossible to know this but we can approximate it by the amount of goods manufactured.

Looked at that way, the picture is quite different. A majority of the total products manufactured in the entire world is manufactured to the inch-pound system. Even that needs to be adjusted upward. An automobile or an automatic washer-dryer requires hundreds of times the measurement of less sophisticated products.

---

*Judson, Lewis V., "Units and Systems of Weights and Measures", Precision Measurement and Calibration, Washington, D. C., U. S. Dept. of Commerce, 1961, NBS Handbook 77, Vol. III, p. 30/7.

**Cattaneo, A., "The Battle of Measurement Systems, Standards Engineering, Vol. XV, No. 9, September, 1963, p. 5.

*35 mm film is a curious "goof". It was established by Thomas A. Edison and is 1 3/8 inches. Metric manufacturers misnamed it and we went along.

**An example: The president favors social security. Communists favor social security. Therefore, the president is a Communist. Such reasoning is particularly damaging to reliable measurement.

Furthermore, by no stretch of the imagination can it be said that the rest of the world is on the metric system. The inch-pound system is far more extensively used in Europe than the metric system is in the inch-pound countries,* excluding scientific work.

Germany used the *zentner* which is 100 pfund (pounds), not 100 kilos. Belgium, Holland, Luxembourg, and even France use hybrid units in trade. Plumbing and heating pipe in Germany is in inches. The inch in Germany is so familiar that it has its own name, *zoll*: Automobile, truck, bicycle and motorcycle tires are sized in inches wherever they are made. The one exception marks their tires in both inches and millimeters. The majority of produce and packaged goods in grocery stores throughout Europe use ounces (converted to grams of course, but often showing both). Canned goods use the American No. 2 1/2 can. In Germany this even has been adopted as a DIN (German official) standard. Shotguns of any source are sized by gage and most rifles are by caliber, both of which are inch measurements. And while at it — organ pipes throughout the world are sized in feet. If amount of measurement is the criterion,

---

*Gilles, Dr. Robert C., "Let's Not Go Metric", F.B.I. Review, London, Federation of British Industries, January, 1963.

the balance swings to keeping the inch-pound system intact.

FACTORS FOR EVALUATING MEASUREMENT SYSTEMS

The natural retort to this is: "So most measurement is in the inch system. If it is not the best, why keep it?" That is a valid argument and the only one on which the ultimate decision should be made. Let's examine which is best.

Before we can do that we must determine what factors are desired. The factors are of three general types: metrological, computational and communicative. The metrological factor concerns the actual act of measurement; for example, comparison of the graduations found on standard instruments in each system. The *computational factor* is the ease with which measurements can be used mathematically. The *communicative factor* is the ease with which information about measurement can be conveyed from one person to another.

Each of these must be considered from four standpoints to determine whether it is good or bad. In each case the goal is the same - *to permit maximum measurement in the minimum time, with the minimum error and at minimum cost.* These factors are compared in Fig. 3-6.

METRIC AND INCH-POUND SYSTEMS COMPARED

Separate factors to be compared:
1. metrological
2. computational
3. communicative

Basis for comparing each factor:
1. maximum measuring ability
2. minimum errors
3. minimum time to measure
4. minimum cost

FACTORS	METRIC	INCH-POUND
METROLOGICAL Science Industry Domestic	Excellent Poor Poor	Poor Good Fair
COMPUTATIONAL Science Industry Domestic	Excellent Good Good	Horrible Poor Poor
COMMUNICATIVE Science Industry	Good Good	Poor Poor

Fig. 3-6 When the two systems are critically analyzed, neither one is all good or all bad.

Maximum measuring ability means that the system has sufficient precision for very fine measurement, and sufficient range for extremely large measurement. The metric system provides uniform increments from very small to very large. The inch system has very dissimilar units (inch, foot, yard, rod, mile) that are not systematically related to each other. This is vividly shown in Fig. 3-1.

The instruments used in both systems are similar. They require about the same time for use and are subject to the same errors. The metric system is somewhat better, when the measurements in a given problem cover a wide range, because only the decimal points are moved. In this situation in the inch system, it is necessary to change in and out of feet, inches, etc. The inch system is slightly better for some small measurements. There is a small but significant range that can be read with an inch steel rule but cannot be read with a metric steel rule. The micrometer of the inch-pound system has four times the discrimination of the metric micrometer. There are some provisos that shade this advantage slightly, as you will find in Chapter 7.

Because the instruments are nearly alike, there is little cost difference, except for one important exception. The inch system requires instruments both for decimal and fractional measurements. This duplication is not required in the metric system.

For science, the metric system is clearly preferred. The quantities involved relate to theoretical considerations rather than existing objects. These range from the minute particles of nuclear physics to the enormous reaches of outer space. Accurate measurements are made painstakingly, but can then be communicated readily or used in complex calculations with relatively small opportunities for error. The inch-pound system in this role is awkward and unsystematic. In fact, it only rarely is used for scientific work any place in the world.*

For industry the picture is different. Compared to science, the measurements have been mostly in a narrow range. *The inch-pound*

system developed from the requirements of practical men producing real products. As unsystematic as the units may be, they are convenient for actual measurement problems. An excellent example is screw threads. Even in metric countries, about half of the threaded products are in inches.

In contrast, the founders of the metric system were concerned only with ideals and theories. They carefully developed an elegant system completely devoid of debasing compromises with anything in the real world of men and machines. Unfortunately, the farmer, the mechanic and the housewife are left out of the lofty forums in which measuring systems are thrashed out. Yet they are stuck with the results. Their choice for mixing feed or baking a cake is a sorry one.

On one side, they have the metric units that bear no resemblance to anything ever known before or to anything encountered in daily life. No where has the metric system found its way completely into the art of cooking, for example. Even in metric countries the old measures have been retained along with reluctant use of the liter and the kilo. The latter were only placed into use by strict laws. As recently as 1961, Japan, where the metric system had been in force for 30 years, found it necessary to pass a law forbidding the calibration of speedometers in miles per hour. Moreover the metric time system has been given up altogether.

On the other side there is the inch-pound system that has units directly traceable to things found in real life: drop, carat, cup, peck, teaspoon, jigger and grain, to name just a few. Never has a more bizarre conglomeration of inconsistent units and terminology been inflicted on man, Fig. 3-7.

From all of this two thoughts should emerge. First, the metric system is by far superior to the inch-pound system as far as its ease of computation and communication. Second, the inch-pound system provides units that are far more compatible to the measurements encountered in practical work. If you accept these statements a question becomes unavoidable. Why not combine the best features of each? In fact, if we take that liberty, then why not also eliminate as many faults as we can at the same time? This is exactly what has been done in the *decimal-inch system.*

---

* Even in this bastion of the metric system there are important exceptions such as "horsepower" and "joule".

TABLE OF ENGLISH SYSTEM WEIGHTS AND MEASURES

Linear Measure or Long Measure

12 inches (in., *pl.* in. *or* ins.; *symbol* ″)	= 1 foot (ft., *sing. & pl.; symbol* ′)
3 feet	= 1 yard (yd., *pl.* yd. *or* yds.)
5½ yards or 16½ feet	= 1 rod (rd.) or pole (p.) or perch (p.)
40 rods	= 1 furlong (fur.)
8 furlongs or 1760 yards or 5280 feet	= 1 mile (m. *or* mi.) (English *statute mile*)
3 miles	= 1 (land) league

Metric Equivalents: 1 in. = 2.54 cm.; 1 ft. = 0.3048 m.; 1 yd. = 0.9144 m.; 1 rd. = 5.029 m.; 1 fur. = 201.17 m.; 1 mi. = 1.6093 km. or 1609.3 m.; 1 league = 4.83 km.

Chain Measure

7.92 inches	= 1 link (li.)
100 links or 66 feet	= 1 chain (ch.)
10 chains	= 1 furlong (fur.)
80 chains	= 1 mile (mi.)

The *engineer's chain* is 100 feet long, with links one foot long (52.8 chains = 1 mile). *Metric Equivalents:* 1 link = 20.12 cm.; 1 chain = 20.12 m.

Square Measure (Area)

144 square inches (sq. in.)	= 1 square foot (sq. ft.)
9 square feet	= 1 square yard (sq. yd.)
30¼ square yards	= 1 square rod (sq. rd.) or square pole or square perch (sq. p.)
160 square rods or 4840 square yards or 43,560 square feet	= 1 acre (A.)

Metric equivalents: 1 sq. in. = 6.452 sq. cm.; 1 sq. ft. = 929 sq. cm. (0.0929 sq. m.); 1 sq. yd. = 0.8361 sq. m.; 1 sq. rd. = 25.29 sq. m.; 1 acre = 40.4687 ares (0.4047 ha.)

Surveyor's Measure (Area)

625 square links (sq. li.)	= 1 square pole (sq. p.)
16 square poles	= 1 square chain (sq. ch.)
10 square chains	= 1 acre (A.)
640 acres	= 1 square mile (sq. mi.) or 1 section (sec.)
36 square miles	= 1 township (tp.)

Metric Equivalents: 1 sq. mi. = 259 ha. (2.59 sq. km.); 1 tp. = 9324.0 ha. (93.24 sq. km.).

Cubic Measure (Volume)

1728 cubic inches (cu. in.)	= 1 cubic foot (cu. ft.)
27 cubic feet	= 1 cubic yard (cu. yd.)
(for measuring cordwood, etc.)	
16 cubic feet	= 1 cord foot (cd. ft.) or 4′ × 4′ × 1′
8 cord feet or 128 cubic feet	= 1 cord (cd.) or 4′ × 4′ × 8′

Metric Equivalents: 1 cu. in. = 16.387 cu. cm.; 1 cu. ft. = 0.0283 cu. m.; 1 cu. yd. = 0.7646 cu. m.; 1 cd. = 3.625 cu. m.

Avoirdupois Weight (Or

16 drams (dr.)
16 ounces or 7000 grains
14 pounds
100 (in Eng. 112) pounds
2000 pounds or 20 hundredweights
2240 pounds or 20 hundredweights

Metric Equivalents: 1 dr. = 1
453.59 g. or 0.4536 kg.; 1 st. = 6
1 short ton = 907.18 kg. or 0.9072
1.0160 M. T.

Troy Weight (Pre

3.086 grains (gr.) = 1 car
24 grains = 1 pen
20 pennyweights = 1 ou
12 ounces or 5760 grains = 1 po

Metric Equivalents: 1 grain
etc., a grain = ¼ carat (0.77 tr

Apothecaries' W

20 grains (gr.) =
3 scruples =
8 drams =
12 ounces =

Metric Equivalents: 1 s. ap

Time Measure

60 seconds (sec. *or* s.; *symbol* ″)	= 1 minute (min. *or* m.; *symbol* ′)
60 minutes	= 1 hour (hr.)
24 hours	= 1 day (da. *or* d.)
7 days	= 1 week (wk.)
30 days (commonly)	= 1 calendar month (mo.)
365 days or 12 calendar months	= 1 common year (yr.)
366 days	= 1 leap year
100 years	= 1 century

The length of the *astronomical year* is about 365¼ days, or 365 days, 5 hours, 48 minutes, 45.51 seconds. As the *common year* is 365 days, it becomes necessary once in every four years to add a day to the year, making the *leap year* of 366 days.

Circular Measure

60 seconds (″)	= 1 minute (′)
60 minutes	= 1 degree (°)
90 degrees	= 1 quadrant
4 quadrants or 360 degrees	= 1 circle

Longitude and Time

1 second of longitude (″)	= 1/15 sec. of time
1 minute " " (′)	= 4 sec. of time
1 degree " " (°)	= 4 min. of time
15 degrees " "	= 1 hour
360 degrees " "	= 24 hours

Nautical Measure

6 feet	= 1 fathom (f. or fm.)
100 fathoms	= 1 cable's length (ordinary). A cable's length, however, is taken variously as: 608 ft. (Br.), 720 ft. or 120 fathoms (U. S. Navy).
10 cables' lengths	= 1 nautical mile, taken as 6080.20 ft. (U. S.) and 6080 ft. (Br.)
1 nautical, or geographical, or sea, mile	= 1.1516 statute miles (cf. *Table* 1, above)
3 nautical miles	= 1 (marine) league = 3.45 statute miles
60 nautical miles	= 1 degree (of a terrestrial great circle)

Metric Equivalents: 1 fathom = 1.829 m.; 1 nautical mile = 1853.248 m. (U. S.) or 1853.2 m. (Br.); 1 (marine) league = 5.56 km.

The nautical, or geographical, mile (called *Admiralty mile* in Great Britain) is the length of a minute of longitude (see *Tables* 7 & 8), or 1/21600 of a great circle of the earth.

Dry Measure (Grain, Fruit, etc.)

2 pints (pt.)	= 1 quart (qt.)	= 67.20 cu. in.	= 1.1012 l.
8 quarts	= 1 peck (pk.)	= 537.61 cu. in.	= 8.8096 l.
4 pecks	= 1 bushel (bu.)	= 2150.42 cu. in.	= 35.2383 l

The Briti... quart = 1.0320 U...

PROPOSED MODIFICATION

TABLE 5

One ounce	= 1/16 pound	= 27/16 cubic inches	(1.6875)		
,, gill	= 4 ounces	= 27/4 ,, ,,	(6.75)		
,, pint	= 1 pound	= 27 ,, ,,		= 3 inches	
,, quart	= 2 pounds	= 54 ,, ,,			
,, gallon	= 8 ,,	= 216 ,, ,,		= 6 inches	
,, peck	= 16 ,,	= 432 ,, ,,			
,, bushel	= 64 ,,	= 1,728 ,, ,,		= 1 foot	
,, barrel	= 256 ,,	= 6,912 ,, ,,		= 4 cu. ft.	
,, hogshead	= 512 ,,	= 13,824 ,, ,,		= (2 feet)	
,, ton	= 2,048 ,,	= 55,296 ,, ,,		= 32 cu. feet	

and { 20 grains = 1 pennyweight; 20 pennyweights = 1 ounce
400 grains = 1 ounce; 6,400 grains = 1 pound }

Weights are in water equivalents

Fig. 3-7 The bizarre conglomeration of the English system can be unified and simplified. The example shown was developed by Dr. Robert C. Gilles and appeared in the article cited on page 27.

BACKGROUND FOR THE DECIMAL-INCH SYSTEM

The idea of using the decimals to simplify the inch-pound system is not new. The decimal foot was used in surveying in the United States before 1856. Up to the Civil War 1/64 in. was considered the finest measurement needed for practical work. (This is not to imply that instrument makers had not achieved precision that has only recently been exceeded.) Thousandths of an inch then became the accepted measure for precision, in spite of the confusion caused by the use of both thousandths and fractions.

The first major decimal-inch breakthrough came when Ford Motor Company adopted the decimal-inch system in 1930. The aircraft industry quickly followed. The Society of Automotive Engineers (SAE) published a manual in 1946 setting forth a complete decimal-inch dimensioning system. From then until 1961 many companies in many industries adopted this system.

In 1959 the American Standards Association (ASA) and American Society of Tool and Manufacturing Engineers (ASTME) decided to jointly urge greater use of the decimal inch. This was to be done by proposing an American Standard for the definition and use of the decimal inch. The committee formed for this purpose was known as ASA Sectional Committee B87. It extensively explored the points made earlier in this chapter, and drafted a proposal. This proposal, known as ASA Standard B87, is being considered for approval at the time of this writing.

Decimals have found wide acceptance for weight in the inch-pound system. They permit prices to be multiplied directly by weight. It must be emphasized that, although this committee studied all phases of the decimal-inch matter, the B87 standard applies only to linear measurement. The terminology applies to inches and their subdivisions. It can be anticipated, however, that at some future date the decimal inch will be extended to larger measurements, area, volume, and possibly weight.

It must be emphasized, also, that both of these associations primarily represent industrial interests. The findings of the committee, therefore, are relevant to industry. Whether or not they are also relevant to other interests, such as science or agriculture, was not investigated. It would be another *non sequitur* to jump to the conclusion that they are.

The balance of the chapter will show how the committee arrived at the decimal-inch standard, and the advantages of the new system.

METROLOGICAL ADVANTAGES OF THE DECIMAL INCH

The size of the millimeter is unfortunate in relation to the resolving power of the human eye.* This refers to the ability of the eye to separate closely-spaced objects which is essential when reading finely-spaced scale divisions.

The millimeter is nearly equal to 0.04 in. This is about twice as large as it should be for the smallest graduations. In the metric system the next smallest unit would be one tenth this size or 0.1 mm. That would be only 0.004 in. which is the thickness of most human hair. Even 0.2 mm would be impossible to resolve except for persons with exceptional eyesight.

The clumsy size causes a tendency to round off measurements to the nearest millimeter or half millimeter. In the next chapter where scales and rules are considered, you will see that rounding off to a half space can produce errors as large as a quarter of that space. In this case, that would be 0.25 mm. This is an error of 0.01 in., and certainly larger than desired.

If, in contrast, the inch is divided into 50 parts the smallest graduations are 0.02 in. These can be resolved easily by normal eyesight without undue eyestrain. If these graduations are rounded off to 0.01 in. the possible error is only 0.005 in., or half of that with the metric scale.

---

*Hastings, Russell, "Decimalized Measure", Engineering Conference of the American Society of Tool and Manufacturing Engineers, May 7, 1962, paper 437, p. 1

FRACTIONAL INCH, METRIC AND DECIMAL INCH SCALE READING COMPARED

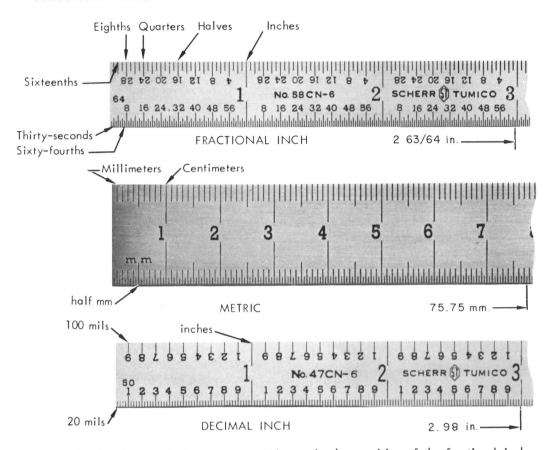

Fig. 3-8 The decimal-inch readings provide nearly the precision of the fractional-inch scale and the simplicity of the metric scale. Steel rules with both scales are now available.

It might be argued that the metric scale could be divided into half-millimeter spaces. These would be nearly the same size as the proposed 0.02-in. graduations. They would, therefore, seem to be as useful. This, unfortunately, gives the metric system one of the supposed defects of the inch system that we would like to eliminate – fractions.

The three alternatives are compared in Fig. 3-8. The top example is the conventional 6-in. steel rule with the 1/64-in. scale at the bottom. There is little argument that the host of fractions required with this scale (halves, quarters, eighths, sixteenths, thirty-seconds, and sixty-fourths) are as difficult to read as they are awkward to use.

The center example is a rule with upper edge graduated in millimeters and lower edge in half millimeters. Rounding off readings

between graduations involves quarter millimeters. Reading that scale involves relatively large whole numbers combined with fractions. On the half-millimeter scale it requires a reading of 75 3/4 mm to correspond to 2 63/64 in. - hardly an improvement.

The same reading on the bottom example, the decimal-inch graduated rule, is 2.98 in. This scale can be read to 0.01 in. by estimating half graduations. The next chapter will convince you, I hope, that this is hazardous. But in comparing scales we must recognize that this will be done. If such an estimation is made, 2.98 in. for example, we have three digits. The corresponding reading on a half-millimeter scale is 75.75 mm. That requires four digits instead of three. Rather than have so many digits it would be natural to "round off" this reading. Thus there is a tendency for error caused by excessive rounding off with the metric scale.

Summing this up: The metric system provides units of length that are either too large or too small for taking advantage of the normal resolution of the unaided eye. The inch system provides good resolution of its smallest divisions but loses this advantage because of the hodge-podge of fractions it requires. The decimal-inch system can take advantage of the resolving power of the eye and still retain the simplicity of decimal divisions.

COMMUNICATIONS ADVANTAGES

In the first chapters we stressed the fact that measurement is a language. We can measure the worth of a language by the ability it gives us to communicate clearly, accurately and simply. The systems of measurements vary considerably in this respect. Because the terminology of the decimal-inch system was devised with this in mind, it would be inexcusable for it not to be superior to either the metric or the inch systems. A distinction must be made between language and vocabulary. The terminology is vocabulary. We are free to invent arbitrary vocabularies any time we desire. In this, teenagers and jazz musicians are particularly fecund. Whether or not a new vocabulary becomes a new language is another matter. Use, not invention or law, makes languages. If the terminology is natural it is easily adapted into a language. This is true regardless of its merit or lack of merit. If it is unnatural, its only use is esoteric. The decimal-inch system is most vulnerable in this area.

In selecting the terminology, the following criteria were used for the new terms:*

1. *Short* - they should preferably use one syllable, like inch, mile.
2. *Pronounceable* - the correct pronunciation should not only be easy, it should be obvious and understandable even with foreign accents.
3. *Distinctive* - to avoid confusion words should not be similar in speech, writing or print.
4. *Pleasant* - terms should avoid unpleasant sounds or connotations. Slang gave clues.
5. *Learnable* - terms should be easy to learn and remembered by logical derivation.

Finding terms that fulfilled all of the above requirements was not easy. Only three terms

---

*Hastings, Russell, "Decimalized Measure", Engineering Conference of the American Society of Tool and Manufacturing Engineers, May 7, 1962, paper 437, p. 5.

are needed for measurements small enough to be expressed in inches: *inch, mil and mike.*

In 1958 an international committee agreed on prefixes. These were adopted by the National Bureau of Standards. Most are too unfamiliar or unwieldy to fulfill the requirements of the decimal-inch system. They did provide some help, however.

The "thousandths" division of the inch is one of the most useful. Unfortunately it is a tongue twister, hard to spell, and easily confused with "thousand". Slang was already shortening it to "thou" which is even more ambiguous. *Mil* was a much better choice. It is similar to the official international prefix "milli" meaning thousandths. The prefix "mil" is from Latin and widely used by nontechnical people. And it is short enough.

In practical work one tenth of a thousandth of an inch is frequently encountered. Alas, it acquired the most unfortunate name possible, the "tenth". Tenth of what? The resulting mistakes that have been made are legend. Furthermore, in the opinion of many who have studied the metric system, if it teaches anything, it is that units that vary only by a factor of ten are an open invitation to trouble. Therefore, this interval is skipped in the decimal-inch system. One-tenth part of one thousandth inch is simply *point one mil* (three syllables) or *one-hundred mike* (four syllables). This may or may not be convenient.

The next unit is one millionth of an inch. Again the word was chosen for much the same reason as was *mil.* "Micro" of the international prefixes was an obvious choice. It is of Greek origin and widely used in many languages. In the decimal-inch system it is simplified to *mike.*

Measurements larger than an inch are described as so-many inches in the decimal-inch system. *The B87 standard does not provide terminology to replace feet, yards, fathoms, rods, furlongs, miles, or leagues.* This was discussed, and it is a good guess that eventually it will come. The larger unit has been suggested as the "hin" taken from Hundred INches or Hecto INches, both of which mean the same. The next unit suggested is 1,000 hin. This would be called a "kay" from the "K" of "kilohin". This would equal 1.578 miles which is convenient because it converts evenly into the metric system. One "kay" would equal 2.54 kilometers.

EXAMPLES OF DECIMAL-INCH TERMINOLOGY

WRITE		SAY (Preferred form is first)	NUMBER OF SYLLABLES (to compare brevity)
For type-set material	On Drawings		
0.002" or 2 mil	.002	two mil	2
0.012" or 12 mil	.012	twelve mil	2
0.02" or 20 mil	.02	twenty mil	3
		point zero two inch	5
0.2"	.20	point two inch	3
2.005"	2.005	two inch five mil	4
		two point zero zero five inch	8
2.00"	2.00	two inch	2
0.000005" or 5×10^{-6} " or 5 mike	.000005 or 5 MK	five mike	2
0.00002" or 20 mike	.00002 or 20 MK	twenty mike	3
0.0002	.0002	point two mil	3
		two hundred mike	4
		point zero zero zero two inch	9
		two tenths *	2
0.0025"	.0025	two point five mil	4
		point zero zero two five inch	8
2.000005"	2.000005	two inch five mike	4
2.0005"	2.0005	two inch point five mil	5
		two point zero zero zero five inch	10

* Colloquialism – not recommended

Fig. 3-9   This is the terminology that is used wherever appropriate in this book. (From ASA B87, American Standards Association, Table 1)

Although these larger units were *not adopted* they demonstrate an important advantage of the decimal-inch system. They get along very well with the metric system. All conversions to metric use the basic number 254 because all units are multiples of an inch.

The way the decimal-inch system works out in general practice is shown in Fig. 3-9. Note that the preferred form is in every case the one with the least syllables. This makes sense, but human preference does not always follow logical lines.

This book is the first major work to use the proposed decimal-inch system. The author's experience has disclosed only three major defects. All are communicative. They are defects of omission, identification and combination.

Much has been said about the superiority of the inch system because its values are natural, or relate to useful things in real life. The reason for omitting the troublesome "tenth" (0.0001 in.) is understood. Unfortunately, it appears to be one of those natural intervals without which discussion of practical work is difficult. It is so unnatural to avoid it that the author has adapted it to the decimal-inch system as a *tenth mil* and used it throughout. There is sufficient agreement on this that it may be added before the final version of ASA B87 is published.

When using the conventional inch system we easily identify the precision we are using, such as "sixty-fourths", or "thousandths". Similarly in the metric system we can refer to "centimeters", "millimeters", etc. But how do you specify that you are working to a precision based on increments of 0.002 or 0.02 in. both of which are basic intervals of the decimal-inch system? "Bob, I want you to make this part to twenty mils." That would be an awkward way to say that steel rule measurement is adequate. It could also be dangerously ambiguous. If the "s" on mils is not heard, Bob might grind the part down to a thickness of *twenty mil.* As a result of the necessity to identify degrees of precision, the author has found it necessary to continually refer to such-and-such "increments", the *"twenty-mil increment"*, for an example. This is seldom necessary with either the conventional inch or the metric terminology. It appears to be a basic defect in the proposed system for which no easy solution is in sight.

The third defect is the difficulty in forming combinations. Unfortunately the proposed standard not even hints at what you should do when you want to state such a dimension as 0.2555 in. Using the recommended terminology it would be "point two inch, fifty five point five mil". Try some other combinations of the recommended forms yourself. They range from preposterous to hysterical. How much easier it is to say "two hundred fifty-five and a half thousandths". And that probably shows what will be done in the decimal-inch system. In this example, the chances are that the natural choice will be to simply replace "thousandths" with "mil" and let it go at that. But what about the "and a half" in that expression? If we try to rule out all frac-

tions we may doom the decimal inch system before it starts.

COMPUTATIONAL ADVANTAGES OF THE DECIMAL INCH

There are two major computational advantages of the decimal-inch system. The first one is quite obvious. It eliminates the difficulty of working with fractions. The second one is more subtle, but just as important. It minimizes the error that is caused when figures are rounded off.

Fractional computations require as much as five times as long as decimal computation. All of the fractions in an addition or subtraction problem must be changed to the same denominator, Fig. 3-10. In multiplication both the numerators and denominators must be dealt with separately. In all cases the results must be reduced to their simplest form before the problem is completed.

Even when it is decided to handle the problem in decimals it remains tedious. In the beginning the fractions must be converted to decimals. Whenever an odd number (smaller than one) is divided by two, the quotient is odd and one place longer than the original number. Successive dividing of one into its fractions results in the 64ths having six-decimal places, Fig. 3-11. Too bad that the dividing could not have begun with an even

COMPUTATION COMPARISON

SYSTEM	ADDITION		SUBTRACTION		
Fractional System	1-1/64 9/32 1-31/32 2-3/4	1-1/64 18/64 1-62/64 2-48/64 4-129/64 6- 1/64	1-5/8 -9/32 -27/64	1-40/64 -18/64 -27/64	104/64 -45/64 59/64
Exact Decimal Equivalents	1.015625 .28125 1.96875 2.75 6.015625		1.625 - .28125 - .421875	1.625000 -.703125 .921875	
Two Place Decimal System	1.02 0.28 1.97 2.75 6.02		1.62 -.28 -.42	1.62 -.70 .92	

Fig. 3-10 The computational advantage of the decimal-inch system is evident from this comparison of an addition problem and a subtraction problem, both solved by the three methods. (Compiled from ASA B87, figs. 5 and 6).

SUCCESSIVE HALVING

1	1.	
1/2	0.5	1 place
1/4	0.25	2 places
1/8	0.125	3 places
1/16	0.0625	4 places
1/32	0.03125	5 places
1/64	0.015625	6 places

Fig. 3-11 Successive halving of one (1) in order to form the common fractions results in the 64ths having six decimal places.

number. If the problem involves 32nds, five place figures must be used. If it involves 64ths, six place figures are necessary. Then at the end of the problem, the long decimal must be converted back into a fraction.

Later in this book the extreme importance of a seemingly obvious point is really hammered upon. Every step in the measurement procedure is a potential source of error. For the most accurate and reliable measurement the total number of steps must be kept at a minimum. This applies to all steps. It is often overlooked that computational steps are just as much distinct measurement steps as those involving setting up or reading of instruments. Therefore, the many extra steps required when computing with fractions must be recognized as a deterrent to reliable measurement.

Much of the entire computation may be meaningless, anyway. A 64th is approximately 0.016 in. If that is the most precise reading, what can an answer such as 1.96875 in. mean? Converted into fractions it is 1 31/32 in. But would not 1.9687 in. also be considered to be 1 31/32 in.? Or even 1.97 in., for that matter?

Beyond the second decimal place the additional numbers have little effect on the selection of the nearest 64th. Yet the accumulative effect of the third to sixth-decimal place in a problem involving several fractions could change the result by a 64th. Therefore, in using fractions they are kept in their unwieldy form, even though the last places are meaningless in each individual quantity.

It has been pointed out that halving an even number is preferable to an odd number, because the number of decimal places is not increased. Thus the successive halving can continue without an increase in the number of decimal places until an odd quotient is reached. For example: 56, 28, 14, 7; but the next halving results in two places, 3.5. As shown in Fig. 3-11, continued halving would then add a decimal place for each division. Obviously, for practical reasons the unnecessary places are rounded off rather than carried along.

ROUNDING-OFF NUMERICAL VALUES

Rounding off refers to the elimination of unnecessary figures in any computational problem. It is useful in business, the crafts, around the home and in nearly every field where numbers are required. Yet, in spite of the serious errors caused by improper rounding off, very few people do it correctly. This discussion should be of value to you in practically anything you do - and it will show the wisdom behind the choice of 0.02 in. as the smallest decimal-inch division.

Suppose, for example, that you desire to know the volume of a block measuring 15.2 x 11.1 x 8.3 in. Multiplying out results in 1,400.376 cubic in. The values in the first place were known only to one-decimal place. Thus, carrying the answer to three places is to presume that the answer is more accurate than the information on which it was based.

If, for example, measured more carefully the block is found to be 15.20 x 11.10 x 8.29 in. Only one of the values is changed and only by 0.01 in., but the change in the volume is startling. It now becomes 1,398.6888 cu. in. If the other measurements had contained errors, the change might have been even greater, or they might have cancelled out. The practical person, recognizing this *area of uncertainty*, would consider 1,400 cu. in. to be the volume. He rounded off the meaningless figures.

Common sense tells us that in rounding off, we should leave the last remaining digit as is if the digit eliminated is less than five, and we should raise it if the digit eliminated is more than five. In this example the reductions would be as follows:

	1,400.376	and		1,398.689
to	1,400.38		to	1,398.69
to	1,400.4		to	1,398.7
to	1,400		to	1,399.
			to	1,400

In both cases, the results are the same. And in neither case did you have to stop and think. The steps were rather obvious. What would the results have been, however, if the two figures had been 1398.6855 and 1398.6845? Now we have choices. In neither case is the last digit greater than or smaller than five.

There are several schools of thought on this. One says that it does not matter if you *round up* or *round down*, Fig. 3-12, as long as you are consistent. Consistency is one of the important means to achieve reliable measurement. In many instances it assures that the inevitable errors have a chance to cancel each other. It is important, however, not to chose techniques that are consistently wrong. And that applies to rounding up and rounding down.

Let us say, to prove this, that instead of one block we are concerned with the total volume of five blocks. We get this by finding their separate volumes and adding them together. Typical results are shown in Fig. 3-13. In this example only three of the values ended in five and were actually "on the fence". Yet the rounded up total was 0.03 in. larger than the rounded down total. If the column of values had been longer, the difference would become progressively greater. If 1,000 blocks had been involved and they were made of one of the new high-strength alloys devel-

oped for space vehicles, the cost represented by this volume difference could be an important factor.

How can you round off to get closer to the average? You could go up half of the time, down the other half. That means keeping track - and there you have introduced another operation. Fortunately there is a better way, so much better in fact that it is incorporated in a standard.* Actually, the indirect advantages of proper rounding off are far more important than the direct savings. They could be a life or death matter when medical data are analyzed, or a high defective rate when production data accumulate errors.

This rule is to raise the remaining last digit if it is odd and to leave it the same if it is even. Thus, the new value will always end up even when the preceding one ended in five. For example: 1398.65 becomes 1398.6, while 1398.75 becomes 1398.8. This was rounding down in the first case, rounding up in the second.

It was on the basis of this method of rounding off that the basic 0.02 division of the decimal-inch system was obtained from the basic 1/64 division of the fractional inch: 0.015625 to 0.01562 to 0.0156 to 0.016 to 0.02. The complete table for conversion of fractions to the decimal-inch system is reproduced in Fig. 3-14.

Whether or not the decimal-inch system is safely convertible to the fractional-inch system, thus, depends upon whether this method of rounding off results in truly average values. Fig. 3-15 shows that it is. In this table the true values and the rounded-off values of all fractions from 1/64 through one are added and averaged. It is seen that the sum of the rounded values, regardless of the places retained, is in each case equal to the sum of the values given to six places. Neither the sums nor the averages have been increased or decreased by the rounding off.

This is not really a proof because the distribution of fives is systematic instead of random. The results will be essentially the same if random values are chosen - providing the series is sufficiently long.

---

*"Rules for Rounding off Numerical Values", <u>ASA Z25.1</u>, American Standards Association, 1940.

ROUNDING UP AND ROUNDING DOWN

VALUE	ROUNDED UP				ROUNDED DOWN			
PLACES	PLACES				PLACES			
5	4	3	2	1	4	3	2	1
2.07550	2.0755	2.076	2.08	2.1	2.0755	2.075	2.07	2.1
2.07551	2.0755	2.076	2.08	2.1	2.0755	2.075	2.07	2.1
2.07552	2.0755	2.076	2.08	2.1	2.0755	2.075	2.07	2.1
2.07553	2.0755	2.076	2.08	2.1	2.0755	2.075	2.07	2.1
2.07554	2.0755	2.076	2.08	2.1	2.0755	2.075	2.07	2.1
2.07555	2.0756	2.076	2.08	2.1	2.0755	2.075	2.07	2.1
2.07556	2.0756	2.076	2.08	2.1	2.0756	2.076	2.08	2.1
2.07557	2.0756	2.076	2.08	2.1	2.0756	2.076	2.08	2.1
2.07558	2.0756	2.076	2.08	2.1	2.0756	2.076	2.08	2.1
2.07559	2.0756	2.076	2.08	2.1	2.0756	2.076	2.08	2.1

Fig. 3-12 Rounding up requires the digits retained to be raised when the digit eliminated to the right is a 5. Rounding down requires that it be kept the same. In the example above, the shaded areas are the values that required the up and down decisions.

ROUNDING UP vs ROUNDING DOWN

	Calculated Volume	Rounded Up	Rounded Down
	1400.375	1400.38	1400.37
	1400.376	**1400.38**	**1400.38**
	1399.995	1400.00	1399.99
	1400.395	1400.40	1400.39
	1399.991	**1399.99**	**1399.99**
Totals	7001.132	7001.15	7001.12
Differences		over by 0.02	under by 0.01
Average T/5	1400.226 (true)	1400.230 (high)	1400.224 (low)

Fig. 3-13 In this group of five values the rounded up column yielded high values, the rounded down, low values. The more items involved, the greater the total error and average error.

Like so many good rules, this one has exceptions, too. *The final rounded value should be obtained from the most precise value obtainable and not from a series of roundings.* For example, 0.5499 should be rounded off successively to 0.550, 0.55 and 0.5. The last one is 0.5, not 0.6 which it would be if you followed the rule. The reason is clear when you study what has taken place. The most precise value is less than 0.55, therefore, the final rounding is to 0.5. Similarly, 0.5501 is rounded off to 0.550, 0.55 and 0.6 because the most precise value obtainable is greater than 0.55.

In the entire 64 conversions from fractions to decimal inch this exception occurs only twice:

29/64 is 0.453125 to 0.45312 to 0.4531 to 0.453 to 0.45 to .5

35/64 is 0.546875 to 0.54688 to 0.5469 to 0.547 to 0.55 to .5

If the two-place values had been exact instead of the result of successive roundings the one-place values would have been 0.4 and 0.6 respectively.

FRACTION TO DECIMAL CONVERSION CHART

4 THS	8 THS	16 THS	32 NDS	64 THS	TO 4 PLACES	TO 3 PLACES	TO 2 PLACES	4 THS	8 THS	16 THS	32 NDS	64 THS	TO 4 PLACES	TO 3 PLACES	TO 2 PLACES
				1/64	0.0156	0.016	0.02					33/64	0.5156	0.516	0.52
			1/32		0.0312	0.031	0.03				17/32		0.5312	0.531	0.53
				3/64	0.0469	0.047	0.05					35/64	0.5469	0.547	0.55
		1/16			0.0625	0.062	0.06			9/16			0.5625	0.562	0.56
				5/64	0.0781	0.078	0.08					37/64	0.5781	0.578	0.58
			3/32		0.0938	0.094	0.09				19/32		0.5938	0.594	0.59
				7/64	0.1094	0.109	0.11					39/64	0.6094	0.609	0.61
	1/8				0.1250	0.125	0.12		5/8				0.6250	0.625	0.62
				9/64	0.1406	0.141	0.14					41/64	0.6406	0.641	0.64
			5/32		0.1562	0.156	0.16				21/32		0.6562	0.656	0.66
				11/64	0.1719	0.172	0.17					43/64	0.6719	0.672	0.67
		3/16			0.1875	0.188	0.19			11/16			0.6875	0.688	0.69
				13/64	0.2031	0.203	0.20					45/64	0.7031	0.703	0.70
			7/32		0.2188	0.219	0.22				23/32		0.7188	0.719	0.72
				15/64	0.2344	0.234	0.23					47/64	0.7344	0.734	0.73
1/4					0.2500	0.250	0.25	3/4					0.7500	0.750	0.75
				17/64	0.2656	0.266	0.27					49/64	0.7656	0.766	0.77
			9/32		0.2812	0.281	0.28				25/32		0.7812	0.781	0.78
				19/64	0.2969	0.297	0.30					51/64	0.7969	0.797	0.80
		5/16			0.3125	0.312	0.31			13/16			0.8125	0.812	0.81
				21/64	0.3281	0.328	0.33					53/64	0.8281	0.828	0.83
			11/32		0.3438	0.344	0.34				27/32		0.8438	0.843	0.84
				23/64	0.3594	0.359	0.36					55/64	0.8594	0.859	0.86
	3/8				0.3750	0.375	0.38		7/8				0.8750	0.875	0.88
				25/64	0.3906	0.391	0.39					57/64	0.8906	0.891	0.89
			13/32		0.4062	0.406	0.41				29/32		0.9062	0.906	0.91
				27/64	0.4219	0.422	0.42					59/64	0.9219	0.922	0.92
		7/16			0.4375	0.438	0.44			15/16			0.9375	0.938	0.94
				29/64	0.4531	0.453	0.45					61/64	0.9531	0.953	0.95
			15/32		0.4688	0.469	0.47				31/64		0.9688	0.969	0.97
				31/64	0.4844	0.484	0.48					63/64	0.9844	0.984	0.98
1/2					0.5000	0.500	0.50	1					1.0000	1.000	1.00

Fig. 3-14 Generally, when converting fractional inches to decimal inch, it is only necessary to convert to two places. If the fractional dimension is the basic size of a final part and a gage is being made to check it, then there is reason to use more than two decimals. (from ASA B 87, page 5).

To assume that your roundings are precise, whenever possible carry the calculation two places beyond the values needed, then round off. The rules for rounding off values are summarized in Fig. 3-16. The result of the application of these rules is the table of equivalents reproduced in Fig. 3-14 which converts fractions into decimal-inch.

1	2	3	4	5	6
6 PLACES	5 PLACES	4 PLACES	3 PLACES	2 PLACES	1 PLACE
0.015625	0.01562	0.0156	0.016	0.02	0.0
.031250	.03125	.0312	.031	.03	.0
.046875	.04688	.0469	.047	.05	.0
.062500	.06250	.0625	.062	.06	.1
.078125	.07812	.0781	.078	.08	.1
.093750	.09375	.0938	.094	.09	.1
.109375	.10938	.1094	.109	.11	.1
.125000	.12500	.1250	.125	.12	.1
.140625	.14062	.1406	.141	.14	.1
.156250	.15625	.1562	.156	.16	.2
.171875	.17188	.1719	.172	.17	.2
.187500	.18750	.1875	.188	.19	.2
.203125	.20312	.2031	.203	.20	.2
.218750	.21875	.2188	.219	.22	.2
.234375	.23438	.2344	.234	.23	.2
.250000	.25000	.2500	.250	.25	.2
.265625	.26562	.2656	.266	.27	.3
.281250	.28125	.2812	.281	.28	.3
.296875	.29688	.2969	.297	.30	.3
.312500	.31250	.3125	.312	.31	.3
.328125	.32812	.3281	.328	.33	.3
.343750	.34375	.3438	.344	.34	.3
.359375	.35938	.3594	.359	.36	.4
.375000	.37500	.3750	.375	.38	.4
.390625	.39062	.3906	.391	.39	.4
.406250	.40625	.4062	.406	.41	.4
.421875	.42188	.4219	.422	.42	.4
.437500	.43750	.4375	.438	.44	.4
.453125	.45312	.4531	.453	.45	.5
.468750	.46875	.4688	.469	.47	.5
.484375	.48438	.4844	.484	.48	.5
.500000	.50000	.5000	.500	.50	.5
.515625	.51562	.5156	.516	.52	.5
.531250	.53125	.5312	.531	.53	.5
.546875	.54688	.5469	.547	.55	.5
.562500	.56250	.5625	.562	.56	.6
.578125	.57812	.5781	.578	.58	.6
.593750	.59375	.5938	.594	.59	.6
.609375	.60938	.6094	.609	.61	.6
.625000	.62500	.6250	.625	.62	.6
.640625	.64062	.6406	.641	.64	.6
.656250	.65625	.6562	.656	.66	.7
.671875	.67188	.6719	.672	.67	.7
.687500	.68750	.6875	.688	.69	.7
.703125	.70312	.7031	.703	.70	.7
.718750	.71875	.7188	.719	.72	.7
.734375	.73438	.7344	.734	.73	.7
.750000	.75000	.7500	.750	.75	.8
.765625	.76562	.7656	.766	.77	.8
.781250	.78125	.7812	.781	.78	.8
.796875	.79688	.7969	.797	.80	.8
.812500	.81250	.8125	.812	.81	.8
.828125	.82812	.8281	.828	.83	.8
.843750	.84375	.8438	.844	.84	.8
.859374	.85938	.8594	.859	.86	.9
.875000	.87500	.8750	.875	.88	.9
.890625	.89062	.8906	.891	.89	.9
.906250	.90625	.9062	.906	.91	.9
.921875	.92188	.9219	.922	.92	.9
.937500	.93750	.9375	.938	.94	.9
.953125	.95312	.9531	.953	.95	1.0
.968750	.96875	.9688	.969	.97	1.0
.984375	.98438	.9844	.984	.98	1.0
1.000000	1.00000	1.0000	1.000	1.00	1.0
32.500000	32.50000	32.5000	32.500	32.50	32.5 (SUM)
0.5078125	0.50781	0.5078	0.508	0.51	0.5 (AVERAGE)

Fig. 3-15 The value of the recommended method for rounding off is demonstrated in this table. Six-place decimal equivalents of all 64 of the 64ths of an inch are successfully rounded off to one place. These vertical columns are added and averaged. When the six-place average is rounded off, the successive results are the same as the averages — This shows that the rounding off did not introduce an error. (from ASA Z25.1 — 1940, page 7).

GENERAL RULES FOR ROUNDING OFF

When a value is to be reduced in number of decimal places, one of the following three rules is followed:

1. When the digit to be dropped is less than 5, there is no change in the preceding figures.
    Examples:
    0.280423 to 0.28042 to 0.2804 to 0.280 to 0.28

2. When the digit to be dropped is greater than 5, the preceding digit is increased by 1.   Examples: 0.046857 to 0.04686 to 0.0469 to 0.047 to 0.05

3. When the digit to be dropped is exactly 5, round off to nearest even number.
    Examples:
    0.09375 to 0.0938 but 0.09385 to 0.0938

Fig. 3-16 Whenever possible, carry the calculation two places beyond the desired value, then round off the last two significant figures.

DANGERS AHEAD

The threat to wide adoption of the decimal-inch system for industrial use is psychological. It is a human tendency to tenaciously extend any system beyond its natural limits. The system becomes a thing to be perfected for its own sake and its original function is forgotten. History abounds with examples. The metric system itself is one. A more recent one is the attempt of telephone companies to force all-number dialing on a public which wants no part of it. And anyone who has ever fought a case in court, except the attorneys, is bound to wonder how the judicial system got so far from the needs of the people who created it. This can happen to the decimal-inch system.

If those responsible for its adoption can keep in mind that it is only a vocabulary and not a language, there is hope. It is those of us who use measurement regularly who will either make it into a new and viable language or will disregard it. Our action will depend upon the degree to which it serves us. For it to serve us well it must be adapted to our needs. If strictly codified it will probably share the fate of the metric standard for time measurement. If allowed to remain pliable it will be usable and valuable.

Summary — Which System to Use

● There is no insurmountable difficulty involved in a total change from the inch-pound system to the metric system, nor is there any convincing proof that such a change is needed or desirable.

● The metric system is unquestionably superior in ease of computation and communication. Popular use of the metric system in science clearly advocates its continued use. The units of the metric system were selected theoretically. It is not surprising that they bear little natural relation to most things in the real world, including the resolving power of the human eye.

● The inch-pound system is handicapped by a bewildering and disorderly assortment of units. This slows computation and clogs communications. However, its basic unit, the inch is of such convenience that it is used to produce a major part of the world's goods. It also finds wide use in places in which the metric system is the legal system.

● Each system has merit and has roles in which it is fully accepted. Rather than force one system into the role of the other it is preferable to combine their best points. This has been done for the inch-pound system. The result is the decimal-inch system.

● The decimal-inch system uses the accepted and convenient unit, the inch. It uses 0.02 in. as its basic scale division. It provides the ease of computation that the metric system boasts. With the terminology chosen for it, communication is much better but leaves something to be desired. Moreover, it can be converted readily in and out of both the metric system and the inch-pound system.

● The reason it converts easily to the fractions of the inch-pound system is the systematic method used to round off values. When the decimal equivalents of the fractions are reduced to two-place decimals, these new values are in the decimal-inch system.

● Acceptance of this system by the American Standards Association and cooperating organizations rapidly will replace the fractional inch-pound system in the United States, and possibly in other places where the inch-pound system is standard. Therefore this text uses decimal inch terminology wherever appropriate. Italics are used in order to call attention to this new terminology. The greatest threat to the decimal-inch system is to consider it a final solution to inch-pound measurement problems.

TERMINOLOGY

Metric system	System of measurement based on the use of the meter, kilo and second as the units of length, mass and time.
Inch-pound system	System of measurement based on the use of the inch, pound and second, often called "English System."
Decimal-inch system	System of linear measurement based on decimal division of the inch and certain specific terminology to assist communication and computation.
Unit of length	Basis upon which the length standard is defined.
Standard of length	Physical embodyment of the unit of length in usable form.
International inch	The reconciled inch standard of major inch-pound system countries, equal to 25.4 mm.
Computational factor	The relative ease with which the measurements can be used.
Communicative factor	The relative ease with which correct information can be passed from one person to another.
Resolution	The ability to distinguish between separate items.
Mil	One thousandth of an inch in decimal-inch terminology.
Mike	One millionth of an inch in decimal-inch terminology.
Round off	To arbitrarily eliminate the last decimal place.
Round up	To raise the next remaining decimal place when rounding off a numeral five in last place.
Round down	To leave intact the next remaining decimal place when rounding off a numeral five in last place.
Area of uncertainty	Those digits in a result that extend to more decimal places than warranted by the information on which calculation was based.

Fig. 3-17   The terms introduced in this chapter have much use in other fields in addition to measurement.

QUESTIONS

1. What was the earliest recorded standard of measurement?
2. How did this ancient standard of measurement recognize metrology principles that apply today?
3. What was the original method used for defining the length of a meter?
4. Why was this original definition of the meter revised?
5. Why did the leading nations finally meet for the purpose of establishing an international metric system?
6. Why did light waves replace metal bars as the international standard?
7. What are the three basic factors that must be considered when making a comparison of the inch and metric systems?
8. Explain each of the three factors.
9. An ideal measurement system must consist of four qualities. What are they?
10. It is obvious that both the metric and the inch-pound systems have advantages and disadvantages. Is it possible to alleviate the problem? Explain.
11. Why is the millimeter difficult to apply to scale divisions?
12. Since measurement can be classified as a language, what criteria should be used for selecting clear and simple measurement terms?
13. What is the basic increment of the decimal-inch system?
14. Can the decimal-inch system be easily applied to the metric system?
15. At the present time, what is the principal objection of opposing factions toward conversion to the metric system?
16. What is meant by rounding off and when is it required?
17. What is the exception to the rule for rounding off?
18. Does rounding off only apply to the inch-pound system?

DISCUSSION TOPICS

1. Man's ability to measure has reflected his way of life, and has been his most serious concern in this struggle to harness and control natural resources. Why has it been so difficult to establish realistic measurement standards?
2. Explain why industry has been primarily responsible for proposing and introducing the decimal-inch system.
3. Enumerate several general terms that you feel could easily be applied to a universal measurement system.
4. A reliable and practical measurement system helps to prevent and reduce errors. Discuss the major elements that contribute to measurement errors.
5. Explain why the decimal-inch system can reduce measurement errors.
6. Explain the significance of the rounding-off matter in a chapter devoted to systems of measurement?

> *The Congress shall have Power . . . To regulate Commerce . . . among the several states . . . to fix the Standard of Weights and Measures . . .*
>
> Constitution of the United States
> Section 8, Article 1

Measurement with Graduated Scales
The Rule for Rules

Gather up all the rulers, yardsticks, machinist's rules and measuring tapes that are handy. Lay them out on a flat surface. Compare their lengths by matching the graduations on one end of two adjacent scales and then noting what happens at the opposite ends. This is shown in Fig. 4-1.

Do not be surprised if no two are alike. Even those that at a glance appear to be in agreement seldom are if a magnifier is used. Perhaps this explains why the cabinet drawer you so carefully made had to be encouraged slightly before it would push into the space it was intended to fill.

Your first contact with linear measurement was probably with a ruler or tape measure. It is appropriate to introduce dimensional metrology with similar instruments. They are still in frequent use and they demonstrate principles that apply even to the most advanced instruments. If nothing else, this chapter will show you how to use rules and scales with greater precision.

THE STEEL RULE

First of all we need to clear up the difference between a scale and a rule. A scale is graduated in *proportion* to a unit of length.

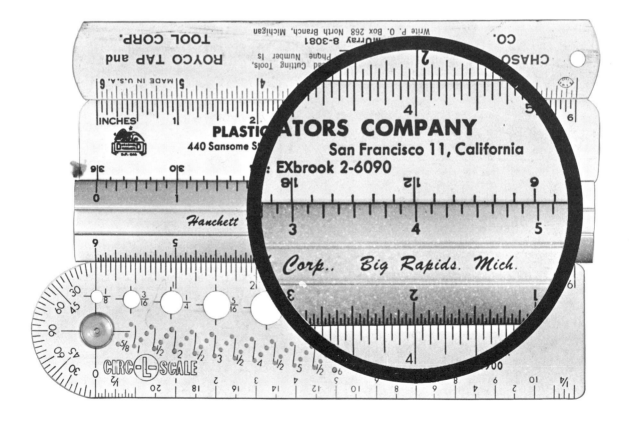

Fig. 4-1 A simple comparison of the scales and rules around you emphasizes the principal problem in measurement — error. Every measurement contains some error. For reliability it must be recognized.

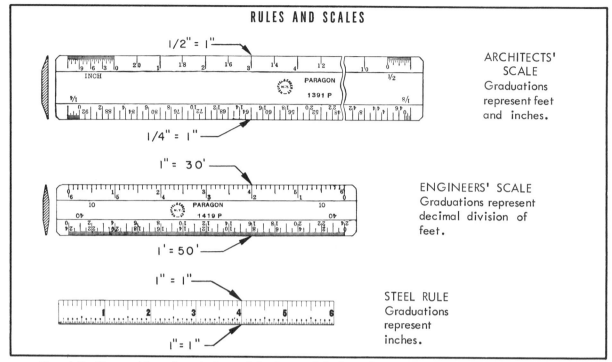

Fig. 4-2 The architects' scale is read in feet and inches, the engineers' in decimals of feet, but the machinists' steel rule reads directly in inches, either fractions or sometimes in decimals.

DISCRIMINATION

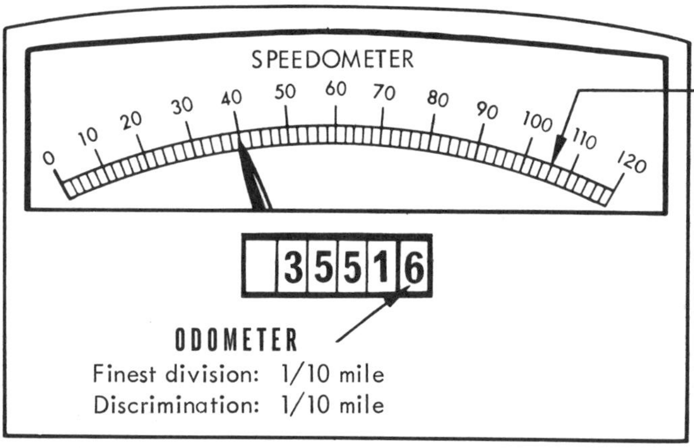

Finest division: 2 mph.
Discrimination: 2 mph.

Finest division: 1/10 mile
Discrimination: 1/10 mile

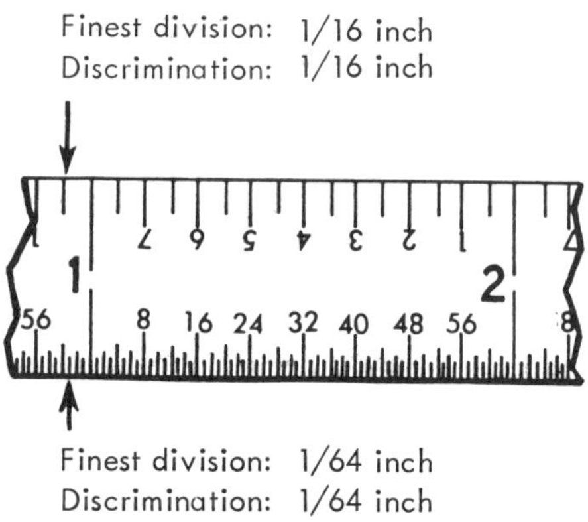

Finest division: 1/16 inch
Discrimination: 1/16 inch

Finest division: 1/64 inch
Discrimination: 1/64 inch

1. Discrimination is the fineness of the scale divisions of an instrument.

2. It is the smallest division of the scale that can be read reliably.

3. It is related to but not a measure of precision or accuracy.

Fig. 4-3 Everyone who has argued with a state highway patrolman knows that the alleged discrimination of his speedometer is not considered proof of its accuracy.

The draftsman uses an architects' scale whose divisions represent feet and inches or an engineers' scale whose graduations are read in decimal divisions of the foot, Fig. 4-2. On the machinists' steel rule, however, the graduations represent full-size inches, hopefully no more, no less. Warning - the old-timer becomes upset if you call a steel rule a scale. Unfortunately, even the old-timer frequently does not understand proper use of his favored instrument. It is best to do it right but not to argue with him.

There are many types of steel rules. All are narrow steel strips with one or more graduated scales. Yes, the graduated edges are properly called scales even though it is not advisable to call the instrument a scale in a machine shop. They vary in length from a fraction of an inch to several feet. The six-inch size is the most popular, perhaps because it can be carried in a pocket.

The degree to which an instrument sub-divides the unit of length is termed its *discrimination*, Fig. 4-3. A wide range of scales is available on rules. These are shown in Fig. 4-4. Most rules in the United States now have the first four scales. It is expected that the decimal-inch scales will rapidly come into wide use.

Using a steel rule requires three important considerations. First, which particular style of *rule* will best do the job? Second, which *scale* should be used? Third, what *method* of holding the rule and the part provides the most reliable measurement?

IT ALL STARTS HERE

The purpose of all three is to establish the most precise relationship between the reference point and measured point on the part with the scale graduations of the rule.

STEEL RULE SCALES

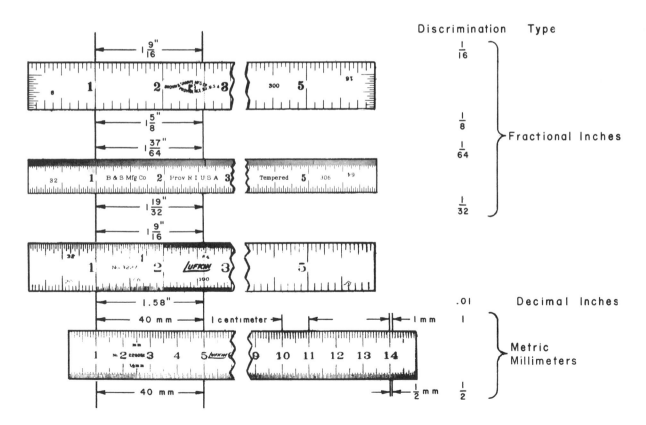

Fig. 4-4 Many scales are available. Top 4 are most popular in U.S. The decimal-inch scale is standard in some plants. Note that two lines 1 37/64 apart read differently on each of the scales.

First of all, consider aligning the reference point with the scale. Figure 4-5 shows several methods. The top two examples rely upon the condition of the end of the rule. Obviously, the corner receives more wear than the balance of the rule and is, therefore, difficult to align with the reference point on the part. Furthermore, the edge of the work may not be sharp which could make the situation even worse, Fig. 4-6.

Why is the last method in Fig. 4-5 preferred when it adds an extra step that requires careful alignment? The alignment of two lines (the edge appears as a line) can be made to great accuracy and visually checked. If a *knee* is used, an inward slope on the shoulder would result in a measurement shorter than the true length. Chips, burrs or poor surface finish could also cause errors. A rule with a hook, similarly, can cause errors unless the hook is frequently checked. *Every attachment used in measurement contributes to accumulated error.*

REFERENCE POINTS

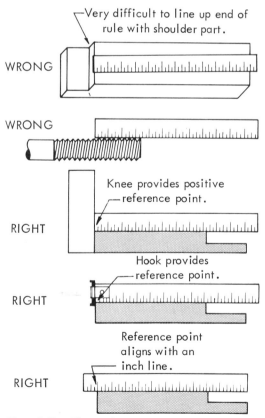

Fig. 4-5   The right way to use a rule is usually the easiest, fastest, and most reliable.

ALIGNMENT

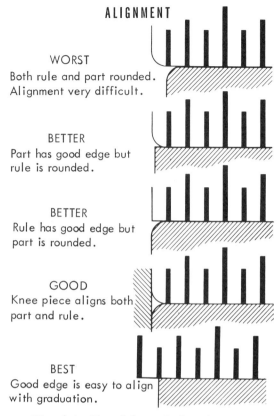

WORST
Both rule and part rounded. Alignment very difficult.

BETTER
Part has good edge but rule is rounded.

BETTER
Rule has good edge but part is rounded.

GOOD
Knee piece aligns both part and rule.

BEST
Good edge is easy to align with graduation.

Fig. 4-6   Use of the end of a rule invites errors. Examination with a magnifying glass will convince you.

SHRINK RULE

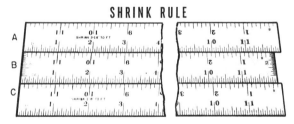

Fig. 4-7   Molten metal shrinks as it cools and solidifies. To compensate for the shrinkage, patternmakers use shrink rules that automatically compensate. A is a rule for brass, B is a standare rule and C is for cast iron. Beware of rules marked "SHRINK" unless you are a patternmaker.

The method needed to align the reference point with the rule is an important consideration in choosing a rule for a particular measurement. A more obvious one is size. If a longer rule than necessary is used, it will be unwieldy. If the dimension to be measured is in a recess, a short rule will be required. If the measurement is in a slot, a narrow one must be used.

DISCRIMINATION AND PRECISION

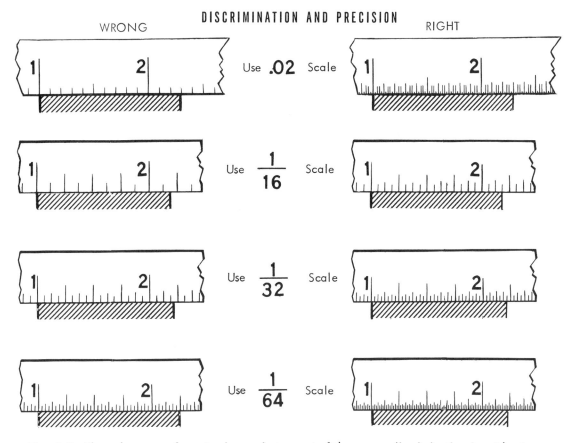

Fig. 4-8 The only excuse for not using an instrument of the proper discrimination is not having one. Unfortunately, this happens often. That is when basic knowledge is invaluable.

THE THINKING END

Any scale, on a rule or any other measurement instrument, should be read to the nearest graduation. There are two reasons for this. First, reliability, Fig. 4-8, is lost when interpolation between graduations is made. Second, there is generally no need for it. If a rule is being used, you can go to a scale with greater discrimination. If the finest scale is not adequate, you should use a more precise instrument. Only when you reach the point that an adequate instrument is not available should you interpolate. Even then, extreme precautions should be taken to minimize all contributing sources of error.

If you begin with a coarse scale and the measured point does not coincide with a graduation, turn to successively finer scales until one of three things happens: (1) The measured point coincides with a graduation. (2) You reach as fine a division as is needed. (3) You reach the finest scale, Fig. 4-8. One of these always happens.

The first situation presents no problems-- if you have carefully aligned both the reference point and measured point. The second is simply one of economy. There is no need to carry a dimension to 1/64 inch if 1/16 inch is all that is needed. But note carefully: if you do carry the measurement of a 1/16 inch measurement to 1/32 inch, you will have increased the reliability.

If you are using the 1/64-inch scale and the measured point falls between two graduations, it is conventional to take the closest one as the measurement. If it appears to equally divide the space, take the one that provides the best *margin of safety*, but keep in mind your bias. This you will hear more about later in the chapter. If you are using a decimal-inch graduated rule you can follow the rounding off instructions given in the last chapter without difficulty.

ERROR IS THE ENEMY

A discussion of the error of steel rule measurements is required to counteract the claims that many make that they can measure to *a mil* (0.001 in.) with one. They can and do--*but not reliably.*

There are four general sources of measurement error when using a steel rule: inherent instrument error, observational error, manipulative error and bias.

The inherent error can be disposed of quickly by using a quality steel rule, not just any advertising giveway with graduations that happens to be handy. Quality steel rules have a maximum inherent error of *point three mil* (0.0003 in.) per inch of length.* That is one-tenth of the usual width of a graduation. Except for very long measurements the inherent error can be ignored.

A fractional-inch rule is not intended to be read closer than 1/64 inch. The maximum error that can result from the width of a graduation is about *three mil* (0.003 in.) or half that (0.0015 in.) each side of the center line of a graduation. That is a potential error of one-tenth of the discrimination. This

amount, one-tenth, has important significance in the selection of measurement instruments and will be discussed later.

The smallest division of the decimal-inch scale is about 28 percent larger than that of the fractional-inch scale. This reduces the discrimination. It increases the ease of reading. And because the width of the graduations are the same in either case, the observational error is decreased with the decimal-inch scale. This is small but significant.

Only when both the reference point and measured point are aligned with graduations might the full width of one line alter the measurement, Fig. 4-9. Quality rules are *engine engraved.* A machine called a *ruling engine* actually cuts each graduation. Low cost rules are stamped or made by printing processes. Engine engraved graduations are sharp V's. For precise work, a magnifier can be used to carefully align the lines on the part with the centers of the graduations.

---

* ASA Z 75.1-1955 for decimal-inch scales limits the overall length tolerance on a 6-in. steel rule to plus 0.004 in. minus 0.002 in. For a 12- or 18-in. scale it is extended to plus 0.005 minus 0.002 inch.

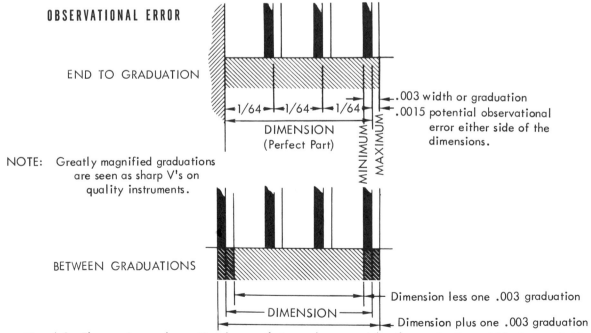

OBSERVATIONAL ERROR

END TO GRADUATION

DIMENSION (Perfect Part)

.003 width or graduation

.0015 potential observational error either side of the dimensions.

NOTE: Greatly magnified graduations are seen as sharp V's on quality instruments.

BETWEEN GRADUATIONS

DIMENSION

Dimension less one .003 graduation

Dimension plus one .003 graduation

Fig. 4-9 The maximum observational error that can be caused by the graduation widths is one-half graduation for an end setting, and one graduation for a setting between two graduations. With decimal-inch scales, the margin of safety is even greater because 0.02 in. is approximately 28 percent wider than 1/64 in.

OBSERVATIONAL ERROR

Parallax is one of the important forms of observational error. It is the apparent shifting of an object caused by the change of position of the observer. For example, the speedometer in an automobile appears to be reading slower to the passenger than to the driver. The only time that parallax does not affect measurement with a scale is when the scale edge of the rule is directly on the line of measurement.

Figure 4-10 is greatly exaggerated to show parallax. In the top drawing, the observer at B would correctly measure X as 16 divisions. A would measure it 15 and C would measure 17. If the scale had been placed directly in contact with the line of measurement as in the lower drawing, all three observers would have correctly measured X as being 16 divisions.

Parallax is combated in two ways. First, always have the scale edge of the rule as close to the line of measurement as possible. Thin steel rules are popular for this reason. Second, align head so that the line of sight is perpendicular to the measured point on the line of measurement, as at B in Fig. 4-10. You can almost always correct for parallax, providing you recognize it.

MANIPULATIVE ERROR

It is not easy to hold both the part and the scale and measure to 0.02 in., let alone to 1/64 in., or even closer. If one attempts to measure any closer, parallax will have to be corrected. The decimal-inch scale helps but even then parallax is a problem. The very act of moving the head to be perpendicular to the measured point will probably cause a shift between the rule and the part. This may cause greater error than the desired results.

Even when measuring as recommended, no closer than 1/64 inch, (or 0.02 with decimal-inch scales), manipulative errors such as the above must be carefully avoided. Some of the common pitfalls are shown in Fig. 4-11. Many of the common errors are due to "cramping." That is the name given to using excessive force. When you squeeze a rule (or any other instrument) tightly, you cannot be sure you are not forcing it against the part. In measurement always use the light touch.

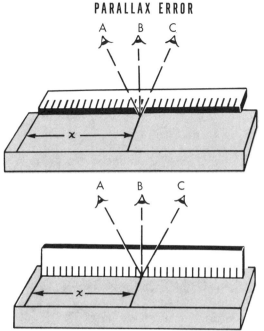

PARALLAX ERROR

Fig. 4-10 Parallax error is minimized by having the line of measurement of the rule as close to the feature being measured as possible.

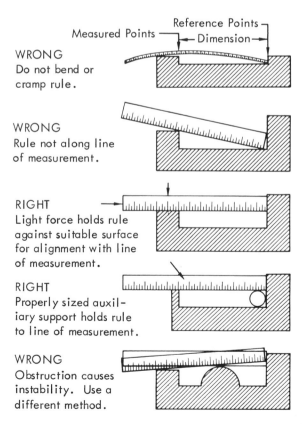

Fig. 4-11 These potential errors are exaggerated. In practice they may be much less obvious.

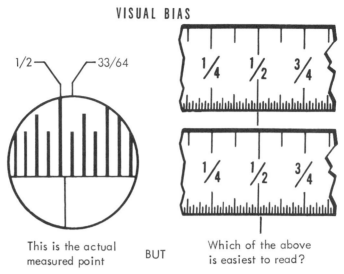

Fig. 4-12   It is a normal tendency to take the easiest route, especially if that produces hoped-for results. The finer the scale, the more serious this problem becomes.

YOU ARE BIASED

Bias has been saved to last. Bringing it up has much the same reaction as admitting a belief in ghosts. *We unconsciously influence each measurement we make.* It is an automatic tendency that exists whenever it is not consciously recognized and combated. Even then some bias usually remains.

For example, assume that a dimension needs to be 5 1/2 inch. It is obviously easier to read that graduation than the 5 31/64 or 5 33/64 graduations, even without bias, Fig. 4-12. Obviously, this will affect the reading of a measured point between the 5 1/2 and 5 33/64 graduations unless a conscious effort is made to fight the normal tendency.

The simple act of moving the line of sight to prevent parallax will usually cause slippage, Fig. 4-13. Strangely enough, these errors will statistically be on the side of the desired results. Add to this the fact that either the part or the scale or (worse) both may be held by hand. Even when they are

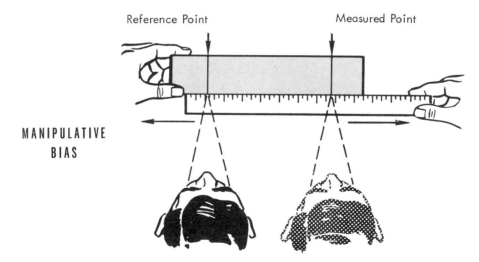

Fig. 4-13   When the line of sight is moved from reference point to measured point, the natural tendency is for the hands to move slightly apart. How much depends upon the bias? Try it and see. The larger the measurement, the more serious this becomes.

lightweight it is easy to have slippage; when heavy, it is difficult not to. If on top of all of this, the operation is a tedious or a repeated one, the reliability declines even more. Note, too, that bias works the opposite as well. This is a persuasive argument for good training and high morale wherever reliable measurements are needed.

Most precise measurements require several readings which are averaged. Obviously wrong ones are thrown out. This is an open invitation to bias. *Never reject any reading unless it is definitely known to be fallacious.* *

*Hume,K.J., Sharp,G.H., Practical Metrology, London, Macdonald & Co. Ltd.,1958, Vol.1,p.19

Summary

● Go back to the oft repeated claim that a rule can be used for measurement to a *mil.* It can, and frequently is-- *but not reliably.* When measurement is not reliable, things get expensive. More hand fitting is required. Pieces are scrapped. Assemblies wear out faster, are harder to repair.

● When you are building parts whose measurements are to be within 1/64 of an inch, you can reliably use a rule. But if you are instructing other people to make parts which must fit together the situation is different. If the measurements required for proper functioning are no finer than 1/32 you may require that they make them to 1/64. Even that would not provide really adequate *control* as you will realize when we discuss the rule of ten to one in Chapter 11.

● Chose a scale that is graduated the same as the part dimensioning, if you are working to a drawing. Use metric scales for metric parts, decimal-inch scales for decimal-inch parts, etc. If the choice is up to you, the decimal-inch scales should be preferred.

● Steel rules of all kinds are among the most useful measurement instruments, but they are not recommended for the control of production, Fig. 4-14.

● The attention box, Fig. 4-15, sums up what you should know about the use of steel rules and similar instruments and Fig. 4-16 reviews the new terms. The next chapter shows some of the things that can be done with a knowledge of steel rules and their associated equipment.

METROLOGICAL DATA FOR RULES AND TAPES								RELIABILITY	
INSTRUMENT	TYPE OF MEASUREMENT	NORMAL PRECISION	DESIGNATED PRECISION	DISCRIMINATION	SENSITIVITY	LINEARITY	PRACTICAL TOLERANCE FOR SKILLED MEASUREMENT	PRACTICAL MANUFACTURING TOLERANCE
Ordinary rulers	direct	12 in.	1/16 in.	1/16 in.	1/16 in.	1/16 in.	1/16 in.	never
Steel rules:								
Decimal inch	direct	6,12,18,24 in.	0.02 in.	0.02 in.	0.02 in.	0.0003/in.	±0.02 in.	±0.04 in.
Fractional inch	direct	6,12,18,24 in.	1/64 in.	1/64 in.	1/64 in.	0.0003/in.	±1/64 in.	±1/32 in.
Steel tapes:								
Decimal inch	direct	100 ft.	0.10 in.	0.10 in.	0.10 in.	0.01 in.	±0.10 in.	± 0.30 in.
Fractional inch	direct	100 ft.	1/8 in.	1/8 in.	1/8 in.	0.01 in.	±1/8 in.	± 3/8 in.

Fig. 4-14   For scaled instruments, some of these columns are repetitive. They will have important meaning, however, for higher precision instruments.

RELIABLE MEASUREMENTS WITH STEEL RULES

1. Choose the proper rule and the proper scale for the measurement.
2. Have both the part and the rule clean.
3. If at all possible have either the part or the rule resting in a stable condition.
4. Align the scale edge of the rule and the line of measurement of the part as closely as possible.
5. Align the reference point to the rule so that unsharp edges on either will not interfere.
6. Read the measured point from a point directly opposite, with a magnifier for greatest precision or to avoid fatigue from repeated measurements.
7. Repeat with higher discrimination scales until a graduation aligns with the measured point, desired discrimination is reached or the finest scale is reached.
8. Remind oneself of bias and repeat operation.
9. Repeat sufficient times for the needed reliability based on your skill.

Fig. 4-15 These seemingly obvious steps can quadruple the chances that your steel rule measurements will be accurate to the full discrimination of the rule.

TERMINOLOGY

Steel rule	Familiar direct-measurement instrument with which the unknown is compared to a graduated scale.
Scale	The graduations on a rule. Also an instrument graduated to read proportions of linear lengths.
Discrimination	The degree to which an instrument subdivides the unit of length on a steel rule. It is the finest division.
Interpolation	Selection of a value between graduations, usually guessing.
Margin of safety	In measurement it means the amount that the measurement process exceeds the required minimum accuracy.
Reliable measurement	Measurement in which the probability of error is small.
Error	The difference between the measured value and the true value.
Observational error	Error formed during the reading of a measurement.
Parallax error	Error caused by apparent shifting of objects when the viewing position is changed.
Manipulative error	Error caused by the handling of the part and instrument.
Bias	Unconscious or conscious influencing of the measurement.
Parallax	Apparent shifting due to changed position of observer. A source of measurement error.

Fig. 4-16 Although steel rules are the most elementary of the measurement instruments, they involve the same factors that are encountered with the most sophisticated instruments, and are subject to the same errors.

TYPES OF RULES

NARROW TEMPERED
RULE

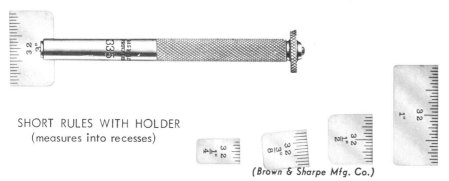

TEMPERED HOOK RULE (integral female reference point)

FLEXIBLE FILLET RULE (measures past obstructions)

SHORT RULES WITH HOLDER
(measures into recesses)

(Brown & Sharpe Mfg. Co.)

ANGLE RULE (Prima-
rily for small parts held
between centers.)

Fig. 4-17 This is a small sampling of many available types of steel rules.

TAPES

STEEL MEASURING TAPE

POCKET STEEL TAPE

REEL MEASURING TAPE   *(The L. S. Starrett Co.)*

Fig. 4-18 Steel tapes are extensions of steel rules and obey the same principles. For
very precise measurement, temperature expansion and contraction must be considered.

QUESTIONS

1. What is the difference between a scale and a rule?
2. How does the architects' scale differ from the engineers' scale?
3. What is meant by the term discrimination as it applies to a measuring instrument?
4. Three requirements must be considered when using a steel rule. What are they?
5. How does discrimination apply to the accuracy of rules?
6. How can attachments contribute to accumulated measurement errors?
7. When can interpolation be applied to a specific measurement application?
8. Why is interpolation between graduations usually objectionable?
9. There are four general sources of measurement error when using a steel rule. What are they?
10. How can we unconsciously influence each measurement that we make?
11. Why are steel rules not recommended for the control of high production?
12. What happens when there is little or no measurement reliability?
13. What is meant by parallax? By parallax error?
14. How can parallax error be eliminated?
15. What length of steel rule is best?

DISCUSSION TOPICS

1. Enumerate some characteristics that can contribute to measurement error when using rules.
2. Explain how several measurement errors with scales and rules can be eliminated.
3. Discuss the various types of scales, rules, tapes, and their proper use and application.
4. Discuss parallax errors in your daily life and the way they could influence actions.
5. Explain interpolation and its proper and improper application.

Man's industrial progress has been linked to his ability to divide and measure the inch and meter finer and finer.

Louis Polk
Sheffield Corporation

Inspection and layout are opposites. Both depend upon measurement. Each is affected by the other. Inspection is examination to determine if material or work performed is according to specifications, Fig. 5-1. Layout is the preparation of material so that work

In the second case the layout is only the guide for the machining. The closer the layout is to the specifications, however, the less work the machines will have to do to bring the part to size. This can be very important when speed or cost is a factor--and when aren't they?

(Northrop Aircraft)

Fig. 5-1 Plate work, or surface plate work, is measurement taken from an auxiliary reference surface called a surface plate. For large or small parts it is an important measuring method.

can be performed according to specifications, Fig. 5-2. Layout usually refers to scribing lines to show the limits for band machining, milling or grinding and the location of the centers for diameters to be drilled, turned or bored to size.

For rough parts, the layout provides the actual features to which the part is machined. For more precise parts, the layout furnishes a guide, but the precision of the machine tool determines the actual dimensions.

In the first case, the layout limits the precision of the machining. A part might be ruined by bad machining regardless of the care used in layout. However, *no matter how carefully machined, if the layout is in error the part must be also.* The layout must be made sufficiently close to the specifications to allow for the error that is bound to occur in machining.

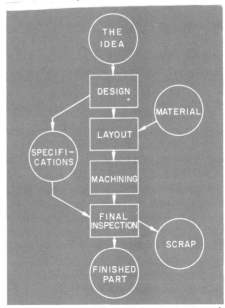

Fig. 5-2 Measurement is used at each step. The inspection step is the use of measurement for <u>verification</u>.

For example, consider a part that is to be band machined to rough size, heat treated and then surface ground to the specified dimension. Band machining removes material in blocks and is generally the least expensive machining method. Grinding removes material in minute chips. While much more precise than band machining, it is also much more expensive. If the part is carefully band machined to within 1/32 inch of the final size, the grinding will be rapid because little material will remain. If, on the other hand, the band machining is careless and an extra 1/16 inch is left, the grinding is unnecessarily prolonged and costly.

The layout requires the same reliability that the inspection requires, but can usually be performed at a lower level of precision. Incorrect location of a hole can scrap a valuable part just as easily as boring the hole oversize. Thus the needs of layout lead to improvements and modifications in the basic steel rule and to auxiliary equipment to use in conjunction with the new instruments.

Layout and inspection instruments may be divided into three groups. First are the mechanical refinements developed to make the steel rule more useful. The combination square is an example. Second are the related devices. Although not rules themselves they are used with rules. Calipers are one of many examples. These two are discussed in this chapter. The third are the highly precise refinements of the steel rule, the vernier instruments, such as the vernier height gage. They are sufficiently important to have their own chapter, Chapter 6.

These instruments in no way change the principles in the last chapter. The mechanical refinements add to convenience. The related devices extend the rule into jobs it could not otherwise perform. Only the last, the verniers, extend the basic measurement considerations beyond the basic rule.

THE DEPTH GAGE

A popular modification of the steel rule is the depth gage, Fig. 5-3. This instrument measures into holes, grooves and recesses. In its simplest form, (on the right) it consists of a T-head, sometimes called the *stock*, but usually referred to as the *head*. Through it a

small rod slides. The head is used to span the shoulders of a recess. Then the rod is pushed into the recess until it bottoms. A screw clamp locks the rod in the head. The depth gage is then withdrawn and the length of the rod protruding from the head is measured to determine the depth of the recess.

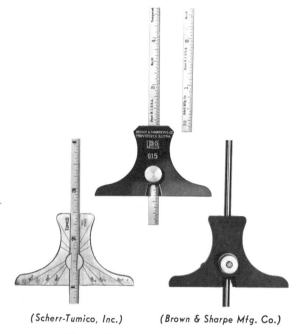

(Scherr-Tumico, Inc.)   (Brown & Sharpe Mfg. Co.)

Fig. 5-3 Types of depth gages.

This is an example of an instrument that simply transfers inaccessible reference and measured points into accessible ones.

Rods are still used in depth gages, particularly for small holes. Both plain and graduated ones are available. The graduated rods have the advantage that they eliminate a transfer of measurement. *Every transfer of measurement reduces reliability.*

It is difficult to read graduations from a small rod. Therefore, the narrow flat scale was a logical development for depth gages. It is properly termed *rule, scale* or *blade*. Most are 6 inches long, providing 5 inches of working length. They are available with hook ends for measuring the depth of through holes, Fig. 5-4.

Some depth gage heads can be turned in respect to their scales. These are usually marked to show left and right 30°, 45° and 60° angles, but cannot be compared to a protractor for angle measurement.

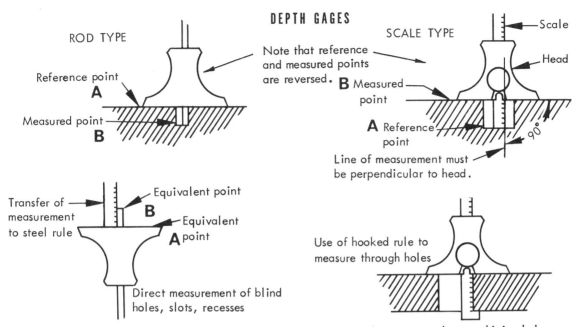

Fig. 5-4   The depth gage is frequently used to check the progress when machining holes and recesses. When close to the finished size a more precise instrument is substituted.

The depth gage has serious limitations. The part size is limited by the width of the head. About 2 inches is the maximum opening that can be spanned. Two precautions must be observed when using the depth gage. The base of the head must be perpendicular to the line of measurement, and the end of the scale must be against the desired reference. The latter is not easy when it is in the bottom of a blind hole.

In most applications the end of the scale contacts the part. This limits the reliability. The inevitable wear is aggravated by the small cross section of the scale. And in view of the low price of replacement scales they should be checked frequently and replaced when worn the width of one graduation line.

THE COMBINATION SQUARE

This is the most useful of the steel rule variations. It is used universally in mechanical work for assembly, layout and work-in-progress inspection. It evolved from the try-square. These were small, fixed-blade squares, once popular with machinists.

The more difficult a dimension is to measure, the lower the reliability of the measurement. Because the combination square makes steel rule measurement far more convenient,

it increases reliability. Thus it may be considered more precise, Fig. 5-5. It does *not* increase accuracy, however. This is still limited by the steel rule itself. The combination square should not be used to measure any finer than its discrimination. That is the same as for steel rules. It is recommended only for limited production inspection such as small lots of rough work.

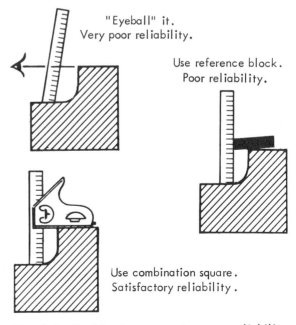

Fig. 5-5   Combination square improves reliability.

The combination square consists of a *blade* and a *square head*, Fig. 5-6. The three sizes available are expressed by the length of the head. Both the most popular combination square (the 4 inch) and the largest combination square use a blade 1-inch wide by 3/32-inch width. The blades are available in lengths from 4 to 24 inches and with a wide combination of fractional-inch, decimal-inch, and metric scales. The smallest combination square uses a narrower blade.

The heads may be either plain or hardened. They have a clamp that locks the blade in any position, a spirit (nonfreezing) level, and a scriber.

SQUARENESS AND HEIGHT MEASUREMENT

Check squareness and measure height.

Measure inaccessible point with right angle attachment.

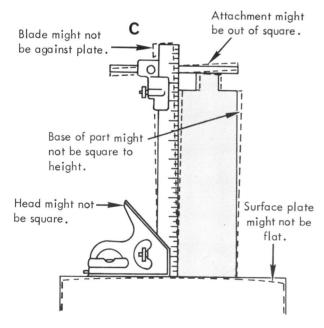

Fig. 5-7 Combining squareness and length measurement in one instrument may combine errors or eliminate errors. It is up to the skill of the measurer.

COMBINATION SETS

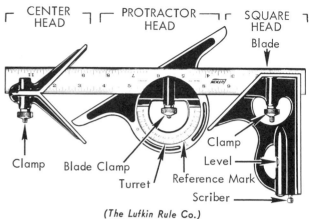

(The Lufkin Rule Co.)

Fig. 5-6 The steel rule and square head are called a combination square. Adding the center head and protractor head changes the name to combination set.

The *square head* contributes two elements missing with the steel rule. First, it provides a right angle reference. Second, it provides a means for transferring either the measured point or the reference point from the work to the rule.

As shown at A in Fig. 5-7, the combination square can be used from a flat reference surface such as a surface plate.* Squareness of a part height to its base and height measurements can be determined. Adapters are available to hold a steel rule at a right angle to the blade as at B. This establishes the reference point and measured point on the work with the blade and improves reliability *if* .... The four big "ifs" that are involved are exaggerated in the figure at C. Even if each one of these factors adds an error no greater than the thickness of one human hair, the total result may be greater than the 1/64-inch *sensitivity* which might be expected.

---

* Surface plates are precisely flat surfaces from which measurements are made. They are sufficiently important to rate Chapter 15.

DEPTH MEASUREMENT

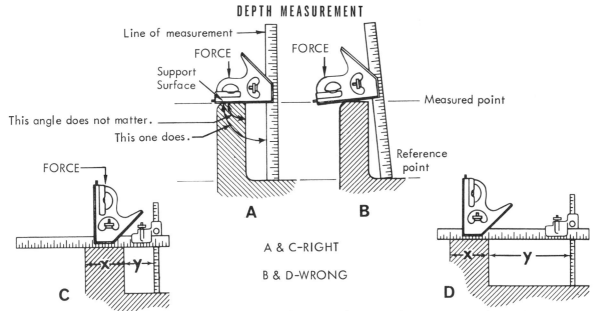

Fig. 5-8 The combination square can be used like a depth gage. Being larger it magnifies its range and the errors.   Overhang is always a problem.

Figure 5-8 shows depth measurement considerations. Used directly as in A the combination square has greater range and better reliability than a depth gage. Note that the support surface must be square to the line of measurement. It is easy to confuse a feature of the part with the real line of measurement. Misplaced holding force (in B) has same effect as out-of-square support.

Attachments extend the range of depth measurements as in examples C and D. *It is important in all gaging to minimize overhang.* In D the ratio of x to y is unfavorable. Any irregularity in the supporting surface will be multiplied. To minimize this tendency, apply holding force at center of support surface as in C, Fig. 5-8.

The 45° angle is met so frequently that one side of the square head is at 45° to the blade. For other angles the *protractor head* is available. This head consists of a rotatable turret within a stock. The scale is on the turret graduated in degrees from 0 to 180 in either direction.   The protractor scale is double. This provides both the angle and its compliment. The advantages of this are discussed in Chapter 16.   The reference mark on the stock is opposite 0° to 90° on the protractor scale.   A level in the turrent corresponds to the 0° and 90° position on the respective scales.

Fig. 5-9   The protractor head with sliding blade forms a versatile instrument for the measurement of angles.

Because the protractor head is easily interchanged with the square head, it provides a convenient means for checking angles no closer than 1°, Fig. 5-9. Its chief advantages are convenience and rugged construction.

When the blade is placed in the *center head*, one edge of the blade bisects the V-angle of the head. It thereby lies along the center line of any circle placed against the faces of the V. This is useful both for measurement and inspection of a wide range of parts.

DIAMETER MEASUREMENT

Hold 1" point at the reference point.

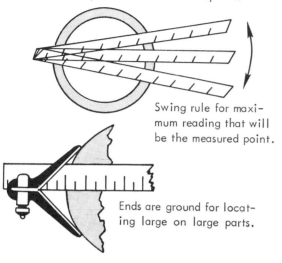

Swing rule for maximum reading that will be the measured point.

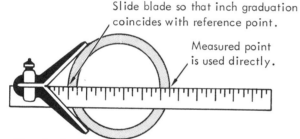

Ends are ground for locating large on large parts.

Slide blade so that inch graduation coincides with reference point.

Measured point is used directly.

Fig. 5-10 The center head speeds the measurement of diameters and improves reliability.

In contrast to the steel rule method for measuring diameters, the center head method is far more reliable, Fig. 5-10. It eliminates the hand juggling of the rule and part while trying to make a reading at the right fleeting moment.   The limitation to the center head method chiefly is the parallax problem. The blade is relatively thick; and both a reference point setting and a measured point reading must be made.   Unless these are properly viewed, the error could be 1/32 inch or more. Care must be taken at all times.

For layout purposes the center head provides one of the most satisfactory methods for finding the centers of shafts. Theoretically the intersection of any two major diameters (A in Fig. 5-11) will be at the center. The reliability of this is greater as the diameters approach right angles to each other (B). If the shaft is slightly irregular, several diameters can be taken. Then the center of the resulting polygon is chosen as the practical center (C) of the shaft.

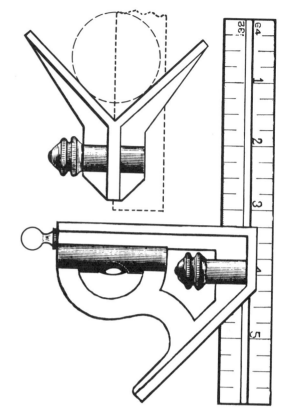

CENTER HEAD AND SQUARE HEAD FROM EARLY PATENT

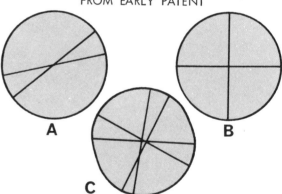

A

B

C

Fig. 5-11 The center head was considered early in the development of the combination square by L.S. Starrett as this early patent drawing shows. In the circles below, B would be a more reliable center than A, while C is an average for an irregular shaft.

When the combination square is accompanied by the protractor and center heads it is called a *combination set*. All components are replaceable.   They should be frequently checked for wear. If subjected to much use, hardened heads are an economy in spite of their higher price.

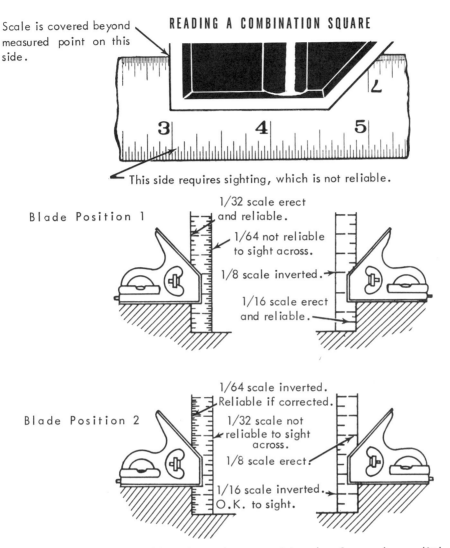

Scale is covered beyond measured point on this side.

READING A COMBINATION SQUARE

3 4 5

This side requires sighting, which is not reliable.

Blade Position 1

1/32 scale erect and reliable.

1/64 not reliable to sight across.

1/8 scale inverted.

1/16 scale erect and reliable.

Blade Position 2

1/64 scale inverted. Reliable if corrected.

1/32 scale not reliable to sight across.

1/8 scale erect.

1/16 scale inverted. O.K. to sight.

Fig. 5-12 When a blade has only two positions but four scales, a little judgment improves reliability. When reading near center of scale, particular care is required.

Special attention to scale selection is needed when using a combination square. The graduations vary among manufacturers, but these general considerations should be remembered. The head covers the portion of one scale beyond the measured point. This makes interpolation even more hazardous than usual, Fig. 5-12.

Furthermore, there are four scales on each blade, but usually they can have only two positions in respect to the head. You must select the position that places the scale you need in the best position for measurement. If you only require the 1/8- or 1/16-inch scale it does not matter because it is safe to sight from the head to the scale. If the 1/32- or

1/64-inch scale is required you cannot reliably sight across and must have the chosen scale on the head side. In one or the other case, the dimensions uncovered by the head will be from large to small in reading. It is necessary to correct for this. Although easy to do, it is also easy to overlook.

This does not cause confusion if you are near either end of the blade. If the reading is between 10 and 11 it is obvious whether or not it should be between 1 and 2 (presuming a 12-inch blade) instead. When the reading is between 6 and 7 it is not obvious and care must be taken. When in doubt, simply overlook the engraved numerals and count from the end.

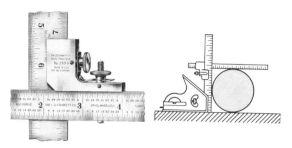

Right angle rule clamp is useful for measuring large diameters.

Rule clamp couples tw rules of same or differ ent sizes.

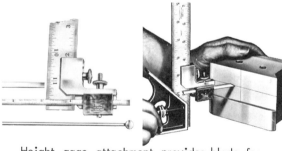

Key seat clamps align rule with axis of cylinder preventing skew condition at right.

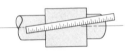

Height gage attachment provides blade for use as pointer or scriber.

(The L. S. Starrett Co.)

Fig. 5-13 Use attachments when their added convenience improves reliability more than their added chance for errors decreases it.

Several accessories are shown in Fig. 5-13. Their uses are largely self-explanatory but one important fact is not. These *attachments in every case add errors.* They must be checked as frequently as the combination square for reliable measurement. These and other considerations for reliable measurement with the combination set are summarized in Fig. 5-14.

RELIABLE MEASUREMENT WITH COMBINATION SETS

1. Observe all applicable precautions for using steel rules.
2. Assure that the heads have not worn excessively.
3. Be sure that the line of measurement is parallel to the blade.
4. Apply holding force to assure stability during measurement.
5. Be sure that the scale is read in proper direction.
6. For work from a plate, be sure the blade is in contact with the plate.
7. Take particular care to minimize parallax error.
8. Check all attachments for wear and squareness.
9. If attachments are used keep overhang as short as possible.

Fig. 5-14 With very little practice, these precautions become automatic.

SIMPLE CALIPER INSTRUMENTS

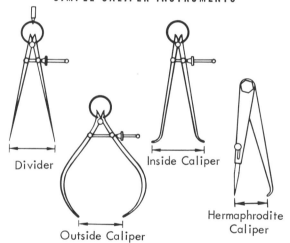

Divider   Inside Caliper

Outside Caliper   Hermaphrodite Caliper

Fig. 5-15 These are simple calipers, although the dividers are not called one. They all have ends that are adjustable to transfer a measurement from part to standard, usually a scale.

CALIPERS — THE ABORIGINAL TRANSFER INSTRUMENTS

Later you will be shown means for reliably measuring to within a few millionths of an inch by transfer of measurement. Here you will meet the forerunner of these methods, the caliper, Fig.

CALIPER DEFINED

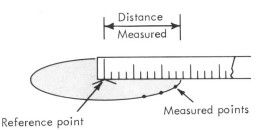

A circle can be approximated with a graduated scale if sufficient measured points are plotted.

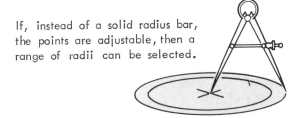

If the distance is made into a radius bar, an infinite number of measured points can be swung around any reference point.

If, instead of a solid radius bar, the points are adjustable, then a range of radii can be selected.

Fig. 5-16 A caliper is an instrument for mechanically duplicating a measurement. Unless otherwise specified, it is an adjustable instrument. Not all calipers are termed that, dividers are an example.

Man discovered the problem of transferring measurement long before he discovered the advantages of a standard of length and a measurement system. Although he probably discovered the method to draw a circle by trial and error, he was actually taking the first step to solve the problem, Fig. 5-16. To draw a circle, he needed a physical embodiment of the desired radius. For convenience in drawing many circles, this physical embodiment needed to be adjustable.

The simple forms of calipers (inside calipers, outside calipers and dividers) are no longer as important in measurement as they once were. They demonstrate some essential measurement considerations. These are easily overlooked but apply equally well to the more sophisticated instruments. Therefore they are discussed at some length.

Calipers are instruments that physically duplicate the separation between the reference point and measured point of any dimension within their range. Many instruments fitting this description are called calipers. The word caliper used alone refers to the simplest instruments of this type, Fig. 5-17. They are used for transfer of measurement only. The more sophisticated calipers incorporate their own scales and standards of length. They are discussed in Chapter 6. Although a member of the caliper family, dividers use only their own name.

TYPES OF CALIPERS

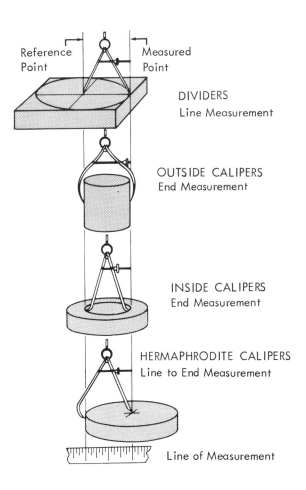

Fig. 5-17 All caliper instruments transfer measurements. These are the basic types. Other versions bear little resemblance, except in principle of operation.

Double calipers for inside
and outside measurement.

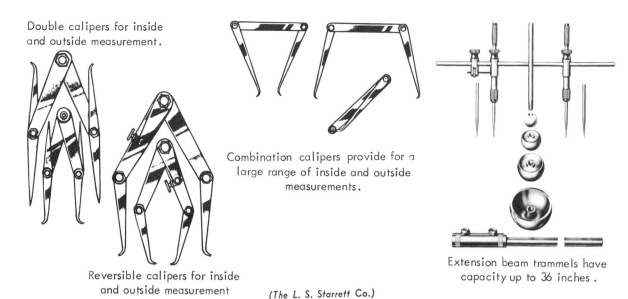

Combination calipers provide for a
large range of inside and outside
measurements.

Reversible calipers for inside
and outside measurement

(The L. S. Starrett Co.)

Extension beam trammels have
capacity up to 36 inches.

Fig. 5-18   Many types of calipers have been devised.

There have been many constructions for calipers. Two predominate today, firm-joint calipers and spring calipers. The firm joint is held to the desired opening by friction between the legs. This may be adjusted, and in some, the legs may be locked. They are adjusted closely for size by tapping a leg. The spring calipers are opened and closed by their screw adjustment. This permits very careful control and no lock is needed. Calipers are

made in a large number of sizes. They are designated not by their measurement range, but by the length of their legs. These range from 2 to 24 inches in length.

There are many good reasons why caliper should be singular and calipers, plural. Both popular use and the catalogs generally use either term to be singular. Dividers also may be either singular or plural. A caliper is frequently spoken of as a "pair of calipers." This makes every bit as much sense as a "pair of trousers."

GEOMETRY BY DIVIDERS

DIVIDERS FOR LINE MEASUREMENT

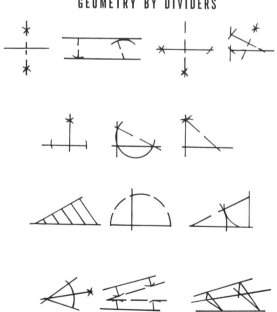

Dividers started as a means to draw circles but eventually lead to the study of geometry. This enabled the premeasurement craftsman to control position and size in the articles he produced. With them he could divide any straight line or arc into convenient spaces, Fig. 5-19. This amounted to each article being its own standard of length. While handy for locating the spokes on a wheel or determining the size of a box, this was not conducive to mass production.

Fig. 5-19   Once man invented dividers he was well on his way towards the mastery of geometry.

As with all calipers, the chief use of dividers is to transfer measurement. They differ from other calipers in that they are used primarily with *line references.* Whenever two planes intersect, a line is formed. Therefore, the features of most parts provide an abundance of lines in the form of shoulders, recesses, O.D.s and I.D.s.

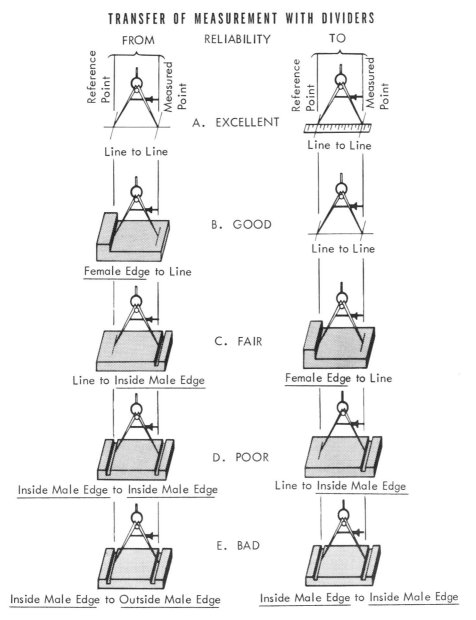

TRANSFER OF MEASUREMENT WITH DIVIDERS

Fig. 5-20 The important thing to remember in transfer of measurements is that two, not one, measurements are involved. Each contributes errors.

Scribing of arcs remains the chief use for dividers. The other important uses are for (1) transferring a dimension of a part to a scale for measurement, (2) transferring a dimension from one part to another part, and (3) for transferring a dimension from a scale to a part for layout of the part. As an instrument for this work, dividers range from excellent to extremely poor. The difference depends upon the lines to which the dividers must be set. Figure 2-7 shows an examination of the edges that in measurement are treated as lines.

The effect these edges have upon the reliability of measurement is shown in Fig. 5-20. The important fact emphasized by this illustration is that in every transfer of measurement, two separate operations are involved: from the unknown to the gage (divider in this case), and from the gage to the known (scale, standard or a part to be duplicated).

It cannot be overemphasized that *the reliability of both operations must be considered.*

To set a divider to a dimension one leg is placed in an even inch graduation of a rule for the reference point. The divider is then adjusted until the other leg falls into the desired graduation for the measured point. Because good graduations are V-shaped this can be done very accurately, providing the divider does not have dull points. A dull point can cause considerable inaccuracy. When a series of dimensions is stepped off this could accumulate and ruin the part. Points can be honed with a stone or returned to the factory for repair.

A form that transfer of measurement frequently takes when using dividers is stepping off series of the same measurements. This is useful in laying out bolt circles. And it ushers in one of the important considerations in measurement, the *accumulation of errors.*

It goes like this. Every measurement contains some error. How much and which way (larger or smaller) is unknown until it is compared with the true measurement. To be safe we must assume that the error is the maximum amount possible: that the instrument, the observer and anything else involved is at its worst. This assumption forces us to agree upon a highly unpalatable axiom of measurement. *Whether measurements are added or subtracted their errors must be assumed to add or reliability is lost,* Fig. 5-21.*

It has been stated that every added operation decreases reliability. This is due to the added errors inherent in the new operation. There can only be two reasons to justify adding an operation: (1) to make a measurement that cannot be made without the added operation and (2) when the errors of the added operation are less than the errors eliminated by the combined measurement operations. The divider provides three cases in which the latter may be true, Fig. 5-22. The one case in which the reverse is true and should be guarded against is shown in Fig. 5-21. These are known as *serial measurements.*

---

* This is unquestionably a safe assumption. It is also highly illogical. It makes as much sense as assuming that your opponent in a poker game receives a royal flush at every deal. After all, it could happen. Modern statistical dimensioning takes a more realistic attitude.

ERRORS ADD
FOR ADDED OR SUBTRACTED MEASUREMENTS

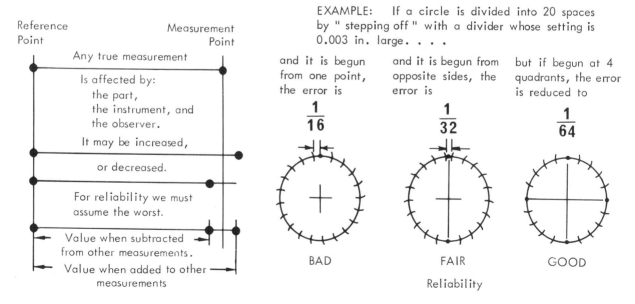

Fig. 5-21   A group of measurements each of which begins where the previous one left off are known as serial measurements. Obviously their errors interact. The accumulation of errors is easily demonstrated by stepping off spaces around a circle with dividers. The results can be startling.

ADVANTAGES OF DIVIDERS

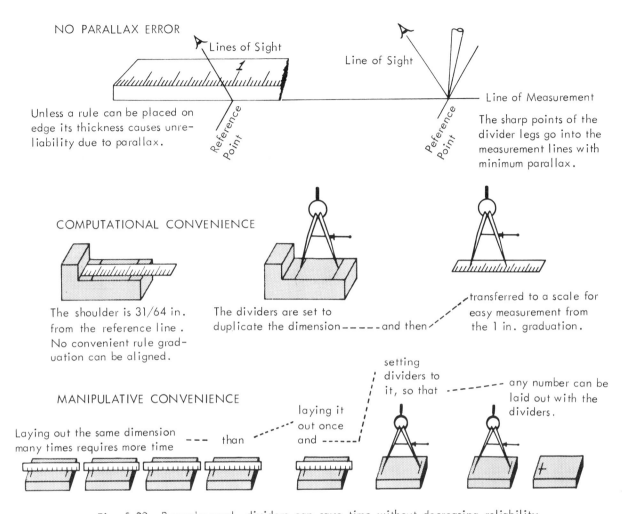

NO PARALLAX ERROR

Lines of Sight

Line of Sight

Line of Measurement

Unless a rule can be placed on edge its thickness causes unreliability due to parallax.

Reference Point

Reference Point

The sharp points of the divider legs go into the measurement lines with minimum parallax.

COMPUTATIONAL CONVENIENCE

The shoulder is 31/64 in. from the reference line. No convenient rule graduation can be aligned.

The dividers are set to duplicate the dimension — — — and then

transferred to a scale for easy measurement from the 1 in. graduation.

MANIPULATIVE CONVENIENCE

setting dividers to it, so that — — — — any number can be laid out with the dividers.

laying it out once and — — — — —

Laying out the same dimension many times requires more time — — — than

Fig. 5-22 Properly used, dividers can save time without decreasing reliability.

INSIDE AND OUTSIDE CALIPERS

Inside and outside calipers perform a similar function to dividers. However they must locate their line of measurement by groping around with areas instead of with clean cut-lines and edges. They are used to measure the features of parts that cannot be reached with a rule but do not require greater precision than afforded by rules. They are geometrically similar to dividers. The thing that matters is the separation of the reference points. All else is simply the mechanical frame required to achieve this.

Measuring with a caliper consists of adjusting the opening so that the reference points of the caliper duplicate those of the feature of the part. Then in a second step, this sepa-

ration is compared with another part or with a measurement instrument. This requires that the reference points of the caliper lie along the same line as the feature of the part, the line of measurement.

This is equally true for steel rules and combination squares. In fact, parallax has been pointed out already as an error that might result from even a small separation of the instrument and part feature. But consider the differences. Aligning the line of measurement of a rule to a feature of the part is easy because the edge of the rule is a line and the flat of the rule is a plane in which the line lies. Simply laying the rule on the surface to be measured places the line of measurement in the same plane as all features of that surface.

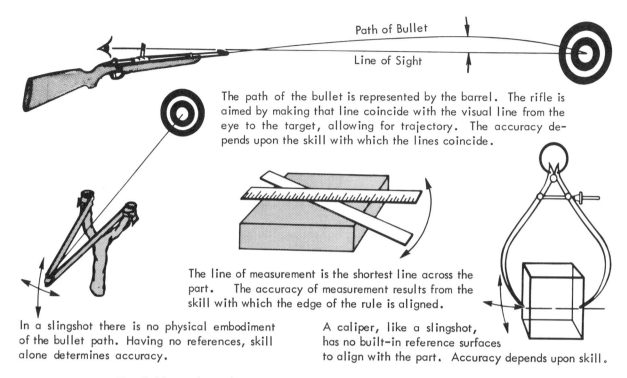

The path of the bullet is represented by the barrel. The rifle is aimed by making that line coincide with the visual line from the eye to the target, allowing for trajectory. The accuracy depends upon the skill with which the lines coincide.

The line of measurement is the shortest line across the part. The accuracy of measurement results from the skill with which the edge of the rule is aligned.

In a slingshot there is no physical embodiment of the bullet path. Having no references, skill alone determines accuracy.

A caliper, like a slingshot, has no built-in reference surfaces to align with the part. Accuracy depends upon skill.

Fig. 5-23   Calipers have more in common with slingshots than looks alone.

Not so with calipers. There is nothing about the instrument that helps align it other than the reference points themselves. It is like a slingshot as compared to a rifle, Fig. 5-23. The correct position for an outside caliper is along the line of minimum separation. This refers to the distances between surfaces that bound the feature being measured, Fig. 5-24. For an inside caliper, it is along the line of maximum distance.

How are these measurement lines found? Mostly by *feel*. And that is the rub. The diffi-

culty varies with the surface. If it is a sphere, there is nothing but *feel* to guide the measurement. If a cylinder, a right-angle plane produces the desired diameter. Often a shoulder or a groove locates the right-angle plane automatically, Fig. 5-25. In these cases, the outside calipers are a relatively precise instrument.

A regular solid whose sides are squares is simply called a cube. Why one whose sides are rectangles should be called a "rectangular parallelepiped" is a puzzle. These solids

LINE OF MEASUREMENT — OUTSIDE CALIPER

The smallest caliper opening that will pass the longest overall line through the sphere. No clues for finding it.

SPHERE

POOREST

The smallest caliper opening that will pass the cylinder. A plane perpendicular to the axis contains the shortest distance.

CYLINDER

BEST
(But that isn't saying much)

The smallest caliper opening that will pass the surfaces. It is along the line formed by any two intersecting planes both of which are perpendicular to the reference surfaces.

RECTANGULAR PARALLELEPIPED

POOR

Fig. 5-24   Seek smallest caliper opening that will pass the part.

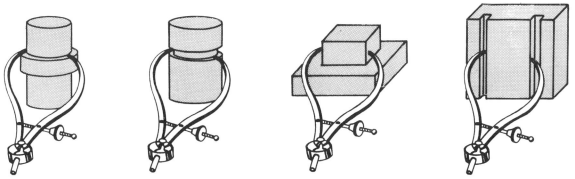

Fig. 5-25 Greatest reliability is achieved when a shoulder or recess on the part restricts the line of measurement to a plane perpendicular to the part.

are far more common than cubes. By any name they present real problems in caliper measurement, because the shortest distance between the faces can be most illusive. Fortunately in practical work they are often accompanied by shoulders that keep the calipers from straying too far, Fig. 5-25. The considerations for inside calipers are similar, Fig. 5-26, but somewhat more reliable. The reason for this is that the inside surfaces help locate the calipers.

USE OF CALIPERS

In use, calipers are (1) set to the feature and that distance transferred to a measuring instrument, or (2) they are set to the desired dimension and used as a fixed gage.

An outside caliper set to the desired dimension shows whether the feature is *too large* or *not too large*. It does not show if it is too small. Similarly, an inside caliper set to the desired dimension shows whether the feature is *too small* or *not too small*. The *something* or *not something* concept is important. It is not really measurement. Instead, it is a form of *sorting*. Under controlled conditions it can provide valuable knowledge. Therefore, it forms an important part of the inspection procedure used by industry.

LINE OF MEASUREMENT — INSIDE CALIPER

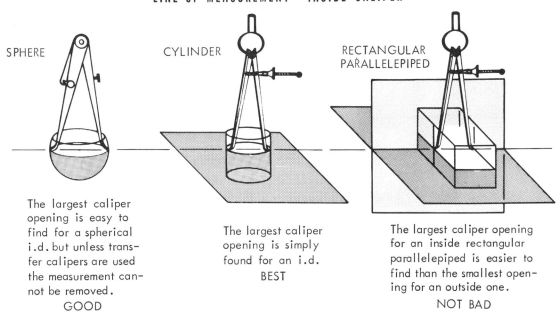

SPHERE

The largest caliper opening is easy to find for a spherical i.d. but unless transfer calipers are used the measurement cannot be removed.
GOOD

CYLINDER

The largest caliper opening is simply found for an i.d.
BEST

RECTANGULAR PARALLELEPIPED

The largest caliper opening for an inside rectangular parallelepiped is easier to find than the smallest opening for an outside one.
NOT BAD

Fig. 5-26 With calipers, inside measurements are somewhat easier to take than outside.

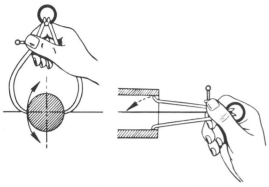

Fig. 5-27 Calipers are carefully rocked over center. The feel provided as they pass center is the limit of their sensitivity.

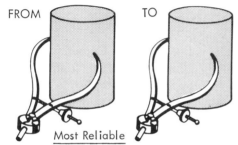

Most Reliable

From a surface to a similar surface, same feel on both.

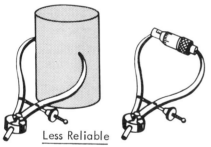

Less Reliable

From a cylindrical surface to a flat surface such a an inside micrometer. Similar feel on both.

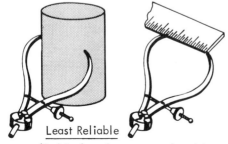

Least Reliable

From a cylindrical surface to a rule. No feel with rule to compare with cylinder.

Fig. 5-28 For greatest reliability with outside calipers, the feel on the work should duplicate the feel on the instrument.

Precision depends upon the *feel* of the caliper rubbing lightly against the part as it is rocked over center, Fig. 5-27. The most reliable situation occurs when both the pickup of the reference points and the comparison with the known distance feel the same, Fig. 5-28. It is infrequent.

IMPORTANCE OF GAGING PRESSURE

One of the important reasons to become familiar with inside and outside calipers is to understand the importance of gaging pressure.* You cannot *feel* the rubbing between the part and the caliper unless something is being distorted. Ordinarily it is the caliper. However, if the part is a thin-walled ring, it may be both.

If the caliper and the part were both absolutely unyielding, there would be no *feel*. They would either go together or not go together. In every gaging situation there will be some yielding of the elements involved. That is distortion--an error. When measuring to a millionth of an inch, even gravity distorts.

The lighter you can keep your *feel*, the more reliable will be your measurements. This presumes that you have already learned the fundamental lesson. The part and the instrument must be scrupulously clean. A coating of dust-laden oil or grease can drastically alter the *feel*. A light *feel* can be acquired with care and practice. The larger the part, the more difficult it becomes. A comfortable position is essential. This point, as well as the importance of cleanliness, will be repeated often through this book.

LAYOUT INSTRUMENTS

Several scribing instruments are used in conjunction with rules. The most important are the beam trammel, the hermaphrodite caliper, the surface gage and, of course, the scriber itself.

The *beam trammel*, Fig. 5-29, is a divider with extended range. Attachments convert it to inside and outside calipers. These are only suitable for flat work because of the shallow throat. Other than its range, the beam trammel has only one unique feature. These

---

* "Gaging pressure" is the popular but incorrect term for gaging force.

(Brown & Sharpe Mfg. Co.)

Fig. 5-29   The beam trammel is useful for laying out large work and as calipers on large flat work.

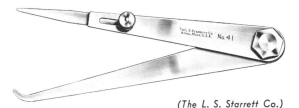

(The L. S. Starrett Co.)

Fig. 5-31   The hermaphrodite caliper transfers measurements bounded by a line and on edge. Its use is limited.

are spherical points that can be centered on holes. They permit arcs to be swung from existing holes, Fig. 5-30.

Hermaphrodite calipers are the odd members of the caliper family, Fig. 5-31. They have one curved caliper leg and one divider leg. The caliper leg is reversible for contact with either inside or outside reference surfaces. This provides a means for picking up a measurement bounded by a line and an edge.

The hermaphrodite caliper is a popular shop instrument and finds wide use for scribing lines parallel with an edge. It is difficult to think of a less reliable way. The reasons are shown in Fig. 5-32, together with a slightly better way to use this instrument. The one recommended role for the hermaphrodite caliper is finding the centers of shafts, Fig. 5-33, for which the centerhead cannot be used.

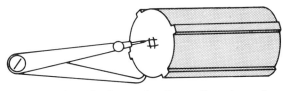

Fig. 5-33   The hermaphrodite caliper is used to find centers of circles. It is particularly useful when part conformation prevents the use of the center head on V-blocks.

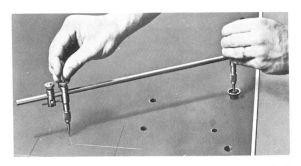

(Brown & Sharpe Mfg. Co.)

Fig. 5-30   The beam trammel is the only convenient layout instrument that can center on existing holes.

HERMAPHRODITE CALIPER

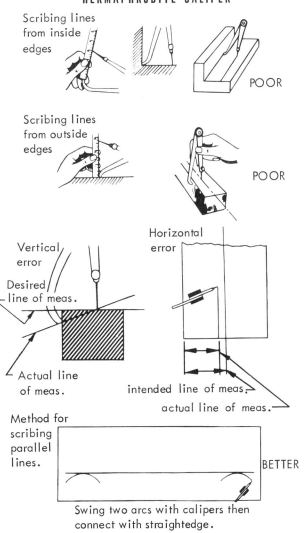

Scribing lines from inside edges

POOR

Scribing lines from outside edges

POOR

Vertical error

Desired line of meas.

Actual line of meas.

Horizontal error

intended line of meas.

actual line of meas.

Method for scribing parallel lines.

BETTER

Swing two arcs with calipers then connect with straightedge.

Fig. 5-32   If you must use this instrument be aware of the chances for positional misalignment and their consequences.

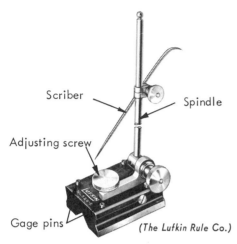

Scriber

Spindle

Adjusting screw

Gage pins

(The Lufkin Rule Co.)

Fig. 5-34 The surface gage is primarily used on a surface plate to transfer height measurements from part to standard.

The *surface gage*, Fig. 5-34, is also a means for transferring measurements, but it has one significant difference. For most applications it transfers height measurements. It usually is used on a surface plate. You will be told later of the potential hazards of vertical measurements taken from surface plates. Why then is the surface gage now being recommended for this use? Because this comparison is between a shaky, hand-held instrument and the stability of the surface plate. More reliability is gained than is lost.

The right half of Fig. 5-35 and Fig. 5-36 show the many setups that can be made using the gage pins to locate edges and the V-base to ride on a shaft. Unlike the hermaphrodite caliper and many other hand-held instruments, the surface gage eliminates many

SURFACE GAGES

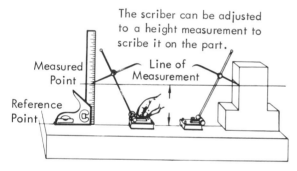

The scriber can be adjusted to a height measurement to scribe it on the part.

Measured Point

Line of Measurement

Reference Point

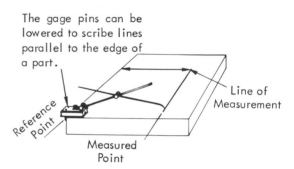

The gage pins can be lowered to scribe lines parallel to the edge of a part.

Reference Point

Line of Measurement

Measured Point

Fig. 5-35 With a surface gage there is stable contact between the part and the instrument along the reference point line.

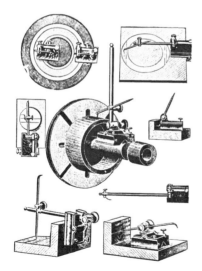

Fig. 5-36 The surface gage has many uses in addition to surface plate layout and inspection. *(The L. S. Starrett Co.)*

alignment problems. The inaccuracy resulting, however, is caused by the vision of the user. This can be minimized for layout by use of a magnifying glass. It is eliminated for inspection by substituting a dial indicator for the scriber. This is discussed later.

The scriber is not a measuring instrument, but is frequently required to transfer measurements. Its handling, although entirely a matter of common sense, deserves the same care that you give other measurement aids.

THE POCKET SLIDE CALIPER RULE

The only thing not handy about this variation of the rule is its name, which you will never see or hear outside of books and catalogs. It is usually called a *slide caliper*. Fig. 5-37 is a typical example.

Slide calipers are made in a wide range of sizes. Three inches is the popular pocket size. It has a two-inch range for both internal and external measurements. Models are made up to 48 inches. Hardly being pocket size, these are called *caliper squares*. A variety of fractional-inch, decimal-inch and metric graduations is available. The *stock* (Fig. 5-37) usually has the coarse scale, the slide the fine scale.

The lock permits the slide caliper to be fixed at any position. It is very handy as a memory device. If you are rough turning a long shaft, you do not have to remember the diameter between passes. You just lock the caliper after the measurement. Before the next measurement you can see what the previous size was before unlocking and rechecking.

The lock can be used also to make the caliper into a gage — but should not. *It is not intended to be a snap gage.* There is no provision to adjust for the excessive wear that snap gaging requires.

The advantages of the slide caliper are threefold. First, it provides positive contact between the instrument and the reference and measured points of the feature being measured for both internal and external measurements, Fig. 5-38. Even the hook rule cannot claim that, Fig. 5-39. Second, it substitutes a line on the instrument for a feature of the part to use as a measured point. Slightly rounded corners do not affect the precision of setting or reading. Third is the built-in memory provided by the slide lock.

POCKET SLIDE CALIPER RULE

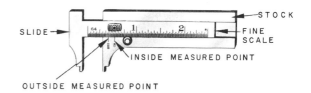

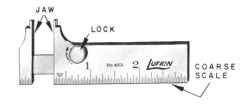

Fig. 5-37   This handy shop instrument is usually called a slide caliper.

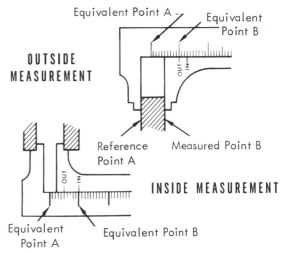

Fig. 5-38   Care must be taken to use the correct equivalent point.

SLIDE CALIPER vs HOOK RULE

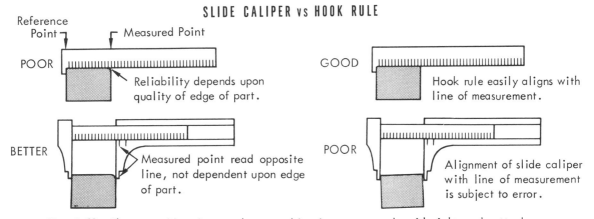

Fig. 5-39   These considerations apply to combination square, rule with right angle attachment and to the hook rule. The slide caliper has disadvantages to match its advantages.

SLIDE CALIPER	
ADVANTAGES	1. Combines rule, inside and outside calipers in one instrument 2. Provides positive contact with reference and measured points. 3. Substitutes line-to-line for line-to-edge readings.' 4. Has built-in memory.
DISADVANTAGES	1. No wear adjustment. 2. Subject to misalignment. 3. Limited discrimination. 4. Cannot caliper inside to outside part features.

Fig. 5-40 When using the slide caliper, remember the adage, " Don't send a boy to do a man's job."

There are two major disadvantages of the slide caliper. Like all caliper instruments, it is subject to positional errors. These are discussed in conjunction with the refined cousin of the slide caliper, the vernier caliper. The other is that there is no provision for wear adjustment. Slide calipers must be rechecked frequently for wear of both the internal and external jaw surfaces. How much wear can be tolerated? If you regularly work to 1/64 of an inch, any wear beyond the width of one graduation is excessive. What should you do when that amount of wear is reached? Forget to lock it up some night and someone else will have to worry about it.

Other disadvantages are its limited discrimination, 1/64, and the fact that it can only measure between two inside or two outside references. It cannot measure from an outside to an inside reference. These points are summarized in Fig. 5-40.

METROLOGICAL DATA FOR SCALED INSTRUMENTS							RELIABILITY	
INSTRUMENT	TYPE OF MEASUREMENT	NORMAL RANGE	DESIGNATED PRECISION	DISCRIMINATION	SENSITIVITY	LINEARITY	PRACTICAL TOLERANCE FOR SKILLED MEASUREMENT	PRACTICAL MANUFACTURING TOLERANCE

Depth gage:
  Decimal inch
  Fractional inch
	direct
	direct
		5 in.
		5 in.
			0.02 in.
			1/64 in.
				0.02 in.
				1/64 in.
					0.02 in.
					1/64 in.
						0.0003/in.
						0.0003/in.
							±0.02 in.
							±1/64 in.
								±0.04 in.
								±1/32 in.

Combination sets:
  Decimal inch
  Fractional inch
	direct
	direct
		12 in.
		12 in.
			0.02 in.
			1/64 in.
				0.02 in.
				1/64 in.
					0.02 in.
					1/64 in.
						0.0003/in.
						0.0003/in.
							±0.02 in.
							±1/64 in.
								±0.04 in.
								±1/32 in.

Calipers:
  Decimal inch
  Fractional inch
	transfer
	transfer
		6 in.
		6 in.
			none
			none
				none
				none
					0.005 in.
					0.005 in.
						none
						none
							±0.02 in.
							±1/64 in.
								±0.08 in.
								±1/16 in.

Slide Calipers:
  Decimal inch
  Fractional inch
	direct
	direct
		3 in.
		3 in.
			0.02 in.
			1/64 in.
				0.02 in.
				1/64 in.
					0.02 in.
					1/64 in.
						0.003/in.
						0.003/in.
							±0.02 in.
							±1/64 in.
								±0.04 in.
								±1/32 in.

Fig. 5-41 Note that although the caliper instruments have relatively good sensitivity, their reliability is low for a skilled operator and very low for manufacturing applications, because two transfers are required for each measurement.

RELIABILITY CHECK LIST FOR SCALED INSTRUMENTS	
INSPECTION OF INSTRUMENT:	1. Set up periodic system for inspection, depending upon use. 2. Inspect contact surfaces with magnifier for wear or abuse. 3. Remove burrs from sliding and contact surfaces. 4. Compare readings against an instrument of higher precision, greater accuracy and with known calibration. 5. Check all mechanical actions for proper functioning. 6. Clean and lubricate internal parts. 7. Check alignment against square of known calibration.
USE:	1. Never use a measuring instrument for a hand tool (scraper, chip digger, burring tool, mallet, screwdriver or sledge hammer). 2. Never use beyond intended size range (don't force open). 3. Never use beyond discrimination or recommended precision. 4. Keep contact force to a minimum. 5. Avoid excessive movements causing wear. 6. Clean both part and instrument before using. 7. Substitute mechanical support for hand support whenever possible. 8. Guard against parallax when reading. 9. Have entire setup rigidly supported. 10. Do not overtighten anything.
CARE:	1. Lubricate instruments before replacing in case. 2. Keep away from moisture. 3. Do not pile instruments together or with other objects. 4. Do not mark tools in any way that interferes with use. 5. Do not hesitate to throw away a worn or defective tool.

Fig. 5-42 Some of these suggestions seem self evident. Overlooking the most minor one can cause an expensive part to be scrapped or a careful experiment to yield incorrect results.

Ordinarily you would expect the sophisticated brother of the slide caliper, the vernier caliper, to be discussed along with it. That instrument moves us out of the 64ths and into thousandths. It rates a new chapter. Before proceeding, however, your attention is called to Fig. 5-42. These suggestions will increase your reliability when using the so-called *non-precision* measuring instruments.

TERMINOLOGY	
Inspection	Verification of conformity to a standard.
Layout	Preparation for machining or assembly.
Transfer of measurement	Operation between taking a measurement and reading the value of the measurement.
Attachment	Anything added to an instrument to increase its usefulness. Always a source of errors.
Overhang	Unsupported portion of instrument. Always to be avoided.
Calipers	Any instrument whose opening is set to a measurement so that it can be transferred.
Accumulation of errors	The assumption that the errors in serial measurements must always add. Safe but sorry.
Serial measurements	An interconnected series of measurements.
Feel	The perception of physical contact between the part and the instrument.
Sorting	Measurement that separates into lots rather than provides specific dimensional information.
Snap gage	A fixed gage used for sorting.
Gaging force	Force holding instrument to part. Incorrectly called "gaging pressure".

Fig. 5-43 The instruments in this chapter have been left out of the terminology box to save room, not out of disrespect for their ancient role in measurement.

Summary

● Most of the basic measurement procedures and measurement instruments derived from the simple steel rule. These instruments were required to lay out parts before machining as well as to check parts afterwards. Among these are the depth gage, combination square, calipers and slide caliper.

● Of these the calipers are transfer instruments. The others contain a scale from which the measurements are read. Measurement with all of these instruments introduces errors. The ends of the scales may be worn. Parallax may exist. The instrument may not be along the line of measurement. The transfer instruments add an extra step. All additional steps add errors.

● Because we need reliability, all steps must be assumed to add their maximum error, although some may cancel out. With sufficient information this can be alleviated.

● Instruments that require contact, such as calipers, depend upon the feel of the user for their precision. Cleanliness will promote reliability, but only skill will provide accuracy. This is a fundamental consideration.

● Any instrument incorporating a scale that does not provide a means for compensating for wear must always be suspect. This applies to the simple steel rule, the slide caliper and to more sophisticated instruments.

QUESTIONS

1. What is the basic difference between layout for rough parts and layout for precise parts?
2. Does layout require mechanical refinements to steel rules? Why?
3. What is the principal use of the depth gage?
4. Why are graduated depth gage rods more reliable than plain rods?
5. Why were rods for depth gages changed to narrow flat rules?
6. Why is the combination square set an extremely valuable all-purpose measurement tool?
7. What advantages does the combination square have over the depth gage for measuring various depths?
8. What is the chief limitation in the use of the center head?
9. How many scales are generally found on the combination square blade?
10. Why are calipers an important layout and machining tool?
11. What is the chief similarity between dividers and calipers?
12. Is an inside caliper more reliable than an outside caliper? Explain.
13. How does gaging pressure affect measurement reliability?
14. What are some additional scribing instruments that are used in conjunction with rules?
15. What is the principal shortcoming of the surface gage?
16. What is the principal advantage for using the surface gage?
17. What are the principal advantages of the slide caliper?

DISCUSSION TOPICS

1. Inspection and layout work depend upon accurate measurement. Explain the principal differences between the two.
2. Explain the importance of measurement reliability when dividers or calipers are used for transferring dimensions.
3. Discuss the principal advantages of dividers.
4. Explain the importance of recognizing the accumulation of measurement errors.
5. Discuss the principal uses of the various scribing instruments.

The Vernier Instruments
Thousandths at Last

Don't sneer at thousandths of an inch. They are by far the most important measurements made today. They do not put space vehicles in orbit but they do provide the money for all far-out projects. They are the dimensions that keep the modern mass producing economy alive and progressing. They are the measurements that you can put to the most immediate use – advancing in a job, developing a hobby, or, for laboratory projects in continued studies. In fact they rate their own name in decimal-inch terminology, *mil.*

The instruments discussed up to now are generally lumped together as "nonprecision" instruments. This makes as much sense as calling ice water, "nonheat" water. Precision, like heat, is quantitative. You do not just turn it on and off. The feature that distinguishes the instruments in the previous chapter from those that follow is not lack of precision but *lack of amplification.* Use of the previous ones was limited by the senses of sight and feel of the user. These senses can be greatly increased by mechanical, electronic and optical means. One of the simplest of these is the vernier. This is a system of scales, invented by Pierre Vernier in 1631. It owes its discrimination to our visual ability to bring two lines accurately into coincidence.

VERNIER INSTRUMENTS

It appears that 220 years had to go by before the vernier principle and the slide caliper were mated. Joseph R. Brown is credited with the union, Fig. 6-1. He is better remembered, together with his father, as a founder of Brown & Sharpe Manufacturing Co.

Fig. 6-1 This historic instrument is the forerunner of the present vernier caliper.
(Brown & Sharpe Mfg. Co.)

In addition to vernier calipers there are vernier height gages, Fig. 6-2, vernier depth gages, gear tooth verniers and vernier protractors. All are simple scaled instruments that use verniers to increase their amplification. This, in turn, increases their discrimination. The vernier also adds precision to instruments that utilize other means of amplification, such as vernier micrometers, Chapter 7.

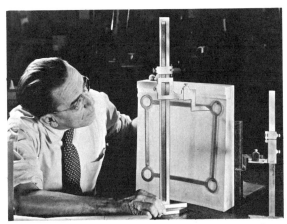

Fig. 6-2 Vernier height gages are among the most popular measurement instruments.

The vernier scale is attached to the instrument so that it may slide in a path parallel to the line of measurement, Fig. 6-3. The main scale is also parallel to the line of measurement. The scales are mounted so that readings can be made from one to the other with minimum parallax error.

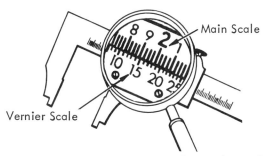

Fig. 6-3 The simple addition of a vernier scale adds amplification to scale reading.

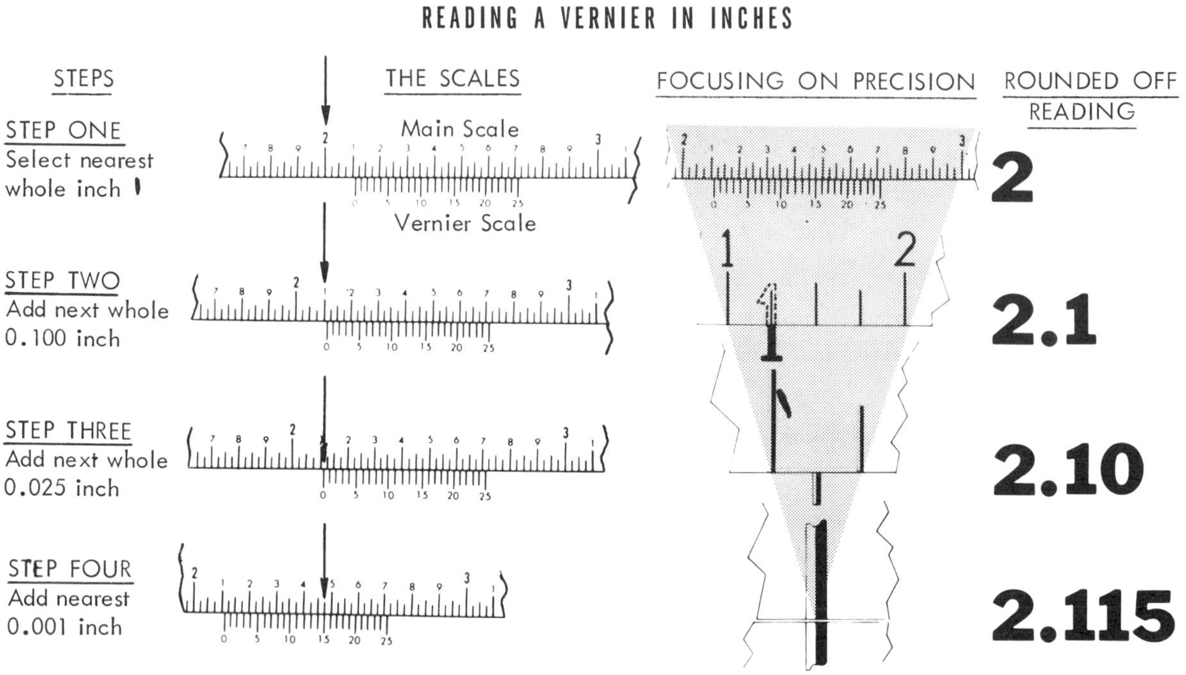

READING A VERNIER IN INCHES

STEPS	THE SCALES	FOCUSING ON PRECISION	ROUNDED OFF READING

STEP ONE
Select nearest
whole inch

Main Scale

Vernier Scale

2

STEP TWO
Add next whole
0.100 inch

2.1

STEP THREE
Add next whole
0.025 inch

2.10

STEP FOUR
Add nearest
0.001 inch

2.115

Fig. 6-4 Each of the steps for reading a vernier raises the results to a higher level of precision, much like focusing a microscope.

Consider a vernier scale measuring length in inches. The main scale is divided into inches. Each inch is divided into ten parts, and each of these divisions is subdivided into quarters, Fig. 6-4. The smallest divisions are thus 1/40 in. One thousand divided by 40 is 25; therefore the smallest divisions are 0.025 in. and read as units of 25 *mil*. Note that from its start, the vernier has been a decimal instrument, not fractional.

In a caliper the vernier scale is attached to the sliding jaw and moves along the main scale. It has 25 divisions in the same length that the main scale has 24. The difference between a main scale division and a vernier division is 1/25 of a main scale division. Hence, 1/25 of 0.025 in. is 0.001 in. or *one mil*. This is the discrimination of the instrument, and is called the *least count*. A vernier is read by adding the total readings from the main scale to the thousandths readings on the vernier scale. This applies to the controls on machines as well as scales on instruments, Figs. 6-4 and 6-5.

The steps are as follows:

1. Read the number of whole inches on the main scale that appear to the left of 0 on the vernier.

2. Read the highest numbered graduation on the main scale that lies to the right of the index (0) on the vernier scale. Read these graduations as even one hundred *mil* (0.100, 0.200, etc.). Add to the whole inches from step one (1.100, 2.100, etc.).

3. Read the highest number of whole minor divisions to the right of the index. Read these graduations as even twenty-five *mil* (0.000, 0.025, 0.050, and 0.075 in.). Add to the sum of steps one and two (1.125, 1.150 in., etc.).

4. Now find the vernier graduation that most perfectly coincides with *any* graduation on the main scale.* That is the nearest *mil* and may be any whole number from zero to 25 (0.000, 0.001, ... 0.024, 0.025). Add this to the sum of the previous three steps.

The principle is the same for metric verniers, Fig. 6-6. Some modern instruments have both inch and metric scales. These provide the best of both worlds.

* Only when the vernier is at zero will two lines, 0 and 25, of the vernier scale coincide with the main scale.

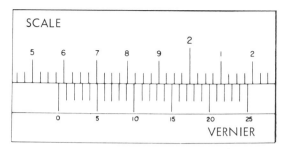

FROM A VERNIER CALIPER 1.582

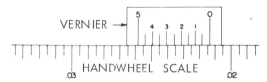

FROM HANDWHEEL OF JIG BORER 0.02135

Fig. 6-5 Practice is required for reliable reading of vernier scales.

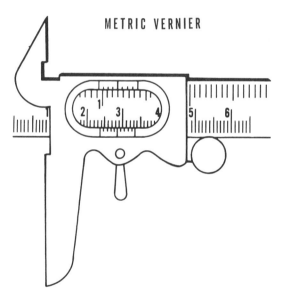

METRIC VERNIER

Fig. 6-6 Nearly all vernier instruments are available with metric scales.

The advantage and disadvantages for vernier instruments are summarized in Fig. 6-8. Here is the essential point. Vernier scales are usually more reliable than the instruments on which they are used. It might be asked why the discrimination is limited? Could not the entire arrangement be expanded so that the discrimination be reduced to, say, 0.0001 instead of 0.001 in.? A vernier designed to read to 0.0001 from an inch scale would have a group of lines so nearly in coincidence that they would be indistinguish-

able. Many modern vernier instruments do have expanded scales. The objective, however, is to achieve greater reliability by improved readability.

An example is the modern 50-division vernier, Fig. 6-7. This vernier makes use of the amazing ability of the eye to bring lines into coincidence. It requires twice the visual acuity of a 25-division vernier, but is quite easy to use. Note that it does not increase the accuracy (that is mostly in the instrument construction), or the discrimination, but does increase the reliability. Another adaptation of the vernier principle is to protractor scales. These are discussed in Chapter 16.

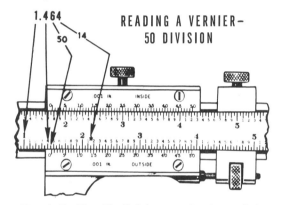

READING A VERNIER-
50 DIVISION

Fig. 6-7 The 50-division vernier is read similarly to the more familar 25-division verniers.

ADVANTAGES AND DISADVANTAGES OF VERNIER SCALES

Advantages:
1. Amplification is achieved by design and is not dependent upon moving parts that may wear or go out of calibration.
2. No interpolation is possible in reading, let alone required.
3. There is no theoretical limit to the scale range (length).
4. Zero setting adjustment is easy.

Disadvantages:
1. The principal disadvantage lies in the instruments on which verniers are used.
2. The reliability of reading depends more upon the observer than most instruments.
3. The discrimination is limited.
4. No way to adjust for any errors other than zero setting.

Fig. 6-8 Important note: These advantages and disadvantages apply to vernier scales, not to complete vernier instruments.

THE VERNIER CALIPER

This is the first and simplest of the vernier instruments. It is no longer as important in shop work as it once was, but still finds many applications because of the long ranges in which it is obtainable (6 to 48 in.), and its economy. One instrument *substitutes* for many outside and inside micrometers. But note, it *does not replace* them.

Vernier calipers are slide calipers with a vernier scale attached. Because this increases their discrimination to *one mil* (0.001 in.) a fine adjustment and a means for zero setting are required, Fig. 6-9. For convenience, on some instruments one side is graduated for inside measurement, the other side for outside measurement. The jaws are reduced to thin *nibs* for small inside measurement on most instruments.

Reliable measurement can be made with the vernier caliper. It is not easy. The reason it is not involves an understanding of a certain point that is fundamental to all highly reliable measurement. This understanding will be far more important to your future use of measurement than anything else in this chapter. You already have been indoctrinated somewhat about the importance of positional relationships in measurement. Now the vernier caliper will show you how much havoc bad alignment can play in measurement. It is all due to . . . . .

ABBE'S LAW

In spite of its Medieval sound, Abbe's law was one of the first fruits of the Industrial Revolution. It came about through the efforts of a truly remarkable physicist who was also a leading industrialist. We can learn much from the life of this distinguished man.

Ernest Abbe stayed on at the university at Jena after his graduation as an instructor. In 1870 he became a professor of physics. In 1878 he was made the director of observatories. At Jena he became acquainted with Carl Zeiss, founder of the famous Zeiss works. When Zeiss died in 1888 he left the entire business to Professor Abbe. With enlightenment uncommon in that age he reorganized the business giving a portion of the profits to the employees and to the university. With his own money he founded Carl Zeiss-Stiftung, an organization for scientific research and social betterment, an organization that is still active and important.

During this period he found time to continue his scientific work. Although his primary efforts were in optics, we are indebted to him for formalizing the relationship we know as "Abbe's law", "Abbe's principle", or more often, "the comparator principle". It is variously stated but may be paraphrased as follows: *maximum accuracy may be obtained only when the standard is in line with the axis of the part being measured.*

VERNIER CALIPER

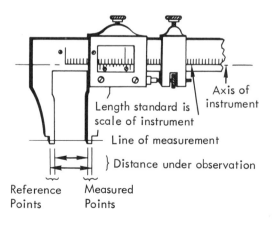

FUNCTIONAL FEATURES

Scale being viewed
(inside on other side)

Divider setting holes

Clamping Screws   Beam

OUTSIDE

Nibs for inside measurement

Fixed Jaw

Jaws for outside measurement

Movable Jaw

Vernier Scale Plate

Adjusting Screws

Vernier Scale

Nut Carrier
Adjusting Nut

METROLOGICAL FEATURES

Axis of instrument

Length standard is scale of instrument

Line of measurement

Distance under observation

Reference Points   Measured Points

Fig. 6-9 The vernier caliper is a slide caliper with trimmings.

THE TWO MEASUREMENT METHODS

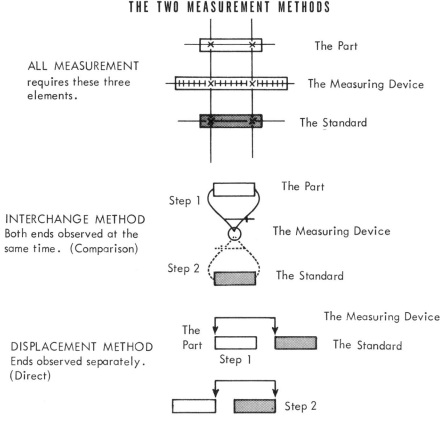

ALL MEASUREMENT requires these three elements.

The Part

The Measuring Device

The Standard

INTERCHANGE METHOD Both ends observed at the same time. (Comparison)

Step 1 — The Part

The Measuring Device

Step 2 — The Standard

DISPLACEMENT METHOD Ends observed separately. (Direct)

The Part — The Measuring Device — The Standard — Step 1

Step 2

Fig. 6-10 All measurement consists of comparison of the unknown with a known. The methods for comparison vary but all fall into one of these two groups. Vernier calipers are of the displacement type.

To understand the implications of Abbe's law you must first consider the fundamental premise of measurement: *If two quantities each equal a third, they are equal to each other.* In all measurement there must be a standard, the part to be measured and a device for relating the two (A in Fig. 6-10). With scaled instruments, such as a vernier caliper, the distinction between the standard and the device is small, but nonetheless, exists. Obviously the results hinge upon the means for relating the standard to the part. There are two general methods, the *interchange method* and *displacement method.*

The interchange method is the method we have discussed up to now. With it both ends of the measured length are *observed* at the same time. An excellent example, is the use of the caliper instruments. The caliper *observes* both ends of the part (B in Fig. 6-10) and then compares the separation to both ends of the standard simultaneously. It is the basis of all *transverse-comparators*, whether they are mechanical, optical, pneumatic or elec-

tronic. These instruments are discussed in Chapters 10 and 11.

The displacement method uses a longitudinal movement which is common to both the part and the standard as the means for relating one to the other. It differs from the interchange method in that only one end of the measured length is *observed* at one time. Both the part and the standard are lined up to an index mark. They are then moved together to see if another set of their reference points line up with the same index mark (C in Fig. 6-10). If they do the lengths are the same. This method is used primarily for positioning in machine tools and measuring machines. The vernier caliper, however, is a simple example of both the method and the problem it has obeying Professor Abbe.*

*Räntsch, Kurt, Machine Tool Optics, International Research in Production Engineering, American Society of Mechanical Engineers, New York, 1963, p. 629.

Figure 6-11 shows the steps required to make a displacement measurement with the vernier caliper. It is similar to Fig. 6-10 except that the *device* is the instrument and contains both the standard and the means for relating the standard to the part.

Moving the jaw is the longitudinal displacement. The positions of both the standard and the part must be observed in step one. Because the vernier caliper is a contact instrument, that simply requires closing the jaws and zero setting. The part is not there; but it is being *observed* in spirit, at least. In step two the other ends are *observed*. In the case of the part, this requires physical contact with the movable jaw. In the case of the standard, a vernier reading suffices. In practice we conclude that the reading is the desired length. But is it?

For the reading to be the exact length the beam would have to be perfectly straight and the jaw absolutely at 90 deg. to it. This is never the case. There is always some lack of straightness of the beam (and its scale standard; more about this in Chapter 18). Even if the jaw was finished very nearly perfectly square to the beam, it must not be forgotten that it is a sliding member. That means there must be "play" between them, and this in itself destroys any hope of near perfection.

For simplicity (what sins have been committed in thy name) let us assume that the total of these errors is represented by x. Let us also call the length error y. At this point I must point out to you that if I were a bonafide metrologist you would not get off so easily. The angularity error would be designated α, and the length error would be $\Delta 1$. Whether the use of Greek letters in these matters is an attempt to be exclusive or to educate us, is beyond the scope of this work.

The goal is that the reading on the instrument equals one, or error y equals zero. Only two conditions can make this possible. First is when the angularity error, x, equals zero. Second is when the distance h equals zero. The first can never happen. That only leaves the second to manipulate. When h does equal zero we have the standard in line with the line of measurement. (When we have this, we have a reliable measurement situation at least as far as geometric or positional relationships are involved.) We find it in the better comparator

VERNIER CALIPER
DISPLACEMENT MEASUREMENT

STEP 1
Observe one end.
(ZERO SET)

The Instrument and the Standard

STEP 2
Observe second end of part simultaneously with observation of standard.

The Part

The Displacement

POSITIONAL ERROR

Axis of Standard

h

Line of Measurement of Part.

x

1

y = error in measurement 1
y = zero when x = zero or h = zero

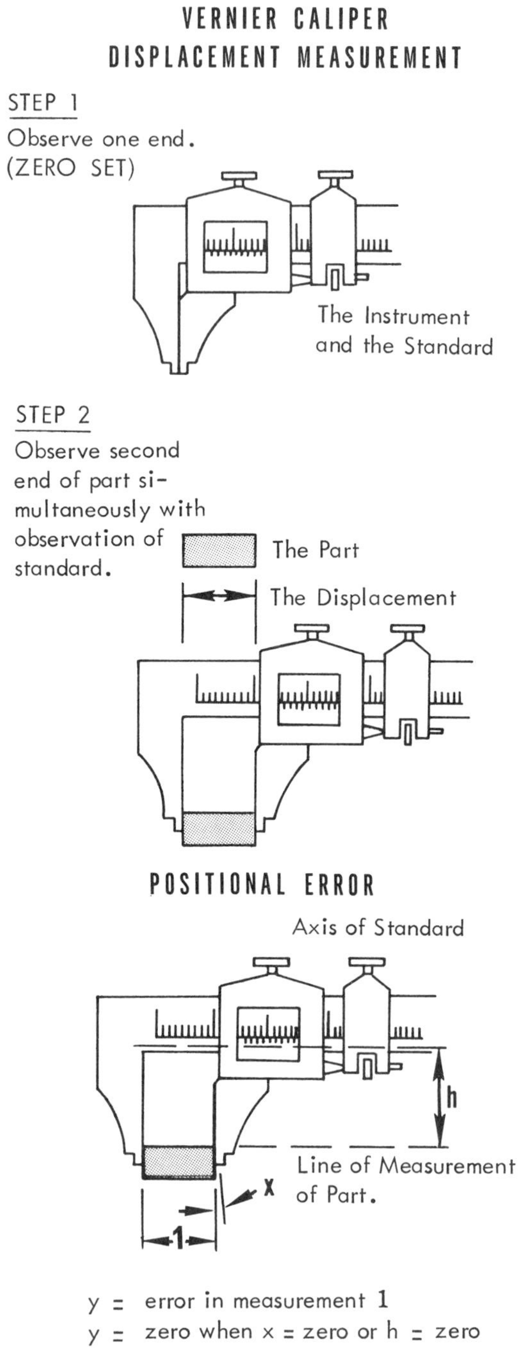

Fig. 6-11 The vernier caliper is shown to be a displacement instrument – but one that does not conform with Abbe's law. x can never be zero and neither can h.

instruments which you will meet in later chapters. You do not find it in vernier calipers. The degree to which an instrument conforms with Abbe's law is the most important single measure of its inherent accuracy.

IMPORTANCE OF THE CLAMPING SCREW

The manufacturers of vernier calipers have tried to minimize the x error by providing a clamping screw that locks the movable jaw to the beam. This brings up a common misunderstanding. If you ask the next one hundred users of vernier calipers whom you meet, the function of the clamp, you will get not more than one correct answer. The majority will state that it is a memory device – a means to retain a reading. The others will state that it is used to prevent a reading from moving between the time it is set to a part feature and the part is removed.

Do not be too severe with these people. The author has been unable to find one single mention of the main function of the clamping screw in any catalog or instruction manual. A government publication* in contrast goes all the way in the other direction. Not only is it suggested that the clamp be locked, but that the reading be ignored entirely and gage blocks used to duplicate the measurement. This then shifts the vernier caliper from a displacement instrument to an interchange instrument – but good.

---

* MIL-STD-120, Gage Inspection, (Washington, D.C., Department of Defense, 1950), 8.4.3.1, p. 81.

There are lamentable schisms among those who earn their living by measurement: the metrologist, the quality control person, the inspector, the skilled worker, and the hosts of engineers of all kinds. This was shown by the example of the clamp screw on vernier calipers. Abbe's law furnishes an even better example. The professional articles on metrology abound with references to it. Yet, of all the reference books in the author's library, including such ponderous tomes as "Mark's Handbook" and "Machinery's Handbook", there is not one single isolated reference to Abbe's law. No wonder, that some estimates of the scrap loss in the United States alone exceeds ten billion dollars per year, and that Soviet scientists have scored most of the firsts in the space race.

MEASUREMENT WITH VERNIER CALIPERS

Reliable measurement can be made with vernier calipers. It is important, however, that you carefully observe certain precautions. Most of these precautions also apply to instruments of even higher amplification. They will be taken up more thoroughly later in the book. For now they are given in the form of a check list, Fig. 6-12. Steps 1, 2, 10, 11, 12 and 14 appear again and again. They are among the most essential contents of this book. They do not take time – they save time. They are your personal measure of reliability.

STEPS FOR USING THE VERNIER CALIPER

1. Determine that this is the best instrument you have available for the particular measurement. Do not use simply because it is handy.
2. Thoroughly clean both the part and the caliper. (Make it a habit to automatically check all contact surfaces of the part and the instrument for burrs or other obstructions.)
3. Loosen the clamping screws on both the movable jaw and the nut carrier.
4. Set movable jaw slightly larger than feature to be measured.
5. Clamp nut carrier to the beam. Snug up but do not lock the clamping screw on the movable jaw.
6. Place the fixed jaw in contact with the reference point of the part feature, Fig. 6-14.
7. Align the beam of the caliper to be as nearly parallel to the line of measurement as possible – in both planes.
8. Turn the adjusting nut so that the movable jaw just touches the part. Tighten clamp screw on the movable jaw without disturbing the feel between the caliper and the part.
9. Read in place without disturbing part of caliper, if possible. If not, remove caliper.
10. Record the reading on scratch paper, chalk on part, or on part drawing. Do not trust memory.
11. Repeat the measurement steps a sufficient number of times to rule out any obviously incorrect readings and average the others for the desired measurement.
12. Loosen both clamps, slide movable jaw open, remove work, if not already done.
13. Clean, lubricate and replace instrument in its box.
14. Ask yourself, what errors may remain in your measurement.

Fig. 6-12 These steps apply to most vernier caliper measurements. Do not slight those that appear obvious.

GEOMETRY OF THE VERNIER CALIPER

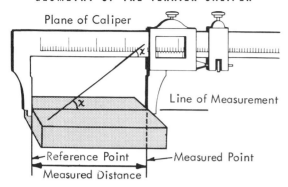

Plane of Caliper

Line of Measurement

Reference Point — Measured Point

Measured Distance

FOUR ESSENTIAL POINTS

1. Line of measurement is shortest distance separating the reference point and measured point. If part is rectangular, this line is perpendicular to the reference edges.

THEREFORE:

2. Line of measurement must lie in plane of caliper.
3. Line of measurement must be parallel to beam of caliper.
4. All deviations from the above will be included as error in the caliper reading.

VERNIER CALIPER ALIGNMENT

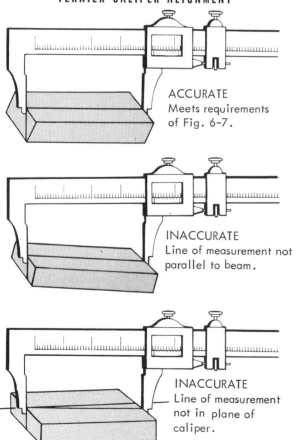

ACCURATE
Meets requirements of Fig. 6-7.

INACCURATE
Line of measurement not parallel to beam.

INACCURATE
Line of measurement not in plane of caliper.

Fig. 6-13 Because it does not comply with Abbe's law, positional considerations are of great importance when using the vernier caliper.

The procedure for inside measurement is similar to outside. It is important to read the proper side of the caliper for the measurement involved.

The most important consideration when using a vernier caliper is the alignment relationship of the instrument and the part. The geometry of these relationships is shown in Fig. 6-13. In the example of alignment, one looks wrong but is right. Two look wrong and are.

Unfortunately when using a vernier caliper you usually cannot locate misalignment by looking. When one *mil* is involved the misalignment will seldom be visible. Generally you will have to rely upon *feel*. That is the reason gaging feel is emphasized.

Measuring a diameter is easier than measuring between flat surfaces, because the diameter is the greatest distance separating the reference and measured points. Parallelism misalignment could result in a plus error reading longer than the true value, but most misalignment error will be minus. Therefore, to measure an outside diameter hold the caliper so that it is balanced and does not have a tendency to upend. Hold the fixed jaw against the edge of the diameter and swing the movable jaw back and forth past center, Fig. 6-14. Slowly close the caliper until the measured point is just felt in passing.

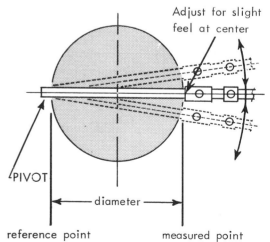

Adjust for slight feel at center

PIVOT

diameter

reference point   measured point

Fig. 6-14 While swinging past center, the beam must be held parallel to the line of measurement.

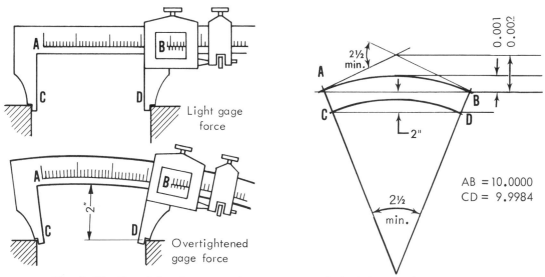

Fig. 6-15 Overtightening causes beams to bow. Only 0.002-in. bow in 10 in. will shorten the reading by 0.016 in.

Even the slightest feel indicates that some metal is being compressed or sprung, however imperceptibly. Unless the part is very thin, it will mostly be the caliper springing. Continued closing of the caliper will only increase the springing. Heavy gaging pressure will cause rapid wear of the jaws, burnish the part, and may actually damage the caliper.

Measuring an inside diameter requires more careful touch, although the principles are the same. When you first feel the work it may be a slight misalignment causing a plus error. This is detected by carefully moving the jaw in all directions, within a small circle. When more nearly on the line of measurement it will feel free again. Then the adjustment is opened until the feel is picked up again. The jaw is again moved about, but this time the circle of movement will be smaller. When the line of measurement is approached another adjustment is made. This is continued until the jaws are squarely on the measured and reference points of the part feature. The name for this technique is *centralizing*. It can best be learned by practice.

Measuring between flat surfaces is much more difficult for the reasons pointed out for outside calipers. Here the problem is greater because the large size of a vernier caliper applies tremendous leverage to compress or spring both itself and the part. Figure 6-15 shows that 0.002-in. curvature in 10 in. shortens the reading by 0.016 in.

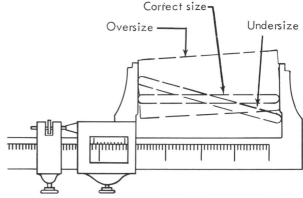

Fig. 6-16 Manipulation of the part and instrument is by far the greatest cause of errors with vernier calipers.

ACCURACY CHECKS FOR VERNIER CALIPERS

Manipulation during measurement is by far the most important cause of unreliable results with vernier calipers, Fig. 6-16. All the more reason to frequently check the instruments so that instrument defects do not make the matter worse. This is known in measurement as *calibration*.

First, check the jaws for wear by closing them tightly and looking through the crack towards a light. No light will be seen if they are in good shape. When the crack is squarely in line of sight, a gap as small as 50 *mike* (0.00005 in.) may be discerned with a little practice.

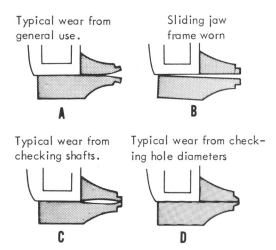

Typical wear from general use.

Sliding jaw frame worn

Typical wear from checking shafts.

Typical wear from checking hole diameters

A   B   C   D

Fig. 6-17 Wear may take very different forms as shown in these exaggerated examples.

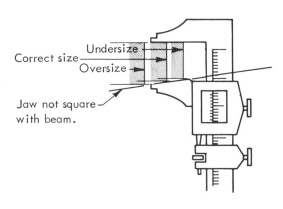

Fig. 6-18 Jaw error can cause plus or minus errors depending upon when the part feature contacts the jaws.

We have now run into one of the little enigmas with which measurement abounds. All the books state that you should check the jaw wear by this method and that when the wear exceeds 0.0002 in., the gage should be returned to the manufacturer for repair. Unfortunately, yet to be found is an explanation of how to tell when an air gap is 0.002-in. wide. Fortunately, there are other ways to measure the amount of wear as you will find later in this book. Optical flats, for example, provide a quick and easy check. To be practical, check periodically with the light gap method. If wear is found, consult an experienced gage maker for a decision. The effect of wear will vary with the use of the calipers. Exaggerated but typical wear patterns are shown in Fig. 6-17. Figure 6-18 shows the effect of either type A or B jaws.

Repeated measurements are frequently used as a check for accuracy. They are a valuable means for checking instruments. However, *simple repetition of the same reading is a partial test of precision and not a test of accuracy at all.* It the true value of the test part is known, a repeatability test does check accuracy — but only for that one measurement, Fig. 6-19.

Incorrect measurements can be repeated with great reliability under certain conditions, such as shown in Fig. 6-20. That is the reason it is emphasized that repeatability tests should be made with standards (gage blocks or master cylinders) of known value.

REPEATABILITY, ACCURACY AND PRECISION

Repeatability of one measurement on one part with the same instrument and observer . . . . . .

1. Does not measure the accuracy of the instrument.

2. Does not measure the precision of the instrument.

3. Does measure the combined precision of the observer and instrument for one portion of the entire range.

4. Can measure the accuracy for the combined observer and instrument for one portion of the entire range if the true value of the test part is known.

Fig. 6-19 Repeatability is extremely important. However, the information it provides is limited.

After checking the jaws for wear the zero setting should be checked. The most obvious thing to do is close the jaws and check to see if the vernier reads 0.000. It is easily adjusted to zero by loosening the adjusting screws, Fig. 6-9, and repositioning. This might not be the best course for reliabile measurement, however. If the majority of measurements are in the, say, 3- to 5-in. range, it would be preferable to zero set at the 4-in. point. This would be done by repeated measurements on a standard of known size (true value) such as a gage block. Chapter 13 considers calibration in greater detail.

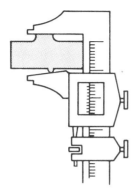

Fig. 6-20 Thanks to the shape of these parts, high repeatability would be obtained, but all measurements could be incorrect as well as correct.

Fig. 6-21 Inspection and quality control offer rewarding careers for women. Some already have risen into the professional ranks of the country's top metrologists. *(Brown & Sharpe Mfg. Co.)*

Formerly vernier calipers only had the outside scale. The cheapest ones still do. The distance across the nibs was "miked*" and added to the outside readings to convert to inside measurements. Many "old timers" in the shop still consider it a mark of distinction to continue this even when using modern instruments with both scales. When asked why, they state that it is more accurate. If pressed they will explain that it eliminates

* "mike" is the popular slang term for a micrometer. "To mike" means to measure with a micrometer. Do not confuse with mike of the decimal-inch terminology, which is not a slang term.

the effect of wear on the outside measuring surfaces. Hogwash. They simply have not caught on to modern instruments, and do not understand one of the fundamental laws of measurement – *every extra operation* (manual, visual, mental, computational or otherwise) *is an additional source of errors and should be eliminated if possible.*

Figure 6-40 sums up the information about vernier calipers. It makes no allowance for good or bad manipulation. These tables will be included for each of the important instruments. They will help you select the best instrument for the job, whether that is a distometer test in the physics lab or inspecting a mountain of refrigerator doors.

VERNIER DEPTH GAGE

The best reason to use a vernier depth gage is the fact dad used it in his shop days and handed it down to you. It is simply a depth gage, as discussed in the last chapter, with a vernier scale hung on, Fig. 6-22.

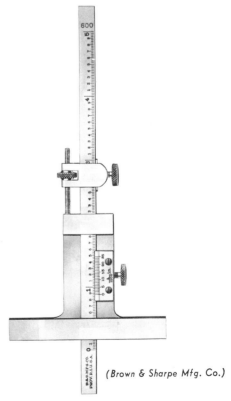

(Brown & Sharpe Mfg. Co.)

Fig. 6-22 If you have a depth vernier, give it the same care (and suspicion) that you reserve for the vernier caliper.

VERNIER DEPTH GAGE ERRORS

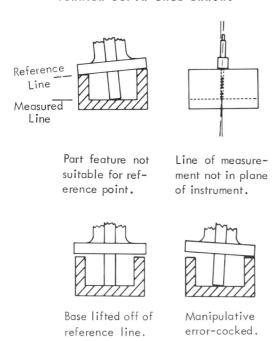

Part feature not suitable for reference point.

Line of measurement not in plane of instrument.

Base lifted off of reference line.

Manipulative error-cocked.

Fig. 6-23 Several ways not to use vernier depth gages.

The justification for an instrument as unreliable as a vernier caliper is that it provides the greatest range for the money. Not so with the vernier depth gage. Depth micrometers would win out in both inches per dollar and reliability per dollar.

The one good point about vernier depth gages is that they conform with Abbe's law. That would be a very important point if depth micrometers did not, but they do also.

The only measurement that justifies a vernier depth gage rather than a depth micrometer is measuring into a slot too narrow for the 0.100-in. rod of the micrometer but wide enough for the 1/32-in. blade to enter. An authority on gaging* states that it is easier to make manipulative errors with the vernier depth gage than any other measuring instrument, Fig. 6-23. They are read the same as vernier calipers and require the same treatment for reliability.

---

*Kennedy, Clifford W., Inspection and Gaging, Industrial Press, 2nd printing, p. 94.

VERNIER HEIGHT GAGE

FUNCTIONAL FEATURES

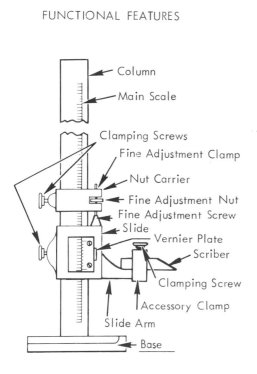

Column
Main Scale
Clamping Screws
Fine Adjustment Clamp
Nut Carrier
Fine Adjustment Nut
Fine Adjustment Screw
Slide
Vernier Plate
Scriber
Clamping Screw
Accessory Clamp
Slide Arm
Base

METROLOGICAL FEATURES

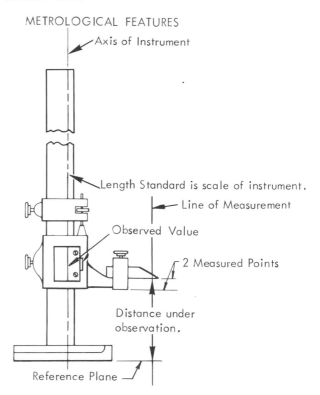

Axis of Instrument
Length Standard is scale of instrument.
Line of Measurement
Observed Value
2 Measured Points
Distance under observation.
Reference Plane

Fig. 6-24 The versatile height gage is associated with surface plate inspection.

REFERENCE POINTS, LINES AND PLANES

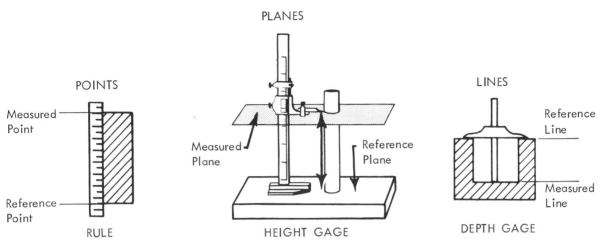

Fig. 6-25 A height gage measures between planes, as a depth gage does between lines and a rule between points.

VERNIER HEIGHT GAGE

This monarch of the surface plate was created with a very unfortunate ratio of base length to height. As a result, it has been a party to more erroneous measurement than any other precision measuring instrument. That view will get you into arguments. Therefore, I suggest that you memorize the following statement from *Military Standard, Gage Inspection*, MIL-STD-120, article 8.4.3.3, page 81 (author's italics have been added for emphasis):

"After a moderate amount of use, the accuracy of this gage is generally slightly *less* than that of ordinary calipers. The *only advantage* in using a vernier height gage in this way (direct height measurement) is that it is very fast."

A person who never does anything, never gets in any trouble. Perhaps that is why the height gage gets into trouble – it does so much.

Essentially the height gage, Fig. 6-24, is a vernier caliper with an entire surface plate for its fixed jaw. In fact, base attachments are available to convert vernier calipers to height gage use. Of course, the surface plate is not part of it, but because the height gage sits on a surface plate it makes effective use of it. The height gage is often used directly on a reference surface of a large part. The

comments based on surface plate use apply equally well to these applications.

When we discussed simple rules we were concerned with points, the reference point and the measured point. Both of these lie on the measurement line. When we added attachments such as a depth gage base, we extended the reference points and thus formed lines. These lines sweep through planes with the height gage, Fig. 6-25. Each instrument may work from the references of the lesser gages, but the reverse is not true. This accounts for the versatility of the height gage.

Height gages are available in sizes (based on maximum measuring height) from 10 to 60 in., in metric and/or inch scales, and with a large selection of refinements. All sizes may be used either as outside or inside vernier calipers, but only the small sizes have a second scale for direct reading of outside dimensions, Fig. 6-26.

Like the vernier caliper it violates Abbe's law – only more so. Therefore, the accuracy that can be expected with a height gage is one *mil* (0.001 in.). It cannot be reliably used to control manufacture of parts with closer tolerance than 5 *mil* (0.005 in.). This accuracy is not usually guaranteed to the upper ends of height gages 18 inches or higher. See Fig. 6-40 for a summary of the measurement information for height gages.

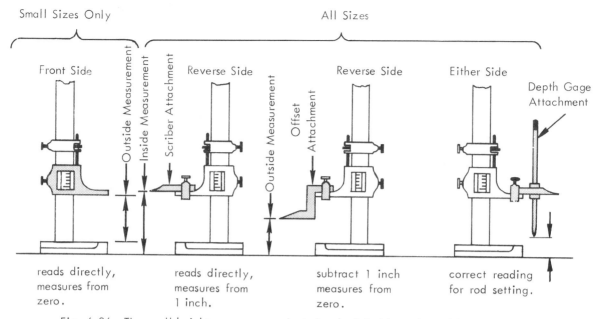

Fig. 6-26 The small height gages are graduated as both inside and outside vernier calipers, the large sizes only as inside vernier calipers.

Figure 6-26 also shows three of the four popular attachments. The *scriber* may be aligned with an edge of a part resting on the surface plate, thereby measuring the height of that edge above the reference base. *This is not recommended.* An indicator used with the height gage is preferred for height measurement. The exception is to measure horizontal lines on the part. A magnifiying glass must be used for accurate alignment of the edge of the scriber to the line. A more common use of the scriber is for layout, Fig. 6-27. The *offset scriber* has similar applications with the added advantage that it can be used from the surface plate, Fig. 6-28.

The *depth gage attachment*, Fig. 6-29, converts the instrument to a depth gage with very large range. It permits relative height differences to be measured in inaccessible and difficult places.

By far the most widely used attachment today is an *indicator head holder*. The indicator substitutes greatly magnified movement of a probe for feel of metal-to-metal contact. The use of the height gage in this fashion is discussed under surface plate work.

USING THE HEIGHT GAGE

The general steps for using the height gage are the same as for the vernier caliper. Step

6 is simplified because the instrument rests on a surface plate, Fig. 6-12. Most measurement with a height gage will involve an indicator making the centralizing a visual step rather than a tricky one of feel.

Use with the depth gage attachment is different from the vernier caliper however. It is used to make depth or height measurements past obstructions. To use it first attend to the general steps 1, 2 and 3 for the vernier caliper. Then proceed as shown in Fig. 6-30.

When several depth measurements are to be taken it is easier to work from a whole inch than a decimal. To do this simply hold the rod against the surface plate and adjust the movable jaw for the nearest inch. After clamping the rod, recheck and carefully adjust as required.

CHECKING THE HEIGHT GAGE

The principles for checking the height gage are similar to those for the vernier caliper with the added complication of the base-to-column squareness check. The checking must be done from a surface plate. The surface plate errors will enter into checking. Therefore height gage inspection and calibration are best left to the gage inspection department or other qualified personnel.

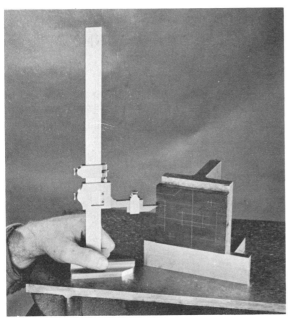

Fig. 6-27 The most important use of the scriber is for layout rather than measurement.

Fig. 6-28 The offset scriber can be used close to the plate.

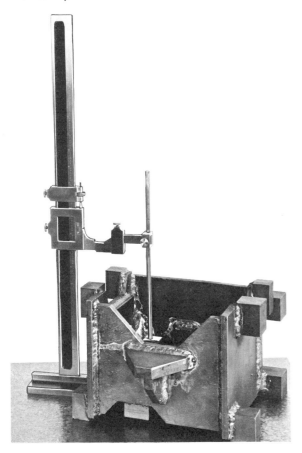

Fig. 6-29 The depth gage attachment gets into otherwise inaccessible places.

STEPS FOR USING DEPTH GAGE WITH VERNIER HEIGHT GAGE

1. Thoroughly clean and inspect the attachment.

2. Fasten it to the movable jaw.

3. Set the depth-gage rod with its end in contact with the surface plate.

4. Write down the reading. Do not rely on memory.

5. Raise the movable jaw to clear the obstruction on the part.

6. Place height gage in position and lower movable jaw until depth-gage rod contacts the measured point of the part feature.

7. Note the reading. The difference between this reading and the reading in step 4 is the height of the measured point above the reference surface.

Fig. 6-30 In general, the steps for using the vernier caliper height gage resemble those for using the vernier caliper. The depth gage attachment is the exception. Proceed as shown in this check list.

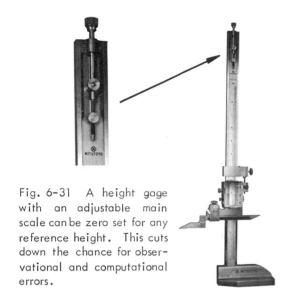

Fig. 6-31 A height gage with an adjustable main scale can be zero set for any reference height. This cuts down the chance for observational and computational errors.

Several tests can be made, however, that will show whether or not the instrument needs to be returned. Visual inspection will reveal much about the previous use of the instrument. Particularly look for burrs which can be removed, and cracks that may indicate severe abuse or damage.

The base should not rock on a flat surface. Apply pressure at various places. There should be no perceptible feel of movement. A really flat base will wring to a fine surface plate. There should at least be a feeling of drag or suction, when sliding the instrument.

The acid test of accuracy and precision is repeated readings of *gage block heights at various points throughout the range.* Note:

we emphasize again that repeatability alone is not sufficient. If after correcting the zero setting by adjusting the vernier plate, the *precision* (dispersion of repeated readings) or the *accuracy* (repeating readings of the gage block height) are more than the *discrimination* of the instrument (0.001 in.), it should be returned for repair. It is essential that the accessories are inspected as carefully as the instrument itself.

THE PROBLEM WITH HEIGHT GAGES

The problem with height gages is height. The great height of these instruments in relation to their small bases creates a lever action, multiplying errors. The higher the measured plane, the greater will be the effect of many of these errors. Furthermore, the height gage is not at all exclusive about the errors it multiplies. It accepts the set-up errors as well as its own. In Fig. 6-32, A shows that a height gage is inherently unstable. The tall, thin column sways like the Eiffel Tower. You cannot see a wobble of 0.001 in., or even 1/32 in. for that matter. Unfortunately it can play havoc with reliable measurement.

Some modern height gages have greatly "beefed-up" columns, Fig. 6-38. This cuts down the wobble due to springiness but some remains. Even in the direction of its greatest strength, a height gage is unstable. This is tested easily as in B of Fig. 6-32. What might be very light gaging pressure when the movable jaw is low, may very well rock the base slightly off of the reference surface, when the jaw is high.

HEIGHT GAGE ERRORS

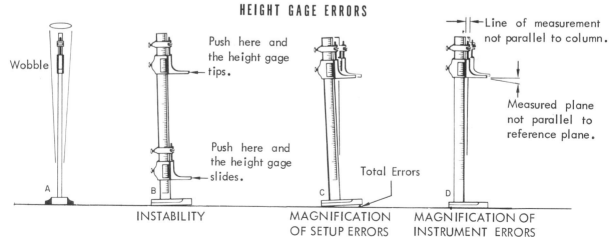

Fig. 6-32 The very height of a height gage creates its measurement problems.

C in Fig. 6-32 shows the result of the total setup errors. These could include dirt, surface plate error, burr on height gage base, etc. A 10 in. height gage has a base of about 3.2 in. length. Therefore, *one human hair under the base will throw the top of the column off by 0.02 in.*

Every instrument that is used will wear. Unfortunately many get abused. This can disturb the squareness of the column to the base and of the movable jaw to the column. The result on measurement relationships is shown in D of Fig. 6-32.

Perhaps even more at fault is the company the height gage keeps. Because it is fast and easy to use, there is a tendency to turn everything possible on end and measure it on a surface plate with a height gage. Remember point one in Fig. 6-12. Always ask yourself if there is a more appropriate instrument before choosing the vernier height gage. This overlooks the fact that the further up you go, the greater the chance for misalignment error, Fig. 6-33. Sometimes the error will cancel out — but can you chance it?

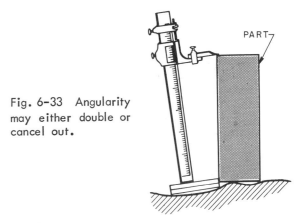

Fig. 6-33 Angularity may either double or cancel out.

You will rarely be free to choose the best reference surface to use as a base for measurement. When you can, it is advisable to keep the height no greater than the base, Fig. 6-34. There are many reasons to deviate from this. Those pertaining to the quality of the surface are shown in Fig. 6-35. *Note that if the reference surface is unknown or poor, all hope of reliable measurement with a height gage is lost.* Following the suggestions in Fig. 6-36 should help.

RATIO OF BASE TO HEIGHT

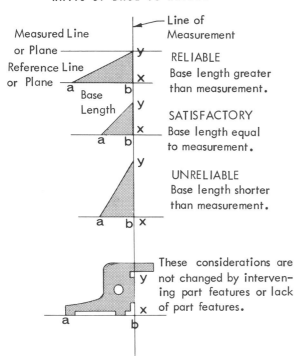

Fig. 6-34 You cannot always select the best surface for a reference. When you can, the above relationships should be remembered.

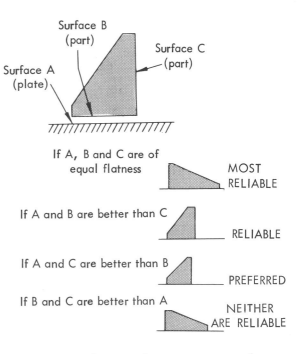

Fig. 6-35 When you choose a support surface, consider its condition as well as the ratio of base to height.

SUGGESTIONS FOR RELIABLE HEIGHT GAGE MEASUREMENT

1. Do not use height gage for measurements requiring greater discrimination (Fig. 6-26).

2. Know that the instrument is in calibration.

3. Observe scrupulous cleanliness.

4. Move height gage by base, not column.

5. Allow height gage, surface plate and part to stabilize (reach same temperature).

6. Prefer low setups to high ones.

7. Repeat each measurement.

8. Question the alignment of instrument, reference surface and part for each use.

9. Know that the surface plate is sufficiently flat for desired measurement.

Fig. 6-36   If these suggestions are followed carefully, they will force you to explore better means for height measurement.

The remaining important vernier instruments are the vernier protractor, and the gear tooth vernier caliper. The protractors are discussed under angular measurement in Chapter 17. The gear tooth vernier caliper is actually two vernier instruments combined, Fig. 6-37. One portion acts as a vernier caliper to measure the tooth thickness at the pitch line. The other portion consists of a *tongue* that travels between the jaws. This measures the distance from the top of the tooth to the pitch line of the tooth.

THREE ELEMENTS OF MEASUREMENT

Every measurement requires three elements. First is the object being measured. Second is a standard of length with which it is to be compared. And third is the means for comparison. This has been true of every instrument we have discussed.

All of them thus far, including the height gage, have had the standard of length built into the means for comparison. It has not been necessary to consider it separately.

(Brown & Sharpe Mfg. Co.)

Fig. 6-37   The gear tooth vernier caliper measures both the tooth thickness and height.

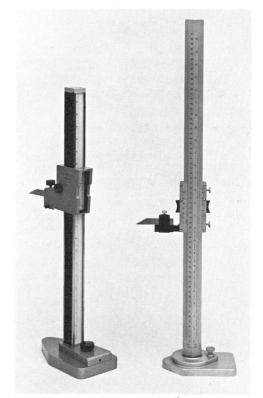

Fig. 6-38   The best modern height gages have large bases and stout columns. *(Scherr-Tumico, Inc.)*

You will be told much more about standards of length and height gages later in the book. The important thing here is that you recognize the existence of the three elements and that *no measurement can be more accurate than the positional relationship of these elements.*

TERMINOLOGY

Vernier instruments	Only those instruments in which the vernier is the outstanding characteristic. Many instruments not so classified use vernier scales to increase their amplification.
Amplification	The amount by which the senses are multiplied in the perception of precision.
Vernier scale	The amplifying scale that has one division more in a given length than the main scale.
Vernier caliper	A slide caliper that utilizes a vernier scale to achieve a discrimination of 0.001 in.
Vernier depth gage	A depth gage that utilizes a vernier scale to achieve a discrimination of 0.001 in.
Vernier height gage	Essentially a vernier caliper whose reference point is on a horizontal base thereby adapting it primarily to height measurement.
Least count	Discrimination of vernier instrument.
Abbe's law	Principle that for maximum reliability the axis of the standard must lie along the line of measurement.
Interchange method	Measurement in which both ends of the part feature are viewed in comparison to a scale or other standard.
Displacement method	Measurement by means of a movement analogous to the feature measured.
Memory device	Means for retaining a reading mechanically so that it will not be lost. Usually a simple clamp.
Feel	The perception of metal-to-metal contact in measurement.
Precision instruments	Those providing amplification of the natural senses of sight and touch.
Scriber	An instrument or attachment that makes a line when drawn across a part by the removal of material.
Centralize	Movement of instrument by feel of contact so that its axis is oriented parallel to the line of measurement.
Centralizing	Movement of the instrument to align it to the line of measurement based on the changes in feel.
Repeatability	The variation among several measurements taken with one instrument on one part feature, a test of precision not of accuracy.

Fig. 6-39 The important terms introduced in this chapter are summarized here. Much is said about repeatability in later chapters.

METROLOGICAL DATA FOR VERNIER INSTRUMENTS

INSTRUMENT	TYPE OF MEASUREMENT	NORMAL RANGE	DESIGNATED PRECISION	DISCRIMINATION	SENSITIVITY	LINEARITY	RELIABILITY	
							PRACTICAL TOLERANCE FOR SKILLED MEASUREMENT	PRACTICAL MANUFAC- TURING TOLERANCE
Vernier caliper	direct	24 in.	0.001 in.	0.001 in.	0.001 in.	0.0003/in.	± 0.002 in.	± 0.010 in.
Vernier depth gage	direct	12 in.	0.001 in.	0.001 in.	0.002 in.	0.0003/in.	± 0.003 in.	± 0.020 in.
Vernier height gage	direct	24 in.	0.001 in.	0.001 in.	0.001 in.	0.0003/in.	± 0.002 in.	± 0.008 in.

Fig. 6-40 Feel free to argue about the reliability data. There is virtually nothing on record to prove you right or wrong. The author's guesses are conservative but practical.

Summary — Much To Do About Something

● The familiar vernier instruments are the vernier caliper and vernier height gage. They discriminate in *mils*, the area in which most practical measurement takes place. Essentially they are steel rules with refinements. The vernier scale provides amplification so that the resolving power of the eye is increased many times.

● The advantages of vernier instruments are their long measuring range and convenience. Their chief disadvantages are reliance on the feel of the user and their susceptibility to misalignment of all kinds. Wherever reliability is essential vernier instruments have been largely supplanted by higher amplification instruments.

● The study of vernier instruments is important because they vividly demonstrate the importance of geometry and positional relationships in measurement. Furthermore, although the vernier instruments have been "phased-out" for the control of production, they remain important tools of the skilled craftsman. It will be a long time before the vernier caliper disappears from the tool box of the master tool maker or the vernier height gage from the surface plate of the model maker. For the latter, Fig. 6-40 will assist with instrument selection.

QUESTIONS

1. What principal advantage does the vernier instrument have over those gages discussed in previous chapters?
2. To what does a vernier owe its discrimination?
3. What are the common types of vernier instruments?
4. What is the principal limitation of verniers?
5. What is the principal advantage of the vernier caliper?
6. Why is it difficult to obtain accurate measurement with the vernier caliper?
7. Why is it simpler to measure a diameter than flat surfaces with a vernier caliper?
8. What mechanical checks should be performed periodically on vernier gages?

9. What is the principal advantage of the vernier depth gage?
10. Why is the vernier height gage an unreliable instrument?
11. What is the principal advantage of the vernier height gage?
12. What are the most popular attachments for the vernier height gage?
13. Of what importance is the reference or support surface upon which height gages are used?
14. How can an operator determine repeatability on verniers?
15. What is the optimum discrimination possible on nearly all verniers?

DISCUSSION TOPICS

1. Explain the steps involved when reading a vernier.
2. Discuss the principal advantages of vernier scales.
3. Discuss the principal disadvantages of vernier scales.
4. Explain the various features of the vernier height gage.
5. Explain the basic three elements of measurement.
6. Discuss the various points that must be considered in order to obtain the most reliable measurements with the vernier height gage.

Old Inspectors Never Die ... They Just Gage Away

If death to the Inspectors came, the smallest loss might be tears;
Production folks out in the plant would lose their several fears.
Inspectors at the Pearly Gates would act like they're expected;
While St. Peter inked a rubber stamp, the one that marks: "Rejected."
At Satan's door they'd have to stand, their heads bowed in dejection;
While pondering their doleful fate, as he read out their rejection.
"Your fellow workers," he might say, "have drawn up your indictment;
You've caused confusion, grief and woe, and all sorts of excitement.
You tried too hard to meet the rules of rigidly tight inspection;
Nothing made in Heaven, I'm sure, could meet your grade of perfection!"
Then Lucifer would send them back: "Take your 'mikes' and your detectors!
You're not needed up above, and my Hell hath no Inspectors!"
So, immortal these rejected are; the reasons why, you now know.
Despite their actions here on Earth, Inspectors all are "no-go."
But, pity not their grievous lot, they're really somewhat wise.
They stand and watch while others work, and get paid to criticize.

Donald Morrison
(From *Quality Assurance*, Feb. 1964)

Vernier instruments differ from rules in that they have amplification. This gives them greater precision. Micrometers are instruments that achieve still higher amplification by the use of screw threads. They would not be able to make full use of this amplification if they had the same geometry as the vernier instruments. Fortunately, micrometers obey Abbe's law. The line of measurement is in line with the axis of the instrument. Like the vernier calipers, they measure by the displacement method as shown in Fig. 6-10.

There are several forms of micrometers: micrometer calipers, inside micrometers, depth micrometers, and others, Fig. 7-2. The word *micrometer* or, in slang, "mike", refers to the 1-in. micrometer caliper. The interesting history of the micrometer, that carries it from astronomy to the apprentice's tool kit, is told in Fig. 7-1.

The designated micrometer size is its *largest opening*, not its *range*. A quick check of catalogs shows that micrometers are available from 1/2 to 168 in. Regardless of overall size, the range is usually limited to 1 in. There are also 1/2-in. and 2-in. micrometer heads, but they are uncommon. One- and 2-in. micrometers are available with vernier scales for *discrimination* (I did not say accuracy) to *one-tenth mil* (0.0001 in.). Other useful features are the ratchet stop and clamping ring.

1638 – William Gascoigne (Yorkshire, England) devised adjustable indicators to measure stars.

Fig. 7-1

1772 – James Watt's micrometer. One dial read in 1/100 inch, the second dial in 1/256 inch.

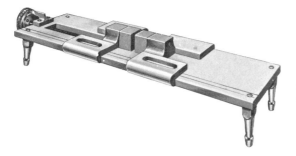

1800 – Henry Maudslay's "Lord Chancellor." Graduations to tenths of an inch were scribed on the base, and measurements to 1/1000 in. were read from the disc head.

1848 – The Systeme Palmer micrometer was patented in France. In 1867, J. R. Brown and Lucian Sharpe saw this device at the Parisian Exposition. The first micrometer sheet metal gage followed and, in 1877, Brown and Sharpe started production of the first micrometer-caliper.

Fig. 7-2 A wide range of types and sizes of micrometers are used for measurement. *(The L. S. Starrett Co.)*

AMPLIFICATION BY SCREW

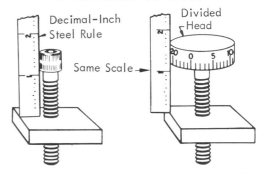

Discrimination = 0.02"   Discrimination = 0.001"

Fig. 7-3 If we use a rule to measure the advance of the thread, discrimination is limited by the rule. If we calibrate the rotation of the screw, then the screw itself becomes the element that limits discrimination.

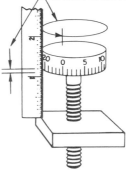

Amplification is ratio of circumference to lateral movement.

Fig. 7-4 The greater the ratio between the circumference and the lead of the screw, the higher the possible amplification.

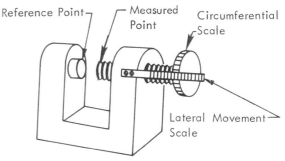

Fig. 7-5 In order to make a screw and nut into a measurement instrument, it is necessary to extend the nut to form a reference point. This creates the frame of the basic micrometer. (Reproduced from "Tools of Metrology, Part II", Quality Assurance, Feb. 1963, p. 46).

All micrometers are based on the relation of a screw's circular movement to its axial movement. Consider a screw threaded through a plate, Fig. 7-3. Each turn of the screw would move the screw a distance that could be measured with a steel rule. The accuracy with which we could measure the change is limited by all of the variables associated with rules. The discrimination is that of the rule, or *20 mil* (0.02 in.) for a decimal-inch rule or scaled instrument.

Instead of the rule, we can place a special scale alongside of the head, and this **can** be calibrated in whole turns. The discrimination is then one turn. If we try to increase this discrimination by subdividing the whole-turn spaces, we again run into the resolving power of the eye as the limit. However, if we divide the head into spaces we have effectively increased the discrimination many times over. Fig. 7-4 shows that the amount of amplification increases with an increase of the circumference and a decrease in screw lead (lateral travel of screw in one revolution).

A screw by itself cannot do much measuring. Even with a nut as in Figs. 7-3 and 7-4 it would be difficult. The screw furnishes just one end of the measurement partnership. It is the measured point. Still needed is the reference point. It is obtained by extending the nut to form a frame, Fig. 7-5.

OUTSIDE MICROMETER CALIPER

FUNCTIONAL FEATURES   METROLOGICAL FEATURES

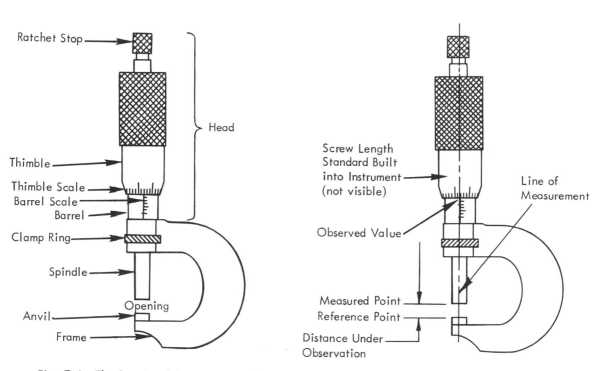

Fig. 7-6 The functional features make the metrological ones usable for practical measurements.

For example, consider the typical micrometer in Fig. 7-6. First of all, learn the nomenclature shown under *functional features*. These terms you will hear whenever micrometers are discussed. The screw (not shown) has 40 threads per in. Each revolution moves the screw 1/40 in. on *25 mil* (0.025 in.). The thimble is divided into 25 spaces. Each division thus represents *one mil* (0.001 in.). Each of these graduations is approximately *60 mil* (0.060 in.) wide. The amplification is therefore *one mil* to *60 mil* (0.001 to 0.060 in.) or 60 times finer than that of the unaided eye.

One important element of reliability of an instrument is its *readability*, Fig. 7-7. The smallest division on a micrometer is four times the size of the smallest division on a 1/64-in. rule, yet discriminates 15 times finer. Obviously it is more readable. Now compare the vernier with the micrometer. The latter is clearly more readable. Therefore, all other things being equal, the micrometer should be more reliable than a vernier.

READABILITY AND DISCRIMINATION

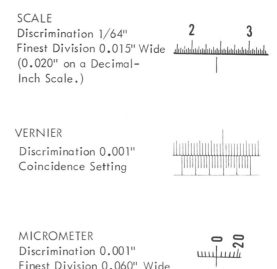

SCALE
Discrimination 1/64"
Finest Division 0.015" Wide
(0.020" on a Decimal-
Inch Scale.)

VERNIER
Discrimination 0.001"
Coincidence Setting

MICROMETER
Discrimination 0.001"
Finest Division 0.060" Wide

Fig. 7-7 The smallest division on the micrometer reads to 0.001 but is four times as wide as a 1/64-in. graduation on a scale. Although the micrometer and vernier have the same discrimination, which is most readable?

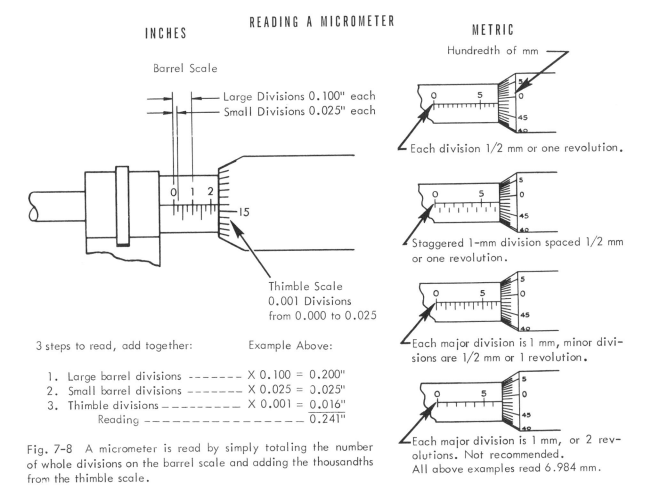

READING A MICROMETER

INCHES   **METRIC**

Barrel Scale

Large Divisions 0.100" each
Small Divisions 0.025" each

Thimble Scale
0.001 Divisions
from 0.000 to 0.025

Hundredth of mm

Each division 1/2 mm or one revolution.

Staggered 1-mm division spaced 1/2 mm
or one revolution.

Each major division is 1 mm, minor divisions are 1/2 mm or 1 revolution.

Each major division is 1 mm, or 2 revolutions. Not recommended.
All above examples read 6.984 mm.

3 steps to read, add together: Example Above:

1. Large barrel divisions ------- X 0.100 = 0.200"
2. Small barrel divisions ------- X 0.025 = 0.025"
3. Thimble divisions --------- X 0.001 = 0.016"
   Reading --------------- $\overline{0.241"}$

Fig. 7-8 A micrometer is read by simply totaling the number
of whole divisions on the barrel scale and adding the thousandths
from the thimble scale.

MICROMETER READING

Fig. 7-6 names the functional parts of a micrometer that you need to know. The barrel of the micrometer is divided into *25 mil* (0.025 in.) graduations. Each fourth one is elongated and numbered for easy reading. These large divisions are *100 mil* (0.100 in.) each. As mentioned, the thimble is divided into 25 graduations of *one mil* (0.001 in.) each.

There are three steps for reading Fig. 7-8 (the examples are for the inch-reading micrometer but apply in principle to the metric as well):

1. Note the highest figure on the barrel that is uncovered by thimble. That is the hundred-mil part of the reading. It is the first figure to the right of the decimal point. For example, 0.300 or 0.700 in., etc.

2. Note the whole number of graduations between the figure and the thimble. It may be 1, 2, or 3. Each of these represent *25 mil* (0.025 in.). Therefore, add 0.025, 0.050 or 0.075 in. to the hundred thousandths found in step 1. For example, 0.325 or 0.775 in.

3. Read the thimble opposite the index on the barrel. This will give the *mil* reading from 0.000 to 0.024 in.(zero and 0.025 in. on the thimble coincide). Add these *mil* to the previous figure. For example, 0.325 plus 0.003 in. equals 0.328 in., or 0.775 plus 0.024 in. equals 0.799 in.

There are only two precautions to observe when reading a micrometer. When reading the thimble be sure you read in the right direction. For example, in Fig. 7-8 the thimble might be carelessly read as 14 instead of 16 *mil*.

BE CAREFUL WHEN PASSING ZERO

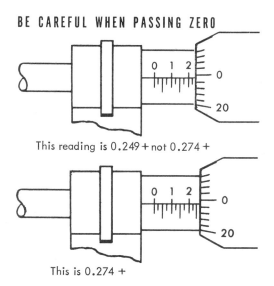

This reading is 0.249 + not 0.274 +

This is 0.274 +

Fig. 7-9   One of the few errors can occur when passing zero on the thimble.

The other precaution concerns the zero position on the thimble. When passing the index line on the barrel there is a chance to read an extra 25 *mil* (0.025 in.). This is caused by the fact that the next barrel graduation has begun to show but has not yet fully appeared, Fig. 7-9.   This is avoided by being careful to read only full divisions on the barrel. This error is particularly dangerous because, once started it is usually continued without detection. Some errors are progressive and if continued will eventually call attention to themselves. Stepping off intervals with dividers, Fig. 5-21, is an example.

Fig. 7-10   The slanting graduations on the barrel of this micrometer reduce the chances of observational errors such as the one shown in Fig. 7-9.

Frequently a machine operator has read his micrometer 25 *mil* (0.025 in.) incorrectly on the first reading of the day and continued to produce scrap the balance of the day. Certainly if he did not suspect his first reading, it would seem that he would have even less reason to suspect his 137th reading. *Suspect every measurement.*

One manufacturer has found a novel method to eliminate this *observational error*. The barrel graduations are at an angle, Fig. 7-10. There is no way a graduation significant to the reading can be completely covered.

The popularity of the micrometer is due to other factors in addition to its improved reliability. These are summarized in Fig. 7-11. Note that end measurement is listed as both an advantage and a disadvantage. It depends upon what you want to do, and emphasizes an important point. *There is no cure-all in measurement.* It requires many different instruments to provide for efficient dimensional measurement.

MICROMETER ADVANTAGES AND DISADVANTAGES

Advantages:

1. More accurate than rules
2. Greater precision than verniers.
3. Better readability than rules or verniers.
4. No parallax error
5. Small, portable and easy to handle

Disadvantages:

1. short measuring range
2. Single-purpose instruments

6. Relatively inexpensive
7. Retains accuracy better than verniers
8. Has wear adjustment *
9. End measurement

* For screw only. Anvil and spindle tip wear remain as a potential source of error.

3. End measurement only
4. Limited wear area of anvil and spindle tip

Fig. 7-11   End measurement is an advantage where it can be used. But it is a disadvantage to have an instrument that can only be used for end measurement.

ACCURACY, PRECISION AND DISCRIMINATION

A review of these terms will show you why some micrometers have vernier scales. By now you must realize that accuracy, precision and discrimination are related. *Accuracy* is the relation between the readings (observed values) and the true values. *Precision* is the "fineness" of the instrument or the dispersion of repeated readings. *Discrimination* is the smallest readable division. The discrimination given to an instrument is arbitrarily selected by the manufacturer. It is based on practical considerations and is a measure of only one thing, the integrity of the manufacturer.

Figure 7-12 analyzes an imaginary instrument. Assume that several readings are taken of a standard of known length. All readings deviate from the true value. They are fanned out in the illustration to show this dispersion. This is the precision, a measurable amount. Note that all of the readings are greater than the true value. Now consider four possible scales for the imaginary instrument, A, B, C and D. The smallest scale divisions are shown at A. This scale would cause serious errors. Although the readings are fine they are virtually meaningless. The total dispersion is greater than the smallest graduation. In fact, as drawn, the minimum error is nearly one graduation.

These graduations measure the instrument instead of the part. The situation would be better with scale B, about as good as flipping a coin. A single reading could be five as easily as six. In either case its value would be reduced by at least the minimum error.

In C the discrimination is sufficient to include all of the possible readings and allow for the error. But that is not saying much. When you measure to five you do not expect that the true value might actually be 5.75 in. It does show that extending the size of graduations in respect to the precision (or increasing the precision in respect to the size of the graduations) increases the *reliability* of correct measurement. An increase to 10 times the width would give the scale at D. Even at the extremes of dispersion and instrument error the readings will be correct 90% of the time. This would be considered a *reliable* scale for the particular instrument in our time. This would be considered a reliable scale for the particular instrument in our example.

If that ratio of scale division to precision (and accuracy) is practical, consider what happens if the graduations are twice as wide as in D. On one hand, the reliability of the instrument is increased. Note: reliability of the instrument is increased, not reliability of the measurement. It is the latter that we ultimately want. On the other hand, the wide graduations invite *interpolation*.

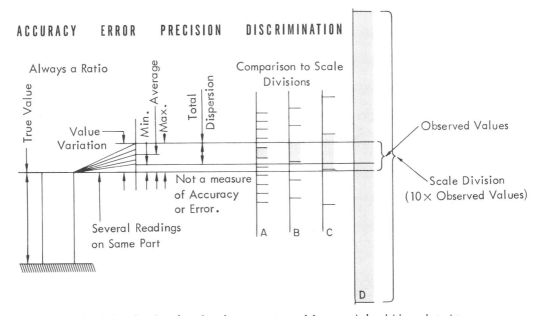

Fig. 7-12 Discrimination is related to instrument precision, and should be related to accuracy.

Interpolation is to instrument reading what *rounding off* is to computation. Like rounding off, the problem occurs when the reading is nearly half way between two graduations. Which one should be chosen? Unfortunately in interpolation you are on your own. It adds human error and only should be resorted to as a last resort. If the scale can be divided into smaller graduations (and the same ratio retained) there will be less need to interpolate. Furthermore, the full practical capability of the instrument will be used, which is of importance whenever cost of measurement is considered. This is the reason that instruments are provided with scales with as high a discrimination as possible. It is understandable that sometimes the manufacturer may be a bit too optimistic.

THE VERNIER MICROMETER

That brings us back to micrometers. The design and manufacture of the measuring portion of these instruments are intended to give them an inherent accuracy of *one-tenth mil* (0.0001 in.) (We have repeatedly observed that accuracy is a ratio. Therefore this means 0.0001 in. to the entire range of 1.0000 in.) Placing a vernier scale on the micrometer permits us to make readings to the limit of the instrument accuracy.

Now here is the point to be emphasized. The *tenth-mil* or *hundred-mike* (0.0001-in.) readings approximately correspond to scale B in Fig. 7-12. They are *not reliable* measurements for *tenth-mil* increments (0.0001 in.). What good are they then? They are extremely valuable; because they provide highly reliable measurements to *mil* increments (0.001 in.) and increased reliability of smaller measurements, such as *five-tenth mil* increments (0.0005 in.)

Having justified the vernier micrometer, here is the way it is read. The principle is exactly the same as for vernier calipers. Only the figures are changed.

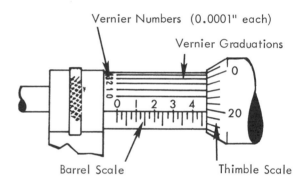

Fig. 7-13   The barrel and thimble scales of a vernier micrometer are the same as those on a plain micrometer.

READING THE VERNIER MICROMETER

4 steps to read, add together:

		Example A	Example B
1. First significant figure large barrel divisions – – – – – – – – X 0.100 =	0.400"	0.400"
2. Second significant figure small barrel divisions – – – – – – – X 0.025 =	0.050"	0.050"
3. Third significant figure thimble divisions – – – – – – – – – X 0.001 =	0.019"	0.019"
4. Fourth significant figure the numbered vernier line that is in coincidence with a thimble line – – – – – – X 0.0001 =	0.0000"	0.0007"
Reading –	0.4690"	0.4697"

Fig. 7-14   The vernier micrometer is read the same as a standard micrometer.

The vernier scale is on the barrel, Fig. 7-13. It consists of 11 equally spaced lines. The first ten are numbered zero to nine. The last one is marked with a second zero. The ten divisions thus created span nine divisions on the thimble, Fig. 7-14. One vernier division, therefore, is $1/10 \times 9/1000$ which equals $9/10000$. The difference between a vernier and a thimble division is $10/10000$ less $9/10000$, or $1/10000$. That is a reading of *one-tenth mil* (0.0001 in.), and is found by matching a vernier graduation to a thimble graduation. The steps for reading a vernier micrometer are shown in Fig. 7-14. They are identical with the steps for a plain micrometer except that the *tenth-mil* decimals are added. Note that when the micrometer is exactly on whole *mil* readings *both* zero graduations on the vernier scale coincide with lines on the thimble. Other examples are shown in Fig. 7-15, and the essential points to remember about vernier micrometers are summarized in Fig. 7-16.

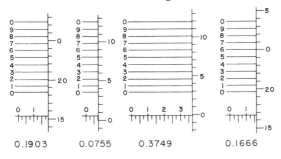

0.1903   0.0755   0.3749   0.1666

Fig. 7-15 Reading micrometer verniers has much in common with reading vernier height gages and calipers, including the need for practice.

METRIC MICROMETERS

Metric micrometers have a discrimination of one hundredth of a millimeter (0.01 mm). This is achieved by a screw with a one-half millimeter pitch. Here is another instance in which the slipshod old inch-pound system happens to fit practical situations better than the metric system. A plain metric micrometer provides better discrimination than a plain inch micrometer. One hundredth of a millimeter is *four-tenth mil* (0.0004 in.) This is less than half of the *one mil* (0.001 in.) discrimination of the inch micrometer. But ..

This is an important "but". A vernier scale can be added to the inch micrometer to raise the discrimination to *one-tenth mil* (0.0001 in.). Although this exceeds the inherent accuracy of the instrument it is usable, as explained earlier. Add a vernier scale to a metric micrometer and the discrimination becomes 0.001 mm or *40 mike* (0.00004 in.). This is the situation shown in A of Fig. 7-12. That discrimination so completely exceeds the instrument capability that it is virtually useless for practical purposes.

Thus, *for practical purposes the inch micrometers may be considered to have four times the discrimination of metric micrometers.* Does this mean that you should use an inch micrometer for metric measurement? Absolutely not. Besides the nuisance, the additional chance for computational errors would eliminate any advantage.

RELIABILITY OF VERNIER MICROMETERS

1. Vernier micrometer is no more accurate than a plain micrometer in spite of its high discrimination. Two tenth mil (0.0002 in.) approx.

2. The structural design of a micrometer is not sufficiently rigid for reliable measurement to one tenth mil (0.0001 in.).

3. The micrometer is a contact instrument dependent upon the user's alignment and feel which is seldom accurate to one tenth mil (0.0001 in.), even when the micrometer is equipped with a ratchet stop.

4. With expert use two tenth mil (0.0002 in.) accuracy may be obtained. However, generally speaking the 1-in. micrometer should be assumed to have an accuracy of plus or minus four tenth mil (0.0004 in.). *

* MIL-STD-120, Section 8.4, 2.1, p.79.

Fig. 7-16 Point 4 will cause arguments but it assures reliability.

READING METRIC MICROMETERS

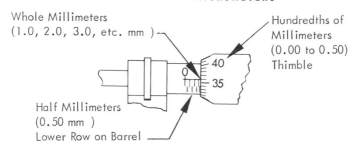

Whole Millimeters
(1.0, 2.0, 3.0, etc. mm)

Hundredths of
Millimeters
(0.00 to 0.50)
Thimble

Half Millimeters
(0.50 mm)
Lower Row on Barrel

Example:

Whole mm lines visible on barrel	3 = 3.00 mm
Additional half mm line (lower) visible on barrel	1 = .50 mm
Lines on thimble which have passed long line on barrel	36 = .36 mm
Reading of measurement Total	= 3.86 mm

Fig. 7-17 The only precaution to be remembered when reading a metric micrometer is to add in the half millimeter graduations of the lower row of graduations when required.

READING THE METRIC MICROMETER

Each complete turn of the thimble moves the spindle one-half millimeter (0.5 mm). Two turns equal one millimeter (1.0 mm). There are two rows of lines on the barrel, Fig. 7-17, each with one-half millimeter graduations. The upper row represents whole millimeters, the lower row half millimeters. The numerals above the parallel line are at each fifth line in the top row representing 5, 10, 15, etc. millimeters. The lower row provides half millimeter (0.5 mm) readings that are added to the whole millimeter readings of the top row. The thimble is divided into 50 parts, thereby dividing the half-millimeter graduations into hundredths of millimeters (.01 mm).

To read the micrometer first note the number of graduations on the top row counting the first as zero. Every fifth one is numbered. These are the whole millimeters. If an additional line is uncovered on the lower row, add 0.50 mm. Then add the thimble reading. If an additional line on the lower row of barrel graduations is not uncovered, add the thimble reading directly.

MICROMETER CONSTRUCTION

Micrometers are very similar in construction, although many manufacturers have developed special features that are described in their literature.

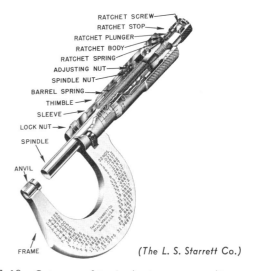

RATCHET SCREW
RATCHET STOP
RATCHET PLUNGER
RATCHET BODY
RATCHET SPRING
ADJUSTING NUT
SPINDLE NUT
BARREL SPRING
THIMBLE
SLEEVE
LOCK NUT
SPINDLE
ANVIL
FRAME
(The L. S. Starrett Co.)

Fig. 7-18 Cutaway of typical micrometer caliper.

Fig. 7-18 shows the typical construction. The frame may be of a number of constructions. Full finished solid frames are popular for the 1- and 2-in. sizes. These have decimal equivalents for 4ths, 8ths, 16ths and 32nds of an inch on one side and 64ths on the other side. It is not clear why, because the most conspicuous things in a machine shop other than the pin-up calendars are the decimal equivalent charts. Perhaps the time is near when metric conversions will be substituted. Drop-forged frames, both plated and black finish, are common, as well as tubular frames that are lightweight and have favorable heat conduction characteristics.

The hardened anvil is pressed or screwed into the frame. It is replaceable in a factory overhaul. The anvil was once adjustable in micrometers. This did permit some wear adjustment. The wear is seldom uniform across the anvil. Therefore, this adjustment seldom completely eliminated the effect of wear, and it always added one more source of error. Little wonder that they are no longer adjustable. Carbide anvils as well as carbide spindle tips are available. These provide a far better answer to wear.

The barrel is fastened to the frame. It carries both the barrel graduations and the nut in which the screw turns. The nut is short compared to the screw. As a result, it wears much more rapidly. To compensate for this the nut is slotted and has a tapered thread on its outside, Fig. 7-19. The adjusting nut draws the slotted nut closed as it is tightened, thereby removing the *play* (term for unwanted clearance).

The thimble is attached to the screw. Not the least of the purposes of the thimble is to protect the screw and nut from abrasive dust. (You will learn later that dust is the enemy in measurement). It also carries the *mil* scale.

Both zero setting and manufacturing problems are solved neatly by making the screw and thimble two separate parts. Zero setting is accomplished by loosening the parts and resetting the thimble scale. The arrangement

for this varies among manufacturers. One typical method is shown in Fig. 7-41.

Two other important construction features are the *ratchet stop* and the *clamp ring*. Both are taken up when discussing the use of the micrometer.

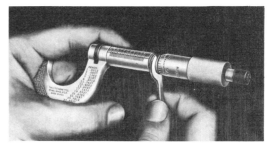

Fig. 7-19 The micrometer screw nut is slotted so that an adjusting nut can be tightened to take up clearance from wear. *(The L. S. Starrett Co.)*

WHAT DOES A MIKE MIKE?

There are four elements in every measurement, *the part*, *the instrument* (or measurement system), *the observer*, (the quality control term for person making the measurement) and *the environment* (temperature, vibration, etc.). The more accurate the desired measurement, the more control there must be over these elements. Although the total control required increases, the emphasis may change as shown in Fig. 7-20. Environment is a constant in these limited examples and has been omitted.

CONTROL OF MEASUREMENT

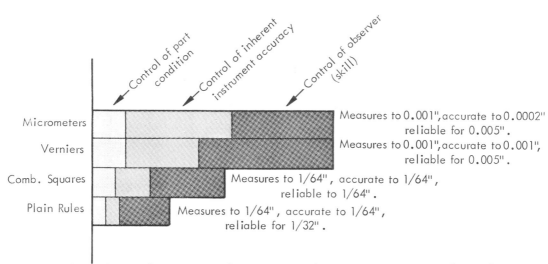

Fig. 7-20 The total control increases as the requirements become more stringent. The emphasis shifts from the instrument user to instrument maker, however.

WHAT A MICROMETER "MIKES"

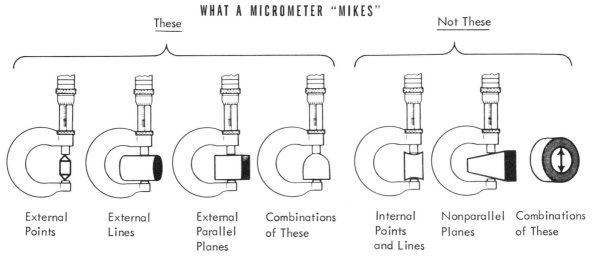

Fig. 7-21 Although versatile, a micrometer is completely inflexible in use. It only produces reliable measurement in the first four cases. Unfortunately, the conditions shown by the next two, cannot be detected by eye.

Control of part condition is the requirement for flatter surfaces, better surface finish and sharper edges. Obviously a measurement cannot be made to greater precision than permitted by these part conditions. Control of inherent instruments accuracy does not imply that greater care is taken in the manufacture of micrometers than of vernier instruments. Satisfactory manufacturing standards are presumed for both. This is primarily a measure of design features. Control of the observer includes all steps from care of the instrument to any computation required to put the measurement in usable form, and there are many.

In the case of the micrometer the manufacturer has built in some of the control that the user would have to furnish with the less precise vernier caliper.

The basic measurement act with a micrometer is the same as with all previous instruments. The instrument is set to duplicate in itself the separation between the reference point and the measured point of the part feature. In previous instruments this required parallel paths for the line of measurement and axis of instrument. Because these can never be perfectly parallel, some serious errors were expected. With the micrometer the axis of the instrument lies along the line of measurement. As predicted by Abbe's law, this minimizes error. Any variation from this caused by faulty use will show up as error in the micrometer reading.

GEOMETRY OF MICROMETER MEASUREMENT

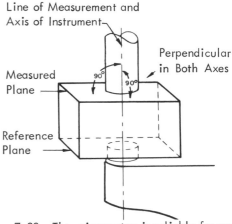

Fig. 7-22 The micrometer is reliable for measurement between planes when these planes are parallel and perpendicular to the axis of the micrometer.

The spindle tip and anvil of the micrometer are *areas*. This provides both the advantage and the problem of micrometer measurement. Accurate measurements are made only between external points, external lines, certain external planes and combinations of these, Fig. 7-21. The certain planes referred to are parallel planes that can be brought into coincidence with the planes of the micrometer's contact surfaces. While these are limitations caused by the area contact, these planes of the micrometer help it locate squarely on the reference surfaces of the part, Fig. 7-22.

USING THE MICROMETER

There is no correct way to hold a micrometer that is suitable for all measurement. By far the most useful method for 1- and 2-in. micrometers is the one shown in Fig. 7-23. The thimble is rotated between the thumb and combined index and third finger of the right hand. (Unfortunately this method cannot be reversed simply for left-handed people. The barrel scale is on the wrong side for them.) The fourth and fifth fingers grip the frame securely against the palm of the hand.

This method is only suitable for "quick-and-dirty" measurement. That slang term refers to checking "wide-open" measurements where the tolerance is plus or minus 0.002 in. or more. Checking size between successive passes of a lathe tool is an example. It is also used for *attribute inspection.*

Attribute inspection* usually refers to the use of a fixed gage to check one part feature on many parts. The parts are either accepted or rejected by the gage, and that's that. No one, opening and closing a micrometer repeatedly on a group of parts, considers this to be attribute gaging; but in most cases it is. The micrometer returns to nearly the same setting, time after time, until a part is met that slipped through the previous operation without being processed. The closing micrometer slams to a halt long before expected. Immediately, the bells ring, the lights flash, the dream about pitching a no-hit game is broken and one part gets tossed into the scrap barrel, after which all goes back to normal.

That brings up the use and abuse of the *clamp ring* or as it is sometimes called, the locknut, Fig. 7-18. The general use for this feature is to make an attribute gage out of a micrometer. It locks the instrument at one measurement so that it can be used as a *snap gage.* That is abuse of a precision instrument. When the micrometer is used as a snap gage, sliding friction results in excessive wear on the contact surfaces.

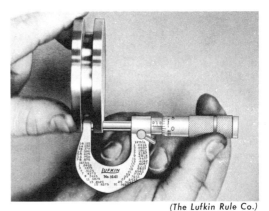

Fig. 7-23 This is the popular way to hold a small micrometer for work not requiring the highest accuracy of which the instrument is capable.

You can be excused for not recognizing this before. The manufacturers of micrometers actually provide stands to hold the instruments for gaging use. Some even go so far as to provide double-headed micrometers to be used as *go* and *not-go* gages. It is hard to imagine a less reliable instrument. It is subject to just about all the errors that plague measurement. Fig. 7-32.

The greatest concentration of metrological knowledge available is held by the manufacturers of precision measurement instruments. Although extremely cooperative in the preparation of books such as this, as a group they have displayed an almost unbelievable lack of interest in the advancement of measurement knowledge - even to the manufacture of ill-conceived instruments. Normally this would be considered the result of high pressure marketing. Strangely enough that, too, has been conspicuously absent. The point is that *the stamp of a reputable manufacture on an instrument should never be the sole justification for using that instrument.*

The clamp ring is a *memory device.* Please repeat until memorized. It retains a reading so that it can be referred to before making the next reading. If you have used micrometers and did not know this, do not be embarrassed. A quick search of manufacturers' literature produced only one useful statement beyond, "Our micrometers feature clamp rings." The statement is "Never tighten the clamp ring when the spindle is withdrawn. To do so injures the clamping mechanism". And, do not eat the daisies, they might make you sick.

---

* MIL-STD-105D, Military Standard Sampling Procedures and Tables for Inspection by Attributes.

Why is the generally used method for holding a small micrometer only suitable for rough measurement? Because only a person with exceptionally long fingers can reach the rachet stop. Some few can be using only the little finger and the palm to hold the frame. Fortunately some micrometers are now being made with the controlled force feature of the ratchet stop lower on the thimble where it can be easily reached. The instrument in Fig. 7-23 is of this type.

THE GREAT EQUALIZER

Some people are stupid enough to purchase micrometers without ratchet stops. Don't be one of them. The micrometer is a *contact instrument.* That means that there must be positive contact between the part and the instrument. And the amount of contact (all-important *feel)* is up to the user. When you are attempting to measure *one-mil* (0.001 in.) reliably - the same true reading time after time - almost imperceptible differences in gaging force can be very important. Because human beings vary so widely this is a source of serious errors in measurement. The "six gun" of the West owed its importance to much the same reason. It was known as the "great equalizer" because the big bully and the undersized whelp were equally persuasive with its help. So it is with the ratchet stop on micrometers, Fig. 7-24.

Fig. 7-24   The modern-day "equalizer" is the ratchet on the trusty mike. *(Hastings Mfg. Co.)*

Basically the ratchet stop is an overriding clutch that "kicks out" at a predetermined torque. This holds the gaging force to the same amount for each measurement regardless of the differences in human physiques.

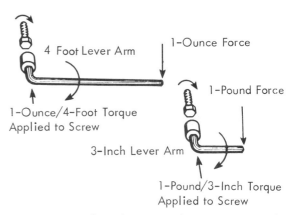

Fig. 7-25   Both of these wrenches are applying the same torque to the screw. 1 pound/4-ft. torque = 16 oz/48-in. torque = 1 oz/3-in. torque.

Torque is the amount of twist. It is based on the principle of the lever, Fig. 7-25. When torque is applied to the rotation of a screw thread, *force* results. Force is the forward push of the thread as it is turned by the torque. In a contact instrument it is the force that develops the *feel* of the contact. This is shown in Fig. 7-26. All material is somewhat elastic. Whenever there is feel of contact, the instrument and/or the part are being distorted. This is symbolized by the spring in the illustration, Fig. 7-26. Instead of fingers turning the ratchet stop, a lever arm has been attached. Weights hanging from the arm create the torque, which generates the force, which squeezes the spring.

With the one-inch lever arm and one-half ounce weight in A of the figure, you should expect a relatively large force because of the amplification of the screw thread. Threads, however, are a very inefficient means for power transmission. Figuring only 10% efficiency, the force against the spring is squeezed down to a length of 0.783 in. as read on the micrometer. As the weight is increased, the increased force closes the micrometer tighter as at B. At one ounce the force is 25 ounces, and the spring is compressed to 0.775 in. If one ounce is the limit at which the ratchet stop is set to disengage, any further addition of weight has no change on the reading. The clutch simply "kicks out" before the spring is compressed any further. The one-pound weight at C results in the same 0.775-in. reading that the one-ounce weight produced.

Note that the term *gaging pressure* is often used to mean *gaging force*. Force is the total push in units of weight. Pressure is force acting upon a specified area, such as pounds per square inch. It is pressure that we are concerned with in measurement, not force. It can be deceiving as shown in Fig. 7-27. The lightweight girl presumably connected with the shoe exerts many times the pressure against the floor than the bruiser whose brogue is shown.

When a micrometer is used, the frame is sprung open and the part is compressed. The ratchet stop on a micrometer is not intended *to eliminate* the compression of the work and expansion of the micrometer frame. It is intended *to equalize* this effect for all measurements. Unfortunately, it does not succeed.

The reasons that the ratchet stop does not assure completely repeatable readings are varied. Of course, there are bound to be some mechanical variations within the ratchet stops themselves.

The friction of the screw in the nut is another important one. A tight adjustment will use more of the torque to overcome friction than to apply to the measurement. This variable should be constant for a given series of measurement performed by one observer with the same micrometer.

Lubrication also affects the action of the ratchet stop. This is a good place to emphasize the great importance of lubrication in measurement. It can be demonstrated easily by unscrewing the thimble and spindle assembly from a micrometer and thoroughly flushing both parts in a solvent to remove all lubrication. Now reassemble. You will find that the screw turns noticeably harder. Disassemble again and apply a few drops of instrument lubricating oil to the screw and nut. Now reassemble and notice how freely the screw turns. This shows that although the film of oil obviously took up space between the screw and the nut, it materially decreases the turning friction.

What happens, however, if the lubrication is dirty or gummy? Then the friction will vary throughout the range and the torque required for the ratchet stop to release will be different from point to point.

TORQUE AND CONTACT FORCE

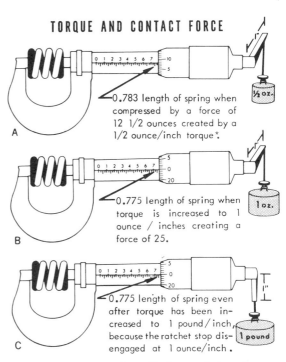

0.783 length of spring when compressed by a force of 12 1/2 ounces created by a 1/2 ounce/inch torque*.

0.775 length of spring when torque is increased to 1 ounce / inches creating a force of 25.

0.775 length of spring even after torque has been increased to 1 pound / inch, because the ratchet stop disengages at 1 ounce/inch.

* Based upon 10% efficiency of thread.

Fig. 7-26 The amount that the spring is compressed depends upon the thimble torque, but is limited by the ratchet stop.

FORCE AND PRESSURE

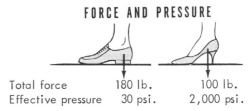

Total force	180 lb.	100 lb.
Effective pressure	30 psi.	2,000 psi.

Fig. 7-27 It is the pressure of a contact point, not the force, that indents the surface against which it is applied.

Another important cause of variation in gaging pressure when using a ratchet stop equipped micrometer is the *momentum* of the turning portion. If the thimble is turned rapidly, energy will be stored up. This energy will be released by giving an extra squeeze to the part being measured, even though the ratchet stop has been used. This is easily demonstrated with any *tenth mil* reading micrometer. Simply take several readings, moving the thimble at different speeds. If you repeat this with several micrometers you will notice the difference that the nut adjustment makes. If it is very tight, the effect will be negligible. If very loose, the readings may vary nearly *one mil* (0.001 in.), hardly desirable for reliability.

Fig. 7-28 Nearly three quarters of a century ago, Carl Johansson , while developing gage block set combinations, modified a micrometer to give better control. He did not use it to measure gage blocks as it has often been stated. Gage blocks are used to measure micrometers , not vice versa. He used it to control manufacturing operations up to the final sizing.

(Brown & Sharpe Mfg. Co.)

Fig. 7-29 The ratchet should be used whenever 0.001 inch or greater precision is required.

These causes for error have been recognized for many years, Fig. 7-28. Unfortunately little has been done about correcting them or even warning the users about them in this country. Much greater attention is given to this in Great Britain and Europe. One text recommends using the ratchet only until you develop the proper feel, because the ratchet stop is considered too heavy.*

There are two important facts that you should remember. First, the ratchet stop is a great help in reliable measurement, Fig. 7-29. Second, if you really put your mind to it, you can certainly defeat its advantages. Dirt, poor lubrication, bad adjustment, careless closing onto the part all destroy the value of the ratchet stop, Fig. 7-30.

---

* Hume, K.J., Sharp, G.H., Practical Metrology, London, Macdonald and Co. Ltd., 1958, Vol. 1, p. 22.

MICROMETER DO'S AND DON'TS

Clamp ring:

Do use it as a memory device to preserve a reading until repeated.

Don't use it to make the micrometer into a snap gage.

Ratchet stop:

Do use it for every measurement between flat surfaces.

Don't expect it to guarantee reliable measurements if

1. the micrometer is dirty.
2. the micrometer is poorly lubricated.
3. the micrometer is poorly adjusted.
4. the micrometer is closed too rapidly.

Don't use it for measuring diameters unless they are very large or very small.

Fig. 7-30 The ratchet stop is similar to an aspirin. It will cure a headache but not a broken arm.

What should you do if the micrometer does not have a ratchet stop? Obviously you must use greater care. Averaging a larger number of readings may help. Many careful workers make it a point to grip the micrometer on the smooth portion instead of the knurled portion. Some have even ground off the knurl of their micrometers.

HOW TO AVOID ALIGNMENT ERRORS WHEN MEASURING WITH A MICROMETER

The first piece of advice is, do not measure with a micrometer, if a more reliable method is available. Except for work-in-progress measurement, there usually is a better way to measure dimensions of four inches or larger. Work-in-progress measurement is the measurement of a part on a machine tool between passes of the cutting tool, and during assembly before final adjustments are made. The portability of the micrometer advocates it for these uses. Although the height gage was thoroughly castigated in the last chapter, its use in conjunction with electronic indicators and gage blocks is usually far more reliable for large dimensions than micrometers. The balance of the advice on alignment errors requires that we distinguish between flat and cylindrical measurement.

CLOSING THE MICROMETER
ON FLAT PARTS

The proper way to approach the part size is shown in Fig. 7-31. Before placing the micrometer on the part bring it to nearly the desired opening. Do this by rolling the thimble along your hand or arm - not by twirling. When placing the micrometer on to the reference plane of the part, hold it firmly in place with one hand. Use the feel of stability (no rock) to show when the axis of the micrometer is perpendicular to the reference plane. Rapidly close the micrometer using the ratchet until the spindle is nearly on the measured plane of the part. This usually can be determined visually. If you hit the part before expected, back off slightly and then slowly and gently close the spindle until the ratchet stop disengages one click.

Note that this procedure requires two hands. If the micrometer is handled with only one hand, the ratchet stop cannot be reached, and reliability will suffer. It works out fine for any part that is large enough to support itself without moving around during measurement. It is awkward for small parts. Tool stands, Fig. 7-32, are available to hold the micrometer so that one hand may be used on the ratchet stop while the other hand is free to handle the part. A tool stand is easily improvised, if not available.

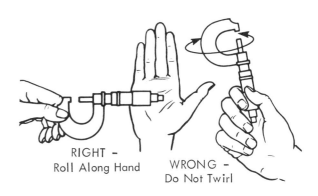

RIGHT -
Roll Along Hand

WRONG -
Do Not Twirl

1. Before placing micrometer on the part, bring it to the approximate opening size.

2. Place micrometer on work. Hold anvil against reference plane and rapidly close micrometer until you cannot see between spindle and part.

3. Slow down. Carefully close spindle, using ratchet stop, onto the part until the ratchet stop releases one click.

Apply pressure to hold anvil against reference plane.

Fig. 7-31   The fastest way to measure reliably is to slow down.

Fig. 7-32   A tool stand permits measurement without supporting the micrometer by hand. This permits small parts to be measured using the ratchet stop (left). It also permits such unreliable practices as the use of micrometers for go and not-go gages (right).

ALIGNMENT ERROR WITH MICROMETER

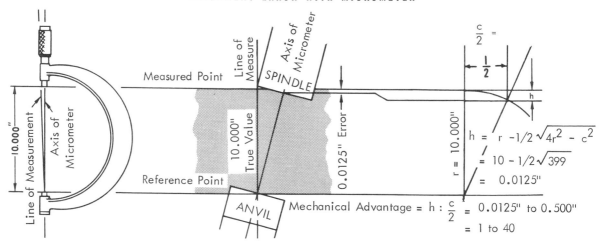

Fig. 7-33 These oversimplified drawings show that a 1/2-in. misalignment when measuring 10 in. has produced an error of one and one-quarter thousandths and the 40 to 1 mechanical advantage easily forces the spindle and the anvil into the part being measured.

When the part is large, the same procedure is followed but greater care must be taken in alignment.   The feel of the anvil "homing" against the reference surface is very deceptive. With a large micrometer the lever arm is so great that an almost imperceptible misalignment can indent the anvil into the part or spring the micrometer frame. A simplified and exaggerated example is shown in Fig.7-33.

MEASURING CYLINDRICAL PARTS

Books on measurement and manufacturers' literature abound with photos of shafts and other diameters being measured with the ratchet stop.   As much as the ratchet stop is to be recommended, it is only suitable for diameters that are very small or very large.

Accurate measurement of cylinders requires that the line of measurement be a diameter. When the diameter of the part is no larger than the diameter of the anvil and spindle, the alignment problem is simple. This is shown as A in Fig. 7-34. If the micrometer is closed and the diameter overhangs the anvil and spindle, the part pops out. Otherwise, the micrometer is closed to the desired feel or until the ratchet stop disengages, no more.

It would be difficult to have the condition shown in B of the figure except for thin wall tubing. Compare this with an odd-shaped part, such as the one in C. It is entirely possible to have severe misalignment and still produce the same feel as in A.

MEASURING SMALL DIAMETERS

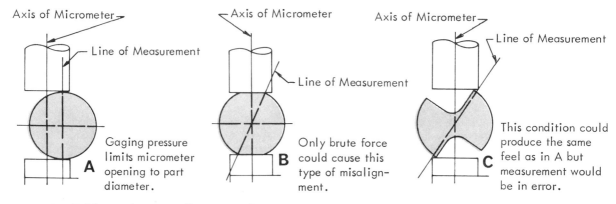

Fig. 7-34 Circles generally provide the same diameter anyplace through center. That eliminates alignment problems for very small cylinders.

DIAMETERS COMPARED

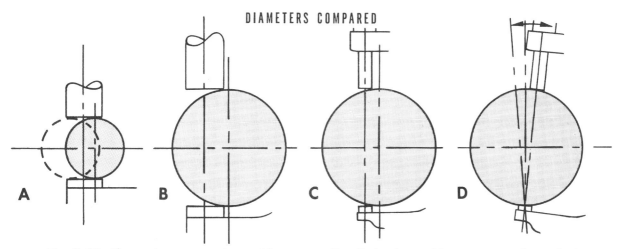

Fig. 7-35   The ratchet stop can be used for very small cylinders but rocking over center is required for large cylinders.

As expected, the problems in cylinder measurement become greater as the diameter increases. In Fig. 7-35, A represents the case for very small diameters. As the diameter approaches one inch, the danger of condition B becomes serious. It is easy to measure a one-inch diameter standard *one mil* (0.001 in.) small. Its fine finish does not give as much of a *feel* as a part with a rougher surface. A relatively small cylinder as in B usually can be aligned by eye to the axis of the micrometer. When the part is quite large, say four inches or greater, the eye is not reliable for alignment as at C.

In these cases the ratchet stop will behave nicely, but unfortunately the ratchet stop has no way of knowing that the micrometer axis is not parallel to the cylinder diameters.

Fortunately, the solution is relatively simple when measuring large cylinders - rock over center. This is shown at D in the figure. All diameter measurement errors will be minus. The reading will be less than the true diameter. Therefore the largest measurement across a circle will be the diameter.

If the micrometer axis is not held perpendicular to the axis of the cylinder being measured a plus error will result. This is removed by *centralizing* as with the vernier caliper. The steps are shown in Fig. 7-36.

ROCKING-OVER-CENTER TECHNIQUE

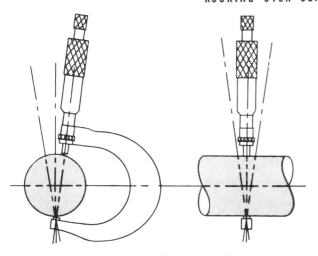

1. Clean the micrometer and the part.
2. Rock across cylinder to find the maximum opening that provides the desired feel – that will be the diameter.
3. Rock along the axis very slightly to find minimum feel ("centralizing") – that will establish micrometer at right angle to axis.
4. Step 3 may have removed all the feel from step 2. Therefore, close micrometer slightly and repeat steps 2 and 3.
5. When feel across the diameter is as desired and the perpendicularity assured, take the reading and write it down.
6. Repeat all steps at different diameter sufficient times to be satisfied with measurement.

Fig. 7-36   These steps are automatic for skilled people.

The most frustrating phrase an inexperienced cook encounters in recipes is "season to taste". Nothing will save the dish at that point except experience or luck. That is the way it is with the *feel* required in step 5. Only experience will teach how much *feel* is the right amount so that the micrometer reading is consistently close to the true value. Some of the factors involved are itemized in

Fig. 7-38 Skill is required to reliably "mike" large diameters. This micrometer is equipped with interchangeable anvils to cover a range of sizes in 1-inch increments. *(Scherr-Tumico, Inc.)*

FEEL IN MEASUREMENT

Factors influencing:

1. Size of the part. If the micrometer is very large it will be awkward and may be heavy. The person using it must be more intent upon support than on <u>feel</u>.

2. Closely related to the first is the position of the measurement. Even a 1-inch micrometer may provide inadequate feet if used at arm's length through a recess in a large machine.

3. Shape of part. If the <u>feel</u> is checked against gage blocks and then duplicated on a cylindrical part of the same size, the reading will vary several <u>mil</u> (0.001 in.)

4. Surface finish affects the <u>feel</u>. A coarse finish will produce a more pronounced <u>feel</u> than a fine finish.

Fig. 7-37 Experience is the best guide for evaluating <u>feel</u> when measuring with contact instruments such as micrometers.

Fig. 7-37. Therefore, allowances must be made. A measurement such as the one in Fig. 7-38, for example, should not be considered reliable to less than plus or minus 5 *mil* (0.005 in.), unless the person making it is highly skilled. The important steps for measuring with micrometers are summarized in Fig. 7-39.

EXTERNAL MICROMETER MEASUREMENT SUMMARIZED

FLAT PARTS	CYLINDRICAL PARTS
1. Clean the contact surfaces of the part and of the micrometer.	1. Same as number 1 at left.
2. Open slightly larger than part feature.	2. Same as number 2 at left.
3. Seat anvil squarely against reference surface of part.	3. Same as number 3 at left.
4. Using ratchet, slowly close the micrometer until the ratchet clicks once.	4. Rock back and forth across diameter, closing micrometer by small steps.
5. Record reading.	5. When first contact is felt, rock sideways to find position of last contact.
6. Repeat entire procedure several times and average readings.	6. Repeat steps 4 and 5 until perpendicular position is found and spindle just contacts the measured point as it passes over center.
	7. Record reading.
	8. Repeat entire procedure several times and average readings.

Fig. 7-39 The method for holding the micrometer might vary considerably for various circumstances but the general procedure will be the same.

CARE OF MICROMETERS

Dirt is the enemy of precision measurement. *Dirt is matter out of place.* White or black, visible or invisible, small particles where they do not belong are dirt. They do not belong in or on micrometers or any other precision instrument. Before each use the micrometer should be wiped free of any dust, cutting fluids and cookie crumbs that may have dropped on it. After each use it should be wiped clean and put in a safe place, preferably its own case. At the end of each day that it is used, it should be wiped clean, visually inspected, oiled and replaced in its case, to await its next use.

This is a good place to mention oil on precision instruments. It is highly essential on steel surfaces for lubrication and to protect against both rust and corrosion. Perspiration can be highly corrosive. It is the main reason that instruments should be cleaned and oiled after use. Cutting fluids also are a source of corrosion to the finely finished surfaces of precision instruments.

The actual rust or corrosion will not be visible unless the instrument has been grievously neglected. That does not mean that it does not exist. Invisible corrosion is rubbed off when the micrometer is used. Although less than a *mike* (0.000001 in.) at each use, this needless loss accumulates and virtually destroys accuracy.

A thin covering of oil has virtually no effect on the measurement until *100 mike* (0.0001 in.) or less is measured. Unfortunately the air is full of dust and the dust collects on oiled surfaces. While the oil film may be of negligible thickness, the buildup of dust can create errors of a *mil* (0.001 in.) or more. Worst of all, most dust is composed of silicon particles. These are highly abrasive and can cause rapid wear of fine surfaces.

Never leave a 1-in. micrometer in the closed position when not in use, or leave any micrometer tightened on its standard. Finely finished steel surfaces corrode rapidly when left in close contact, probably due to electrolytic action.

If the micrometer sticks do not dunk it in solvent and hope for the best. This has three objections. First, it washes away needed lubrication. Second, the solvent may carry the foreign matter further in between the sliding surfaces. An abrasive particle could then become imbedded and difficult to remove. Third, this may becloud the real reason for sticking which could be more serious.

If a micrometer sticks, first inspect carefully for evidence of damage. Then disassemble, clean all the parts in *clean* solvent, lubricate and reassemble. If the sticking persists, loosen the adjusting nut slightly. If that does not correct it, return the micrometer to the manufacturer for repair.

Fig. 7-40 The contact surfaces can be cleaned by closing on a soft piece of paper. Then pull the paper out and blow away any remaining fuzz or dust.

A handy method to clean the contact surfaces is to close the micrometer lightly on a piece of soft paper, Fig. 7-40. It is then withdrawn. The paper will probably leave fuzz and lint on the surfaces. Blow this out with lung power - not the air hose. *Never use compressed air to clean any precision instrument.* The high velocity forces abrasive particles into the mechanism as well as away from it.

Micrometers are very rugged for precision instruments, but they can be sprung or bent. Never "cramp" or force a micrometer when aligning it on a part. There is a human tendency to make things fit. This is the bias mentioned in Chapter 4   *For reliable measurement you must recognize that you have bias in your actions and consciously correct for it.* Every time a micrometer is cramped there is danger of a small permanent set that could eventually destroy the reliability yet escape detection.

Overtightening is perhaps the worst of all sins of micrometer use. There are three reasons for it: ignorance, bias and carelessness. Many people using micrometers daily are not aware of the errors caused by overtightening. Some are not sufficiently sensitive to *feel* for correct use. Certainly these people should use only ratchet stop equipped micrometers. Unfortunately, many of these people stand in front of boring mills and lathes where nearly all the measurements are diameters and ratchet stops do not help. That is one reason indicator type instruments are often preferred to micrometers.

Bias has been discussed as a cause of cramping. It is also the largest cause for overtightening. Be completely impersonal when measuring. Forget everything about the dimension desired. Seek only one thing - the correct measurement. After you are certain that you have measured reliably, and only then, compare the measurement with the desired dimension.

Carelessness causes errors whenever it creeps into measurement. For accuracy, do not simultaneously observe the micrometer along with the mail girl as she passes through. The greatest chance for serious overtightening occurs when measuring diameters. Do not have the opening too large when beginning the measurement. Then there is a tendency to close it in too large stops. At one rock over center there is no feel. And the next time you have closed it too much but have forced it past center before your brain overruled your reflexes.

INSPECTION OF THE MICROMETER

Calibration of an instrument refers an important inspection process. The instrument is compared with standards of known accuracy to determine its condition. *Out of calibration* means two things. Most frequently it means that calibration has shown that the instrument no longer conforms with the specified standards that apply to that particular class of instruments. It also means that the stipulated period of time between calibrations has been exceeded and the instrument can no longer be considered reliable.

In a manufacturing operation the quality control department establishes the period for calibration of the instruments. Contrary to common belief, they do not use a crystal ball. The period is found by applying statistics to past experience records. In the laboratory or small shop, calibration is equally important, but frequently overlooked until something goes wrong. Protect your measurement data whether they are for progressive education or progressive dies. Calibrate all instruments before undertaking critical work. The closer the measurement requirements are to the specified accuracy of the instrument, the more frequently calibration is needed.

Gage blocks and optical flats are now in general use wherever precision measurement is employed. They provide reliable means for calibrating micrometers. The devices in Fig. 7-42 provide quick shop tests. These techniques are discussed in Chapters 13 and 14, respectively. The primary features that require calibration are the condition of the contact surfaces and the actual accuracy of the measurement.

The contact surfaces should be checked for parallelism and flatness before the accuracy check. There are several easy methods that can be applied in the shop.* A ball, of course, presents nearly point contact. It therefore provides a convenient means for checking flatness and parallelism of the contact surfaces. The ball "explores" the surfaces as the micrometer is opened and closed on it. Great care must be taken to apply uniform gaging pressure each time and to read the micrometer carefully.

Note: Tests such as this show when something is wrong and requires further attention. *They do not show that nothing is wrong.* They should not be substituted for thorough, periodic calibration. The method used requires an autocollimator, an optical device of almost uncanny ability to track down errors of parallelism.

A thorough check of accuracy requires gage blocks. However, a partial but useful check of a one-inch micrometer can be made at zero and one inch in the shop.

After checking for and removing burrs and surface defects, the micrometer is closed

---

* Kennedy, Clifford W., Inspection and Gaging, Industrial Press, 2nd printing, p. 102.

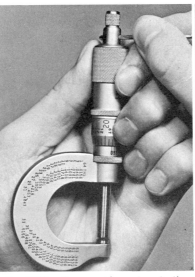

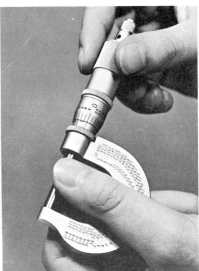

1. Carefully clean the measuring surfaces by pulling a piece of soft paper between the surfaces, while they are in light contact with the paper. Do not use a hard paper.

2. With the anvil and spindle apart, unlock cap with spanner wrench: then tighten cap lightly with fingers to bring slight tension between thimble and spindle.

3. Bring anvil and spindle together by turning spindle and set zero line on thimble to coincide with line on barrel.

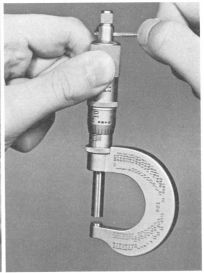

4. Move spindle away from anvil by turning spindle, not by turning thimble.

5. Holding thimble only, tighten cap with fingers. Be careful not to touch frame.

6. Lock cap with wrench still holding thimble only, and adjustment is complete.

(Brown & Sharpe Mfg. Co.)

Fig. 7-41   This is a typical procedure to zero set a micrometer. Be sure to grip micrometer as shown.

tightly using the ratchet stop. If the micrometer does not have a ratchet stop, close it with the same gaging pressure that you use when measuring. The micrometer should read zero, right on the nose. If not, reset according to the manufacturer's instructions, as shown in Fig. 7-41.

During the zero setting, check the adjustment nut for the desired degree of friction as has been discussed. Also check for excessive looseness that would warn of a worn-out instrument. Be sure that the final zero setting is made after adjustment. These precautions will certainly improve reliability.

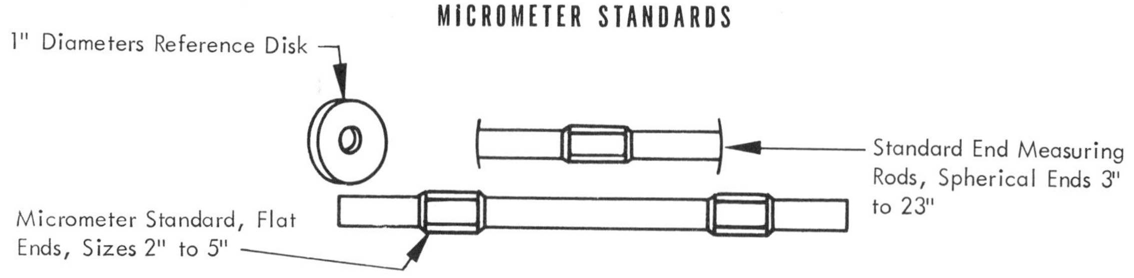

MICROMETER STANDARDS

1" Diameters Reference Disk

Standard End Measuring Rods, Spherical Ends 3" to 23"

Micrometer Standard, Flat Ends, Sizes 2" to 5"

Fig. 7-42   Standards are used to check the calibration of micrometers. The end surfaces of the standard end measuring rods are portions of spheres. If flat parts are being measured the flat ended micrometer standard is preferred for calibration because the feel is nearly duplicated. Similarly the spherical ends more nearly duplicate the feel of measurements made on cylindrical parts. Gage blocks are frequently used in place of these standards.

Accurate masters, called *setting standards*, are frequently purchased with micrometers and are usually furnished in sets of micrometers, Fig. 7-42. For one- and two-inch micrometers, these are cylindrical shaped. For large sizes, they are rods. After checking the closed setting of the micrometer (against the anvil with a one-inch size, against a setting standard for larger sizes), the open setting should be checked against a setting standard. If the reading is more than *one tenth mil* (0.0001 in.) in very carefully controlled operations, or *two-tenth mil* (0.0002 in.) in ordinary operations, the micrometer should receive expert inspection and repair, if needed.

Whether you send a micrometer in for service because of *one-tenth* or *two-tenth mil* (0.0001 or 0.0002 in.) error depends upon two factors. First, unless you can reliably repeat measurements to *one-tenth mil* (0.0001 in.) *without helping them along*, there is no reason to accuse the beleaguered micrometer. Second, if the measurements you are making are plus or minus *one mil* (0.001 in.) or coarser, it is desirable to have *one-tenth mil* (0.0001 in.) accuracy but *two-tenth mil* (0.0002 in.) should be sufficient. A better solution in cases such as these is to select an instrument of higher precision. The use of gage blocks and comparator instruments is often the answer, as you will find in later chapters.

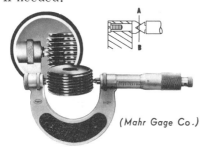

(Mahr Gage Co.)

Fig. 7-43   Screw thread micrometers measure pitch diameters directly. When set at zero the pitch lines of spindle and anvil coincide as AB. Reading corresponds to distance between pitch lines. With fixed anvil instruments, seven micrometers are required to cover pitches from 3 to 64 threads per inch. The instrument shown here has interchangeable anvils to cover a wide range of thread pitches as well as other special requirements. It is available in 0 to 1-inch size and 1 to 2-inch size.

(Scherr-Tumico, Inc.)

Fig. 7-44   Disc micrometers have narrow discs for contacts. This enables them to enter slots as narrow as 1/32 inch. They are useful for measuring to one mil (0.001 in.), three sizes to 3-inch opening. A similar appearing micrometer but with thicker discs is used for measuring paper and other soft materials that require large contact areas.

(Brown & Sharpe Mfg. Co.)

Fig. 7-45   The blade micrometer has 30 mil (0.030 in.) thick anvil and spindle end to reach into recesses. The spindle does not rotate. They are available in one mil (0.001 in.) reading sizes to six inches. All require careful checking because of the small contact area subjected to wear, and of wear from the nonturning spindle working against the rotating screw.

VARIATIONS OF MICROMETERS

Because this book gives priority to principles instead of hardware, no attempt will be made to discuss all of the many types of micrometer calipers. Some notes will acquaint you with a few of the more familiar types that you might encounter.

In Fig. 7-38 is a micrometer whose anvils are interchangeable to cover a range of sizes. This is not as reliable as a separate micrometer for each one-inch range required, but it is much less expensive. Recognizing the potential error (to say nothing about additional sales), manufacturers usually limit the range obtainable to 4 or 6 in. Heating of the frame while handling a large micrometer could cause errors. Many of the very large instruments have tubular frames to minimize this effect as well as to lighten.

The value of the thread micrometer, Fig. 7-43, will be apparent when thread inspection is attempted. The disc micrometer and blade micrometer, Figs. 7-44 and 7-45, permit micrometer measurements of part features that cannot otherwise be reached. The hub micrometer is primarily used for in-process measurement, Fig. 7-46. Measuring wall thickness of tubing is frequently required and not surprisingly there are several ways to "mike" it, Fig. 7-47. The instrument that bridges the gap between shop instruments and gage room instruments is the bench micrometer, shown in Fig. 7-48. When not available, the next best thing is the bench stand in Fig. 7-32.

The indicating micrometer is mentioned in Chapter 10 because its features are more like comparators than micrometers.

The second disadvantage of micrometers listed in Fig. 7-11 is the fact that they are single purpose instruments. Where a vernier height gage may be used for inside or outside measurement plus as a depth gage, separate micrometers are needed for these roles. Fortunately, the principles remain nearly the same and an understanding of the other micrometer instruments will come easily now that we have covered the micrometer caliper.

There are three principal micrometer instruments for making inside measurements: inside micrometer calipers, inside micrometers, and micrometer plug gages.

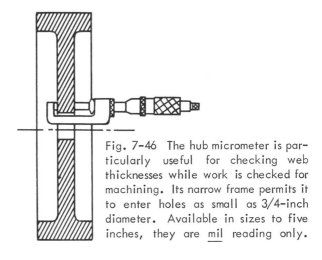

Fig. 7-46 The hub micrometer is particularly useful for checking web thicknesses while work is checked for machining. Its narrow frame permits it to enter holes as small as 3/4-inch diameter. Available in sizes to five inches, they are <u>mil</u> reading only.

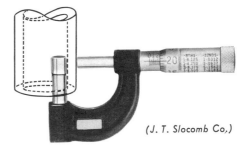

(J. T. Slocomb Co.)

Fig. 7-47 Several micrometers are specially adapted for measuring tubing. A round anvil micrometer is made in zero to half-inch range, reading to one <u>mil</u>. A ball anvil attachment may be added to standard micrometers. The above micrometer has the advantages that it can enter a hole of only 3/8-inch diameter and can reach a depth of 3/4 inch.

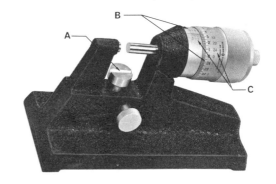

A — ADJUSTABLE HEIGHT WORK REST.
B — BLACK FIG'S READ OPENING FROM ANVIL.
C — RED FIG'S READ CLOSING FROM MAX. OPENING.
Fig. 7-48 The bench micrometer is by far the most reliable micrometer for measuring to tenths of <u>mil</u> (0.0001 in.). Its discrimination is directly in tenths <u>mil</u>. No vernier is required. It owes greater reliability to its superior stability as well as the large diameter of the head. *(Brown & Sharpe Mfg. Co.)*

THE INSIDE MICROMETER CALIPER

This instrument is shown in Fig. 7-49. The inside micrometer caliper is intended for relatively small inside measurements, from point *two inch* (0.200 in.). The *inside micrometer* is intended for larger inside measurements, from one inch.

Note in Fig. 7-50 the similarity between the inside micrometer caliper and the plain micrometer. To adapt the micrometer principle to inside measurement of small dimensions, the axis of the instrument had to be separated from the line of measurement. This destroyed the adherence to Abbe's law that provides micrometers with much of their reliability. Furthermore, while the flat contact surfaces of the plain micrometer are suitable for flat, or cylindrical part features, the reverse is true for the inside micrometer caliper. The *nibs,* as the contacts are called, are ground to a radius. This radius is smaller than the smallest radius the instrument might measure. Therefore all measurements are made with line contact.

Fig. 7-49 Available in three sizes, the inside micrometer calipers all have one-inch range in mils. The smallest closes to point two inch (0.200 in.). The largest opens to two inches. *(Scherr-Tumico, Inc.)*

From these two characteristics arise the potential problems with the inside micrometer caliper. Because the axis of the instrument is separated from the line of measurement, the alignment problems of the vernier caliper are encountered. The shoulders behind the nibs assist in proper alignment, but can only be relied upon if the part is faced sufficiently square to the line of measurement, Fig. 7-51. The moral is, know your part.

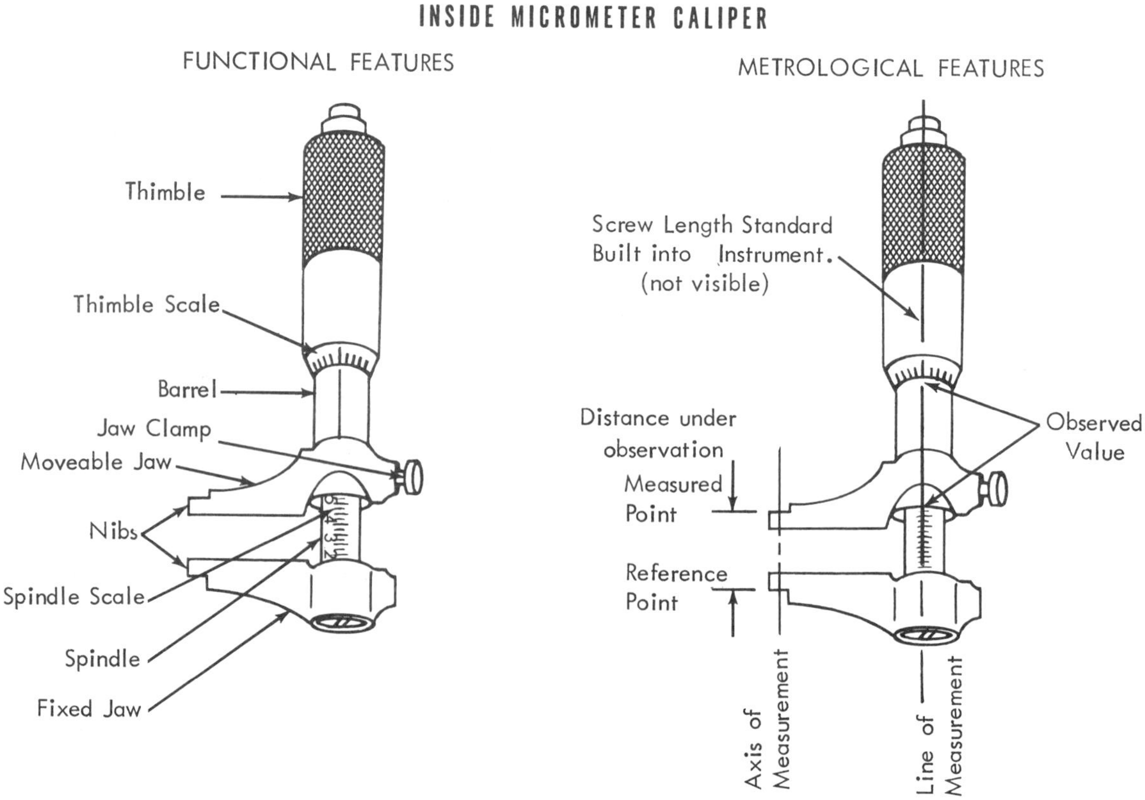

INSIDE MICROMETER CALIPER

FUNCTIONAL FEATURES

- Thimble
- Thimble Scale
- Barrel
- Jaw Clamp
- Moveable Jaw
- Nibs
- Spindle Scale
- Spindle
- Fixed Jaw

METROLOGICAL FEATURES

- Screw Length Standard Built into Instrument. (not visible)
- Distance under observation
- Measured Point
- Reference Point
- Observed Value
- Axis of Measurement
- Line of Measurement

Fig. 7-50 Both the functional and metrological features of the inside micrometer caliper correspond to those of the plain micrometer.

GEOMETRY OF INSIDE MICROMETER CALIPER INSTRUMENT

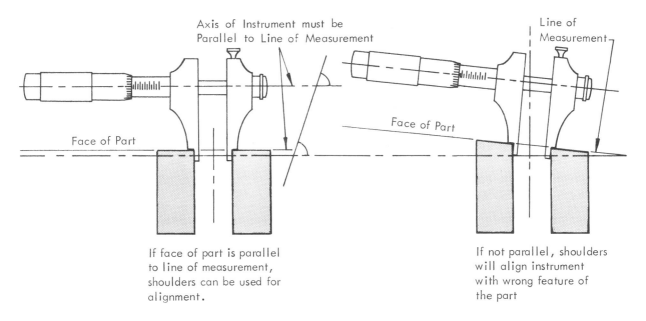

If face of part is parallel
to line of measurement,
shoulders can be used for
alignment.

If not parallel, shoulders
will align instrument
with wrong feature of
the part

Fig. 7-51   The shoulders on the jaws may either help or hinder accurate measurement.

All the usual rules for rocking and *centralizing* need to be used to assure that the axis of the instrument is parallel to the line of measurement. Of course, this action creates sliding friction. Because the inside micrometer caliper has line contact instead of plane contact, the wear is accelerated. This instrument must be frequently checked for zero setting accuracy.

The inside micrometer caliper is read exactly the same as the plain micrometer, except that the eye must jump a distance between scales. Why was the scale for the first and second significant figures placed on the spindle instead of the barrel? To avoid reading from large to small.

The same general instructions for care and calibration apply to the inside micrometer caliper as to the plain micrometer with the happy exception that we do not have *to pretend* that we are working to *tenth mil* with it.

Unfortunately telescope gages have the same range and are substituted for the inside micrometer caliper. This has little to recommend it as you shall see within a few pages.

Fig. 7-52 The inside micrometer is simply an adjustable rod that is opened until it duplicates the distance separating the reference points being measured.

THE INSIDE MICROMETER

This is one of the most straightforward instruments in that the instrument itself duplicates the distance to be measured, Fig. 7-54. The axis of the instrument *is* the line of measurement. You cannot comply with Abbe's law any closer than that, of course. The inside micrometer is considerably more common than the inside micrometer caliper because, fortunately, telescope gages can only substitute for a limited part of the range covered by an inside micrometer set. Unlike

INSIDE MICROMETER SET

Rods: 6 to 7 inch

1/2-inch Spacing Collar

5 to 6 inch

4 to 5 inch

2 to 3inch rod

Micrometer Head

3 to 4 inch

Micrometer Handle

(The L. S. Starrett Co.)

Fig. 7-53 Only one head is required to cover a large measurement range.

the previous instruments (except the combination set), inside micrometers are usually purchased in sets, Fig. 7-53.

The practice of using extension anvils to add range to the plain micrometer was frowned upon. With inside micrometers it is the only way. The smallest sizes of inside micrometers have movements as short as one-fourth inch. The largest sizes have one inch. The most popular have one-half inch of movement.

The inside micrometer is very similar to a plain micrometer, except that it has no frame and the short spindle is equipped with a chuck for the attachment of extensions, Fig. 7-54. The spherical contacts have a radius shorter than the smallest radius for which the micrometer measures. The chuck permits either the collar or the rods to be added to the micrometer head to make up the required length as shown in Fig. 7-55. The rods available on the micrometer shown provide up to 32 in. of measurement. Even greater lengths are available on some models, notably those

that use hollow tubes instead of rods for greater rigidity.

It is a common practice to set the micrometer to size on the part then remove it for reading. For that reason the nut is adjusted for a relatively tight fit on the screw. If it is used as shown in Fig. 7-52 that would be necessary, because when adjusted for the proper feel the index on the thimble may be turned away from the observer. This method is preferred by some because the heaviest part of the micrometer is at the bottom.

ADDING EXTENSION RODS

Example: 2 inch inside micrometer with 1/2-inch movement.

2 to 3-inch rod (the shortest). Reading is 2.125 inch.

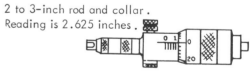

2 to 3-inch rod and collar. Reading is 2.625 inches.

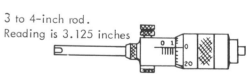

3 to 4-inch rod. Reading is 3.125 inches

3 to 4-inch rod and collar. Reading is 3.625 inches.

INSIDE MICROMETER

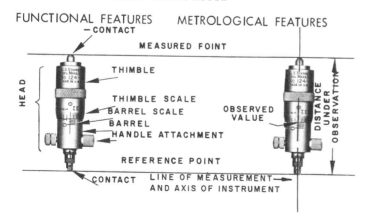

FUNCTIONAL FEATURES   METROLOGICAL FEATURES
— CONTACT

MEASURED POINT

THIMBLE

HEAD

THIMBLE SCALE
BARREL SCALE
BARREL
HANDLE ATTACHMENT

OBSERVED VALUE

DISTANCE UNDER OBSERVATION

REFERENCE POINT

CONTACT   LINE OF MEASUREMENT
AND AXIS OF INSTRUMENT

Fig. 7-54 An inside micrometer is basically a plain micrometer without a frame.

Fig. 7-55 The closed length is 2 inches. The screw provides 1/2 inch, a collar adds 1/2 inch and the rods add whole inches.

A better method is to hold the reference point against one side of the part, Fig. 7-56, and adjust for fit while moving the head end. If the hands or forearms can be rested on the part all the better.

Warming of the long instrument from handling can substantially increase its length thus resulting in a minus error. Some long rods are provided with insulating grips to minimize this effect.

The handle aids in the use under the opposite extreme. It is obviously helpful to reach nearly inaccessible parts of large machinery. It also is helpful when measuring parts so small that the fingers would be in the way, Fig. 7-57.

Fig. 7-56 The inside micrometer must be carefully centralized to provide the desired feel.

(Brown & Sharpe Mfg. Co.)

Fig. 7-57 The handle for the inside micrometer gets the fingers out of the way when measuring small diameters.

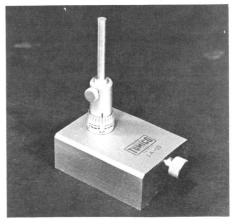

(Scherr-Tumico, Inc.)

Fig. 7-58 There are bases available that convert inside micrometers to height gages.

The rod ends are spherical and present nearly point contact to the part. They should be frequently checked for wear which would result in plus errors.

Bases are available that permit inside micrometers to be used from flat surfaces, Fig. 7-58. This is an inexpensive way to make height gage length comparisons from a surface plate. In no case can it compare with gage blocks for accuracy. In low setups it may have some advantages over vernier height gages. It is certainly easier to read, but has very limited measuring range and takes longer to set up. Economy rather than reliability warrants these applications.

Calibration of the inside micrometer is more difficult than for the plain micrometer.

It cannot be used to measure a standard and then compare its reading to size of the standard, as outside measuring instruments can. Both the standard and the micrometer must be compared to each other using an indicator instrument. This type of calibration is discussed later.

All of the precautions about cleanliness, calibration and alignment that have been mentioned before apply to the inside micrometers. An additional precaution is required. Care must be taken that there is no dirt between the shoulders of the rods and the micrometer. Furthermore, these joints must be tight so that they will not work loose. That is why wrenches are furnished with the sets. A reminder list is in Fig. 7-59.

INSIDE MICROMETER DO'S AND DON'TS

DO:

Support the instrument in a comfortable position.

Hold rod firmly against reference point.

Thoroughly clean rod joints when assembling.

Lock the rods securely in their sockets.

DON'T:

Handle long enough that heating of the rods expands them.

Measure one place too long. The point contact will burnish a groove.

Forget to add collar length, if used.

Fig. 7-59   Any one of these points can cause a five-mil (0.005-inch) error in measurement or more.

THE DEPTH MICROMETER

Some snide remarks were made about the vernier depth gage. Now we meet the reason, Fig. 7-60. Generally speaking, the depth micrometer provides as great a measurement range, much better reliability and is easier to use than a vernier depth gage. Best of all, it is less expensive.

Depth micrometers have two peculiarities. They measure from a plane to a point, Fig. 7-61. A very small area such as the end of the rod may be considered to be a point when analyzing an instrument. The vernier height

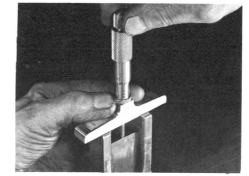

(The L. S. Starrett Co.)

Fig. 7-60   One of the most useful measurement instruments is the micrometer depth gage.

MICROMETER DEPTH GAGE

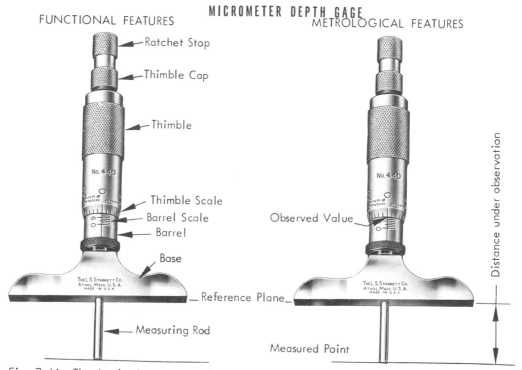

FUNCTIONAL FEATURES

- Ratchet Stop
- Thimble Cap
- Thimble
- Thimble Scale
- Barrel Scale
- Barrel
- Base
- Reference Plane
- Measuring Rod

METROLOGICAL FEATURES

- Observed Value
- Distance under observation
- Reference Plane
- Measured Point

Fig. 7-61   The depth micrometer, unlike all other micrometers, measures from a plane to a point.

gage and the vernier depth gage are the other instruments thus far with that feature. The other peculiarity is that it reads in reverse from other micrometers, Fig. 7-62. This is also true with the vernier depth gage. With it the entire scale is visible and for that reason it is not confusing. The confusion of the reversed scale of the depth micrometer is more than compensated for by the easier readability of a micrometer scale as compared to the poor readability of a vernier scale.

All of the positional relationships discussed for vernier depth gages, Fig. 6-23, apply to depth micrometers. It complies very thoroughly with Abbe's law. The problem is not whether or not the line of measurement lies along the instrument axis, but whether or not the line of measurement being picked up is the part feature desired.

Reliable measurement can be expected to *one mil* (0.001 in.) providing two points are observed. First, the base must be held firmly against the part feature so that it is not disturbed by the rod bottoming against the measured point. The ratchet stop should be used for this purpose. Second, the part feature used to support and locate the base must be accurately square with the line of measurement. This cannot be overemphasized.

The head movement of depth micrometers is one inch. They achieve their range by interchangeable measuring rods. Rods for measurements up to 10-in. long are available. They are held against their seat by the thimble cap. The same care that is taken when changing inside micrometer rods must be used when changing the rods in the depth micrometer.

READING THE DEPTH MICROMETER

.300"	.600"	1st Significant Figure
.000"	.075"	2nd Significant Figure
.024"	.001"	3rd Significant Figure
.324"	.676"	Total

Fig. 7-62 A depth micrometer measures in reverse of other micrometers. Remember to read the covered graduations on the barrel and that the thimble reads clockwise instead of counterclockwise.

No *centralizing* or other moving around to find proper feel is required with the depth micrometer. In fact, it is to be avoided. It would not contribute to accuracy and would cause accelerated wear of the small rod ends. Several readings should be taken at slightly different positions, but the rod should be raised when shifting positions.

Because of the possibility of wear the depth gage should be calibrated frequently. Note that zero setting check is made when the thimble is screwed up, not down as with a micrometer, Fig. 7-63. Accuracy, along the scale, is checked by measuring over gage blocks. This in general is discussed in Chapter 13 on calibration.

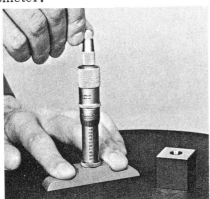
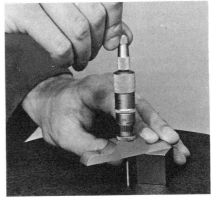

Fig. 7-63 The micrometer depth gage is checked for zero setting against a flat plate. Accuracy along its range is checked by measuring over gage blocks. *(The L. S. Starrett Co.)*

Micrometer instruments are expected to provide greater accuracy and reliability than the previous instruments discussed. For this reason the manipulation of these instruments has been separated from the reliability check list, Figs. 7-64 and 7-66, so that the latter can be emphasized.

SMALL HOLE GAGES

The accurate measurement of small holes is one of the perplexing problems encountered in inspection work. Small hole gages are certainly not the answer. They are not reliable and require a high degree of skill. They are convenient and inexpensive, and are widely used by skilled model makers, tool makers, machinists and maintenance people.

STEPS FOR MEASUREMENT WITH MICROMETER INSTRUMENTS

1. With a rule determine the approximate size instrument required.
2. Select the instrument and attachments, if required.
3. Thoroughly clean the part and the instrument components.
4. Assemble instrument.
5. Set instrument to slightly clear expected size.
6. Place instrument on part with the line of measurement nearly parallel to axis of instrument.
7. Find comfortable position.
8. Hold the reference point firmly in place.
9. Locate measured point by carefully adjusting instrument, centralizing, if necessary.
10. Take reading.
11. Back off instrument for removal from part.
12. Repeat as required for reliable results.
13. Clean, lubricate, disassemble (if necessary) and replace instrument.

Fig. 7-64   These general steps apply to all micrometer instruments.

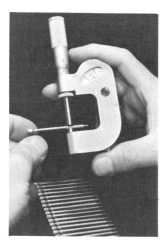

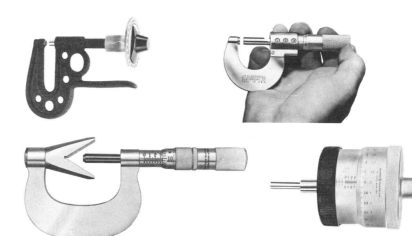

Fig. 7-65   Many micrometers have been developed, some to increase specialization, some to increase versatility.

RELIABILITY CHECK LIST FOR MICROMETER INSTRUMENTS

1. Determine whether you have available an instrument that would be more suitable for the measurement.

2. When cleaning instrument, visually inspect it.

3. Know (by records, memory or actual check) that the instrument is in calibration.

4. Examine the geometry of the part to determine the best method to assure alignment.

5. Use the ratchet stop, if possible.

6. Take precautions to prevent excessive heating.

7. Read the tool in place, if possible.

8. Write down the reading.

9. Question the observational error in reading.
   a. Was scale read from right direction?
   b. Were 0.025" divisions read correctly?
   c. Were the scales added correctly?

10. Consciously look for bias in manipulation and reading.

11. Recognize the extra precautions needed for measurements whose tolerances are close to the specified accuracy of the instrument.

12. Repeat until satisfied with measurement.

Fig. 7-66   Note that these steps repeat what you have been told about lower precision instruments.

They are not used in production measurement. There are excellent production inspection methods for measuring small holes such as air gages. However, there are virtually none for single application measurements unless considerable time can be expended. Even the smallest inside calipers are too

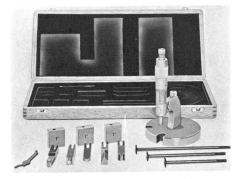

Fig. 7-67 The use of interchangeable parts has greatly extended the versatility of micrometer instruments. However, the attachments must be scrupulously clean and checked frequently.

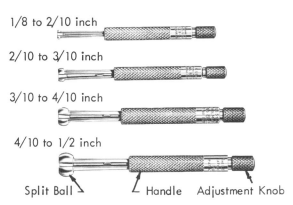

1/8 to 2/10 inch

2/10 to 3/10 inch

3/10 to 4/10 inch

4/10 to 1/2 inch

Split Ball ⌐   Handle   Adjustment Knob

Fig. 7-68 The small hole gage is included here because it is used with micrometers to measure into small holes and recesses.

Fig. 7-69 After setting the small hole gage to the desired feel in the hole, it is measured with a micrometer. *(The L. S. Starrett Co.)*

large for holes smaller than about *point two inch* (0.2 in.), which is not very small.

Small hole gages are available in sets, Fig. 7-68. These consist of four tools covering a range from 1/8 to 1/2 in. They consist of a split-ball gaging number. By turning the knurled adjustment the ball can be opened for the desired *feel* with the hole. The gage is then removed, and measured with a micrometer, Fig. 7-69.

The principal problem with the small hole gage is its dependence upon *feel* - not just one *feel* per measurement, but two.

A second problem is in the geometry of the instrument. The adjustment does not *expand* the ball as the catalogs and the books state. It merely opens it up. The line of measurement is not clear cut and the axis of the handle may or *may not* be perpendicular to the line of measurement, Fig. 7-70.

GEOMETRY OF A SMALL HOLE GAGE

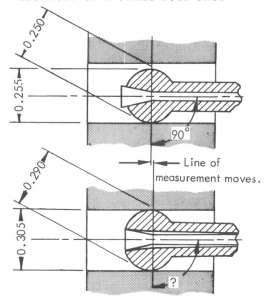

Size depends upon where on the ball you measure and does not change uniformly with adjustment.

Perpendicularity of line of measurement to handle is not definite due to unreliable spring of sides.

Fig. 7-70 Nothing is very positively fixed in relation to anything else in a small hole gage. This makes reliability depend upon the user's skill.

Nevertheless, with care, a skilled technician can measure reliably to *one mil* (0.001 in.) with small hole gages. If he is really skilled, he will not attempt greater precision.

The main justification for small hole gages is that there is nothing better that is generally available for 1/4 in. or smaller. That is still a good argument. Fortunately other instrumentation that might help is coming into wider use. The measuring microscope, the coordinate measuring machine and air gages that can be quickly set up, are examples.

TELESCOPE GAGES

The author checked all of the technical books in his library and manufacturer's catalogs in the files to find the preferred name for these instruments. The manufacturer's catalogs unanimously agreed on "telescoping gages". All of the book authors except one agreed upon "telescope gages". Never *once* having heard them called anything else is good enough reason to standardize on the latter as the preferred form.

Telescope gages are T-shaped instruments. They consist of a pair of telescoping tubes connected to a handle, Fig. 7-71. The tubes are called legs or plungers. They are closed at their outer ends and spring loaded to push them apart. The knob on the end of the handle provides careful adjustment of the locking action. The spacing of the outer ends duplicates the distance between the reference and measured points of the part feature. They have spherical surfaces with radius ground for their smallest hole setting.

Telescope gages are made in sizes to measure holes from one-half to six in. They are usually sold in sets. They have been heralded because they are more reliable for inside measurement than calipers. This is much like praising the stagecoach as a means of cross-country transportation because it is undeniably better than horseback or walking.

When *precise inside measurement* is required look first to a comparison method using gage blocks and an indicator instrument. If these are not available consider a height gage setup or an inside micrometer. The next choice would be the rather unfamiliar, tapered, parallel gages (up to one inch). If none of these is available, then, and only then, use the telescope gages.

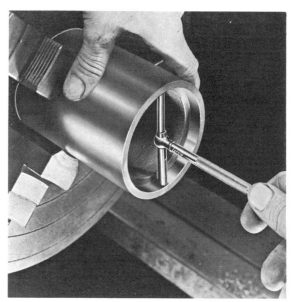

Fig. 7-71 Telescope gages provide a quick way to make rough measurements of inside diameters.

Fig. 7-72 Telescope gages can be very useful when measuring odd-shaped holes. *(The Lufkin Rule Co.)*

On the other hand when *rapid, in-process inspection* is required between cuts of a machine tool, a telescope gage may well be the best method. It also is helpful in measuring odd-shaped holes into which an inside micrometer might not fit, Fig. 7-72.

To use, the legs are telescoped together smaller than hole or recess, and the knob tightened. The instrument is inserted, canted at a slight angle as at A in Fig. 7-73, and the knob loosened. This permits the legs to open against the walls. The knob is then carefully tightened so that when the legs are pushed closed they will remain. This is tested while tightening by rocking forward slightly.

The tightening should be sufficient so that the instrument is firm, yet not so much that it might be sprung or bent. If too tight or too loose, go back to position A and try again. When satisfied, rock over center, B. That will be the shortest distance separating the reference and measured points and will occur when the axis of the instrument (along the legs) coincides with the line of measurement. In position C the instrument is removed, the knob securely tightened and the leg span measured with a micrometer.

Alignment is somewhat easier for inside diameters than for flat-sided recesses. However, especially for large diameters, care must be taken that the axis of the legs will pass through the line of measurement.

When measuring, the *feel* must be adjusted to suit the surface finish of the part. If it was as good as the micrometer spindle tip, use the same *feel*. Use heavier *feel* to compensate for the greater drag of poorer finishes on the part.

When rocking past the center, pivot about the lower leg. Do not slide both. The reason that the telescope gage is rocked from back to front is that there seems to be firmer control when pulling down than when pushing up.

The fundamental difference between the telescope gage and the inside caliper is found in the picking-up-the-measurement stage. Once the clamping pressure is set, the telescope

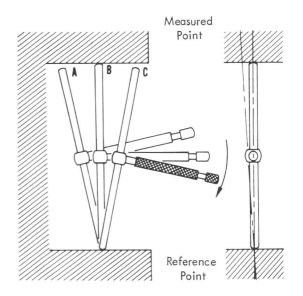

Measured Point

Reference Point

Telescope gage is rocked over center and squeezed to length that separates measured and reference points.

Axis of instrument must be aligned with line of measurement.

Fig. 7-73 The feel is determined by the adjustment, but technicians must still assure proper alignment.

gage thinks for itself. With an inside caliper, you're on your own all the way. You do the adjusting.

The steps for using the telescope gage are listed in Fig. 7-74.

STEPS FOR USING THE TELESCOPE GAGE

1. Determine that this method is adequate.

2. Select proper size of gage.

3. Clean both gage and part.

4. Close gage slightly smaller than measurement required, and clamp legs.

5. Insert and with handle canted up slightly, release legs.

6. Adjust clamp for firm but not overly stiff contraction of legs. Check by rocking slightly, starting over, if necessary.

7. Align gage so that axis of legs and line of measurement are in same plane.

8. Rock over center, pivoting on one leg, pulling handle down.

9. Secure legs with clamp knob.

10. "Mike" the legs, and write down the measurement.

11. Repeat steps 4 through 10 until satisfied with reliability of measurement.

Fig. 7-74 Even these precautions do not make the telescope method reliable except for "wide open" tolerances of $\pm$ 0.003 in. or more.

(The L. S. Starrett Co.)

Fig. 7-75 This is a job for tapered parallel gages, not for the more familiar adjustable parallels.

TAPERED PARALLEL GAGES

When the hole or recess is accessible from both sides, tapered parallel gages can be used for measurement. Let us begin with a warning. These are not to be confused with *adjustable parallels* which they resemble. Adjustable parallels are generally used for supporting parts in machine setup or in gaging set up. They have flat contact surfaces and are inexpensive.

The tapered parallel gages are used for measurement. They have crowned contact surfaces, and they are not inexpensive. They are usually furnished in sets of 10 pieces that have a measuring range from 1/4 in. to 1 in. A plate in their case shows how to combine the pieces for size.

Tapered parallel gages are inserted in the hole to be measured. The two parts slide together until their contact surfaces are stopped by the walls of the part. The protruding part is measured with a micrometer, Fig. 7-75.

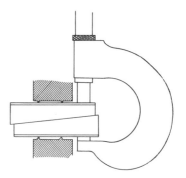

Fig. 7-76 The tapered parallels provide line contact. Any obstruction will change the measurement.

There are two essential differences between tapered parallel gages and telescope gages. Telescope gages are *point contact*. The parallel gages are *line contact*. Because they are line contact they will give a measurement caused by the two highest points on each wall. They do not permit exploring the hole. Conversely, they are not dependent upon the vagaries of two points to express the real measurement. Another advantage of the line contact is simplification of alignment. Care is still necessary for reliable work.

The other principal difference is that both required measurements take place simultaneously, Fig. 7-76. The parallels are held against the hole surfaces while the micrometer measures the parallels. It is always desirable to avoid separate operations. When that is not possible, it is desirable to bring separate operations as close together in time and place as possible.

Skilled technicians claim reliable measurement to precision as fine as *point five mil* (0.0005 in.) with tapered parallels. The author is unable to measure closer than *one mil* (0.001 in.) with any degree of regularity.

SUMMARY

● Knowledge of micrometers is important because they are among the most frequently encountered instruments. It is also important because micrometers demonstrate six important considerations.

● Relatively simple and inexpensive instruments can provide high accuracy when they conform with Abbe's law, as well as other good design principles.

● Whenever you gain something you must give up something. Micrometers provide better reliability than vernier instruments, but at a sacrifice in range and versatility. Think of the range you will be restricted to when you use instruments that measure to *mike increments* (millionths of an inch.)

● Micrometers show the relationship between discrimination and accuracy. At the same discrimination as vernier instruments they provide greater accuracy (less error). However, if a vernier scale is added to a micrometer its discrimination then exceeds its accuracy. The vernier micrometer provides high reliability for measurements in the plain micrometer range but is unreliable at its maximum discrimination.

● Micrometers are contact instruments. This means that there must be some distortion, however minute, for every measurement. This must be controlled either by skill (the *feel* of the contact) or by mechanical limiting of the force of contact (the ratchet stop).

● The proper geometry for accurate measurement can only be achieved by adjusting the method to the conformation of the part. For nearly all cylindrical and spherical part features, contact instruments must rely on *feel*. One technique is *centralizing*, but even that ultimately depends upon skill. For parallel plane part features the ratchet stop may substitute for most of the skill of *feel*. However, no matter how well designed the instrument, careless or stupid handling will result in measurement errors.

● Much measurement with micrometers involves the transfer of measurement through an intermediate step. This doubles the measurement problem and greatly reduces the reliability. If a more suitable instrument is not obtainable, the only course open is to reduce the level of accuracy that you expect in your result.

TERMINOLOGY

Readability	The relative ease with which the measurement can be distinguished. For example: both a plain micrometer and a Vernier caliper have the same discrimination, but the micrometer is more <u>readable</u>.
Interpolation	The selection of the nearest graduation when a measurement lies between. The observational equivalent to the <u>rounding off</u> process in computation.
Contact instrument	An instrument that depends upon physical contact with the part. Hence distortion is introduced.
Feel	The preception of the distortion that results from physical contact between the instrument and the part or standard.
Torque	The amount of turning moment, or twist. It increases with the length of the lever arm and with the force applied to the end of the arm.
Force	Simply the amount of push exerted.
Pressure	Force spread out over an area.
Rocking	The pivoting of an instrument around one or the other of its contact points.

Fig. 7-77 Most of these terms take on increased importance for higher amplification instruments.

METROLOGICAL DATA FOR MICROMETERS

INSTRUMENT	TYPE OF MEASUREMENT	NORMAL RANGE	DESIGNATED PRECISION	DISCRIMINATION	SENSITIVITY	LINEARITY	RELIABILITY	
							PRACTICAL TOLERANCE for SKILLED MEASUREMENT	PRACTICAL MANUFACTURING TOLERANCE
Micrometer, bench	direct	1 inch	0.0001 in.	0.0001 in.	0.0001 in.	0.0001 in.	± 0.0001 in.	± 0.0005 in.
Micrometers, with Verniers and ratchets 1 and 2 inch	direct	to size	0.0001 in.	0.0001 in.	0.0001 in.	0.0002 in.	± 0.0001 in.	± 0.001 in.
Micrometers, Plain to 6 in.	direct	to size	0.001 in.	0.001 in.	0.0005 in.	0.0002 in.	± 0.001 in.	± 0.005 in.
6 to 12 in.	direct	to size	0.001 in.	0.001 in.	0.001 in.	0.0002 in.	± 0.002 in.	± 0.010 in.
over 12 in.	direct	to size	0.001 in.	0.001 in.	0.0001 in.	0.0002 in.	± 0.005 in.	± 0.015 in.
Micrometers, inside	direct	to size	0.001 in.	0.001 in.	0.001 in.	0.0002 in.	± 0.001 in.	± 0.005 in.
Micrometers, depth	direct	to size	0.001 in.	0.001 in.	0.001 in.	0.0002 in.	± 0.001 in.	± 0.005 in.
Small hole gages	transfer	$\frac{1}{8}$ to $\frac{1}{2}$ in.	none	none	0.001 in.	none	±0.001 in.	± 0.004 in.
Telescope gages	transfer	$\frac{1}{2}$ to 6 in.	none	none	0.002 in.	none	±0.002 in.	± 0.005 in.

Fig. 7-78 Regardless of its size, the range of a micrometer is limited by its screw travel. That is the reason the linearity is constant and not stated as so much per inch as it was for previous instruments.

QUESTIONS

1. What is the main advantage the micrometer has over the standard vernier instruments?
2. What is the typical range of a micrometer?
3. How is higher amplification achieved on the micrometer?
4. What are the basic limitations of micrometers?
5. Why is a micrometer easier to read than a vernier instrument?
6. How is a micrometer read?
7. What is "degree of accuracy" as related to a micrometer?
8. How is the precision of a micrometer related to its accuracy?
9. How can precision increase the reliability of an instrument?
10. What is the maximum inherent accuracy that is possible on a micrometer?
11. Are tenth-mil readings reliable on a micrometer?
12. Why is a ratchet stop a valuable addition to the micrometer?
13. Of what value is the clamp ring?
14. What should you do if a micrometer does not have a ratchet stop or clamp ring?
15. As diameters and lengths increase, the problems of measurement increase. Why?

16. What points should be considered for adequate micrometer care?
17. Why is it more difficult to calibrate an inside micrometer than a plain micrometer?
18. What additional precaution must be observed when using the inside micrometer?
19. What are the principal problems involved in the use of small hole gages?
20. What is the measuring range of telescope gages?
21. What are the principal advantages of telescope gages?
22. What is the principal disadvantage when using tapered parallel gages?

DISCUSSION TOPICS

1. Explain the basic advantages of the micrometer over vernier instruments.
2. Explain the principal features of the micrometer.
3. Explain the steps for reading the vernier micrometer.
4. Discuss the reasons why micrometers are not suitable for reliable measurement to tenth-mil increments.
5. Explain how the ratchet stop on a micrometer can help to obtain more reliable measurements.
6. Explain the steps to be followed for locating the diameter of round parts with a micrometer.
7. Discuss some of the factors that influence "feel" in measurement.
8. Explain the important steps to be followed when calibrating a micrometer.
9. Explain the types and uses of various "special" micrometers.
10. Explain the important points to be considered to provide reliable measurements with micrometer instruments.
11. What considerations are involved in transfer measurements using small hole gages, telescope gages and tapered parallels?

> *As a rule, inventions — which are truly solutions — are not arrived at quickly. They seem to appear suddenly, but the groundwork has usually been long in preparing. It is the essence of this philosophy that man's needs are balanced by his powers. That as his needs increase, the powers increase . . .*
>
> Louis Sullivan
> (architect who pioneered the skyscraper)

Measuring *in* millionths of an inch and measuring *to* millionths of an inch accuracy are two entirely different matters. Although the number is growing rapidly, few have need to measure *in* millionths. But everyone concerned with reliable measurement is concerned with measurement *to* millionths accuracy, whether he knows it or not. In explaining this, it will become apparent why gage blocks by far are one of the most important measurement tools with which you can become acquainted.

Gage blocks are the only means by which the standard of measurement is given physical form and made available for practical measurement use. They provide *tenth-mil* (0.0001-in.) precision, *mike* (0.00000-in.) accuracy and a measuring range from *ten mil* (0.010 in.) to several feet, Fig. 8-1. They are a direct link between the measurer and international length standards.

DEVELOPMENT OF GAGE BLOCKS

By now the point has been made that precision - the ability to divide into small increments - means little unless it is accompanied by accuracy. Accuracy means the ad- herence to a standard, so that all parts made to that standard can be used in predictable relationships to each other. Both precision and accuracy are needed so that scientists can pass on quantitative knowledge to each other about the measurements required in their research.

Prior to the Industrial Revolution, vernier instruments extended the precision of the craftsman's own sight and feel. His concern, however, remained with complete functioning units, not with individual parts. Machine tools greatly extended the ease with which parts could be made. But parts are useless unless they can be assembled into functioning units. A transference of measurement skill from the craftsman and to the manufacturing operation was needed.

Eli Whitney is credited with the introduction of interchangeable manufacture. This required that all the parts within one operation be made to one standard of length. Often the standard was a handmade model of the device rather than a measurement system. It is said that Henry Ford still used this method for producing the Model T. Anyone who has ever worked on his car will have a hard time believing that it died with that venerable auto.

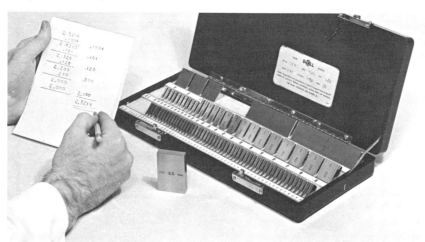

Fig. 8-1 Over 200,000 size combinations can be made from a standard set of gage blocks - all accurate to millionths of an inch. *(DoALL Co.)*

Within individual plants this method developed to a high level. By 1900 all of the principles of precision measurement discussed in this book and most of the instrumentation was already well on its way. It took World War I, however, to show that accuracy must accompany precision when you are really serious about killing off your neighbors. Plant A might make really wonderful antipersonnel shells and so might plant B; but unless fuses from plant A could explode charges from plant B, there just was not much point in being at war.

The more extensively goods are produced and used, the greater is the need for universally accepted standards. This is a good place to relate standard of length with accuracy. *The* standard is the ultimate in accuracy *by definition*. Any deviation from *the* standard is error. This applies to *the* standard, meaning the absolute master from which all others are judged. *A* standard means something else. It is any replica of *the* standard or even a replica of a replica.

This basic need for standards has increased tremendously because three other factors have increased: speeds of mechanisms, number of components and cost of manufacture. Whenever any of these go up, either standardization must improve or the reliability will go down. No wonder that missiles and space exploration have put the greatest strain on our standards to date. They involve tremendous increases in all three of the factors.

It is logical that the development of international standards should have progressed hand in hand with the development of precision manufacturing methods. Too bad it was not that way. Probably we would be having commuter service to the moon by now if it had been.

Great progress was made in measuring the international standard and correlating the lengths of the various national standards. Unfortunately, this all involved *line standards;* and had little relation to the problems of the men who were trying to make more things that would run faster and do it for less money.

These men were working along an entirely different line. They were developing *end standards*. As early as the beginning of the 18th century, "gage sticks" were known to have been used in Sweden. About 1890, go and not-go gages were introduced for the manufacture of rifles, Fig. 8-2. Obviously such gages had to be made for each and every part feature whose size must be controlled. Because the gages were bound to wear, separate sets of setting gages for each and every working gage were required. And because these too were bound to wear..... seldom did the cost allow this to be carried as far as reliability required.

C. E. Johansson is not the inventor of gage blocks as popular opinion insists. They were used long before his time – his actual contribution is far more important. And while on errors, the way you tell an amateur from

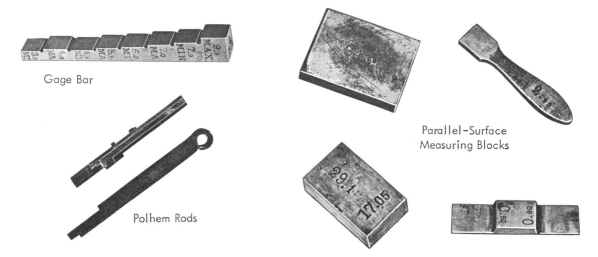

Gage Bar

Polhem Rods

Parallel-Surface
Measuring Blocks

Fig. 8-2 Long before systematized gage block sets were developed, end standards had shown their value in the dimensional control of manufacturing. *(DoALL Co.)*

a "pro" in measurement is that the amateur calls gage blocks "Jo-Blocks". You know what you would think of a mechanic tuning up your car if he asked you to hand him his "monkey wrench".

Around 1890, Hjalmar Ellstrom, chief mechanic in a Swedish arms factory, manufactured parallel-surfaced steel measuring blocks. Though they could be used together in measuring combinations, a major problem was the maintenance of a sufficient quantity to fill all measuring requirements.

Johansson worked with Ellstrom on this development. Why not, he reasoned, make a set of master gages that could be combined for all sizes within its range?

Johansson's genius lay beyond recognizing this need and working out a system of block sizes to satisfy the need. He went further. He recognized that the resulting set of gage blocks, to be of universal value, must be calibrated to the international standard.

We owe an everlasting debt to him for patiently solving the latter problem. As a result of his efforts, along with others such as Ellstrom, of course, end standard replicas of the international standards became generally available to the major metrology laboratories of the world. It was Johansson, for example, who proposed 20 deg. C. as the temperature for standard measurement. This was the average temperature of European machine shops and it converted to a whole number in the Fahrenheit scale.*

Much of the credit for present day use of gage blocks goes to Major William E. Hoke of the U.S. Bureau of Standards. He developed mechanical methods for lapping (final finishing) which made the mass production of gage blocks possible.

FROM LIGHT WAVES TO SOLID STEEL

We all know that the metric system was a result of the French Revolution. That is not surprising when you realize that the lack of a systematic system of measurement was one of the *causes* of that revolution. Prior to that

time, measurement was oppressively used in levying taxes and rent payments. A peasant paid his bills with large measures and bought his necessities with short measures.

In the great tidal wave of liberation that swept the world, standards based on anything other than physical phenomena seemed inappropriate. The fact that the French scientists failed to measure the earth exactly when establishing the meter, is unimportant. The world had a length standard on which all could agree. As explained in Chapter 3, in 1866 the United States accepted this standard, making it nearly universal among the industrial nations.

Unfortunately the international meter was a line standard which made difficult the manufacture and verification of the replica meters for the various countries. It was nearly a quarter of a century, for example, before a replica was received in the United States. Why? *Whenever work is performed requiring a subdivision of the standard that is smaller than the error that is known to exist in the standard, then that work is not in conformance with the standard and cannot be considered accurate.* Great effort has been extended to reduce the error. This finally resolved in

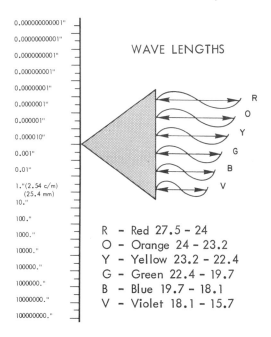

Fig. 8-3 When Michelson determined the number of cadmium red wave lengths corresponded to the length of the meter, he simultaneously established light as a basis for length standards.

---

* Moody, J.C.,"New Standards for Old", <u>Ordnance</u>, May-June, 1962, p. 4.

UNITS OF LENGTH AND STANDARDS OF LENGTH

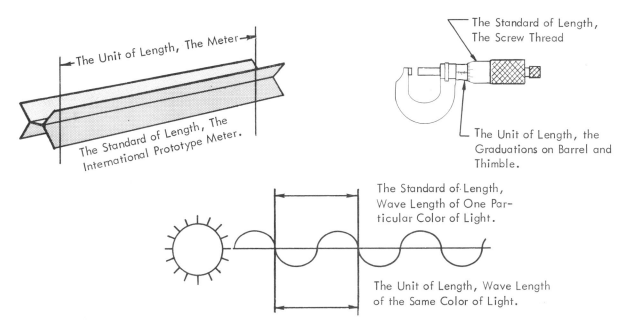

Fig. 8-4   Only for light does the unit of length and standard of length become one. Therefore light waves are the standard for both accuracy and precision.

expressing the standard in an unchanging physical phenomenon, whose units should be smaller than any conceivable error that might be encountered. Light was an obvious choice.

The idea that the standard of length should be defined in terms of light was not new. In 1827 there is a record that this was suggested by a French philosopher. Much later Albert A. Michelson and W. L. Morley had constructed an interferometer to test the theory of ether drift. Then in 1892-3 Michelson used the same principles to determine the wave length of the red cadmium spectral line, Fig. 8-3.

By now, I am sure, no reader will argue that measurement is relative. One thing is always measured in terms of another. Therefore *in measuring the wave length of a particular color of light, Michelson also did the reverse. He defined the unit of length of the standard that he was using in relation to that particular wave length.*

We need to re-emphasize a point made in Chapter 2. A *unit of length* is an arbitrary unit, such as inch or meter, that is used to describe lengths as either fractions or multiples. *A standard of length* is the physical thing that defines the unit of length. To illustrate, the International Prototype Meter is a standard length but the distance separating the scribed lines on it is the unit of length, the meter, Fig. 8-4.

With all standards prior to the use of light waves, the unit of length and standard of length were two separate things. For the first time, with light, the unit and the standard become one and the same. Equally important, given the necessary laboratory equipment, the standard can be reproduced anywhere and at any time with complete assurance that it will be in agreement with all similar reconstructions of it.

Of course, light cannot be handled and used as a direct-measuring instrument. You will learn in Chapter 14, however, that it can be used to measure distances between flat surfaces by a method called interferometry. Gage blocks are simply flat, parallel surfaces whose separation has been established to light wave precision and accuracy, Fig. 8-5. Precision is assured because the discrimination of the light waves is much smaller than the tolerance of the gage block; and accuracy, because light waves *are* the standard of accuracy.

ARITHMETICAL PROGRESSIONS
TO THE RESCUE

Before the turn of the century the trains leaving the Mauser plant in Germany often carried visionaries who were so possessed with their plans that they saw little of the surrounding country. What strategies were thought out, we can only guess. Johansson's characteristic precision saves us the necessity for conjecture in his case. He recorded his thoughts.* While on the train in 1896 he conceived the idea for *arithmetical combinations for gage blocks.*

This was a greater feat than might be realized because it went contrary to the entire concept of tolerances that was then accepted for all measurement from shop work to the national bureau of standards. In the old system a constant tolerance was specified for both a reference tolerance and all of its subdivisions. If a precision device was

---

* Torsten, K. W. Althin, C. E. Johansson, The Master of Measurement, Stockholm, Nordisk Rotogravyn, 1948, p. 51.

being manufactured to plus or minus *two mil* (0.002 in.), all of its parts would have tnat tolerance. A groove that contained ten parts would have the same tolerance as each of its individual parts. The reasoning was that with everything at its basic size there would be no problem. True, but unrealistic.

Johansson did not apply for a patent on his system of block sizes until he thoroughly investigated his system of tolerances. He later patented it. It is the basic system used today.

> "When two or more combinations of a certain measurement are laid together the combination measurement thus obtained will lie within the tolerance fixed from the beginning for this measurement."

The idea for using arithmetical progressions to combine into any desired length is the subject of Johansson's invention. Johansson fully understood a matter that has required an additional half century and untold billions of dollars in scrap losses for the industries of the world to accept. Gage blocks

GAGE BLOCKS

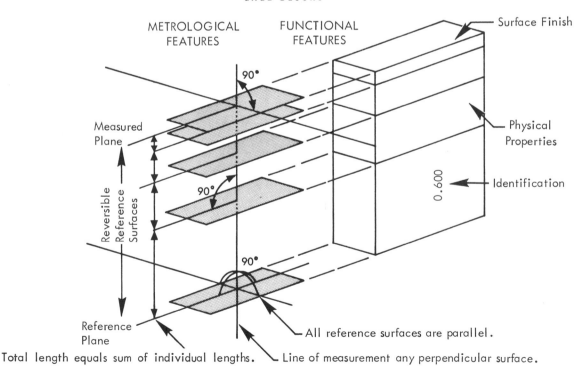

Fig. 8-5 Both the metrological and the functional features of gage blocks suit them for exacting measurement.

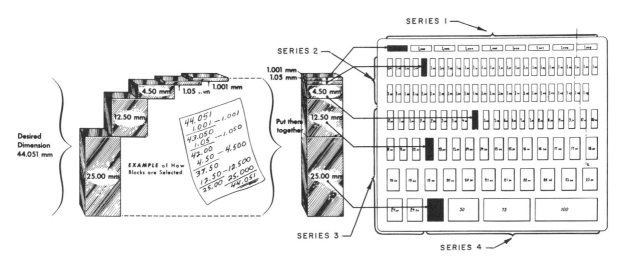

Fig. 8-6   In Johansson's patent, 111 blocks in 4 series combine to form any dimension from 2 mm to 202 mm in steps of 0.001 mm, a total of 200,000 combinations.

COMPARISON OF ENGLISH AND METRIC GAGE BLOCK SETS

	JOHANSSON'S PATENT (111 pieces)				PRESENT INCH SETS (81 pieces)
			SAME DIGITS		
1ST SERIES	9 Blocks from thru Increments of	1.001 mm. 1.009 mm. 0.001 mm.	INCHES	0.039439 0.039755 0.000039	1ST SERIES 9 Blocks from 0.1001" thru 0.1009" Increments of 0.0001
			BUT DIFFERENT RESULTS		
2ND SERIES	49 Blocks from thru Increments of	1.01 mm. 1.49 mm. 0.01 mm.		0.039794 0.058661 0.000394	2ND SERIES 49 Blocks from 0.101" thru 0.149" Increments of 0.001
3RD SERIES	49 Blocks from thru Increments of	0.50 mm. 24.50 mm. 0.50 mm.		0.0197 0.9646 0.0197	3RD SERIES 19 Blocks from 0.050" thru 0.950" Increments of 0.050
4TH SERIES	4 Blocks from thru Increments of	25.00 mm. 100.00 mm. 25.25 mm.		0.9842 3.9370 0.9843	4TH SERIES 4 Blocks from 1.000" thru 4.000" Increments of 1.0000

Fig. 8-7   Inch sets copied the form but not the intent of Johansson's standard set. Note the sameness of the digits but difference in the actual sizes.

are needed to establish accuracy as close to the actual manufacturing operation as possible. Here are his own words:

"So it was mainly in France that Johansson standard gage block sets were acquired, even though they often ended up in a glass case in the laboratory *instead of in the workshop where they belonged.*"

The author's italics clearly shows his intention. It was further shown when he modified his original set so that it could cover fully the range required for practical shop use.

This set and the way it combined into the desired lengths is shown in Fig. 8-6. Continuing his drive to get the sets into the shop, he designed the end standards and holders required for practical use of the blocks. These are covered in the next chapter.

FIRST INCH SETS INTRODUCED

As a result of an almost inexplicable lack of understanding we were deprived of much of the benefit that was intended from the use of gage blocks. The inch sets slavishly imitate the form but not the intention of Johansson's metric sets. This is shown in Fig. 8-7. The smallest block in the metric set (1.001 mm) is equal to 0.0394 in. This is compared with the smallest block in the inch set which is 0.1001 in. or *2 1/2 times larger*. The discrimination of the metric set is 0.001 mm which is equal to 0.0000394 in. In contrast the discrimination of the inch set is 0.0001 in., again *2 1/2 times larger*.

If Johansson's ideas about the practical range were correct, then the inch sets badly missed their mark. The best proof that this was the case is the tremendous number of sets that gage block manufacturers have since placed on the market to compensate for the inadequacy of the basic inch set.

The explanation is clear. The original inch set was made by Johansson in 1906 as part of a program to reconcile official differences between the length standard of the English and metric systems. There is no account in his voluminous records, to the author's knowledge, that Johansson ever intended that set for any other purpose. For its purpose the lack of small sizes and fine discrimination was not as important as the accuracy of the set, which no doubt was as good as his metric sets.

Much of Johansson's remaining time and effort was spent in the rarified atmosphere of international standards and in administrative activity. The problem of developing the best possible combination for inch sets probably never came to his attention. Perhaps one of you will accomplish it.

THE JUMPING OFF PLACE

Creating a gage block set from scratch seems like lifting oneself by one's bootstraps. Where do you get the standard from which to start? Johansson's method is given here because there are many applications you can make of the principle he used.

His method was to make a 100 mm gage block as close to that dimension as he could. Figure 7-28 shows one of the instruments he developed for this work. He then sent it to the International Bureau of Standards at Sevres, France. Instead of asking them to measure it, he asked that they determine the exact temperature at which it would be 100 mm long. This did not make the gentlemen at the laboratory particularly happy, but eventually they finally came through with the information. The block was 100 mm long at 20.63 deg. C.

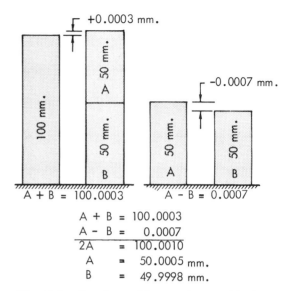

$$A + B = 100.0003$$
$$A - B = 0.0007$$
$$\overline{2A = 100.0010}$$
$$A = 50.0005 \text{ mm.}$$
$$B = 49.9998 \text{ mm.}$$

Fig. 8-8   Starting with one known block, Johansson used simultaneous equations to derive all of the other blocks in the set.

Johansson's next step was to make a block that would measure 100 mm at 20 deg. By calculation based on the coefficient of expansion, he determined that the block should be 0.0007 mm longer than the first one.

He was now ready for the step that may have practical value to you. *He used the single master to calibrate every block in the set.* He used the method of subdivision based on simultaneous equations. If both the sum and difference of two or more quantities are known, the individual quantities may be found. This simple method is shown in Fig. 8-8.

THE STUFF THAT MILLIONTHS ARE MADE OF

We will not delve into gage block manufacture further than necessary to show you that it's one thing to talk about millionths of an inch and another to make them behave in a block of metal.

What functional characteristics are wanted in gage blocks? First, they must be made from something that can be accurately sized and finished. Second, they must stay put and not change in size of their own accord. Third, they must be able to withstand considerable wear. Fourth, they must be practical. The latter covers a host of things, such as ruling out gold, platinum and iridium alloys, to hoping that they will be reasonably corrosion-resistant.

What material best provides these characteristics? Each manufacturer claims his product does. Undoubtedly each is best in some particular area. Unbiased testing of blocks accompanied by accurate records will certainly show which one is best for any particular situation. No attempt will be made to evaluate them here. They simply will be listed along with the claims made for them.

By far the vast majority of blocks are made from hardened alloy steel. The hardening is obviously required to resist wear. Hardening also traps stresses in the material which are slowly relieved as it ages. When this happens the blocks lose their calibrated size. Blocks subject to this are termed *unstable*. This is minimized by elaborate heat-treatment processes which involve cooling the blocks at temperatures 200 deg. F. below zero and lower.

You might wonder why surface hardening is not used so that the block itself can be left soft and free from stresses. For sufficient ability to withstand the pounding of comparator points and normal use, the hard case must be supported by a tougher backing material than the soft steel would provide.

There are two large advantages for the steel blocks. They have very nearly the same coefficient of temperature expansion as the majority of the steels with which they are used. The second advantage is their low cost.

At the opposite extreme are the carbide blocks. These blocks are very wear resistant, very expensive and have very different coefficients of expansion than most materials. They will give the longest service life, and if they are used in standard temperature laboratories, their low coefficient of expansion is no problem.

If they are used at ordinary temperatures great care must be taken to correct for the expansion rate. This is one of the reasons they never should be mixed with other blocks. If, for example, aluminum parts are to be measured at ordinary temperatures to *tenth mil* (0.0001 in.) or closer, the difference in expansion rate will have to be calculated with any gage block. In that circumstance no extra work is required for carbide blocks.

If a burr is thrown up on a carbide block it will seriously damage other fine surfaces.

This is another reason why they should not be used with other gage blocks. Burrs can be quickly and easily found as you will be shown in Chapter 14.

Carbide blocks are made by powdered metallurgy. That involves compacting powdered particles under pressure and heat until they join. Because the material in its natural state is hard, no heat treatment is necessary. This means that internal stresses are not a problem and the blocks have good stability.

Some say carbide blocks have minute pits into which the comparator point may fall. It is said also that they may be porous and trap moisture. Others state that both of these are absurd. Aside from advertising claims, the author has found nothing to prove or disprove either side. One thing is clear – carbide *wear blocks* are recommended for use with all other types of blocks. Wear blocks are placed at the reference ends of gage block stacks. That is the one exception to mixing blocks.

The other blocks fall between ordinary steel and carbide in wear resistance and price. Important among these are the chromium-plated blocks and the stainless steel blocks. The latter are surface hardened. Unlike ordinary steel blocks, the nonheat-treated interior is sufficiently hard and tough to support adequately the tough outer case. An important advantage of these two blocks as well as the carbide blocks is resistance to rust and corrosion. Recognizing the need for very stable gage blocks at a reasonable price, the Bureau of Standards has been investigating stainless steel blocks for a number of years under the sponsonship of the leading gage block manufacturers.

MACROGEOMETRY AND MICROGEOMETRY

"Macro" means large and "micro" means small. *Macrogeometry* generally refers to the easily determined general shape of a part – the features that are measured with the instruments we have discussed thus far. *Microgeometry* refers to the very minute analysis of shape.* When we attempt to measure to

_____

*Moody, J. C., "Geometrical and Physical Limitations in Metrology," International Research in Production Engineering, New York, American Society Mechanical Engineers, 1963, p. 568

very high precision or accuracy, micro-geometry becomes an important consideration. It includes flatness, parallelism, straightness, roundness and surface finish.

When using end standards, length measurements depend upon the condition of the contacting surfaces. *Surface finish* is a general catchall term used to describe these conditions. It is unfortunately both ambiguous and redundant at the same time. The term *surface texture* is coming into use, but promises to be little better.

The complete descriptive term is *surface topography*. This is an involved subject in itself. Only enough will be included here to avoid pitfalls in the use of gage blocks. Surface topography recognizes that the surface condition is made up of several separate factors: roughness, waviness, flaws and lay, Fig. 8 - 9. * Too often, only the roughness is considered, because it is this factor that is most

---

* "Surface Roughness, Waviness and Lay," <u>ASA Bulletin B46.1</u>, New York, American Standards Association, 1947.

SURFACE FINISH CONSIDERATIONS

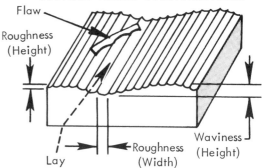

<u>Roughness</u> is finely-spaced surface irregularities, in a consistent pattern, produced by machining or processing.

Waviness is an irregular surface condition of greater spacing than roughness. Usually caused by deflections or vibrations, not by the cutting edge.

<u>Flaws</u> are irregularities that do not appear in a consistent pattern.

<u>Lay</u> is the predominant direction of surface pattern.

Fig. 8-9   These terms are usually used to describe surface conditions.

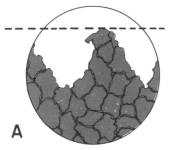

A

A crystalline metal surface that has been machined on a lathe may be visualized as having high peaks and deep valleys.

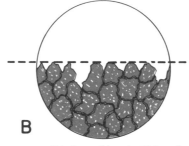

B

Grinding reduces the high peaks, but produces new and smaller peaks and valleys, the dimensions depending on the grit size.

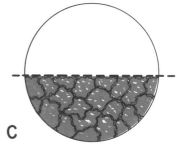

C

Honing, due to the fineness of the cutting grits and irregularity of cutting motion, reduces a surface to one of fine scratches.

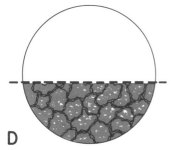

D

Lapping normally leaves a finish of minute scratches. The scratch pitch is finer than that left by a hone.

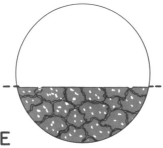

E

Gage block surfaces have a high degree of flatness, as well as surface finish. Only occasional extremely shallow scratches are left.

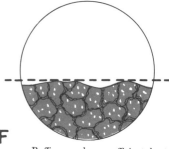

F

Buffing produces sufficient heat to bring about "plastic flow" and, while aiding reflectibility, will reduce the quality of surface flatness.

Fig. 8-10   Comparison of surface finishes shows the exacting requirements for gage blocks.

responsible for the appearance of the surface. This can be very misleading. Several surfaces are shown in the photomicrographs of Fig. 8-10. Precise gages and instruments usually require all of the treatments from A through E. A quick buffing as in F would produce a surface that would look as good as the one in E, or even better. Yet, as you can see by its waviness, it is unsuitable for precision measurement.

"Smooth as a mirror" is a poor compliment in metrology. A buffed or polished surface may have good reflective qualities. This is often mistaken for good surface finish. The

difference is shown in Fig. 8-12. Caution – the reverse is not true. Never assume that surface finish is good because its reflectivity is poor. In fact, when the very best surface finishes are attained, the reflectivity is excellent for nearly all hard materials.

Surface finish is now specified by arithmetical average, abbreviated as AA. This is simply the distance between the mean line and the crests or peaks. Formerly it was referred to as the root mean square (rms), a more complicated method that is no longer standard. Figure 8-11 shows that this measure is not entirely adequate.

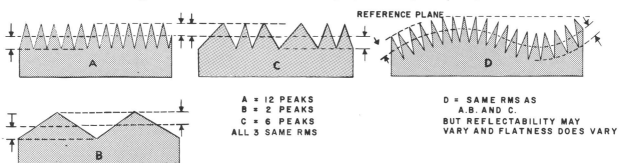

Fig. 8-11 The top drawings show that an irregular surface or poor surface finish may have better reflectivity than a good surface finish. The bottom drawings show the reason for this.

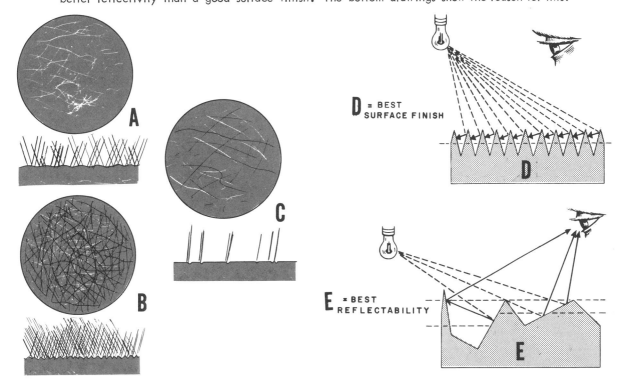

Fig. 8-12 Surface finish is measured by the height of the crests from the average. These examples show that this does not fully describe the surface.

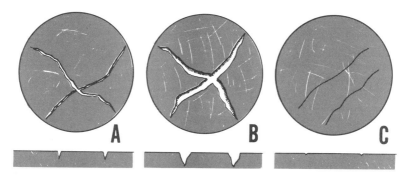

Fig. 8-13 The top width of scratches reduces the load-supporting area and thus reduces the wear life.

Generally, wear is proportional to surface finish. The larger the bearing area, the more evenly the load is distributed and the lower the pressure.   This was demonstrated in Fig. 7-27. A woman exerts a heel pressure many times greater than a man although she weighs less. The reason is the small heel area. Each minute scratch reduces the contact area and places a higher pressure on the remaining area, Fig. 8-13. Surface finish affects wear life for another reason. Irregularities trap abrasive dust which then wears away both surfaces when they are wrung.

Fine surface finish produces some problems.   For example, one manufacturer has developed a means to produce a surface finish so nearly perfect that ordinary, surface-finish measurement   instruments   are   unable   to measure it. They have repeatedly had blocks rejected because of scratches.   The actual scratches in some cases were less deep or wide than the normal finishing marks on some other blocks.   They looked bad only because the finish was so good.

SIZES AND SHAPES

There have been a variety of gage block sizes.   Two are in general use and covered by Federal Specification GGG-G-15. These are the rectangular and the square blocks. The square blocks are made both with and without a center hole. The popularity of the blocks is in that order. Both square and rectangular blocks are available in thicknesses from *ten mil* (0.0010 in.) to 20.00 in. in several series. Blocks are usually purchased in sets consisting of one or more series.

A series is a group of blocks all of one nominal size in which one digit changes consecutively through the series. For example, one of the *one-mil* (0.001-in.) series consists of blocks from 0.101 through 0.149 in. Another *one-mil* (0.001-in.) series consists of blocks from 0.021 through 0.029 in.

The exact number of blocks in a set varies among manufacturers. The following is typical: 121, 86, 56, 38 and 35. There are also sets to extend the range of these sets either larger or smaller.   Long block sets have blocks up to 20.000 in. in length. Thin block sets consist of blocks starting at *ten mil* (0.010 in.) and increasing by increments of *ten mil* (0.010 in.) to *90 mil* (0.090 in.).

Much of the choice between square and rectangular blocks is a matter of personal preference, although some very technical reasons are given. The square blocks have greater surface area. They last longer, are less likely to be toppled over because of their greater mass. And they are much more expensive than rectangular blocks.   There are efficient holders, end standards and accessories for both types.

The choice of set size is another matter. There are three considerations that must be balanced. The range is the first one and the only arbitrary one. You either need a given range or you don't and that's that. The second one is the number of combinations that requires more than three blocks to form.

There are usually several ways to form any given length within the set range. It is always desirable to use the one that requires the fewest blocks. Not only is it fastest but it eliminates the change for additional error. *A large set will make more lengths with fewer combined blocks than a small set.* Therefore, in some circumstances a 38-piece set might not be an economy even if it has sufficient range for most applications. The third consideration is the number of people who will use the set at the same time. The longer the number the larger the set should be to avoid delays.

GAGE BLOCK GRADES

The federal specification establishes three grades for gage blocks, AA, A and B. These stipulate minimum requirements. They are exceeded by some manufacturers. Blocks are graded in several categories in addition to

the length tolerance. These are shown in Fig. 8-14 and are self-explanatory.

Grade AAA blocks are not included in federal specification GGG-G-15 but are manufactured as grand master sets for top-level industrial calibration. They are accurate to within plus or minus *one mike* (0.000001 in.) from nominal size, flatness and parallelism. Blocks of this accuracy are obtained by selecting the best blocks from the production of grade AA blocks.

Grade AA blocks are frequently called *laboratory grade*. They are accurate to plus or minus *two mike* (0.000002 in.). They are reserved generally for use as laboratory working standards. This usually means that they will be used only for the calibration of other gage blocks. Rarely will they be used for other high-precision calibration.

TOLERANCES ON FLATNESS, PARALLELISM, AND SURFACE FINISH OF WRINGING SURFACES

Grade	Tolerances on flatness and parallelism for span of wringing surface				Maximum predominant average surface roughness	Maximum depth of individual scratches[3]	Permissible number of stray scratches per block[4]
	½ inch or less		Over ½ to 1¼ inches				
	Blocks to 1 inch in length	Blocks over 1 inch in length	Blocks to 1 inch in length	Blocks over 1 inch in length			
	Micro-inches	*Micro-inches*	*Micro-inches*	*Micro-inches*	*Micro-inches*	*Micro-inches*	*Micro-inches*
AA gage blocks	3	4	4	5	0.5	2	0
A gage blocks	4	5	4	6	0.8	3	4
B gage blocks	5	6	6	7	1.2	5	10
Attachments and accessories	5	6	6	7	1.0	4	10

TOLERANCES ON LENGTHS OF GAGE BLOCKS

Grade	One inch and less				Each additional inch			
	English		Metric		English		Metric	
	Plus	Minus	Plus	Minus	Plus	Minus	Plus	Minus
	Inch	*Inch*	*Mm.*	*Mm.*	*Inch*	*Inch*	*Mm.*	*Mm.*
AA	0.000002	0.000002	0.00005	0.00005	0.000002	0.000002	0.00005	0.00005
A	0.000006	0.000002	0.00015	0.00005	0.000006	0.000002	0.00015	0.00005
B	0.000010	0.000006	0.00025	0.00015	0.000010	0.000006	0.00025	0.00015

Fig. 8-14 These tables from federal specification GGG-G-15 establish the grades for gage blocks. They are exceeded by some manufacturers.

Grade A blocks are considered to be *inspection* grade and are frequently termed that. They are accurate to within plus six and minus *two mike* (+0.000006, -0.000002 in.). Grade B blocks are accurate to plus ten and minus six *mike* (+0.000010, -0.000006 in.).

All of these tolerances are based on one inch of length or any fraction smaller. For example, a 2-inch nominal length block in grade AA could vary ± 4 *mike* in size, twice the specified ± 2 *mike*. But a half-inch nominal block in the same grade would not have its tolerance reduced. It would remain ± 2 *mike*. The manufacturers of gage blocks have made substantial improvements beyond the minimum standards. This clearly shows the general awareness of the need for better standards of length.

BUILT-IN WEAR ALLOWANCE

From time to time gage blocks are marketed that claim to have a wonderful feature, a built-in wear allowance. It is achieved by shifting the tolerance entirely over to the plus side. The usual plus or minus *two mike* (0.000002 in.) tolerance of a AA grade block becomes plus *four mike* (0.000004 in.), minus nothing. Thus the expected wear only brings the block closer in to its basic size. Feel free to quote the author's retort to this -- "hogwash!"

Perhaps the proponents of this strategem think that you do not know better. Or is it possible that they do not? It is illogical for two reasons. It misses completely the *real meaning* of tolerance, and it effectively *doubles* the amount that the block can deviate from its intended size.

A dimension consists of the basic size and the tolerance. The tolerance is not part of the size. It is a rule about that size. A part feature 1.000, + 0.002, - 0.000 in. certainly does not mean that the part feature measures 1.002 in. If it does, watch out. It is on the border line beyond which the parts will be rejects. That is the function of tolerance. It is not size. It is size limit. Therefore a gage block of 1.000000, + 0.000004, - 0.000000 in. certainly should not be 1.000004 in.

Now let's see what it does when blocks are used. In most cases blocks are combined. With bilateral tolerances (plus or minus), the uncertainty in each block at least has a chance to cancel out. It may not but it could. With unilateral tolerances (all one way) uncertainty cannot cancel out and must accumulate.

If a manufacturer could consistently hold the blocks exactly at the high limit, he could also hold them exactly at the basic size. This would be utopia -- no tolerances would be needed. Of course, no one can. The tolerance gives the needed "leeway." The manufacturer choosing the all-plus tolerance can miss his mark by twice as much as usual and still not scrap the block. This is the real reason and has nothing to do with any altruistic desire to give you "built-in wear allowance."

This built-in wear allowance can be further "debunked." Of blocks returned to laboratories for recliabration only a very small number are rejected because of wear. One laboratory states 10%. Wringing of *clean* blocks results in negligible wear. Setting of comparators causes some wear. Most defective blocks are rejected because of nicks, scratches and corrosion.

Refering to Fig. 8-14 you might wonder why gage block tolerances are not split equally plus and minus except for AA grade. Tolerances reflect practical considerations as well as theoretical ones. They must reflect manufacturing processes as well as product intentions. They not only may be adjusted, but should be. You must assure yourself that it has been done honestly. Even in metrology . . . "false prophets will arise and show great signs and wonders."

CALIBRATION OF GAGE BLOCKS

The fact that gage blocks are used for calibration of other instruments does not excuse them from calibration. Quite the reverse. If a setting standard wears out of calibration, it may add uncertainty to the readings of several gages that it sets. This is serious enough and will show up in increased scrap losses, assembly costs, and field servicing. But if the master set of gage blocks wears out of calibration, the entire plant's performance may suffer. This is not a theoretical or academic matter. It does happen, and at times the costs total millions of dollars. Cases cannot be cited, of course, because stockholders and customers both take a very dim view of such high placed carelessness.

All precision instruments should be on a systematic program of periodic inspection. The length of the period will depend upon the amount and severity of the instrument use. Calibration is expensive and unnecessary calibration should be avoided. Quality control people use statistical methods to set up calibration programs so that a desired level of assurance is maintained without excessive cost. Federal specifications require at least annual calibration for grade AA sets, quarterly for grade A sets and semiannually for grade B sets.

Progressive manufacturers and research laboratories use gage blocks directly for measurement. These sets should receive calibration no less frequently than required by the federal specifications. They should be sent into the metrology laboratory whenever any of the following conditions exist:

1. Visual inspection discloses abnormal wear based on past experience.

2. Wringing becomes difficult.

3. An increase in rejects casts suspicion on the measurement equipment.

A plant using many sets of gage blocks usually will calibrate its working blocks against its own master sets. Conservative practice requires at least two sets of master blocks, so that one is always on hand when the other is away being calibrated by a metrology laboratory. Smaller users send their working sets out for calibration.

While the working blocks may be safely calibrated within the using organization, master blocks must be calibrated against *grand masters* that are directly traceable to the national standard of length. This can be performed by the National Bureau of Standards, by independent metrology laboratories and by the manufacturers of gage blocks.

This calibration is not a proper function of the National Bureau of Standards. By law that department is required to maintain our national standards. As a necessary service the bureau has calibrated gage blocks for many years. The other two sources are preferred, leaving the bureau free for far more important work.

All manufacturers of gage blocks provide calibration service, not only for their own blocks but for competitive blocks. When a manufacturer calibrates sets of blocks he usually is expected to replace those worn out of the tolerance for the grade of the set. Some people feel that a manufacturer will be more severe with competitive blocks than those of his own manufacture. While this seems logical, the author knows of no verified instances of it.

Independent metrology laboratories are relatively new and unfamiliar in this country. They are independent laboratories that sell nothing other than calibration service. Their calibrations are, therefore, above suspicion. While gage block calibration is only one of their functions, it promises to become an increasingly important one.

GAGE BLOCK CALIBRATION

What is it? Verification that the block is within the tolerance specified.

How is it done? By comparison with master standards of known calibration and known traceability to the national length standards.

When is it done?
1. Periodically according to a systematic program based on statistical quality control.
2. Not less often than annually for grade AA, quarterly for grade A, and semiannually for grade B.
3. When required by visual inspection, wringability decline or slipping quality control level.

Who calibrates working sets? The user's gage department or any of the facilities for calibrating master sets.

Who calibrates master sets?
1. Independent metrology laboratories
2. Manufacturers of gage blocks
3. The Bureau of Standards

What is done with worn blocks? They are replaced.

What is done with worn sets? They are regraded for a lower level of precision.

Fig. 8-15 Considering the critical use of gage blocks, their calibration is of crucial importance.

Calibration does more than indicate whether the blocks are in or out of tolerance. The deviation from the nominal lengths is given. For the highest level of measurement, the results are adjusted for this deviation.

Many plants progressively replace gage block sets. The new sets purchased are of the highest grade required for their needs. When a set is worn past that grade it is moved down to a lower precision level, say inspection. When the periodic calibration shows that it is no longer suitable for that level, it is moved down to the next one, probably for measurement at the machine or on the as-sembly floor, in this example. This practice is recommended providing all sets are under control and providing that a suitable means of identification is used.

The only permissible way identification may be added to blocks is by color dot or band on the side surfaces. Never etch or in any way deface the contact surfaces.

Calibration considerations are summed up in Fig. 8-15. It cannot be emphasized enough that each situation is different, but that modern statistical quality control can develop a suitable plan for each situation.

TERMINOLOGY

in millionths (or in mikes)	Measurements carried to six decimal places of precision.
to millionths (or to mikes)	Measurements whose accuracy is known as a specific amount of error in one million parts.
gage blocks	The mass produced end standards that combine arithmetically to form usable length combinations.
the standard	The ultimate physical embodiment of the unit of length.
a standard	A copy traceable to the standard.
interferometry	The use of interference phenomena of light waves for measurement.
arithmetical progression	A series in which the elements change by a constant difference.
stability	Inherent ability of a material to retain its size over a period of time.
wear blocks	Carbide blocks used on the ends of gage block combinations to minimize wear.
macrogeometry	Determination of shape to the precision of standard inspection instruments.
microgeometry	Minute analysis of shape to a precision in mike increments, including the evaluation of surface topography.
surface topography	The combined characteristics that define a surface such as waviness and surface finish.
AA (arithmetical average)	The mean between the highs and lows.
RMS (root mean square)	The average of the squares of the deviation of the highs and lows.
reflectivity	Mirror-like quality having little relation to surface finish or flatness.
gage block grades	The accuracy grades as set forth by government specifications and expanded by the manufacturer's specifications.
AAA grade	A manufacturer's standard for master calibration blocks.
AA grade	Government standard for laboratory grade blocks.
A + grade	A manufacturer's standard for improved A grade blocks.
A grade	Government standard for inspection grade blocks.
B grade	Government standard for working grade blocks. No longer made by some manufacturers.
metrology laboratory	A laboratory for the calibration of standards. May be a department within a company or an outside service.

Fig. 8-16 Of these terms, an understanding of the first two is of paramount importance.

Summary — Slips and Gauges

● No matter how much we owe to the British metrologists they certainly do not speak good English -- at least not our kind of English. They consider us rustics because we drop the unnecessary "u" in gauges. Yet they have permitted the colloquial "slips" or "slip blocks" to become the proper name for gage blocks. "Whilst" on this point, their contribution to metrology has been great. This is shown by the number of references in this text to British sources.

● Summing up this chapter, never assume that a measurement to millionths means anything more than that there are six places to the right of the decimal point. Unless based on standards the accuracy is doubtful. Gage blocks provide the most practical standard for linear measurement. They are readily available. A small number forms thousands of dimensions. The errors involved with them are known and therefore predictable. Well organized channels verify their accuracy. The size of a gage block set required depends not only on the smallest and largest dimensions to be formed, but also on the average number of blocks required to form combinations. A large set will require fewer, thereby saving time and reducing the wringing errors. The material and shape selected will depend upon both economic and metrological factors. The cost per use will be far lower for a set of hard but expensive material -- if extensively used. Cost per use will soar if expensive blocks are rarely used. The probability of wear will be reduced for any given number of combinations. This decision must be tempered by environmental conditions. Abrasive dust calls for the hardest blocks while temperature expansion problems may recommend the lower priced steel blocks. While all gage blocks adhere to federal specifications, caution must be applied to their purchase.

QUESTIONS

1. Why are gage blocks referred to as end standards?
2. Why are gage blocks extremely accurate?
3. Who is usually credited as being responsible for the introduction of interchangeable manufacture?
4. Why did interchangeable manufacturing hasten the need for gage blocks?
5. What was C. E. Johansson's most important contribution to the development of gage blocks?
6. What happens when work is performed that requires a subdivision of the standard which is smaller than the error existing in the standard?
7. How were the existing errors in standards reduced?
8. Are gage blocks exactly the size they are marked?
9. What has The National Bureau of Standards to do with gage blocks?
10. Why should gage blocks be calibrated periodically?
11. How are gage blocks calibrated?
12. When should worn gage blocks be replaced?
13. What are the principle considerations when selecting a gage block set?

DISCUSSION TOPICS

1. Explain the functional characteristics that are wanted in gage blocks.
2. Explain the specifications for gage blocks that are required by the federal government.
3. Explain the important considerations relative to gage block calibration.

9

Use of Gage Blocks
Putting Millionths to Work

Why use gage blocks? The most common reason is to achieve greater precision than is obtainable with other instruments. Of course, in this situation they should be used. But this situation is rare compared to the tremendous scope of opportunities for gage blocks in practical measurement.

There is no reason to use gage blocks for measurements that can be made with *sufficient reliability* and *less expensively* by other methods. Generally, the need for gage blocks increases when the (1) precision *increases*, (2) length of part *increases*, (3) importance of value of the part *increases*, and (4) skill of the personnel *decreases*.

The last point is not an error. It is one of the strongest arguments for a greater understanding and use of gage blocks. It takes a highly skilled person to use gage blocks to their fullest capacity, achieving millionths precision combined with millionths accuracy. But a less skilled person can measure to much greater reliability with gage blocks than by any other method.

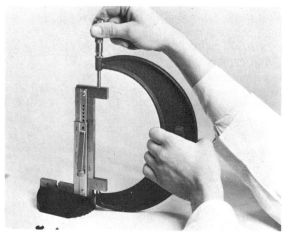

Fig. 9-1 Gage blocks provide the accepted method for calibration of measurement instruments. When calibrating an instrument, such as a micrometer, ask this question: "Could the measurement be made directly to gage blocks instead of through this middleman?"

As paradoxical as this seems at first, it is easy to understand. Three things explain it. First of all, the difficulty in measurement is many times greater at the upper end where *mike* measurements (0.000001 in.) are involved than at the *mil* (0.001 in.) end. Second, gage blocks measure through the intervention of comparators of one kind or another, and these instruments eliminate the problem of feel. Third, and most important from a metrology standpoint, gage blocks have a built-in cushion of accuracy. For measurement in *mil* increments (0.001 in.) there is so little error carried into the measurement from the gage blocks that the measurer has a margin of safety for his own lack of skill.

Now, let us examine where gage blocks should and should not be used. At the lowest level of precision, gage blocks seldom substitute economically for rules. They are not even required for calibration. Buy good rules in the beginning and pass them on to Junior when the ends wear.

At the next level, vernier instruments, gage blocks are preferred under some circumstances even if the tolerance is within the discrimination of the instrument. They are preferred to vernier calipers for parts over approximately 5 in. in length, and to vernier height gages for parts over 10 in. Gage block measurement is preferred to vernier measurement whenever the measurer is less than highly skilled.

The actual sizes mentioned in this discussion are the author's. They make wonderful arguments. Your conservative friends will say they should be smaller, the liberals will suggest larger ones. The Q.C. (quality control) people will say that each case must be decided separately based on consideration of the variables. Q.C. people can be so logical.

Micrometers have the same discrimination as vernier instruments, but have higher reliability. Remember Chapter 7? Therefore, the points at which gage blocks are preferred are the same — but it takes less skill to produce comparable results. Micrometers are reasonably safe instruments to use for measurements no finer than *one-mil* (0.001-in.) discrimination up to about 5 in. Beyond that, even at this low precision, gage block methods may be preferred. Certainly for any measurement over 10 in., gage block methods should be used. Of course, practical considerations must influence the final decision.

These remarks about both micrometers and verniers are based on one extremely important assumption. The instrument must be calibrated by gage blocks, Fig. 9-1.

For measurements to *tenth-mil* (0.0001-in.) precision, vernier instruments cannot be considered. One- and two-inch micrometers can. Convenience may advocate their use — providing you have a margin of safety. The margin of safety is the width of the tolerance. At

plus or minus *three-tenth mil* (0.0003 in.), little problem will be encountered. Even at plus or minus *five tenth mil* (0.0005 in.), you will have to be very careful. At any closer than that, the care you will have to use with a vernier micrometer will usually take more time than making a gage block setup.

If you want to measure to *tenth-mil* (0.0001 in.) precision larger than two inches, your best choice generally will be a gage block comparison method. Today, a large part of the measurements required both by industry and science is in that category. What do you need to know to begin? First, you need to know the rudiments of gage block care. You need to know how to figure block combinations. Then you need to know how to wring blocks reliably. And finally you need to know how to use the packaged millionths you have at hand after you have wrung together the needed stack of blocks. Some generalities about the use of gage blocks are shown in Fig. 9-2. Figure 9-3 explains the terms you are about to meet.

THE PLACE FOR GAGE BLOCK MEASUREMENT

Consider Gage Blocks when:

1. Precision increases
2. Length increases
3. Importance of reliability increases
4. Skill of measurer decreases

The General Rule: Use gage blocks for every measurement unless adequate reliability can be more economically obtained by another method.

Rules by precision required: (alternatives listed in order of preference). For 0.002 in. or finer, use only high amplification comparison instruments set to gage blocks. Check the gage blocks with gage blocks.

For 0.001 to 0.0002 in.:

1. Use only high amplification instruments set to gage blocks for all lengths.
2. Over 2 inch, use only comparison instruments set to gage blocks.
3. Under 2 inch, use vernier micrometer checked with gage blocks.

For 0.002 to 0.001 in.:

1. Use comparison instruments (such as dial indicators), set to gage blocks for all lengths.
2. Use micrometers to approximately 2 in. Check with gage blocks.

For 0.005 to 0.002 in.:

1. Use comparison instruments (such as dial indicators), set to gage blocks for all lengths.
2. Use micrometer instruments to approximately 5 in. Check with gage blocks.
3. Use vernier instruments to approximately 5 in. Check with gage blocks.

For 64th to 0.005 in.:

Use micrometers or vernier instruments. Check with gage blocks.

For precision coarser than 1/64 in. gage blocks are rarely preferred.

Fig. 9-2   Gage blocks are needed at many levels. These rules are based on average measurements, equipment and skill. Availability of a measuring microscope or an interferometer obviously alters them and they cannot be applied to extremely large or small measurements.

GAGE BLOCK NOMENCLATURE

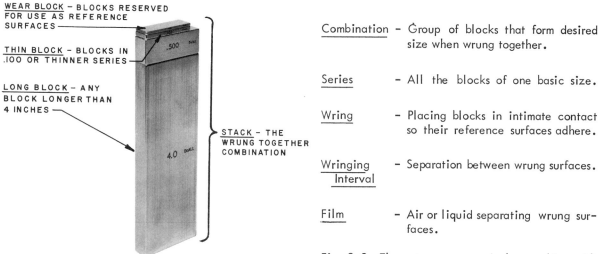

WEAR BLOCK – BLOCKS RESERVED FOR USE AS REFERENCE SURFACES

THIN BLOCK – BLOCKS IN .100 OR THINNER SERIES

LONG BLOCK – ANY BLOCK LONGER THAN 4 INCHES

STACK – THE WRUNG TOGETHER COMBINATION

Combination – Group of blocks that form desired size when wrung together.

Series – All the blocks of one basic size.

Wring – Placing blocks in intimate contact so their reference surfaces adhere.

Wringing Interval – Separation between wrung surfaces.

Film – Air or liquid separating wrung surfaces.

Fig. 9-3   These terms are used when working with gage blocks.

CARE OF GAGE BLOCKS

A few rules will extend the life of gage blocks, reduce time required to wring combinations, and most important, enhance the reliability of gage block measurement. For easy reference they are shown in Fig. 9-4.

Point two is of utmost importance, but a word of warning is needed. The evaporation of solvent cools blocks severely. This causes them to shrink. If you are working to *50 mike*

(0.00005 in.) or finer, try to keep the blocks sufficiently clean all the time, so that only a small amount of solvent will be required for cleaning before use. ·Do not immerse in solvent, keep aerosol cleaning sprays to a minimum, and do not use compressed air to remove excessive solvent. To clean, simply apply solvent and quickly wipe dry with a lint-free tissue. To apply solvent you may dip, squirt or spray. The important thing is to get

---

RULES FOR GAGE BLOCK CARE

1. Never attempt to wring or otherwise use gage blocks that have been in contact with chips, dust- or dirt-laden cutting fluids.

2. Before using, clean blocks with a high-grade solvent or commercial gage block cleaner. Wipe dry with a lint-free tissue.

3. Do not allow blocks to remain wrung together for long periods. Separate daily.

4. When not in use, place blocks in a safe place where they will not be damaged, preferably in their case.

5. Before putting blocks away, clean the blocks and cover with a noncorrosive oil, or grease or commercial preservative.

6. Be on constant guard for burrs. If anything has been placed on a block or if it does not wring readily, use a conditioning stone immediately.

7. Thoroughly clean gage block case periodically.

Fig. 9-4   Obeying these rules will improve reliability, speed measurement and lower the cost of measurement.

BLACK GRANITE CONDITIONING STONE

CERAMIC CONDITIONING STONE

Fig. 9-5 The conditioning stone is one of your best allies to assure reliable measurement and long life from gage blocks. The blocks are firmly wrung along its smooth, crystalline surface to remove burrs and nicks without abrading the surfaces.

it off rapidly to minimize cooling. Repeat as often as necessary to remove any visual signs of dirt, stains, or oil. In the United States most authorities recommend the use of disposable tissues. British authorities prefer chamois, however. *

Point three is a good precaution for all blocks. The more nearly perfect the gage block surface, the greater the chance for damage from extended contact. This problem is greater for steel blocks than for the plated, carbide, or stainless steel types.

Point six is extremely important. Burrs so minute that they cannot be seen unaided can seriously damage the fine surfaces of gage blocks and of the precision instruments they contact. Whenever anything hard has been placed accidentally on a gage block, suspect that burrs may have been formed. Whenever a block resists wringing, investigate. Check first for burrs. If that is not the reason, the block probably is worn badly and should be calibrated.

In Chapter 14 you learn that burrs can be detected easily with optical flats. The use of a conditioning stone, Fig. 9-5, is so quick and easy that it is usually quicker to use it than look for the burr. The block is rubbed firmly on the stone until it begins to wring. The crystalline structure of the stone is smooth and nonabrasive but shears away any burrs or other minute particles adhering to the block surface.

* Hume, K. J., Sharp, G. H., Practical Metrology, London, Macdonald & Co., Ltd., 1958, Vol. 1, p. 17.

All blocks are subject to corrosion but differ considerably in their resistance to it, ordinary steel blocks being most seriously affected. Human beings differ considerably in the corrosion they cause. Some people can virtually destroy a block with a single handling unless it is promptly cleaned afterwards. Others have very little effect upon them. If proper cleanliness is observed, there will be no corrosion problem for any type of gage blocks – or from the most corrosive humans.

WEAR BLOCKS

The use of wear blocks will increase the life of a set. Wear blocks are special blocks reserved for use as reference surfaces at the ends of gage block stacks. They are recommended when blocks are used for direct comparison. They are not needed when gage block holders are used because these provide end standards for the reference surfaces.

By means of wear blocks the total wear that ordinarily would be distributed over numerous blocks is concentrated on two. The useful life of the set is extended. The wear blocks are economically replaced when worn. Always use the same face of a wear block as the reference surface. If they are etched on an end surface, a good practice is to always have that surface showing.

Wear blocks are available in both steel and carbide. The carbide blocks are recommended although more expensive because of their longer life. The difference in coefficient of expansion is not a serious problem because the wear blocks are thin.

The admonition about the damage that can be caused by burrs with carbide blocks should be remembered. It is good practice to pass them over a conditioning block for nearly every use.

WRINGING — THE PRIVILEGE OF PERFECTION

If two sufficiently flat and smooth surfaces are brought intimately together, they will adhere, Fig. 9-6. This remarkable phenomenon is called *wringing* and is important in the use of gage blocks. It is such a convenience to be able to handle a stack of blocks as one unit, that the most important benefit is overlooked.

Whenever any two surfaces are brought together, some space must exist between them. For most parts the peaks in surface finish will probably hold them apart, Fig. 9-7. For better finished surfaces, oil or grease films will occupy the space between. If this is cleaned away a minute air gap will still separate the surfaces. When the surfaces are nearly perfect, the air gap becomes so fine that it acts much the same as a liquid film, hence its name, *air film*. This film may be forced almost completely out but some separation always remains. This separation is called the *wringing interval*. If it were missing, instead of two parts there would be only one, a solid piece.

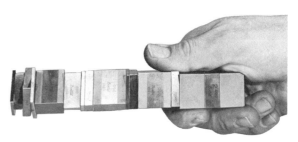

(DoALL Co.)

Fig. 9-6 These blocks show how tenaciously finely finished surfaces will adhere to each other.

The wringing interval can be reduced to a fraction of a *mike* (0.000001 in.), although this is rarely achieved. Even under ordinary measurement conditions, it can be reduced to one or two *mike* (0.000001 to 0.000002 in.). If you learn to wring blocks properly, the wringing interval can be ignored for most practical measurement. *Therefore, a stack of gage blocks is assumed to have a length equal to the sum of the individual block lengths*, as shown in Fig. 9-8.

Why do surfaces adhere? Occasionally one hears that it is magnetic attraction. Demonstrably it is not, at least not in the conventional sense. Frequently the reason given is atmospheric pressure. This certainly appears logical, but, as Fig. 9-9 shows, logical or not it is incorrect.

SURFACE SEPARATION

POOR FINISH	WAVINESS	LIQUID FILM	AIR FILM	WRUNG

Metal to metal contact between the high points holds surfaces apart. Even heavy grease will not eliminate friction if surfaces are very rough.	Metal to metal contact may be prevented by lubrication or air film for wavy surfaces.	Oil or grease may hold the surfaces apart by several thousandths, but they can be forced together to within a few millionths.	Air film will separate the surfaces until squeezed out. The last few millionths are difficult to remove.	Although apparently metal to metal contact some film remains.

Fig. 9-7 Air, oil, grease and surface irregularities hold surfaces apart.

Fig. 9-8 Three stacks of varying numbers of blocks, all of which total the same length, have been wrung to a toolmaker's flat. Other blocks have been wrung across the three stacks so that the final block supports them all. This shows that the lengths of stacks are self checking.

The reason for adherence is still not completely understood but is generally agreed to involve molecular attraction. An over simplified explanation is that in the final lapping some portions of some molecules are knocked off. This leaves an unsatisfied condition. They are always seeking replacements. When brought close to another surface in the same condition, the unsatisfied molecules on each side mistake their equals on the other side as their missing parts and clamp onto them.

There also has been much controversy over the thickness of the wringing interval. Some investigators have claimed that it is negative; that is, actually intertwined so that the whole is less than the sum of the parts. In 1921 two Frenchmen stated that it was a negative 2.4 *mike* (0.0000024 in.). The following year the National Bureau of Standards in the U.S. reported that it varied between 1 and 2.8 *mike* and was positive. In 1927 two Englishmen announced that it was about 0.2 *mike* (0.0000002 in.) positive. The most recent report was in 1956 from the National Standards Laboratory of Australia. They concluded that the separation between two wrung surfaces was not consistently positive or negative; but that it was very small, never more than 0.4 *mike*. Present work at The Bureau in Washington indicates that it is even smaller. Further experimentation is needed.

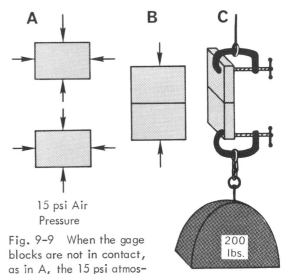

15 psi Air
Pressure

Fig. 9-9 When the gage blocks are not in contact, as in A, the 15 psi atmospheric pressure acts on all sides and balances. When placed together in B, the effect on sides balance but the pressure is now closed off between the blocks. This leaves air pressure on both ends holding the blocks together. But C shows the blocks can resist 200 pounds pull. Based on 15 psi atmospheric pressure and a block area of 0.525 inch, the force holding them together would be 7.7 pounds – a far cry from 200 pounds. *(Pratt & Whitney Co., Inc.)*

In an experiment at The Bureau a thick application of oil was used, almost "gluing" two blocks together. More and more pressure thinned the film until wringing just began to occur. The separation at this point was found to be about 0.7 *mike* more than the final wrung separation. If mechanical manipulation was stopped after the wringing began, the surfaces would pull together by themselves. This would require from 20 min. to 2 hrs. depending upon the film material.

The better the surface topograpy (finish and flatness) of blocks, the easier they are to wring together, and the smaller is the wringing interval. However, after a certain point further improvement in finish does not increase the uniformity of the wringing interval. In a series of tests at the National Bureau of Standards the wringing interval repeated to less than *two mike* (0.000002 in.) 90% of the time.

The main cause of gage block wear is wringing of dirty blocks. When thoroughly cleaned, proper wringing produces negligible wear. In a series of tests performed at The Bureau, two stainless steel gage blocks were wrung 100 times without any measurable wear.

WRINGING RECTANGULAR BLOCKS

A Be sure gaging surfaces are clean.

B To start – overlap gaging surfaces about 1-8 inch.

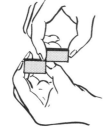

D Blocks will now adhere.

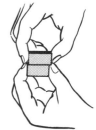

C While pressing blocks lightly together, slip one over the other.

E Slip blocks smoothly until gaging surfaces are fully mated.

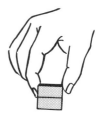

Fig. 9-10 These steps need to be practiced to develop the feel for wringing.

TECHNIQUES FOR WRINGING GAGE BLOCKS

When you wring gage blocks, you are controlling accuracy to within *mike* increments (millionths of an inch). When proficient at wringing, the reliability of all of your measurement with gage blocks is improved. Once the technique has been learned it is no harder to wring well than it is to wring poorly.

The steps for wringing are shown in Fig. 9-10. Note that the first one is to clean the blocks. That needs to be done *immediately* before wringing. If there is a delay of even a minute, dust will settle on them. A camel's hair brush may be used to remove the dust.

Many times a user of gage blocks will clean them carefully and then methodically and deliberately dirty them before wringing. I am referring to the old idea that an application of "palm oil" or "nose grease" was needed to assure wringing. As blocks wear they become more difficult to wring unless an oil film is furnished. It is quite permissible to apply a small amount of oil providing that it is noncorrosive and does not carry microscopic dust particles onto the blocks, as "palm oil" will.   A dropper is used often. The oil spreads itself during the wringing.

The instructions for wringing minimize the effect of dust. The action between steps B and C should push away dust from the contact surfaces rather than trap it between. The more the blocks are worn, the firmer the blocks must be pushed together. One pass is not always sufficient to start the wring. In that case move back from C to B, but not all the way off, and begin over again.

When resistance is felt the wringing action has begun. As you continue to slide the blocks, you will feel it get stronger. The fact that the blocks adhere does not mean that you have achieved a satisfactory wring. It should be continued until further movement does not increase the adhesion. Caution is usually given not to use a circular action because this might cause serious wear or even damage from abrasive dust trapped between the surfaces. This is a sound precaution until the blocks are wrung. After that it may be more convenient for you to wring the blocks the rest of the way by moving one through a small arc. *Remember that blocks are not fully wrung just because they adhere.*

The general steps for wringing square blocks are similar to those for rectangular ones. They are shown in Fig. 9-11. Consistency is important for uniform wringing. That may be the reason for the adoption of a specific procedure by the National Bureau of Standards, Fig. 9-12.

WRINGING SQUARE BLOCKS

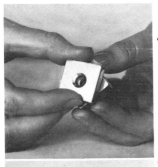

1 Place the blocks together as shown. This minimizes the danger of the corner of one block scratching the other.

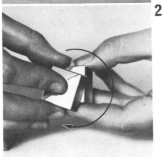

2 Slide the upper block over the lower with a slight circular motion. If they are clean, they will begin to take hold rapidly. Foreign matter can be detected readily and should be removed before it damages either block. Some authorities object to the circular motion.

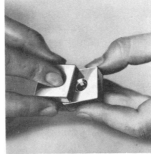

3 Slide the upper block half out of engagement using light but firm pressure.

4 Slide the upper block back into full engagement under light pressure, at which point they should be wrung together ready for use. Worn or dirty blocks will not wring.

Fig. 9-11 Square blocks are wrung in much the same manner as rectangular blocks. *(Pratt & Whitney Co., Inc.)*

Deposit a few drops of alcohol on the flat and wipe dry with one of the cloths.

Repeat this procedure on one of the gaging surfaces of each block.

Deposit a drop or two of kerosene on the flat.

Using the remaining cloth, wipe the surface in increasingly larger circles until there is a thin, evenly distributed yellow cast remaining.

Quickly wring the previously cleaned surface of either block to that flat. Using light pressure, slide the block onto the flat. By slight rotation, it should wring firmly.

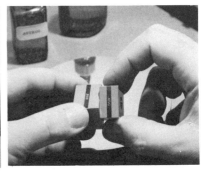

Remove the block promptly and, again with light pressure, slide it onto the previously cleaned surface of the remaining block. With a circular motion, slide it half out of engagement and back into a mating position.

Fig. 9-12 This is the method approved and used by the National Bureau of Standards. Required are 2 lint-free cotton wipers, pure grain alcohol, filtered kerosene, and a toolmaker's flat. Practice will dictate the slight amount of kerosene required. *(Quality Assurance, Hitchcock Pub. Co.)*

How can you tell that you have wrung gage blocks together properly? Generally speaking, with a little practice, you can repeat reliably the wringing interval to *one mike* (0.000001 in.). That is sufficiently small so that you need not worry about the uncertainty added to your measurements as a result of wringing interval. But until you are wringing reliably, you may introduce large errors. For example, in a stack of five blocks each interval if poorly wrung may be 10 *mike* or more. That is a total of *40-mike* (0.00004 in.) total interval. (No error. There is always one less wring than the number of gage blocks involved.)

While learning you can periodically check your progress with a high-amplification comparator. You would do this only between several practice wrings because it requires time that would be better spent practicing than checking.

To check your wringing, take two blocks, and when they just begin to wring, place them on a *heat sink* to lose the heat they have picked up from your hands. The smaller the blocks, the less time they will require to *normalize*. If you are using blocks in the 0.100 series, 15 min. should be adequate. Place them on the anvil of a high-amplification comparator and zero the instrument. Now remove the blocks and continue until you consider them

adequately wrung. Again normalize, perhaps for a longer time if you handled them extensively. Replace them in the comparator and note the minus reading. That is the reduction in wringing interval. If it is as much as *six mike*, you are probably as far as you can go. Now remove, wring some more, normalize again, recheck. If there has been no measurable change, your wringing was adequate.

There is one catch to all of this. Measurement to *mike* precision (millionths of an inch) involves many subtle variables which could produce as large a reading as the wringing interval. These are discussed in Chapter 13. Until you have considerable experience, it would be best to have a skilled technician perform the measurements to check your wringing interval.

Now comes the real test of your wringing ability — and there is no use in making it until you have had plenty of practice. Select six blocks that total the whole size of one large gage block, for example 2.0000, 1.0000, 0.4000, 0.3000, 0.2000, and 0.1000 in. Wring these blocks, normalize together with a 4.0000-in. block and carefully measure both the stack and the single block. Dividing the difference by five will give you a clue to your ability, Fig. 9-13. This will not be a true measure of your average wringing, because the sample is too small but it will guide your continued efforts.

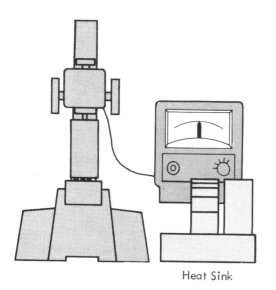

Heat Sink

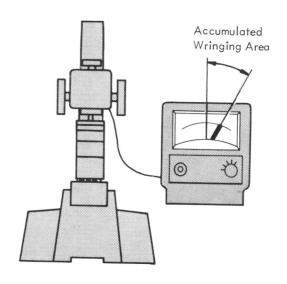

Accumulated Wringing Area

Fig. 9-13 Patience and a high amplification comparator are all that is needed to test your wringing ability. The stack of several wrung blocks is normalized with a single block of the same nominal length. The comparator is zeroed on the single block. That block is replaced with the stack. The plus reading will be your total accumulated wringing interval.

COMBINING GAGE BLOCKS

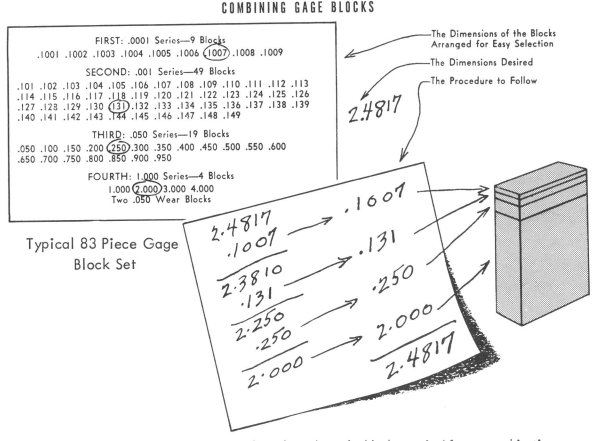

Fig. 9-14   Elimination of digits from the right, selects the blocks required for any combination.

COMBINING FOR A GIVEN DIMENSION

As you have observed by now, gage blocks are produced in several series. The 0.0001-in. series, for example, is made up of nine blocks each different in size by *one-tenth mil* (0.0001 in.). This is the system devised by Johansson that makes possible the great number of dimensional combinations. However, when figuring the blocks required to make up a given dimension, a procedure must be followed to save time, to reduce the chance of error, to use the least number of blocks, and thereby to leave the maximum number for duplicate combinations.

When using this system, the first and principal concern is with the figures to the right of the decimal point. (With the exception of the fractional series, gage block dimensions are designated in decimals.) The first step is to eliminate the last figure, and proceed to eliminate figures from right to left. Here is a practical application.

How can it be set up with the least number of blocks? Before reading further, examine Fig. 9-14. Sizes of blocks constituting a typical gage block set are tabulated in an order corresponding with their arrangement in the box. Always work with the smallest increment first (from the top down).

Assume a dimension of 2.4817 in. The first step is to eliminate the 7 by choosing block designated 0.1007. When this figure is subtracted from the dimension, the result will be as shown in Fig. 9-14. Next eliminate the 1 with block designated 0.131, then the 5 and 2 with the 0.250 block. Finally, eliminate the figure to the left of the decimal with the 2.000 block. The addition of the sizes of various blocks selected will prove the accuracy of selection. And, when wrung together, the unit dimension will be 2.4817 in. It is just that simple as a trial will show.

ALTERNATE COMBINATIONS

Finding a second combination choice:		
Dimension sought	2.4817"	Proof
To eliminate the 7, add .1004 and .1003	0.2007	0.2007
Result	2.281	
To eliminate the 1	0.141	0.141
Result	2.140	
To eliminate the 140	0.140	0.140
Result	2.000	
Adding combinations to 2" (1", .600" and .400")	2.000	2.000
Result		2.4817"
Combination with wear blocks		
Dimension sought	5.6879"	Proof
Eliminate the wear blocks (.050 + .050)	0.100	0.100"
Result	5.5879	
Eliminate the 9 at the right	0.1009	0.1009
Result	5.4870	
Eliminate the 7	0.147	0.147
Result	5.340	
Eliminate the 4	0.140	0.140
Result	5.200	
Eliminate the 2	0.200	0.200
Result	5.000	
Add 1.000 and 4.000	5.000	5.000
Result		5.6879"

Fig. 9-15 Standard block sets provide great versatility in forming combinations.

Now suppose a duplicate combination is desired. Since the only block in the 0.0001-in. series with a size designation ending with 7 is in use, a combination totaling 7 is chosen, 0.1003 and 0.1004, or 0.1005 and 0.1002. This leaves a 1 to be eliminated. Since block 0.131 is in use, the 0.141 or 0.121 block is selected, preferably block 0.141. Now select the 0.140 block to eliminate the 4 and the 1. Then blocks that total 2 in. complete the combination. This is the top example in Fig. 9-15.

Excessive wear on commonly used blocks can be avoided by this method. It often is possible to select a combination that will substitute less-frequently used blocks for the more popular ones. For example, to make up a 1.200-in. stack, 1.000- and 0.200-in. blocks would usually be selected. Instead, 0.900- and 0.300-in. blocks or 0.800- and 0.400-in. blocks could be selected.

The bottom example is a case in which wear blocks are used. Note that their sizes are fixed. You have no choice, so you must remove the digits for them at the beginning.

DIMENSIONS THAT CANNOT BE COMBINED DIRECTLY

It is possible to set up dimensions of less than 0.100 in. The procedure is simple. Suppose, for instance, that a part with a nominal dimension of 0.0236 in. is to be gaged. In cases such as this, it is necessary to use a combination of blocks that will total a dimension that is greater than the dimension designated. Example: 0.1006 + 0.103 + 0.120 in. = 0.3236 in. Hence, 0.300 in. has to be subtracted to leave the desired dimension of 0.0236 in. This is done in one of two manners.

A. When setting a comparator, master micrometer, dial indicator, etc., the complete stack of blocks is used (0.1006, 0.103, 0.120 in.). Then a block or blocks of the dimension subtracted (in this case a 0.300 in. block) is wrung to the anvil or base. This serves as a reference plane from which the nominal dimension of a part is measured. An example is Fig. 9-16.

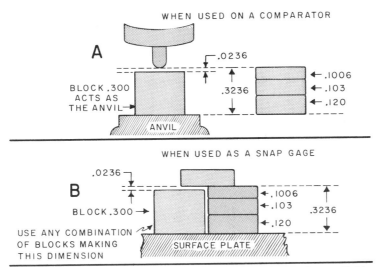

WHEN USED ON A COMPARATOR

A

BLOCK .300
ACTS AS
THE ANVIL

ANVIL

.0236

.3236

← .1006
← .103
← .120

WHEN USED AS A SNAP GAGE

B

.0236

BLOCK .300 →

USE ANY COMBINATION
OF BLOCKS MAKING
THIS DIMENSION

SURFACE PLATE

← .1006
← .103
← .120

.3236

Fig. 9-16 Dimensions too small to be combined with the available
set may be formed as the difference between two stacks of blocks.

B. For a snap gage. a block or blocks
which equal the subtracted dimension is used
as an anvil, example B in the figure. In this
case the blocks should be wrung to a master
flat or surface plate in order to establish a
uniform reference plane, and the overhanging
block should be wrung in position as shown.
When this is done parts may be gaged by feel,
Fig. 9-17. Should a part be oversize, the
wring of the overhanging block will be broken.

COMBINING TO FIND A DIMENSION

The previous discussion showed how to combine gage blocks to make a stack of specified
length. It is included in nearly every domes-
tically published book in the author's library
that even touches on gage blocks. It appears in
no familiar British metrology books. However,
a consideration does appear in them that generally is missing from United States books.
That is the combining of gage blocks *to find
an unknown dimension.* That is as near to direct
measurement as you can get with gage blocks.
The glib explanation is that we have more comparators and therefore, do not have to rely on
feel to transfer measurements from part features to gage blocks. To debunk this, the
example chosen is a common one that cannot
easily be performed with an ordinary indicator
or comparator.

Consider the measurement of a slot approximately one-half inch wide. This could

Fig. 9-17 A feeler gage, which is thinner than
any gage block, is calibrated by the difference
between two stacks. *(DoALL Co.)*

be the opening of a snap gage or an opening on
a part. First, use a scale to find the approximate dimension. The usual practice is to then
build up a stack of blocks and try it in the opening. It will be too large or too small. *It will
not be the correct size – so do not force it.*
At this point haphazard changing of the stack
will waste time and rub thin the patience of
everyone involved.

The technique to speed selection is the
same *bridging* that you will meet soon in the
chapter on calibration. This consists of choosing
successive values on both sides of the desired
unknown value. At each selection, the spread
is reduced, until you are *zeroed in.* To do this
with gage blocks, you must use in the trial stack,
blocks from more than one series so that you
can make orderly changes either of large or
small increments.

The way to begin is to build up a stack smaller than the opening, say 0.2-in. smaller. To this are added a 0.1005-in. and a 0.125-in. block. If this is too small, change the 0.125-in. block for 0.150 in. If this is still too small, change one of the larger blocks by 0.05 in. and begin again. If the 0.150-in. block made the stack too large, the correct size is known to lie between the stack with the 0.125-in. block and the stack with the 0.150-in. block.

To begin changing at this point by *mil* (0.001-in.) increments is fatal.* You could very well go through the entire 25 blocks in that range before the last one fits. By the law of averages you can expect to go through half of the blocks. Continue the bridging. Select a block about halfway between, say 0.137 in. Again, it is either too large or too small. Thus, this cuts the range of uncertainty in half. The process is continued until a stack to the correct *mil* is found. In this example, only six changes are required, which is typical.

When the nearest *mil* is determined, the same process may be continued to find the nearest *tenth mil* (0.0001 in.). This is begun by substituting a 0.100-in. or 0.101-in. block for the original 0.1005-in. block. Successive halving of the remainder continues until a stack of the correct *tenth-mil* size results - or until your combining ability exceeds your feel. The correct feel can only be determined by much practice. It should be tight enough so that there is no perceptible wobble whatsoever. Yet the blocks should slide quite easily through the entire opening.

This introduces a precaution. If the opening has a good surface finish, the stack might wring to one or the other surface and falsely appear to be a snug fit. Check this by repeating. Furthermore, the effect of heat transfer must be remembered. When you think you have duplicated the size of the opening in the stack of blocks, lay both the part and the stack aside to normalize, preferably on a heat sink. After an hour or more, try the fit again. The chances are that you will have changed the stack by *one-* or *two-tenth mil* (0.0001 or 0.0002 in.).

<hr />

* Hume, K. J., Sharp, G. H., <u>Practical Metrology,</u> London, Macdonald and Company, Ltd., 1958, Vol. 1, p. 19.

Fig. 9-18   Remember that body temperature is 98.6 degrees F. even on a cold day. The expansion co-efficient of standard gage block steel is 6.4 micro-inches, per degree, per inch, which means that for a 1 degree rise in temperature, a 4-inch stack of blocks will expand 25.6 mike (0.0000256 in.). Therefore, insulating gloves and tweezers should be used for all precision work.

GAGE BLOCK HOLDERS

The use of gage blocks is improved in both convenience and reliability by well-designed gage block holders. Two of their four basic advantages are metrological. They assist the formation of combinations with uniform wringing intervals, and they reduce the length change due to heat transfer. The other two are functional. They are a decided convenience in handling stacks of blocks, and they are an extremely valuable means to prevent damage. Some very valuable gage block techniques are not possible without holders. Of course, there must be two kinds of holders, those for square blocks and those for rectangular blocks.

Gage block holders of all types provide a frame into which the blocks are constrained. The constraining force in every case known to the author is applied by screw thread. Because of the tremendous mechanical advantage of screw threads, the compressive force can be carefully controlled. While no substitute for good wringing, this clamping force contributes an additional measure of reliability to gage block measurement. The holders shown in most of the examples provide a means for actually controlling the clamping force.

Heat has been mentioned earlier as an important problem in high-precision measurement. Whenever a gage block stack is handled, it absorbs heat from the hands. The holders for rectangular blocks stand between the stack and the hands, and thereby delay heat transfer. Note: they do not eliminate it, they only give you more time. Insulating gloves or plastic insulated tweezers should be used for all work to a precision of 25 *mike* or finer and/or for all lengths 6 in. or greater, Fig. 9-18. Why 6 in.? You have to pick some point in a book like this. The intelligent technician thinks out each case for himself.

There is another way in which holders for both square and rectangular blocks reduce heat transfer. They enable the blocks to be combined in easier-to-use, self-supporting assemblies. These, of course, do not require as much handling as blocks without holders.

Just as in the beginning of this book we found every opportunity to make a point about cleanliness, we are now emphasizing thermal problems of measurement. Figure 9-19 exaggerates the effect of temperature expansion so that you will have a clear idea of its relative effects.

As important as these metrological considerations are, the principal reasons that gage block holders are sold are functional ones. Of the host of functional advantages one stands out. This is safety from falling. Whenever a high stack of blocks tips over you have trouble. The blocks cannot ever again be considered reliable— until recalibrated. Furthermore, the precision surface onto which they fell will be suspect, too, until it has been calibrated.

SQUARE GAGE BLOCK HOLDERS

Square blocks are far less likely to fall over than rectangular, but they, too, are in danger. Gage block holders for square or rectangular blocks on one hand decrease the likelihood of toppling over, on the other hand — they reduce the damage when it happens. In the case of the gage blocks, there is usually no damage. In the case of the part surface or precision reference surface it is usually minimized.

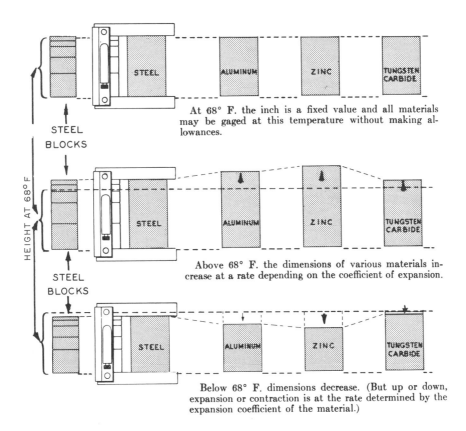

At 68° F. the inch is a fixed value and all materials may be gaged at this temperature without making allowances.

Above 68° F. the dimensions of various materials increase at a rate depending on the coefficient of expansion.

Below 68° F. dimensions decrease. (But up or down, expansion or contraction is at the rate determined by the expansion coefficient of the material.)

Fig. 9-19   These exaggerated drawings show the effect of temperature changes on measurement.

The holders for square blocks are very similar, Fig. 9 - 20. They consist essentially of expandable rods that pass through the center holes of the blocks. They have threaded ends so that they can be secured against the end standards. To use these holders in the most reliable method, fully wring the blocks together with the end standards, assemble with the tie rods, and then firmly draw up the end screw with a torque screw driver. This will provide an assembly with a repeatably uniform total wringing interval and sufficient rigidity for practical measurement.

One of the advantages of these holders, particularly when used with the torque screw driver, Fig. 9 - 21, is that sufficiently reliable assemblies can be made for most measurements without wringing the blocks. They are simply threaded onto the tie rods and clamped with the end screws. This is satisfactory for most measurement – but only if all of the usual precautions about clean surfaces are observed. Actually, the only way to assure that dust has not found its way between the surfaces is by using the wringing procedure as described earlier. Therefore, the compromise procedure is to wring the blocks lightly, and then rely on the clamping force to squeeze out the excess film for uniform wringing intervals.

RECTANGULAR GAGE BLOCK HOLDERS

In spite of the strong partisans for square gage blocks, the vast majority of blocks are rectangular. This is due to their lower price. The inadequacies of the rectangular blocks can be largely compensated for by well-designed holders.

These narrow blocks are very unstable when a stack is stood vertically. Held to a base by holders, they are as stable as square blocks. Their large, flat sides, together with small mass, make them very subject to heat expansion. This, too, is alleviated by block holders. An assembled stack is handled by the holders rather than directly by the blocks.

Unlike the holders for square blocks, holders for rectangular blocks vary considerably in design. Basically, they consist of a channel to contain the gage blocks and the end standards with a fixed end and a movable end. The movable end applies the required force to the

SQUARE GAGE BLOCK HOLDERS

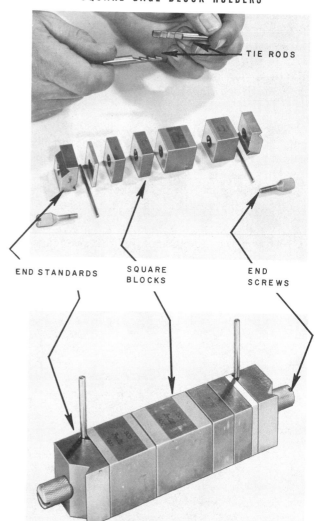

Fig. 9-20 This pin gage assembly shows the use of square gage block holders. *(DoALL Co.)*

Fig. 9-21 The torque screwdriver assures uniform clamping force. *(DoALL Co.)*

assembly. The differences stem from the fact that a large range of stack sizes must be accommodated. Each manufacturer of holders has met this in his own way. The holder shown in Fig. 9 - 22, for example, uses a very long clamping screw. Others have worked out channels with stops for the fixed end so that the screw movement does not need to be long. Still others have used telescoping channels.

The holder selected for this discussion is of the telescoping type, and has several other features that should be mentioned, Fig. 9 - 23. To provide a long range of sizes the channels are made up of components that telescope

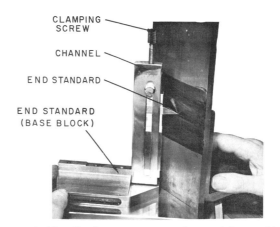

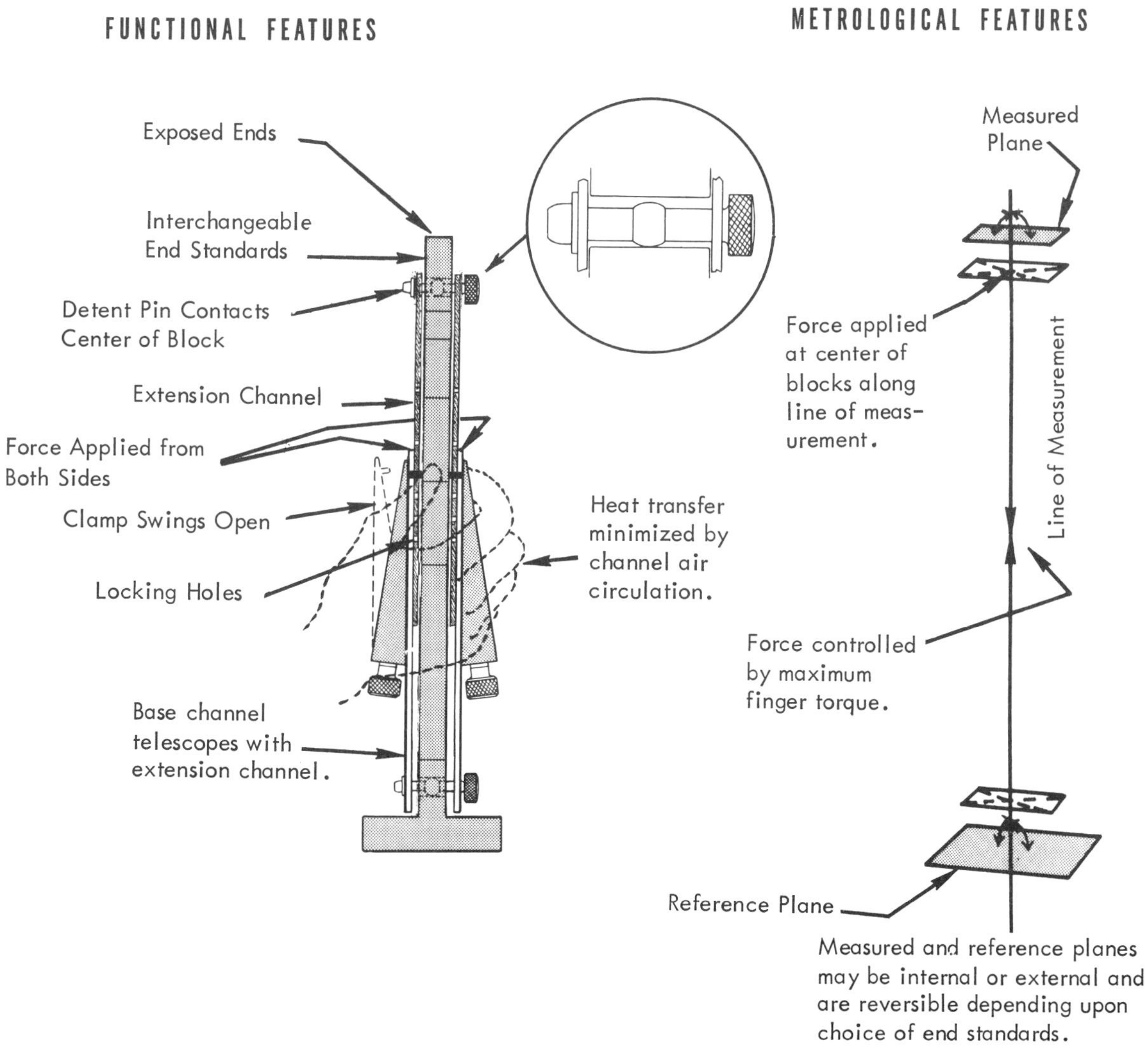

CLAMPING SCREW

CHANNEL

END STANDARD

END STANDARD (BASE BLOCK)

Fig. 9-22   This basic holder can be used for a wide range of setups by changing the end standards.

FUNCTIONAL FEATURES

Exposed Ends

Interchangeable End Standards

Detent Pin Contacts Center of Block

Extension Channel

Force Applied from Both Sides

Clamp Swings Open

Locking Holes

Base channel telescopes with extension channel.

Heat transfer minimized by channel air circulation.

METROLOGICAL FEATURES

Measured Plane

Force applied at center of blocks along line of measurement.

Line of Measurement

Force controlled by maximum finger torque.

Reference Plane

Measured and reference planes may be internal or external and are reversible depending upon choice of end standards.

Fig. 9-23   Not all gage block holders provide all of these desirable features.

together. The base channels have clamps on both sides. The pins in these clamps engage holes in the extension channels and have knurled screws which pull the telescoped parts together. The components have been combined to hold gage blocks up to 12 ft. in length. Such practices are very questionable, of course, but preferred to most practical alternatives.

The channels have an open construction that minimizes heat transfer. Their design permits the end standards to be exposed fully on end. This has the distinct metrological advantage shown in Fig. 9-24. Another metrological advantage to the design in Fig. 9-23 is that the force controlling the wringing interval is along the axis of the stack of blocks. It does not tend to bend the stack.

One of the most important functions of gage block holders is to adapt gage blocks for measurements other than simply setting comparators. This is achieved by means of end standards, our next topic.

END STANDARDS

Let us go back a ways. In the first chapter it was shown that a dimension is the size that the designer feels a part feature should be for perfect parts. It is a specified separation between two points that lie along the line of measurement. Tolerance is the allowable deviation from the dimension, arrived at after much compromise between the designer, the production people, and, in an up-to-date situation, by the quality control people. The latter have the task of verifying whether or not actual part features conform with the dimensions and the tolerances.

In order to verify, they must recreate the separation between the reference point and measured point within a measurement system. The system compares the unknown distance with a standard and provides a reading. This reading shows whether or not there is conformation.

The reliability of the measurement depends upon how nearly the system duplicates the separation between the reference and measured points, both in position and in the character of the reference surfaces. And that brings us up to the need for end standards.

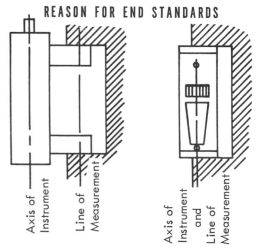

REASON FOR END STANDARDS

When the outer ends of the end standards are exposed, inside measurement can be performed in conformity to Abbe's law.

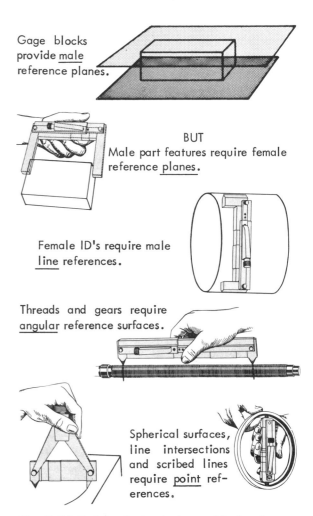

Gage blocks provide male reference planes.

BUT
Male part features require female reference planes.

Female ID's require male line references.

Threads and gears require angular reference surfaces.

Spherical surfaces, line intersections and scribed lines require point references.

Fig. 9-24 End standards adapt gage block reference surfaces to part features.

Gage blocks themselves present only one type of references. They are planes and only male planes, at that, Fig. 9 - 24. Measurement of all outside part features requires male reference surfaces, inside dimensions require female references, but lines not planes. To scribe an arc or measure a scribed arc, line references are required instead of planes. The measurement of threads requires lines at angles. To carry it to the full limit, the measurement of distance between two line intersections requires the use of point references. This then is the function of end standards, to convert male plane reference surfaces to all the others required for practical measurements, Fig. 9 - 24.

Figure 9 - 25 shows the end standards for the particular gage block holders that have been used for examples in Figs. 9 - 23 and 9 - 24. Similar end standards are available for other styles of gage block holders. Although they differ in appearance the functions are the same.

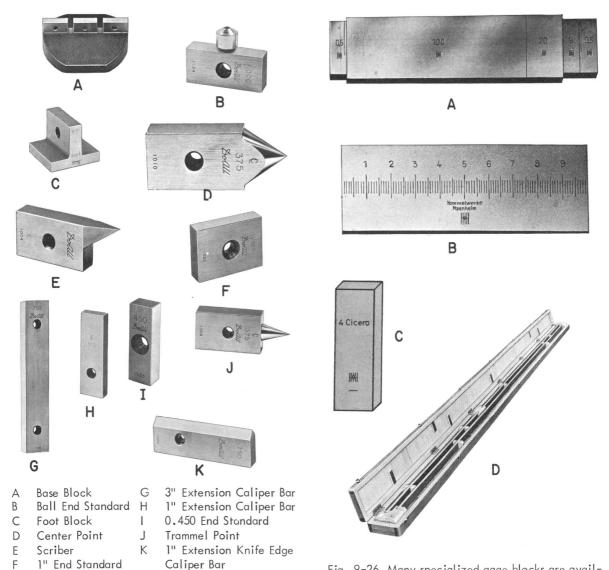

A	Base Block	G	3" Extension Caliper Bar
B	Ball End Standard	H	1" Extension Caliper Bar
C	Foot Block	I	0.450 End Standard
D	Center Point	J	Trammel Point
E	Scriber	K	1" Extension Knife Edge
F	1" End Standard		Caliper Bar

Fig. 9-25 Design of end standards varies according to the types of gage block holders with which they are used. These types are typical of the references provided by end standards.

Fig. 9-26 Many specialized gage blocks are available. A shows blocks with scribed lines for converting end measurement to line measurement. B is provided with a graduated scale for typographical measurement. C is also for typographical measurement. D shows extremely long blocks. (Hommelwerke)

Gage blocks with and without end standards and holders are used in several distinct manners, Fig. 9-27. The basic objective in each case is reliable measurement. Beyond that they have little in common. They are performed to very different levels of precision and under widely divergent circumstances. Only the first and the second listed uses are properly within the scope of a book on the basic principles of measurement. They are so completely in keeping that the next chapter is devoted to them.

The two things that all of the uses do have in common are the need for end standards and efficient holders. The principle behind these gage block accessories and the advantages are just as valuable for attribute gaging as they are for precision comparator measurement. Therefore the three lower precision applications will be reviewed briefly before moving on to more sophisticated applications.

SETUP, LAYOUT AND ASSEMBLY

Some gage block measurements are slower than other methods but the improved precision, accuracy and reliability recommend them anyway. Not so with machine setups. Even when high precision is not the main objective, they may well be the fastest and most convenient method.

Spacing relationships are checked quickly before operation with gage blocks as shown in Fig. 9-28. The old way would be to make an approximate setup. Measuring between two milling cutters is not easy because of the compound cutting angles. A cut would then be made. This would be measured and the spacing altered. This would be continued until within tolerance. The flat reference planes of the blocks are ideal for such locating. Note the absolute necessity of wear blocks for work of this type.

When the dimension is longer, holders are advocated, as in Fig. 9-29. In this case the distance between the table and the cutter is being established on a milling machine. Because the tool block makes this a free-standing gage, the setup man has both hands free for machine adjustments. Even with such a rigid setup, reasonable care should be taken. A wear block should be wrung to the top end standard. The spindle should be disengaged.

PRINCIPAL USES FOR GAGE BLOCKS

1. Calibration of other instruments and lesser standards.

2. Setting of comparators and indicator-type instruments.

3. Attribute gaging.

4. Machine set-up and precision assembly.

5. Layout.

Fig. 9-27 These are the five principal uses for gage blocks.

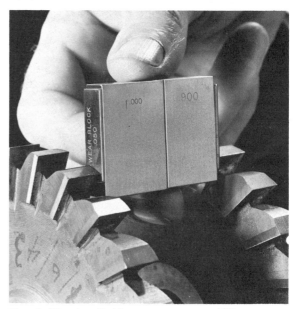

Fig. 9-28 A reliable way to space milling cutters is with gage blocks. Note importance of wear blocks for work of this type.

If the table must be moved very far and a rapid traverse is used, place a piece of wood on the gage block assembly. If daydreaming, the crunching wood will wake you up before the cutter or gage block assembly is damaged. As you approach the final size, rotate the cutter by hand through a small arc. You will become aware of slight rubbing on the wear block when the dimension is reached.

For very long machine adjustments such as shown in Fig. 9-30, no other method can compare with a gage block, end standards and holder assembly. Only for fractional tolerances might other methods be as quick and convenient. When the tolerance is *one mil* (0.001 in.) or closer, only this method permits direct setting without time-consuming trial-and-error methods, or the previous manufacture of expensive setting gages. Obviously, such measurements would not be practical without gage block holders.

Layout applications are methods for precisely scribing lines that show the extent of part features for future machining. The scribed lines are the measured ends of dimensions. The reference ends may be points such as punch marks, lines such as other scribed lines, intersections of scribed lines, or plane surfaces.

Three types of end standards have been developed for these purposes: center point, trammel point and scriber. Their appearance varies among manufacturers, but those in **Fig. 9 - 25** are illustrative. The scriber is used primarily when the reference end of the dimension is a plane. In the example in Fig. 9-22, a surface plate is the reference. It could have been a part feature and could have been performed horizontally as well as vertically.

When an arc is swung around a punch mark or the intersection of two scribed lines, the center point is used as the reference point, **Fig. 9 - 24.** The scriber is used to make the line. The trammel points are more delicate than the center point or scriber. They are usually reserved for checking scribed lines rather than to actually make the lines.

Note that the scriber does not require an addition or subtraction to the length of the block combination if the scribing face is wrung to the end of the stack. For both the trammel and center points, however, an allowance must be made for half of the block width. This should not result in confusion, because in every case known to the author the size etched on the block is the center-to-face dimension.

There sometime must be a reason to use gage blocks to set conventional trammels and dividers, as in Fig. 9-31. The only one that the author can think of is when the desired end standards are already in use. With the end standards shown in Fig. 9-26, cross lines for this purpose are included on the one-inch extension caliper bar. This feature is useful even more in making transfer of measurements from lines by means of the measuring microscope.

Fig. 9-29 The tool block and holders make this a stable setup for setting the depth of a milling cut.

Fig. 9-30 No other method can compare with a gage block and holder assembly for setting large machine dimensions such as this.

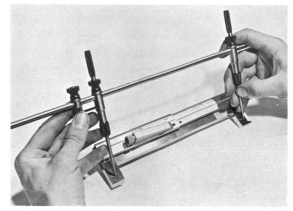

Fig. 9-31 Trammels and dividers may be set to the scribed lines provided on some end standards, such as the 1-inch extension bar in Fig. 9-25.

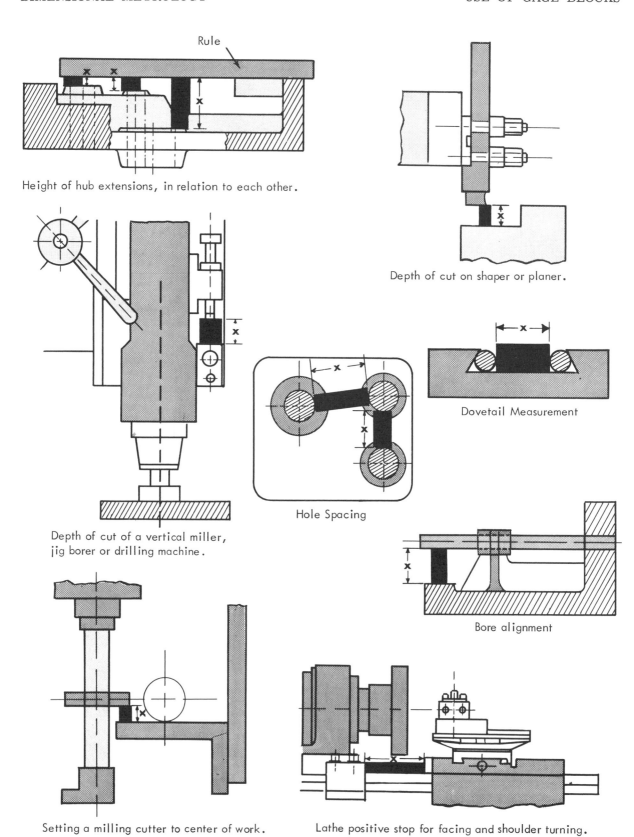

Rule

Height of hub extensions, in relation to each other.

Depth of cut on shaper or planer.

Depth of cut of a vertical miller, jig borer or drilling machine.

Hole Spacing

Dovetail Measurement

Bore alignment

Setting a milling cutter to center of work.

Lathe positive stop for facing and shoulder turning.

Fig. 9-32   Direct use of gage blocks to set desired dimension (x) eliminates errors from transfer of measurement.

GAGE BLOCKS FOR ATTRIBUTE GAGING

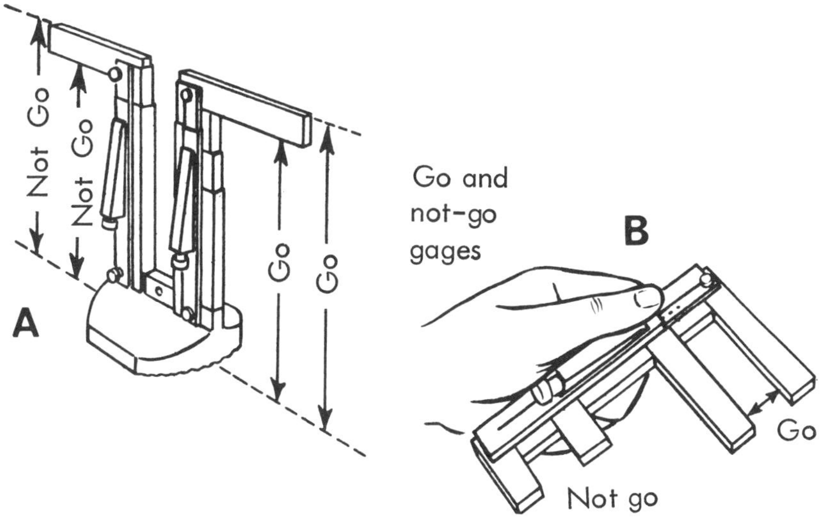

Fig. 9-33 Gage blocks with end standards and holders are valuable as attribute gages. The most important reason is that they eliminate one or more intermediate steps between the gage and the standards. Another reason is that one set of blocks, end standard, and holders can be made up into thousands of gages, saving a tremendous inventory.

Assembly applications are too obvious to require elaboration. When two parts have play between them but must be secured to a fixed dimension, gage blocks may well be the fastest and most reliable method for establishing the dimension. Don't forget to remove the blocks. That sounds funny but it is no joke.

ATTRIBUTE GAGING WITH GAGE BLOCKS

Attribute gaging is not within the scope of this chapter because it is a sorting operation rather than measurement in the sense we are considering it. It involves much measurement on the part of everyone up to the inspector. At that point, however, all decisions have been reduced to good and bad, go and not-go, etc. Those of you who continue into manufacturing will get to know attribute gaging well, which is studied along with inspection and quality control. Gage blocks are not widely used for attribute gaging. They certainly should be for one very sound metrological reason and an equally sound functional reason.

Figure 9-33 shows gage blocks made up into two types of go and not-go gages. The type at A is used from a flat surface to a feature above the surface. The reference surface may either be a surface plate or on the

part itself, if large. This particular gage could be used either for inside or outside measurements. If it is used for outside measurement, the part feature could be either a plane surface or a curved surface. The lower face of the extension caliper bars could pass over it equally well--or not pass over it in the case of the not-go side.

If it is used for inside measurement it is restricted, however, by the flat top surface of the extension bar. It could measure to the bottom of flat surfaces or outside diameters. It could not measure to an inside diameter for the reasons shown in Fig. 9-24. For this purpose a knife-edge caliper bar would be used with the knife edge upward.

The example in B of Fig. 9-33 is similar except that the gage provides both reference surfaces instead of standing on one as in the previous example. In a typical case the go side may have 1.001-in. gage blocks separating the extension bars, and the not-go may have 1.000 in. Then if a part fails to pass through the not-go side, it is rejected as too small. If it fails to pass through the not-go and does pass through the go, it is satisfactory.

Now that you know all about go and not-go gages, what is so important about the use of gage blocks for them? Gage blocks provide

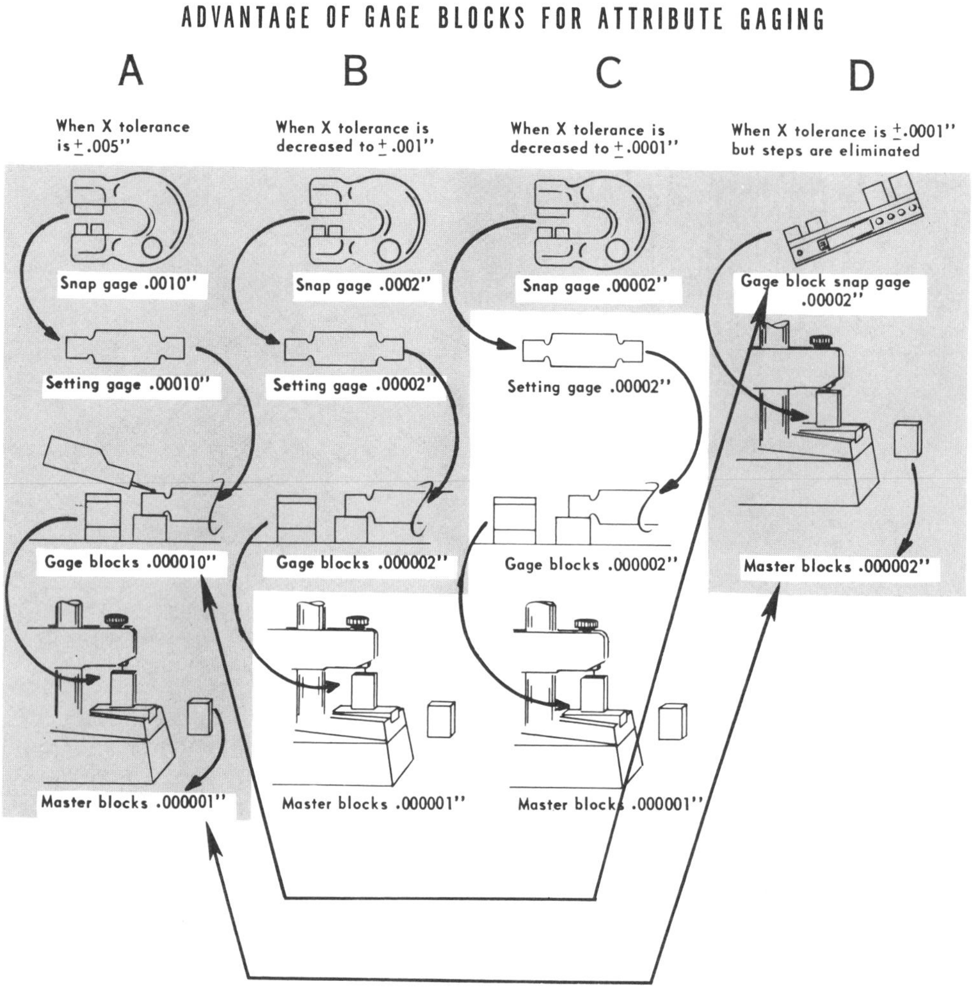

ADVANTAGE OF GAGE BLOCKS FOR ATTRIBUTE GAGING

Fig. 9-34   In example D, a tolerance,1/50th as large as the one in example A, is being controlled with double the final margin of safety. This is achieved by the direct use of gage blocks for attribute gaging.

discrimination in *tenth-mil* increments (0.0001 in.). *One set of gage blocks with suitable holders and end standards is equivalent to ten-thousand different sizes of fixed gages for each inch of its range.* If production is large, a fixed, single-purpose, attribute gage may pay for itself. Otherwise there is a tremendous saving in the cost of the gages when gage blocks are used.

It seemed wise to mention the direct saving first, but the indirect saving that arises from the metrological advantage may be far more important. Figure 9-34 shows the reason. This is an example of what the ten-to-one rule means to gaging. In example A the tolerance is ± 0.005 in., considered "wide open". The snap gage was accordingly set to

one tenth of that tolerance or 0.0010 in. This required a setting gage of 0.0001 in. To calibrate that gage required gage blocks with an uncertainty no greater than one part in ten million or 10 mike (0.00001 in.). Even presuming that the comparator used to calibrate the working set of gage blocks to the master blocks is perfect (which it is not), the master blocks would have to be accurate to one part in one million, or one mike (0.000001 in.). This is not an uncommon situation.

In example B, the tolerance is closed down to ±0.001 in., which know-it-all types still call "wide open". Then the demand for discrimination exceeds the capability of existing equipment, as shown by the shaded areas.

Fortunately, it is exceeded in the final calibration stage where it will be in the hands of experts. In example C, however, the tolerance of ± 0.0001 in., which is not at all uncommon, throws us a real curve. The equipment capability is exceeded at a lower level where the personnel may not be fully aware of the uncertainty that is certainly being added to the gaging.

Example D shows a reliable way to solve this problem. The gage blocks are simply moved closer to the actual gaging. Remembering that each step represents a tenfold increase in required discrimination, moving them two steps helps the situation by a factor of 100. The results are startling. *In example D a tolerance one fiftieth the size of the one in example A is being held with twice the final margin of safety.* In A, *one-mike* blocks were required; in D, *two mike.* What stronger case could be made for gage blocks?

INSTANT ACCURACY

In Fig. 9-27 the five principal uses of gage blocks are listed. Many uses are combinations of these. A strikingly valuable one is the *precalibrated indicator technique.* When the author first encountered this years ago, he immediately made himself sound foolish by condemning it. It does appear that a gaging setup such as shown in Fig. 9-35 immobilizes a set of gage blocks for no better reason than to furnish a chassis for the gage. But what a chassis! It builds into the gage its connection with the international standard of length. It provides instant accuracy, being recalibrated in seconds whenever the slightest doubt arises. And if these were not enough, it eliminates errors because the dimension that it is set to, can always be read directly from the sizes marked on the gage blocks.

Back in Fig. 6-10 you were shown that there are two basic measurement methods, interchange and displacement. Vernier and micrometer instruments are of the displacement method. Importance of interchange method is discussed in the next two chapters. The interchange instruments provide a numerical comparison between the known and unknown lengths. The most common of these is the dial indicator. You will meet it now in conjunction with the precalibrated indicator technique.

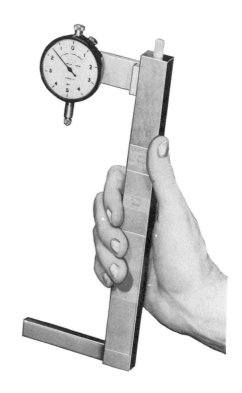

Fig. 9-35 This could well be a demonstration of the effect of heat upon gage block length, but it is not. It is an example of the precalibrated indicator.

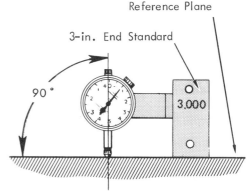

Fig. 9-36 For precalibrated indicator measurement, an end standard is needed that will hold a transducer at a right angle to a reference plane and at a suitable distance from that plane to permit zeroing.

All that is needed is an end standard that will hold a dial indicator (or any other type of transducer, mechanical, pneumatic or electronic) perpendicular to a plane to which it can be zeroed, Fig. 9-36. Several of these

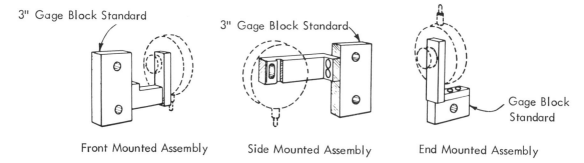

3" Gage Block Standard

3" Gage Block Standard

Gage Block
Standard

Front Mounted Assembly   Side Mounted Assembly   End Mounted Assembly

Fig. 9-37   If mechanical indicators are used, both right- and left-hand
side-mounted brackets are required. This is simplified for remote reading electronic and pneumatic instruments.

PRECALIBRATED INDICATOR GAGE

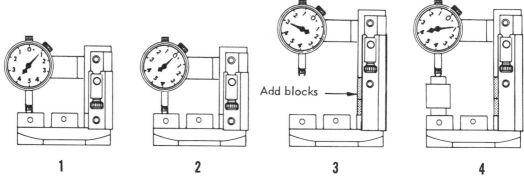

Add blocks

1

Step 1 : Assemble gage
without any gage blocks.

2

Step 2: Zero indicator.

3

Step 3: Separate gage,
add blocks for desired
dimension, reassemble.

4

Step 4: Inspect parts. Any
deviation from gage –block
length can be read on the
indicator.

are available commercially, Fig. 9-37, and they can be improvised easily. Any reliable reference surface then can be used to zero the indicator. This is the nexus. *When you zero the indicator you calibrate a point in space.* A point not to be forgotten.

Having this datum, all you need to do is stack gage blocks upon it and you automatically know the length you have created. Because gage blocks provide hundreds of thousands of lengths with *mike* (millionths of an inch) accuracy, this method allows you to make up hundreds of thousands of gages using only standard components.

The way this is done is shown in Fig. 9-38. In this example a comparator base is shown.

The principle would be exactly the same to form a snap gage, Fig. 9-35. Quite obviously all that is needed to recheck this gage is to remove the blocks, repeat step 2 and see if the instrument reads zero. Actually, it is easier than that -- simply insert an extension bar end standard, Fig. 9-39. After clamping the reading is taken, the indicator zero set if necessary, and the gage returned to its original condition. This eliminates the need to disassemble the gage.

The precalibrated indicator technique is not limited to these examples. Figure 9-40 shows that it can be used effectively to make up comparators. The effective height is the difference between gage block stacks A and B. Therefore, this setup can be used to measure dimensions that are ordinarily too small for gage block methods. For inside measurement the technique can also be used. It requires an additional transfer of measurement thereby losing a part of its advantage. The precalibrating step is shown in Fig. 9-41. After this step, gage blocks and an end standard making up the desired dimension less 4.500 in. (in this case) are simply clamped to the precalibrated indicator. An extreme example is shown in Fig. 9-42.

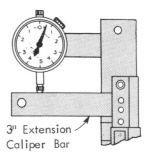

Fig. 9-39 An extension bar is inserted to recalibrate the gage.

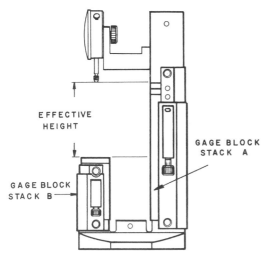

Fig. 9-40 The use of standard holders and end standards gives great versatility to the precalibrated indicator method.

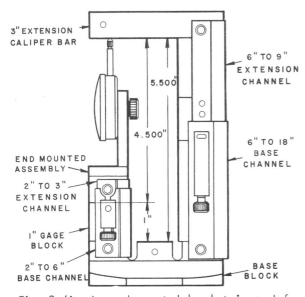

Fig. 9-41 An end-mounted bracket is used for inside measurement. The indicator is zeroed on a plane a fixed distance from the plane established by the end standard.

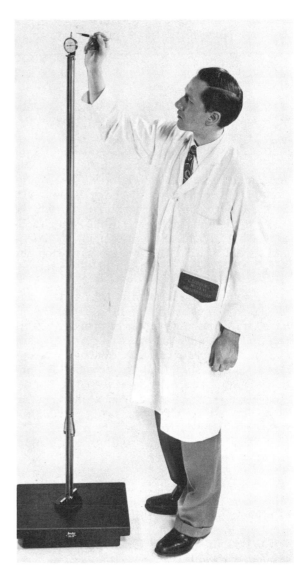

Fig. 9-42 Although very unusual, this setup shows the versatility of the precalibrated indicator gage. In this case, it is set up for internal measurement.

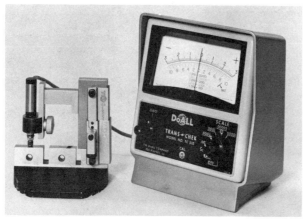

Fig. 9-43 The precalibrated indicator technique only requires one bracket when remote-reading transducers are used.

For simplicity these illustrations all showed dial indicators. Figure 9-43 shows that the principle applies even better to electronic instruments. Because they read remotely only one bracket is required.

Summary — Playing With Blocks

● This is not kid's play. In fact, it is probably expertise with gage blocks that really separates the men from the boys in linear metrology, with due apologies to the distaff metrologists. With them you have reliable access to measurement in *mike* (millionths of an inch) accuracy. Without them, even with *mike* reading comparators you will be lucky to achieve *tenth-mil* (0.0001-in) accuracy. They should be used for every measurement from *mil* (0.001-in.) discrimination and finer unless there is a specific reason not to.

● To achieve the reliability inherent in gage blocks you must practice proper gage block care and learn to wring them properly. This requires practice as well as knowledge; and can be attained by anyone of above average intelligence, patience and dedication. Beyond that you do not need any special qualifications. Just because blocks adhere does not mean that they are fully wrung. When fully wrung the wringing interval will be about *one mike* (0.000001 in.) which is negligible for most practical measurement.

● To combine blocks to form a given dimension, you select blocks that successively eliminate the last digits from the right. To combine blocks to find an unknown dimension, you bridge the unknown, reducing the interval by one-half with each new combination until the stack duplicates the unknown dimension. Very small dimensions are formed by having one stack subtract from a larger one.

● To achieve real versatility and reliability from gage blocks, holders and end standards must be used. These are in many forms but serve certain basic functions. They adapt the gage blocks which are only male end standards to other references. They provide stable setups, enhance uniform wringing and insulate from heat. The latter becomes an all abiding consideration for most parts whenever the tolerance is *one-tenth mil* (0.0001 in.) or smaller. It becomes important even for *one mil* (0.001-in.) tolerances for large parts, say over 10 in.

● The dictum is to use gage block setups unless there is a good reason not to. Therefore they unquestionably are required for calibration of linear instruments. They should be considered also for layout, machine setup, assembly and attribute gaging. One set of gage blocks and end standards will make up nearly every type of attribute gage required and in nearly any size. The application of the precalibrated indicator technique brings laboratory standards right to the line inspection operation, thus improving reliability.

● All in all, gage blocks can be the greatest single tool for accurate measurement -- if they are accurate ones. If inaccurate they can do irreparable harm. It is the function of calibration to determine this. First however, we must consider the means by which gage blocks are used to measure, measurement by comparison.

METROLOGICAL DATA FOR GAGE BLOCK MEASUREMENT

INSTRUMENT	TYPE OF MEASUREMENT	NORMAL RANGE	DESIGNATED PRECISION	DISCRIMINATION	SENSITIVITY	LINEARITY	RELIABILITY — PRACTICAL TOLERANCE FOR SKILLED MEASUREMENT	RELIABILITY — PRACTICAL MANUFACTURING TOLERANCE
Gage blocks, A grade: 0.001" series	end stand.	0.1000 to 4.0000	+ 0.000006 − 0.000002	0.0001	not applic.	0.000008/in.	0.000010	0.0001
0.000025" series	same	same	same	0.000025	same	same	same	same
long blocks	same	4.0000 to 24.0000	same	0.0001	same	same	0.00010	0.001
0.0001" series with holders	same	0.1000 to 4.0000	same	0.0001	same	same	0.000010	0.00005
long blocks with holders	same	4.0000 to 24.0000	same	0.0001	same	same	0.00005	0.0005
Precalibrated Indicator: 0.001" dial indicator	comparison	0.1000 to 6.0000	0.001	0.001	0.0001	limited by indicator	0.001	0.001
0.001" dial indicator	same	same	0.0001	0.0001	0.00005	same	0.0001	0.0002
0.00005" mechanical dial indicator	same	same	0.00005	0.00005	0.00002	same	0.00007	0.0001
0.00001" electric dial indicator	same	same	0.00001	0.00001	0.000005	same	0.00002	0.00008
0.000001" electric dial indicator	same	same	0.000001	0.000001	0.000001	same	0.00001	0.0001

Fig. 9-44 Note that the gain in reliability does not increase in amplification and that it may actually fall off if carried beyond practical limits. This will become clear in the next two chapters.

TERMINOLOGY

Stack of blocks Combination of blocks	Two or more blocks wrung together to form a length dimension.
Burrs	Minute particles of metal above the finished surface that scratch other contacting surfaces.
Conditioning block Conditioning stone Deburring stone	Special blocks that remove burrs from surfaces rubbed against them.
Wring	Adhesion of very nearly perfectly flat surface when brought into intimate contact.
Wringing interval	The separation of two wrung surfaces.
Film	Substance separating wrung surfaces. May be air and/or oil.
Heat sink	A surface with rapid heat transfer on which objects may be placed to rapidly reach ambient temperature.
Normalize	Act of being warmed or cooled to ambient temperature.
Bridging	Finding a dimension by testing of values spanning unknown and successively closing the distance between them until the unknown is approximated.
Zeroed in	Bridging condition when very close to unknown value.
Gage block holders	The hardware developed to secure end standards and gage blocks into one unit.
End standards	Special blocks that are placed on ends of stack to convert to other measurements than the normal external end measurements of the stacks. See Fig. 9-26 for specific terminology.
Attribute gage	Sorting measurement that separates into size groups.
Go gage	Attribute gage that passes all parts within size limits.
Not-go gage	Attribute gage that rejects all parts out of size limits.
Precalibrated indicator technique	Method for measurement in which the standard forms a frame member of a comparison instrument.
Dial indicator	Most familiar measuring instrument of the interchange or comparison type.
High-amplification comparator	Similar in purpose to dial indicator but capable of much greater precision.

Fig. 9-45   The precision with which gage blocks can be used to define lengths is not matched by our language precision.

QUESTIONS

1. When does the need for gage blocks become increasingly important?
2. What is meant by wringing blocks together?
3. What is meant by wringing error?
4. Why are wear blocks used?
5. What is a gage block conditioning stone?
6. What methods have been devised to help eliminate wringing error?
7. What is the principal cause of gage block wear?
8. How can you tell that you have wrung gage blocks together properly?
9. What procedure is used to select combinations of blocks for obtaining a given number?
10. How are desired dimensions obtained that are too small to be combined in the available set?
11. Why is heat such an important problem in high-precision measurement?
12. What can be used other than gage block holders to reduce heat transfer?
13. What basic advantages do square gage blocks have over rectangular blocks?
14. Why are there more rectangular blocks sold than square blocks?
15. Why were end standards developed for use with gage blocks?
16. Although gage block measurements are relatively slower than many other methods, why are they recommended for many applications?
17. What types of end standards have been developed for layout applications?
18. Why is it possible for gage blocks to replace thousands of other gages?

DISCUSSION TOPICS

1. Explain the important maintenance considerations for gage blocks.
2. Discuss the principal rules which concern the proper place for gage block use in measurement.
3. Explain the proper steps that should be used for wringing blocks together.
4. Explain the basic advantages for using gage block holders.
5. Discuss the principal uses for gage blocks.
6. Explain how end standards adapt gage blocks to various part features.
7. Explain the advantages of gage blocks for attribute gaging.

Ode To A Gage Block

Hardened tool steel, polished and lapped,
Size to a millionth, parallel and flat.
Measured by light waves or electronically
Used to build planes that fly supersonically.
Basis of production interchangeability,
For laboratory instruments — great versatility.
Simple in design, true and dependable,
Economical if cared for, definitely commendable.

Mary E. Hoskins
Sheffield Corporation

10

Measurement by Comparison
Mechanics Extend the Senses

All measurement requires the unknown quantity to be compared with a known quantity, called a standard. This involves three elements, the measurement system being the third. Fig. 10-1 shows that these three elements are required in all of the fundamental measurements, time, mass and length.

The standard in rules, vernier and micrometers is not only built in, it is calibrated. Another way to measure is to have the standard separate from the instrument, Fig. 10-2. In both cases measurement is made by comparing the unknown length with the standard. Both methods require comparison. When the instrument is the standard, however, nothing extra is needed. It is therefore called *direct measurement*.

In direct measurement precision is dependent upon the discrimination of the scale and the means for reading it. In *comparison measurement* it is dependent upon the discrimination of the standard and the means for comparing.

Accuracy, in contrast, is dependent upon other factors. Among the most important are the geometrical considerations. In Fig. 6-10 it was shown that all measurement is either by the interchange method or the displacement method. The verniers and micrometers are measured by the displacement method. Comparison measurements use the interchange method. This enables them to have a very favorable geometry. Most comply closely to Abbe's law. This becomes increasingly important as amplification increases.

In the direct-measurement instruments discussed thus far we have increased precision by amplification. This can go only so far in the race for greater and greater precision. As the amplification is increased, a point is reached at which the standard gets in the way. By separating the standard from the measuring instrument each can be perfected independently of the other. Before exploring the things that can be done after the divorce, let's look into the cause of the rift between instrument and standard.

THREE ELEMENTS OF MEASUREMENT

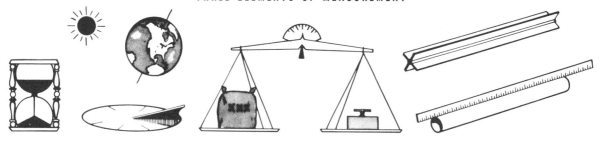

	TIME	MASS	LENGTH
THE UNKNOWN:	Hour Glass	Bag of Flour	Tubing
THE STANDARD:	Earth's Rotation	Weights (copies of International mass standard.)	Yardstick (copy of international length standard.)
THE MEASUREMENT SYSTEM:	Sundial	The Balance	Also the Yardstick

Fig. 10-1 All measurement is made up of measurement of time, mass and length. In each of these cases, three elements are involved, the unknown, the standard, and a system for comparing them. Only in crude length comparison are the standard and system the same object.

THE TWO TYPES OF MEASUREMENT

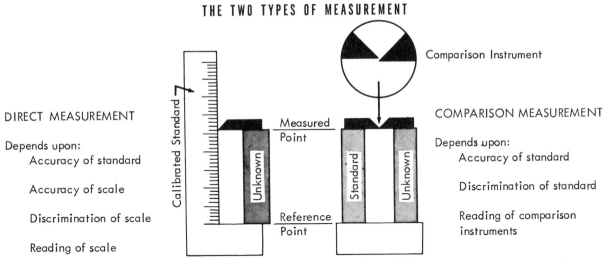

Comparison Instrument

DIRECT MEASUREMENT

Depends upon:
  Accuracy of standard

  Accuracy of scale

  Discrimination of scale

  Reading of scale

COMPARISON MEASUREMENT

Depends upon:
  Accuracy of standard

  Discrimination of standard

  Reading of comparison instruments

Fig. 10-2   All measurement requires comparison. Rules, verniers and micrometer instruments are called direct measurement because they do not require the intercession of another element.

THE RACE FOR POWER

Steel rules have one power (written 1X), or an amplification factor of one, or a mechanical advantage of one, or a leverage ratio of one to one. All these expressions mean that they do not increase or decrease the apparent size of the object. A quarter inch on the part feature reads one quarter inch on the rule. The smallest division on the rule that can be read (its discrimination) reliably is 1/64 in. or 0.02 in. on a decimal-inch scale. Thus the smallest change in a part feature that can be discerned is also 1/64 in. or 0.02 in. respectively.

If we want to observe smaller changes (higher precision) we must enlarge the *apparent size* of the part feature. If we use lenses for this it is called *magnification*. If we use a mechanical or electronic system, it is called *amplification*. In either case the amount of the enlargement is called the *power* of the instrument.

A lever arm is perhaps the simplest means for amplification and is used in Fig. 10-3 to illustrate the relationship between amplification and discrimination. As the pivot point of the pointer arm moves to the right, the ratio of the long end to the short end becomes greater. This increases the amplification or power.

The one-to-one relationship that is equivalent to a rule is shown at the top. Nothing

AMPLIFICATION IN MEASUREMENT

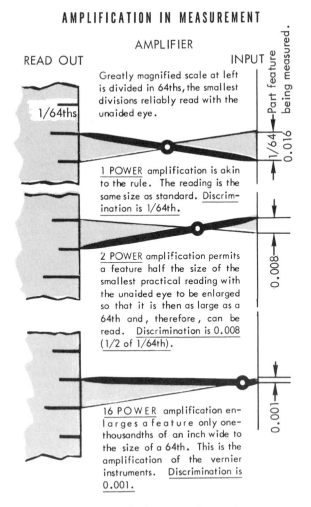

AMPLIFIER

READ OUT   INPUT

Greatly magnified scale at left is divided in 64ths, the smallest divisions reliably read with the unaided eye.

1/64ths

1 POWER amplification is akin to the rule. The reading is the same size as standard. Discrimination is 1/64th.

2 POWER amplification permits a feature half the size of the smallest practical reading with the unaided eye to be enlarged so that it is then as large as a 64th and, therefore, can be read. Discrimination is 0.008 (1/2 of 1/64th).

16 POWER amplification enlarges a feature only one-thousandths of an inch wide to the size of a 64th. This is the amplification of the vernier instruments. Discrimination is 0.001.

Fig. 10-3   A simple lever can be used to represent amplification in measurement systems.

lost, nothing gained. In the center drawing the 2X amplification enlarges a distance of 0.008 in. sufficiently to span 1/64 in. on the scale. Hence, it then can be read.

The bottom drawing represents the amplification gained by adding a vernier scale to a rule. Although the method for reading is different, it enables us to observe a change of only 0.001 in. Compared to the plain rule, this is an amplification of 16X. Note that to achieve this the right end of the pointer has shrunk to a nubbin.

There is a limit to how far this can go, and we meet it in the micrometer. In order to get us back in the ball park, the pointer has been redrawn in Fig. 10-4. In the top drawing the graduations on the left represent one division of the thimble scale. Far, far away at the right end, the invisible movement of the pointer would actually be 0.001 in. Therefore, the thimble divisions represent, as you know, 0.001 in. The thimble graduations vary in separation because they are on a bevel. At the reading end they measure about 0.060 in. giving 60X power amplification.

If one minuscule *mil* (0.001 in.) is blown up to *60 mil* (0.060 in.) so that it can be read reliably, then the one thousand of these that it takes to make up an inch would stretch out to 5 ft in length. How do we cram a 5-ft. scale into the confines of a small micrometer? By wrapping it around. Again the principle of the screw has come to the rescue.

The maximum reading on a micrometer is 1.025 in. but the manufacturers modestly throw in the last 0.025 in. as a bonus, and only specify the range as an even one inch. We will also.

Trace the path of 0 on the thimble from full closed to full open. It makes a long spiral crossing the index line every time the spindle has moved 0.025 in., a total of 40 times. Each time past, it has rotated through 25 spaces of approximately 0.060 in. each. It has paced off a scale of 1 1/2 in. in length. After the 40th time around, a scale of 5 ft. has been created. (It actually figures closer to 62 1/2 in. for most micrometers.) So we have stretched 1 in. into 5 ft.

This is a neat trick, but there are practical limits to how far we can carry it. As

EFFECT OF HIGH AMPLIFICATION

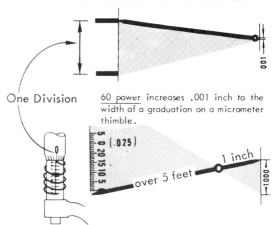

One Division

60 power increases .001 inch to the width of a graduation on a micrometer thimble.

The actual length of a micrometer scale is the spiral path made by zero on the thimble as it progresses from a zero reading to 1.000 inch. Unwound, it would extend over 5 feet long. The ratio of this path to the total travel of the spindle is the amplification.

Fig. 10-4 The micrometer provides the equivalent of an enormous lever arm.

predicted, the limit is caused by the built-in standard. You simply cannot get something for nothing. Every increase in power brings a decrease in some other feature; scope, range, convenience, economy and others.

The rule may have only 1X but it is carried in the shirt pocket and gets into everything. The vernier scale increases its amplification to 16X. But that pampered instrument spends its time locked in a velvet-lined case when not in duty. The micrometer jumps up the amplification to over 60X. But the measuring range shrinks to one inch. Even for this, a wrap-around scale five feet long is required.

If, in our quest for greater precision, we continue along the same route, we run into trouble. Suppose we want an instrument with ten times the discrimination of the present micrometers. Easy, you say; simply add a vernier scale to a micrometer. It already has been shown that while a vernier improves the precision, it does not improve the inherent accuracy.

So, let's try another route. Why not simply scale down a micrometer to one-tenth size? That would result in a 400-pitch thread (impractical in the large version) and each revolution of the thimble would close the micrometer 0.0025 instead of 0.025 in. Even if we had to decrease the number of divisions on the drum, in order to read them clearly, we still would have an instrument of great precision, accuracy and reliability.

It is obvious that this instrument has a serious fault. In our trading of range for power, we have exceeded practical limits. The range is now 0.1 in., that being the length of the screw standard. The short range of the "micro-micrometer" would limit it to very few actual applications. Remember that even when a large frame is put on a micrometer, the *size* is increased, *not the range.*

When an instrument incorporates its own standard, the measuring range is limited by the length of the standard. Its discrimination is limited by the system used to divide the standard into units and enlarge them to a readable size. And its accuracy is limited by the care in manufacture.

THE PARTING OF THE WAYS

If we could achieve high amplification precision without sacrificing the length of the standard, we would have a breakthrough that would permit us to go to any power--well, almost any power.

This has been done. It is achieved simply by separating the standard from the measuring instrument. Without the standard, what is left? The instrument then is simply a device that compares the standard with the unknown length. Every linear measurement, even with a rule, requires a measurement system to compare the standard with the unknown distance. When the standard and the instrument (major part of the measurement system) are combined, changes in one affect the other. When they are separate, each can be refined irrespective of the other.

As the rule evolved into the micrometer, range was sacrificed, Fig. 10-5. However, at every step along the way, the greater precision provided readings for a host of hitherto impossible measurements. This continues to happen even after the separation of instrument and standard. At the first plateau, for example, we find that the new comparison instrument can do far more than just assist in linear measurement. It can provide checks of concentricity, flatness, roundness and position. No wonder that a large family of comparison instruments have been developed. All ordinary means to achieve amplification are used; mechanical, electronic, optical and pneumatic. The family is so large that no attempt will be made to describe them all.

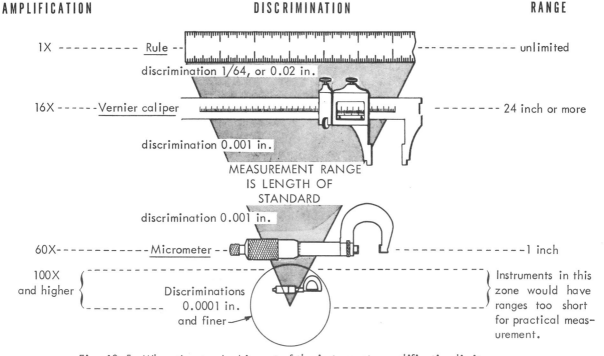

AMPLIFICATION   DISCRIMINATION   RANGE

1X --------- Rule --   ------------------ unlimited

discrimination 1/64, or 0.02 in.

16X ------Vernier caliper   -------- 24 inch or more

discrimination 0.001 in.

MEASUREMENT RANGE
IS LENGTH OF
STANDARD

discrimination 0.001 in.

60X----------- Micrometer --   -----------------1 inch

100X
and higher   Discriminations 0.0001 in. and finer   Instruments in this zone would have ranges too short for practical measurement.

Fig. 10-5 When the standard is part of the instrument, amplification limits range.

If you understand the principles, you will readily understand the instruments when you meet them, from the rough to the highly precise, Fig. 10-6.

THE DIAL INDICATOR

In order to understand measurement by comparison it will help to first become acquainted with the most familiar of the comparison instruments, the *dial indicators.* They were not the first comparison instruments. They cannot compare with the versatility, discrimination and reliability of the electronic instruments. The dial indicators cannot be used at all in some roles for which pneumatic comparators (air gages) and optical instruments excel. However, they are familiar and they are suited admirably for teaching the principles of comparison measurement.

The balance of this chapter discusses aspects of dial indicators over which you have no control--at least at the moment. Those of you who go into quality control, instrument design or purchasing will most certainly use these facts. And, those who simply want to know how to measure reliably will have a better understanding of what they are doing.

The simplest indicator is a lever. This is shown in A of Fig. 10-7. The earliest ones actually looked like this. Note that like the lever examples used earlier, this lever is shown as a pointer. It has, however, a ball on the short end so that distance changes may

(B. C. Ames Co.)

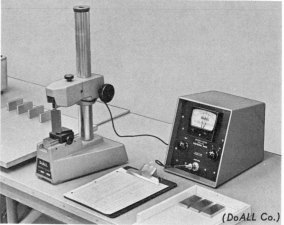

(DoALL Co.)

Fig. 10-6 Indicator instruments range from low amplification test indicators, used for setup and in-process inspection, to highly precise electronic instruments with discrimination in millionths of an inch.

AMPLIFICATION BY LEVER

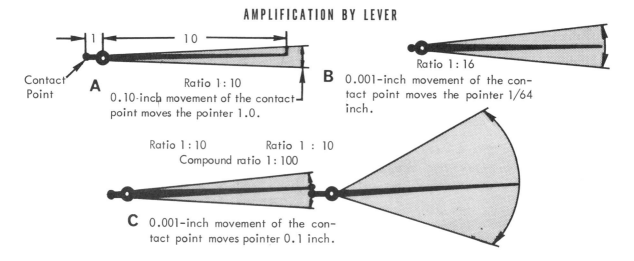

A Ratio 1:10
0.10-inch movement of the contact point moves the pointer 1.0.

Contact Point

B Ratio 1:16
0.001-inch movement of the contact point moves the pointer 1/64 inch.

C Ratio 1:10   Ratio 1:10
Compound ratio 1:100
0.001-inch movement of the contact point moves pointer 0.1 inch.

Fig. 10-7 By means of compound levers, very small movements can be greatly enlarged.

be picked up by contact with them. The ratio shown is 1 to 10 (written 1:10) which provides an amplification of 10X.

If greater amplification is required the ratio of the ends could be changed to say 1:16. This is shown at B, and as you already know, would enlarge a 0.001-in. change to 1/64 in. The proportion of the lever may make this impractical. Compounding levers is a better method for gaining high amplification. In C of the figure an amplification factor of 100 is achieved.

Note that we have achieved greater precision by lengthening the scale through which the pointer swings in respect to the change of length being observed. If the swing of the pointer in A is the maximum practical amount, then the range of measurement in B would have to be less, and the range in C less yet. In fact, the range in C could only be 1/10 that of A. This would reduce greatly the practical applications of a measuring instrument.

Gearing provides the solution. With gears we can achieve the same mechanical advantage as levers but not be as limited in scale range. Consider two gears in mesh. The angular movement of the *driven gear* is determined by the amount the *driving gear* turns, and by the ratio of the gear teeth. This is illustrated at A in Fig. 10-8. The driving gear has 10 times as many teeth as the driver gear. Every revolution of the driving gear rotates the driven gear 10 revolutions. In B two more gears have been connected. Now the ratio is 100 to 1.

To use this amplification for measurement we need two things. First, we must have a way to cause the driving gear to rotate in proportion to changes in length. Second, we must have a means of reading the amount of rotation of the last driven gear. At C a spindle has been added. It has a *rack* that engages the driving gear and a contact that engages the part whose action is to be observed. Note, I did not say that it engages the part to be measured. To measure we must have a standard, and all we have here is the action--a football game without the gridiron.

In C, black dots show the "at-rest" position of the gears. At D the contact has been raised slightly. The black dots show the ro-

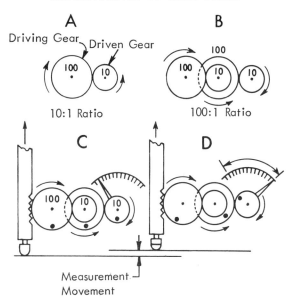

Fig. 10-8 Gear trains provide the mechanical advantage of levers but without the limitation of range.

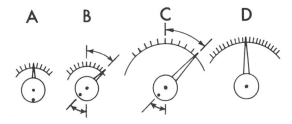

Fig. 10-9 The length of the pointer enters into the overall amplification. Increasing amplification permits the discrimination to be increased. In a dial indicator, the pointer often is called the hand.

tation of the gears caused by the displacement of the spindle. Although the rotation of the first gear was slight, the final gear has rotated 100 times as much.

Does this mean that the pointer has moved exactly 100 times as far as the spindle was displaced? No, because the pointer itself provides amplification. A point on the pitch diameter of the last driven gear will have moved 100 times as far as the spindle moved; but along an arc, of course.

In A of Fig. 10-9 the gear is shown at rest. B is the position after the spindle has been displaced. The pointer has turned through

FUNCTIONAL FEATURES

METROLOGICAL FEATURES

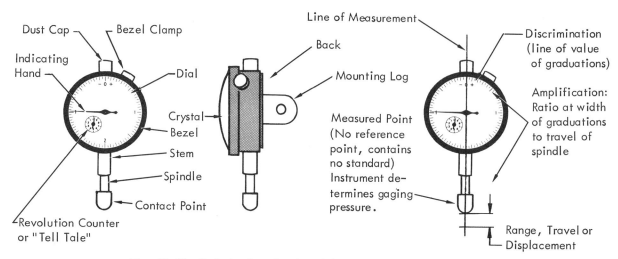

Fig. 10-10 Both the functional and the metrological features of the dial indicators are very different from previous instruments.

the same angle as the point on the pitch line of the gear, but has moved twice as far. It has moved twice as far because it is twice as far from the center of rotation. It has doubled the amplification of the gear train. This can be carried as far as practical. In C, for example, the pointer is four times as far from the center as the pitch diameter, resulting in an amplification factor of 4X. *The amplification of a dial indicator is the amplification of the gear train multiplied by the amplification contributed by the pointer.*

SCALE DISCRIMINATION

Now is the time to distinguish between *read* and *discriminate*. If I wrote that you are a "dir-y d-g", you would *read* it correctly. If you were in a hurry, you might not notice the missing letters. Your past reading experiences fill them in automatically.

That will not do in precision measurement. All we want are the facts. The facts when reading a scale mean the exact values, not conclusions based on hunches, intuition or carelessness. When reading a scale you must be able to separate each division from the next one. That is called *resolution*. It explains why you were told earlier in the book that the discrimination of an instrument is the smallest graduation on its scale. Smaller divisions than can be resolved would be useless.

FUNCTIONAL FEATURES

Going back to Fig. 10-9, the graduations on A and B are satisfactory for reliable discrimination. Therefore the same spacing could have been used for C. This is shown at D, which is an instrument with twice the discrimination as the one at C. Thus increasing the amplification permitted finer discrimination.

The general sizes, dials and mounting dimensions of dial indicators are standardized*. Fig. 10-10 shows a typical instrument. There has been a general similarity of features among all of the instruments discussed up to the dial indicator. This instrument is quite different. The internal construction of dial indicators varies considerably among manufacturers. A typical one is shown in the phantom view, Fig. 10-11.

The dial is attached to the bezel, the outer edge of which is knurled. The dial can thus be turned in respect to the hand. The bezel clamp locks the dial in position.

The spindle and rack are usually one piece. The rack spring resists the measurement movement. It thereby applies the gaging pressure rather than leaving it up to the tech-

---

* National Bureau of Standards Bulletin CS (E) 119-45, Dept. of Commerce. Agreement reached by American Gage Design (AGD) Com.

nician. It also returns the mechanism to the "at-rest" position after each measurement. The hairspring (not to be confused with the rack spring) loads all of the gears in the train against the direction of the gaging movement. This eliminates backlash that would be caused by gear wear. Various indicator backs are available. The instrument is sometimes attached by the stem instead of the back. This may cause damage. The *telltale* is simply a revolution counter for indicators with long range to avoid errors of whole turns.

METROLOGICAL FEATURES

The metrological features of dial indicators bear even less resemblance to the previous instruments than do the functional features. Note, for example, that there is no reference point. The dial indicator amounts to nothing more than a mechanized measured point looking for a place to be from.

In all the previous discussion it has been emphasized that the dial indicator has no standard. This is substantially true in that there is no standard that relates to the *overall length* of the part feature. There is a standard that relates to the observed *change in length*.

The first indicators were simple levers. They only showed that a length was or was not the same size as another part, Fig.10-12. They gave no information about the size of the deviation. Modern indicators measure the amount of this deviation. There can be no measurement without a standard. Therefore, it must be admitted, that in this limited sense, indicators do incorporate a standard. The standard in the indicator measures change in length. It cannot measure length unless it is less than the total range of the indicator.

Three terms can now be clarified. *Test indicators* are the least precise of these instruments. They do not measure, but only indicate whether or not a point moves in one axis. Most test indicators have graduations and some even have dial faces, Fig. 10-6 top, but these are *convenience features* not *metrological features*. Test indicators are used for machine setup and in-process checking.

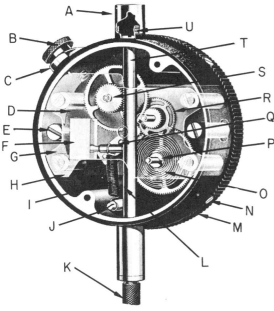

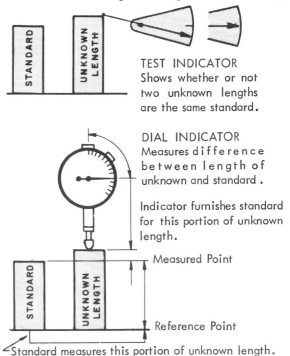

TEST INDICATOR
Shows whether or not two unknown lengths are the same standard.

DIAL INDICATOR
Measures difference between length of unknown and standard.

Indicator furnishes standard for this portion of unknown length.

Measured Point

Reference Point

Standard measures this portion of unknown length.

A.	Dust cap	H.	Bottom plate	P.	Gear assembly takeup
B.	Bezel-screw clamp	I.	Rack spring	Q.	Bearing center pinion
C.	Bezel clamp	J.	Rack spring stud		
D.	Top plate	K.	Lower point	R.	Intermediate gear assembly
E.	Screw movement	L.	Pin rack slide		
F.	Rack slide	J.	Bezel	S.	Rack gear assembly
G.	Screw top plate	N.	Bezel screw	T.	Rack
		O.	Hairspring	U.	Top screw

Fig. 10-11 Phantom view of a typical dial indicator resembles clockwork. *(Federal Products Corp.)*

Fig. 10-12 The dial indicator measures change in length, not length itself.

Dial indicators are refinements of test indicators. The design and construction provide greater precision, accuracy and reliability than test indicators. Discriminations range from one *mil* to *point five mil* (0.001 to 0.0005 in.).

Both test indicators and dial indicators can be used for comparison measurement, hence the term *comparator*. They also have functions, such as concentricity checking, in which they do not act as comparators but simply as indicators of change.

There are other instruments using other principles for amplification. The electronic ones are rapidly becoming more familiar. These are particularly valuable for higher amplifications than can be used satisfactorily in geared instruments. These instruments are usually referred to as *comparators* not as indicators.

COMPARATOR INSTRUMENTS			
Name	Alternate Name	Use	Discrimination
Test indicator	Dial test indicator	Set up and in-process checking	Not intended for measuring
Dial indicator	Dial comparator, Dial gage	Comparison measurement Alignment and positional measurement	0.001 to 0.0001 in. (some to 0.00005 in.)
Comparator	Mechanical comparator, Electronic comparator, Air gages, and many are referred to by trade name.	Comparison measurement of precise parts and for gage calibration	0.0001 to 0.000001 in. (some have low ranges)

Fig. 10-13 These are the terms as generally used. They are ambiguous. For example, all are comparators, yet, comparator alone usually refers to the higher precision instruments. The test indicator and dial indicator are mechanical comparators, yet, that term is generally used only for the more precise types.

COMPARATOR COMPARISON

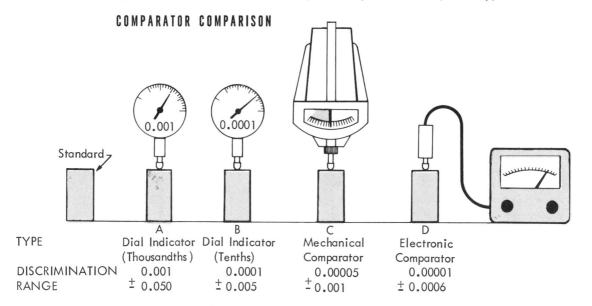

TYPE	A Dial Indicator (Thousandths)	B Dial Indicator (Tenths)	C Mechanical Comparator	D Electronic Comparator
DISCRIMINATION	0.001	0.0001	0.00005	0.00001
RANGE	± 0.050	± 0.005	± 0.001	± 0.0006

Fig. 10-14 To grasp the significance of these comparisons, see Fig. 10-15. (Note: Both C and D instruments are available in other ranges, D in multirange models.)

DISCRIMINATION AND RANGE COMPARISON

	Range	Discrimination

◄ 6' 6.12"

A

± 0.050   0.001 (one mil)

7.81"

B

± 0.005   0.0001 (one-tenth mil)

1.56"

C

± 0.001   0.00005 (fifty mike)

The total ranges of the instruments
are represented by these lengths.

0.94"

D

± 0.0006   0.00001 (ten mike)

One graduation represents these lengths.

Fig. 10-15 These bars, drawn to a scale of 10 millionths of an inch to 1/64 inch, show the tremendous differences between the comparators in Fig. 10-14.

The various instruments are separated in Fig. 10-13 and the ambiguity in terminology pointed out. Fig. 10-14 compares their principal metrological features, discrimination and range. Because the figures are hard to visualize, Fig. 10-15 does that for you.

The bars above the base line in Fig. 10-15 compare the discrimination of the four instruments in Fig. 10-14. The bars below compare their ranges. They have been drawn to scale to show their relative proportions. At this scale the smallest reading (discrimination) of the highest amplification instrument, D, is hardly visible - that is the electronic comparator. The next most powerful one, C, a mechanical comparator, has one-fifth as much discrimination but greater range.

When we jump to the rather commonplace dial indicator in B, the discrimination is reduced another fifth, but the range now is extended greatly. Drawn to the same scale it is over seven inches. The least precise instrument is the *mil*-reading dial indicator. Lowering the amplification here reduces the discrimination to one *mil* but pushes the range out to over six feet in comparison to the range of the others.

This illustrates the tremendous sacrifice that must be made in range in order to gain precision. The length of the ranges in the illustration may be misleading, however, for reasons you will discover when you investigate comparator accuracy.

MOVEMENT IN MEASUREMENT

All measurement requires movement. Of course, there is the movement required to achieve the correct alignment of the standard and the part. This is the same in principle for both direct measurement and comparison movement for each of the two methods of measurement. In direct measurement it consists of aligning the reference and measured points of the part feature with a scale. In some cases, such as a rule, this movement is visual, the eye counting the minor graduations between the reference point and the next major division, Fig. 10-16. In the case of the

MOVEMENT FOR DIRECT MEASUREMENT

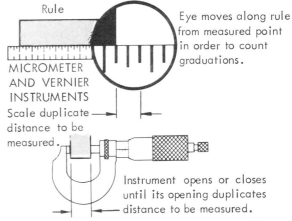

Rule

Eye moves along rule from measured point in order to count graduations.

MICROMETER AND VERNIER INSTRUMENTS

Scale duplicate distance to be measured.

Instrument opens or closes until its opening duplicates distance to be measured.

Fig. 10-16 In direct measurement, movement is required. It consists primarily of correlating the distance to be measured with the instruments, after which the instrument is read.

vernier and micrometer instruments, it consists of physically changing the instrument to duplicate the distance being measured. The change in the instrument is then read from the scale or scales. In all of these cases, it is not the movement that is read, but the results of the movement.

When measuring by comparison we start with a standard whose length is known. All we need to measure is the amount that the part deviates from the standard, Fig. 10-17. We measure not a length, but a *change in length. Direct measurement is static,* but *comparison measurement is dynamic.* The ability to detect and measure the change is the *sensitivity* of the instrument.

SENSITIVITY

Many comparator instruments have relatively low amplification yielding no greater discrimination than vernier or micrometer instruments. This type of instrument, however, is capable of enormous amplification, in the order of 20,000X and greater. At such amplification two factors, not previously of concern, become very important. They are *sensitivity* and *resolution.*

Sensitivity is defined as the minimum input that will result in a discernible change in output*. High sensitivity is desirable for high precision. From the definition it is seen that sensitivity is made up of two factors. One is mechanical. It is the system's ability to provide a proportional output for the change in input. This must be done at low-input magnitudes This is called the *loading* of the system. Second is metrological. It is the *readability* of the system.

These factors work against each other. If the amplification is sufficiently high, even a very small change in input will *theoretically* move the hand a discernible amount. But the higher the amplification, the harder the input signal will have to work to push its way through the maze of gears, shafts and springs to reach the hand.

---

* "Selecting Dimensional Gages," <u>Metalworking Magazine</u>, Boston, Cahners Publishing Co., Nov. 1960, p. 25.

MOVEMENT FOR COMPARISON MEASUREMENT

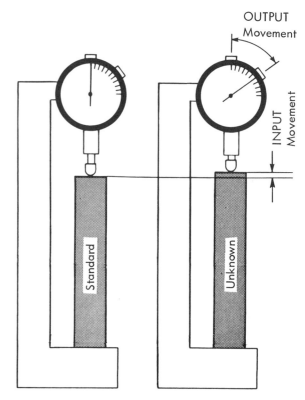

Fig. 10-17 Comparison measurement requires movement because it is change of length that is desired.

In Fig. 10-18, A shows an indicator with a very small input signal being applied to the spindle. Nothing happens--or so it appears. Actually the hand has moved but the movement is not discernible.

The solution seems simple. Increase the amplification tenfold and then the movement of the hand should certainly be seen. Drawing B of Fig. 10-18 shows no improvement, however. What are the reasons? Consider just one as an example, the hairspring that loads the gear train against backlash. Part of the input must be energy to create a torque that will counteract the torque of the spring. The mechanical advantage that gives the large swing of the hand works the reverse also. The force required to counteract the torque of the spring in A is multiplied 100 times by the gear train (assuming 4X amplification from the hand). When the extra gearing is added in B, the required force goes up to 1,000 times. As minute as this is, it causes "wind up" of the shafts and additional frictional loads on the bearings, rack and gears.

That is just one reason the hand did not move as expected. Adding all of them together, including inertia and compressibility, the sensitivity was not discernibly improved by the increased amplification.

Now in C, David comes upon the battlefield and his trim lines are no match for the Goliath-like graduations of the instruments in A and B. Without any more amplification than in the first case, A, the movement of the

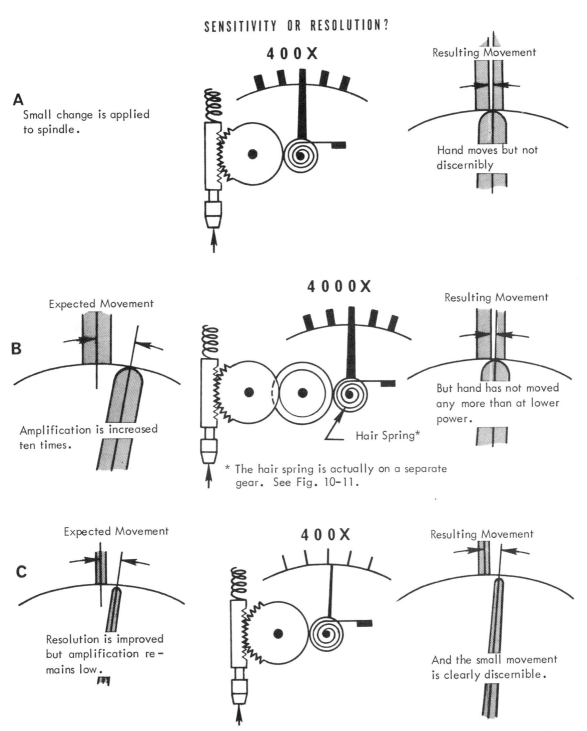

SENSITIVITY OR RESOLUTION?

400X

A Small change is applied to spindle.

Resulting Movement

Hand moves but not discernibly

4000X

B Amplification is increased ten times.

Expected Movement

Hair Spring*

* The hair spring is actually on a separate gear. See Fig. 10-11.

Resulting Movement

But hand has not moved any more than at lower power.

400X

C Resolution is improved but amplification remains low.

Expected Movement

Resulting Movement

And the small movement is clearly discernible.

Fig. 10-18   Increasing amplification may not be as effective for improving sensitivity as better resolution.

hand is discernible. The reason is the improved *resolution.*

When high sensitivity and high resolution are combined in a low-amplification instrument, it will in many cases provide more accurate and more reliable measurement than a high amplification instrument. [*]

In popular use the term, sensitivity, is usually used instead of precision when comparing dial indicators. This is not without some justification, but is inaccurate. When amplification is increased, the potential precision and sensitivity both are increased. The sensitivity, for the reasons mentioned, may not be increased in proportion to the increase in precision. In spite of popular use, this book will use precision to designate differences made in indicators by changes in amplification. For example, a *tenth-mil* (0.0001-in.) indicator will be spoken of as higher precision than a *fifty-mil* (0.0005-in.) indicator-- and we hope that the sensitivity lives up to our faith in the principal manufacturers of these instruments.

---

[*] Hume, K.J., Engineering Metrology, London, Macdonald and Co., 1960, p. 110.

RESOLUTION

Resolution is the ratio of the width of one scale division (one output unit) to the width of the hand (the read-out element). [**]

Resolution is not the same as readability. It is the most important of the many factors that make readability. Among the others are size of graduations, dial contrast, and our old enemy, parallax. The width of the graduations should be the same as the width of the hand. The definition permits a numerical value to be assigned to resolution, Fig.10-19.

In Fig.10-19, B has five times the resolution of A. This might slightly reduce some of the other readability factors but the improved resolution will more than compensate.

In fact, with the graduations and hand as fine as shown at B it would be possible to divide the scale into smaller divisions, thus raising the discrimination.

[**] "Selecting Dimensional Gages," Metalworking Magazine, Boston, Cahners Publishing Co., Nov. 1960, p. 25.

RESOLUTION

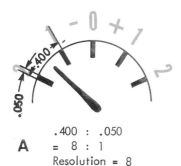

.400 : .050
A   = 8 : 1
Resolution = 8

.400 : .010
B   = 40 : 1
Resolution = 40

Ratio of the width of one scale division to the width of the hand.

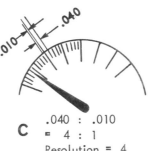

.040 : .010
C   = 4 : 1
Resolution = 4

Fig. 10-19 The larger the resolution factor (without impairment of other readability factors) the greater the reliability.

This is shown at C. Because resolution is a function of the space between graduations, it must be refigured. Now the resolution has dropped to four, only one half of that in A. This is a numerical value for something we already know. Scale C is more difficult to read than scale A or B. The net result is that scale C has ten times the discrimination but reduced readability.

What determines the number of graduations that can be put on a dial, then? The precision of the instrument. Precision here is not simply synonymous for "mighty accurate" but has the meaning explained in Chapter 2. *Precision is the measure of the dispersement of repeated measurements.* To review, remember that precision is essential for accuracy, but not a measure of accuracy. And precision plus accuracy are essential for reliability but are not alone a measure of reliability.

In Fig. 10-20 at A is an enlargement of one division from scale A in Fig. 10-19, but instead of one reading, five are shown. The

precision leaves something to be desired, but all five readings would clearly be associated with just one graduation. Being highly readable the reliability of the observed reading would be high.

At B the scale has ten times the discrimination, but what would the measurement be, 0.0018, 0.0019 or 0.0020 in.? Having five readings to observe we can choose reliably 0.0019 in. What would the reliability be if there was only one reading to go by?

What was poor (but usable) precision when the discrimination was poor, becomes completely unacceptable at higher precision. Generally, resolution should increase with an increase in amplification and/or discrimination. Low resolution is used in shoddy instruments to conceal the fact that they do not have expected precision accuracy, sensitivity or reliability.

COMPARATOR ACCURACY

This is a good place to clear up two terms used in conjunction with indicators, *repeat accuracy* and *calibrated accuracy*. It is unfortunate that repeat accuracy is in general use. It has nothing to do with accuracy. Forget it—the word is *repeatability*. Calibrated accuracy is a redundancy. Accuracy is provable only by calibration. Therefore the word calibration is unnecessary. The preferred term is simply *accuracy*.

Accuracy is usually stated differently for comparators than for the previously discussed instruments. Instead of so much per unit of measurement, as *two-tenth mil* per inch (0.0002 in. per in.), it is usually stated in terms of dial graduations. A frequently stated accuracy for dial indicators is plus or minus one dial division with repeatability within 1/4 of one dial division.

These are very inaccurate expressions of accuracy and precision. They fail to take into consideration that the *accuracy decreases as the distance traveled increases.* To take this into account the preferred method of stating accuracy is as a percentage of the full scale range.

The way this works is shown in Fig. 10-21. A comparator with discrimination of *one mil*

RESOLUTION AND PRECISION

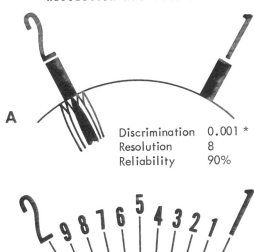

Discrimination	0.001 *
Resolution	8
Reliability	90%

Discrimination	0.0001
Resolution	4
Reliability	20%

* As an example. Could be 1.0, 0.1, 0.0001, etc.

Fig. 10-20 High resolution will not substitute for high precision.

(0.001 in.) and range of *twenty mil* (0.020 in.) has been used for illustration. In A the comparator is zeroed on a 1.0000-in. gage block standard. In B, that standard has been replaced with a 1.0200-in. block in order to move the hand through the full range. (Range is not this simple, but that will be discussed later.) Instead of the correct reading of 1.0200 in. the comparator reads 1.0190 in., or *one mil* (0.001 in.) short.

The error (0.001 in.) divided by the range (0.020 in.) and multiplied by 100 is the percentage of error. This works out to 5% error.

Note that this percentage is error, not accuracy. The frequently heard statement that a given comparator has "an accuracy of ±2% of full scale" is incorrect--it had better be! The thought should be stated (author's italics for emphasis): "Accuracy of an instrument is the *limit*, usually expressed as a percentage of full-scale value, which errors will not exceed when the instrument is used under referenced conditions."* Accu-

* American Standards Association, ASC C39.1 - 1959.

racy is thus defined as the limit of permissible error, which makes sense.

In A and B of Fig. 10-21 we have shown that the error is five percent. In C and D unknown parts have been substituted for the standard. The advantage of basing accuracy on a percentage is shown. The adjustment in C is substantial, but so is the distance measured. At D where only *one mil* (0.001 in.) has been measured, the correction is very small.

By using the percentage method, it is seen that the error in the measurement reduces more rapidly as the distance is decreased than does the distance itself--at zero distance there is (theoretically) zero error. This is an extremely important point. You can improve the reliability of your comparator measurements by remembering that *a comparator has its greatest accuracy at zero and the accuracy decreases in proportion to the distance from zero.*

Fig. 10-22 shows that this does not apply to just one point on the dial, but that zero may be any point within the range. The rotating dial provides this feature.

COMPARATOR ACCURACY

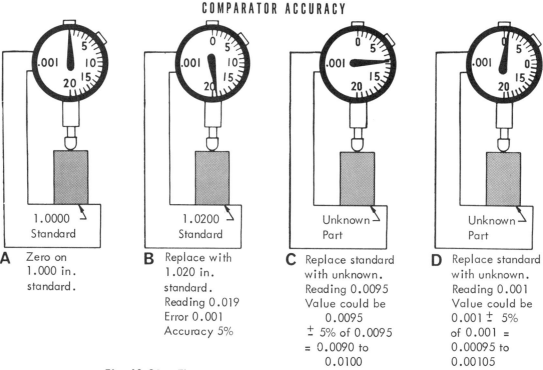

A Zero on 1.000 in. standard.

B Replace with 1.020 in. standard.
Reading 0.019
Error 0.001
Accuracy 5%

C Replace standard with unknown.
Reading 0.0095
Value could be
   0.0095
± 5% of 0.0095
= 0.0090 to
   0.0100

D Replace standard with unknown.
Reading 0.001
Value could be
0.001 ± 5%
of 0.001 =
0.00095 to
0.00105

Fig. 10-21   The percentage method of expressing accuracy, proportions the error in measurement by the distance measured.

ZERO SETTING

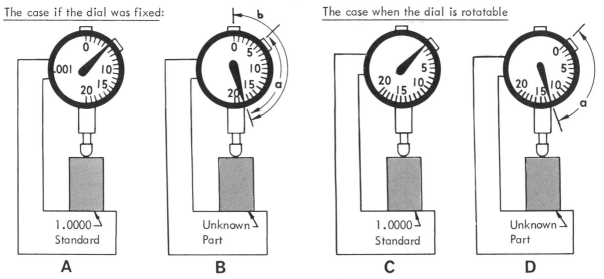

The case if the dial was fixed: The case when the dial is rotatable

A.   In this case, the 1.0000 standard provides a reading of 0.005.
B.   The unknown (a) reading is 0.018. The difference is 0.013. Adding 0.013 to 1.0000 gives the length of the unknown as 1.013.
C.   In this case, the dial is turned to provide a zero reading on the 1.0000 standard.
D.   The unknown part provides an 0.013 reading directly. The part size is 1.013. There is no question that it is the travel of the hand to which the error correction applies.

NOTE: Assuming same 5% full-scale accuracy as in Fig. 10-21, the correction is applied to 0.013 travel of the hand (a) not to the 0.018 travel (b).

Fig. 10-22   The dial rotation permits zero setting. This eliminates possible computational errors.

Zero setting is important because it eliminates two steps, (remembering the setting and computing the correction). Every step in the measurement sequence is a potential source of errors. These steps are particularly troublesome when they must be repeated over and over. These are random errors and should not happen--however, some are bound to occur. In eliminating these sources of error, another source has been added, the zero setting error. *Any error in zero setting will be contained in every measurement made,* Fig. 10-23. The overall result of a sufficiently large, systematic error such as a zero setting error is many times more serious than that of a random error, because it applies to every measurement made until it is detected and corrected.

INDICATOR DIALS

Most of the remarks in this chapter apply to all comparator instruments. In order to discuss comparator *range*, however, it is necessary to first consider the scales. The

---

ZERO SETTING ADJUSTMENT

Errors Eliminated:

1. Memory error.

2. Computational error.

    These are random errors and <u>will</u> not be contained in all measurements.

Error Added:

Zero setting error.

    This is a systematic error and <u>will</u> be contained in all measurements.

Fig. 10-23   Every act in measurement is a source of error. When the errors created are less than those reduced, the act is justified. These are the errors added and eliminated by zero setting.

scales for dial indicators, conveniently called *dials*, are necessarily quite different from those of other comparator instruments.

STANDARD SIZES

3 5/8
Group 4

2 3/4
Group 3

2 1/4
Group 2

1 11/16
Group 1

Monogram of the American
Gage Design Standards

Group AGD	Bezel Diameter		Discrimination	
	Above	To and Inc.	Inch	Mm.
1	1 3/8	2	0.0001 0.0005 0.001	0.005 0.01
2	2	2 3/8	0.00005 0.0001 0.0005 0.001	0.001 0.002 0.005 0.01
3	2 3/8	3	0.0001 0.0005 0.001	0.001 0.002 0.005 0.01
4	3	3 3/4	0.00005 0.0001 0.0005 0.001	0.001 0.002 0.005 0.01

Fig. 10-24 These four standardized dial sizes are available in a wide variety of graduations.

Dials have been standardized for dial indicators as well as overall dimensions and mountings.* The standard dials have four discriminations, *mil* (0.001 in.), *five-tenth mil* (0.0005 in.), *two hundred fifty mike* (0.00025 in.), and *one hundred mike* (0.0001 in.). This means, of course, that the indicators must be made in a range of amplifications. Closely related to the discrimination are two other factors, dial size and range.

* American Gage Design Specifications, GS-E-119-45 (designated as AGD or AD).

The sizes are shown in Fig. 10-24. If a given spacing is required for readability, the larger the dial and the longer the range. Conversely, the higher the amplification, the higher the discrimination that is possible, providing that the dial is large enough to provide readability, Fig. 10-25.

The dials are of two general types, *balanced* and *continuous*. The other comparators, particularly those of high amplification, generally have only the balanced type.

The continuous dial has graduations starting at zero and extending to the end of the recommended range, Fig. 10-26. They are made both clockwise and counterclockwise. These dials correspond to the unilateral (one-way) tolerances used in dimensioning. The balance dials have their graduations running both ways from zero. These dials correspond to the use of bilateral (both way) tolerances.

UNILATERAL AND BILATERAL TOLERANCES

Briefly, every design is conceived in perfection.** The designer knows that it cannot be manufactured perfectly, at any cost, let alone a practical cost. Therefore, tolerances are given to the dimensions. *A tolerance is the amount that a part feature may differ in size from the basic size and still be acceptable.*

There are two types of tolerances, *unilateral* and *bilateral*. Fig. 10-27. A unilateral tolerance is permissible variation in one direction. It is specified by stating the two limits, such as 0.250 to 0.256 in. That provides a range of *five mil* (0.005 in.) in which all parts would be acceptable. A bilateral tolerance states the basic size and the permissible variation in each direction. For example, 0.253 ±0.003 in. means that the basic dimension of 0.253 in. may be departed from by three mil (0.003 in.) in either the plus-or-minus directions. The total acceptable range is the same as in the previous example. If the dimension has been (0.250 in. + 0.006 in. -0.000 in) or (0.256 in. + 0.000 in. - 0.006 in.)

** Wolowicz, C. S., Rindone, C. P., and Wilcox, R. J., "Axioms of Drafting Technology - VII," American Machinist/Metalworking Manufacturing, New York, McGraw-Hill, September 3, 1962, p. 101.

DESIGN AND METROLOGICAL FACTORS

An increase in these factors: ↓	Has the following permissible effect on these:				
	Amplification	Dial Size	Discrimination	Readability	Range
Amplification		Increases	Increases	Decreases	Decreases
Dial Size	Decreases		Increases	Increases	Increases
Discrimination	Increases	Increases		Decreases	Decreases
Readability	Increases	increases	Decreases		Decreases
Range	Decreases	Increases	Increases	Increases	

Fig. 10-25 Standard indicators compromise these factors for the greatest range of practical applications.

TYPES OF DIALS

CONTINUOUS
CLOCKWISE
0-90

CONTINUOUS
COUNTERCLOCKWISE
0-90

BALANCED
0-50-0

METHOD FOR DESIGNATING NUMBERS

Fig. 10-26 These are the three types of dials used on dial indicators, and the method for designating.

DIMENSIONAL TOLERANCES

UNILATERAL   BILATERAL

$\dfrac{0.256}{0.250}$ 0.250 ± 0.003 $0.250 \begin{array}{l} + 0.006 \\ - 0.000 \end{array}$ $0.250 \begin{array}{l} + 0.000 \\ - 0.006 \end{array}$

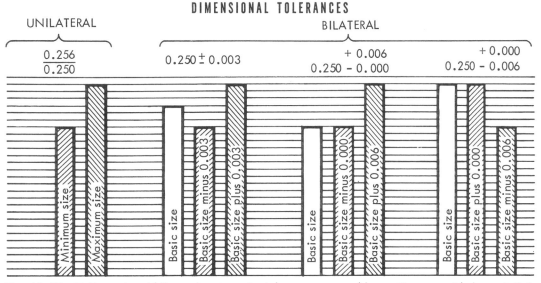

Fig. 10-27 Unilateral and bilateral systems for tolerances, resemble continuous and balanced dials.

all permissible deviation would have been in one direction, but the tolerance would still be bilateral.

The word frequently paired up with tolerance is *allowance*. It is mentioned here just to make sure that you do not fall into the common error of thinking these are the same or even very close relatives. Allowance is the space between mating parts. Tolerances are required to machine parts so that they will have allowances, and that is where the similarity ends.

When we compare indicator dials with tolerance systems, continuous dials resemble unilateral tolerances, and balanced dials resemble bilateral tolerances. This is given as a reason for the existence of the two dial types. It is true that you must use a balanced scale to directly read bilateral tolerances. But you can measure a unilateral tolerance just as easily with a balanced dial as with a continuous dial. In fact you have to at times,

because high amplification instruments are only available with balanced scales. Long over-all range seems to be a better explanation for the existence of the continuous dial.

BALANCED INDICATOR DIALS

There is a tremendous variety of dial sizes and graduations available for both balanced and continuous dials. The general principles of reading all balanced dials are shown in Fig. 10-28. The value of the smallest division is always shown on the dial. This is never the value of the numerals around the dial. These are simply aids in counting small divisions. They usually represent 10 divisions but also may represent 2, 5 or 20 division steps around the dial.

In reading a balanced dial indicator remember that you might be measuring but the indicator is comparing. The indicator reading in itself means nothing. To have mean-

READING BALANCED DIALS

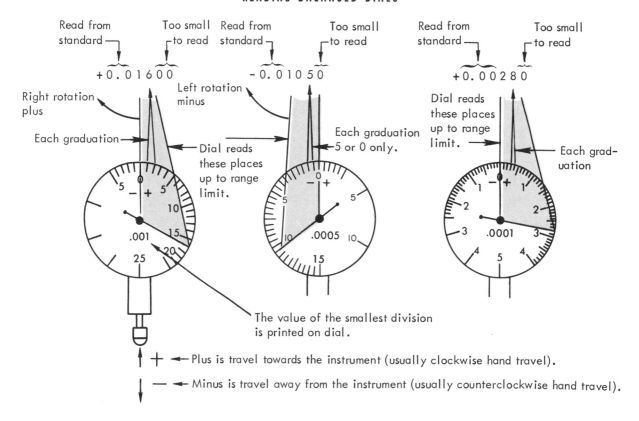

Fig. 10-28   Although the dials may vary in size and graduations, these general principles apply.

REVOLUTION COUNTERS ON BALANCED DIAL INDICATORS

NOT RELIABLE for measurement.

RELIABLE for:

    1. Showing that you are past the overtravel and within the measurement range.

    2. Warning that an indicator with too high an amplification is being used.

    3. Warning that something in the setup has shifted.

    4. Warning that the part being measured is <u>out</u> of <u>control</u>. *

* Out of control means that the previous operation has been completely missed or that the machine is completely out of adjustment.

Fig. 10-29   These warnings are not calculated to make friends. There are those who trust "tell-tales" for measurement, but, a word to the wise . . . .

ing it must be added to or subtracted from the size of the standard to which the instrument has been calibrated.

Revolution counters might seem odd on balanced indicators. *They are.* The conventional indicator has a total travel of two and one-half turns. The hand could revolve completely through the plus range, through the minus range backwards and into the plus range again. The explanation for the revolution counter, or "telltale" as it is called, is to warn you when this has happened.

That warning is often misinterpreted. If you remember what happens to accuracy with long travel you will understand why. The correct message that you should get from a reading on the revolution counter of a balanced dial indicator is that you are using the wrong indicator for the measurement requirements. There are two exceptions to this. One occurs when the long travel is required to get *to the point* from which the measurement is referenced. That is discussed in the next section. The other is to show starting tension. That, too, is discussed further on.

This does not mean that revolution counters are not recommended on balanced indicators. They are, but *use them as warning signals.* Use them to show that something in the setup has shifted, that something is radically out of control, or that an overly sensitive indicator has been chosen for the measurement, Fig. 10-29. The exception comes later.

CONTINUOUS INDICATOR DIALS

A continuous dial has double the range in one direction of an equivalent balanced dial. When even greater range is desired, (note, I did not say necessary) indicators are available with very long racks that pass completely through the cases, Fig. 10-30.

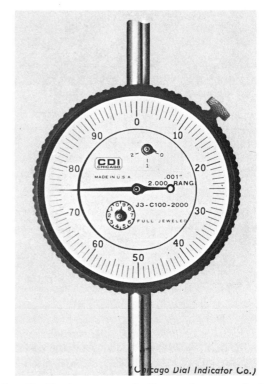

(Chicago Dial Indicator Co.)

Fig. 10-30   Long-range indicators have revolution counters and very long-range ones have inch counters.

These indicators are read similarly to balanced scale indicators. The main differences are no minus readings and larger numbers. A typical example is shown in Fig. 10-31.

The great ranges of the continuous dial indicators are extremely valuable in certain circumstances. Unfortunately, the same long ranges make possible the abuses that have given these instruments their bad reputation. It is easy to forget that they are comparators and not direct-measurement instruments. All too often the relationship between the percent of full-scale accuracy to the length traveled is forgotten.

This point is developed in Fig. 10-32. In the oversimplified example, a long-range indicator is being used to control the tool travel on a lathe to cut a shoulder of accurate length. On the left is shown the popular but wrong use. After roughing cuts, the lathe faces the shoulder. This defines one end of the part feature. The tool is backed off to the finished diameter of the next cut. The long-range in-

READING CONTINUOUS DIAL INDICATORS

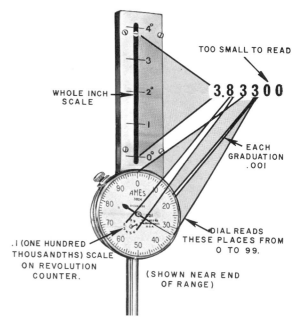

Fig. 10-31   Continuous dial indicators have ranges up to 10 inches. *(B. C. Ames Co.)*

USE OF LONG RANGE INDICATORS

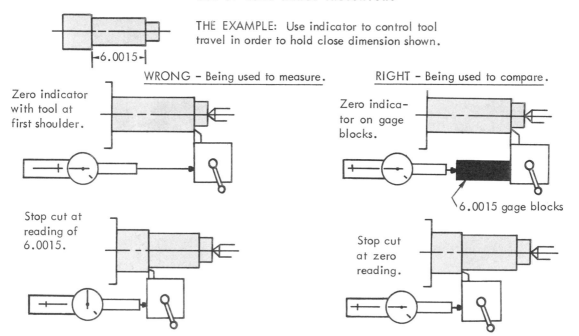

THE EXAMPLE: Use indicator to control tool travel in order to hold close dimension shown.

WRONG – Being used to measure.

Zero indicator with tool at first shoulder.

Stop cut at reading of 6.0015.

If the indicator has 2%" full scale accuracy" the part feature produced may vary from 5.8815 to 6.1215.

RIGHT – Being used to compare.

Zero indicator on gage blocks.

6.0015 gage blocks

Stop cut at zero reading.

Even if indicator has 2% "full scale accuracy" the part feature will be the desired 6.0015 (less error inherent in machining).

Fig. 10-32   When used properly, the long-range indicator is not only convenient but reliable as well.

dicator which is attached to the lathe bed is zeroed somewhere against the carriage. The long cut is made, the indicator warning the machinist when he is approaching the desired measurement. Unfortunately, the small percentage of error in the indicator when multiplied by the long travel becomes a serious error. A 2% error could result in the part being off any amount from 5.8815 to 6.1215 in. Why the bad results? Because *the comparator was used as a direct-measurement instrument instead of as a comparator.*

The drawings on the right of Fig. 10-32 show the proper way to use the long-range comparator for the same operation. After the shoulder on the small diameter at the right end of the part has been turned, the tool is withdrawn to the diameter of the next cut. So far the same as before. Now the difference. Instead of zero setting the indicator against the carriage, gage blocks for the dimension are inserted between the spindle and the carriage. This represents the desired position of the carriage when the tool is at the end of the cut. The indicator is zeroed. This then permits the machinist to follow carefully the progress of his cut and stop it accurately at the desired place.

In the latter example, the errors in the final result are primarily a result of the machining (overtravel of the carriage, etc.). The reason for the accuracy is that the distance traveled by the lathe carriage was *measured by the gage block standards.* The indicator only showed when that distance was reached. The indicator acted in its intended function--as a comparator.

COUNTERCLOCKWISE DIAL INDICATORS

This example also explains why continuous dials are available to read in either direction. In the example shown, it would be more convenient to use a counterclockwise indicator. In that case, the so-called zeroing should consist of setting the indicator to read the finished dimension when the gage blocks are in place. As the cut progresses the indicator should read the distance traveled directly, as shown.

The reason for counterclockwise dials is shown in Fig. 10-33. In this example the part feature being measured has a *unilateral tolerance* of 2.000 to 2.005 in. At A, a conven-

CLOCKWISE AND COUNTERCLOCKWISE DIALS

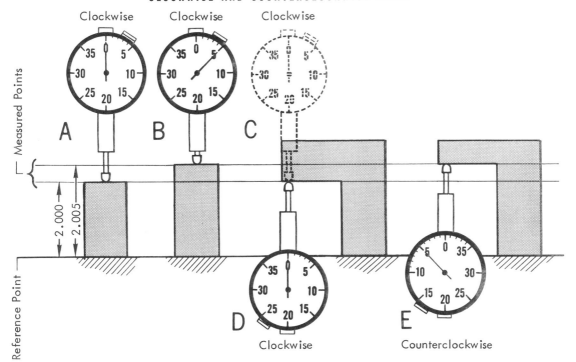

Fig. 10-33 The dial rotation depends upon the side of the measured point that the indicator is on.

tional indicator with a clockwise dial is zeroed on a 2.000-in. standard. At B, the part that replaces the standard is at the high limit of the tolerance. As expected, the indicator reads plus *five mil* (0.005 in.). That reading is added to the original setting as established by the standard to provide the measurement of 2.005 in.

Suppose, however, that the shape of the part is very different but that the dimension and tolerance are the same. This is shown at C, D and E. The phantom indicator at C could theoretically be exactly the same as in A and B. The readings also would behave the same. The only trouble is that you cannot punch holes in parts in order to measure them.

Therefore, the indicator is run in from the bottom as at D. Remember that the bezel permits the dial to be turned. As far as the indicator is concerned it does not matter how much the dial is turned. So the dial position at D would be normal. The zero setting is correct. There is just one trouble now. As the part feature becomes longer the indicator travels backwards. The extreme reading would be *thirty-five mil* (0.035 in.) instead of *five mil* (0.005 in.). If this is kept in mind a simple computation could be made to adjust for this.

The counterclockwise dial provides a better solution. The hand in E travels the

same as it normally would but the scale is reversed. Now the readings can be added directly to base dimension as at A and B.

The rule is that a clockwise indicator should be used when the indicator is on the opposite side of the measured point from the reference point -- and a counterclockwise indicator should be used when the indicator is on the same side of the measured point as the reference point.

Returning to Fig. 10-32, it is seen that this works. When using the indicator the wrong way, the reference and measured points are interchangeable. In doing it the right way with the gage blocks, the reference point becomes crystal clear in the first step. Then it is seen that the indicator and the reference point both are on the same side of the measured point, and a counterclockwise indicator should be used.

INDICATOR RANGE

Before discussing indicator range, we need a method to describe the position of the hand on an indicator. It is confusing to speak of hand position by the reading because of the tremendous number of dial sizes and types. The method often used is to state the position of the hand as if it were the hour hand on a clock, Fig. 10-34.

The very long ranges of some indicators have been mentioned. Most indicators, however, have a working range of two and one-half turns. The full range is even greater because of the necessity for overtravel. At rest, the hand is usually at the 9 o'clock position shown in Fig. 10-34. The measuring position of a continuous indicator should not begin without at least one-quarter clockwise rotation of the hand, at 12 o'clock or beyond. For balanced indicators, the measuring should not begin until the hand has been rotated to the center of its working range, or one and a quarter revolutions. Note that this initial travel is independent of the zero position on the dial.

Overtravel on most instruments is a safety precaution for the occasional measurement that may run beyond the scale. There is a more important reason in the case of dial indicators. It is to preload the mechanical

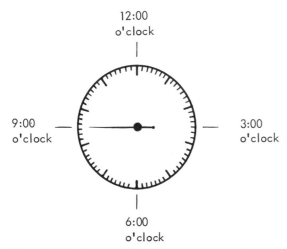

TIME ANALOGY

12:00
o'clock

9:00
o'clock

3:00
o'clock

6:00
o'clock

Fig. 10-34 Because dial faces vary, the time analogy often is used to state hand positions.

COMPARATOR PERFORMANCE FACTORS

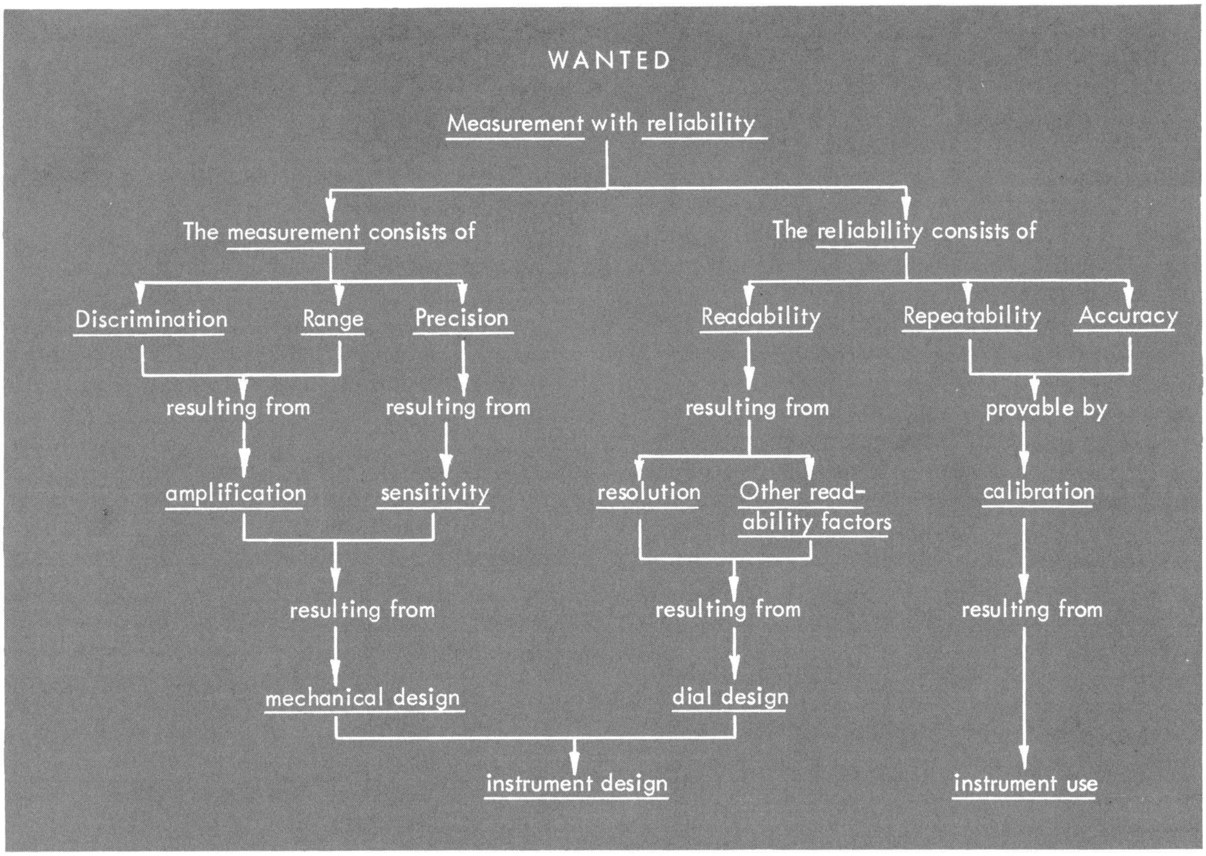

Fig. 10-35 Although oversimplified, this chart shows the maze of interconnecting relationships that govern dial indicator design, selection and use.

movement for *starting tension.* As explained earlier, the hairspring shown in Fig. 10-18 preloads the gear train to take up backlash.

If Fig. 10-11 is examined closely, it will be seen that this spring is actually on a separate large gear and that it is in mesh with a small gear on the spindle. This means that only a small amount of spring tightening is required for the entire range. This provides nearly uniform torque.

It is to get within that range of uniform torque and correspondingly uniform preloading that overtravel is required. This spring is also one of the factors limiting the indicator range. For any given proportion of spring, a point is reached after which the force for further tightening will increase rapidly. That would contribute still greater errors to the ones that already exist as a result of inevitable inaccuracies in the rack and gear train.

The range is independent of dial size or discrimination. It is a function of the amplification. It is not affected by direction of rotation. Remember that a large dial indicator is not selected in order to achieve greater range (or accuracy) than a small one.

SELECTION OF A DIAL INDICATOR

There are two groups of decisions that must be balanced--and often must be compromised. One group consists of functional decisions concerning size, attachment and accessories. The other group consists of the metrological considerations. Some of the items overlap. For example, the range is both a functional and metrological factor in selection. The interconnection of these factors is shown in Fig. 10-35. Fig. 10-25 shows the effect that a change in one factor may have on the others. Furthermore, there may be little choice. When you only have one indicator available, it is more important to know how to get the best results from it in spite of its limitations, than to know that another one would be better.

First, select the precision. That is variously referred to as the sensitivity, the magnification, and the minimum graduation. In a dial indicator, discrimination is the measure of precision and refers to the four general classifications; *one mil* (0.001 in.), *five-tenth mil* (0.005 in.), *two hundred fifty mike* (0.00025 in.) and *one hundred mike* (0.0001 in.), Fig. 10-36.

The general rule for selecting the precision of a comparator instrument is based upon the range requirement. It should be selected with the shortest range that will contain the tolerance. The shorter the range, the higher the permissible amplification, therefore the higher the precision. This rule is used for the very precise electronic and mechanical comparators that you will find in Chapter 11. It must be modified for dial indicators because of two previously mentioned

characteristics of dial indicators: (1) the loss of accuracy that accompanies the relatively long travel, and (2) for the loss of sensitivity that accompanies high amplifications.

The practical method for selecting dial indicator precision and range is shown in Fig. 10-37. From the available precision classes (minimum graduations) select the indicator that most nearly spreads the tolerance over 10 dial divisions. In other words, the discrimination should be 10 percent of the tolerance. The example in Fig. 10-37 is a tolerance of *six mil* (0.006 in. or ± 0.003 in.). The discrimination should be *five-tenth mil* (0.0005 in.).

Select the range so that the total tolerance spread occupies at least 1/10, but not more than 1/4 of the total dial. Note: The spread is over the *dial*, not the total indicator *range*. This rule for selecting the range drives home a point that has been made repeatedly before. The indicator is a comparator instrument. The closer to the setting position (zero as established by setting to a standard), the more accurate the measurement will be.

DIAL INDICATOR RANGES						
Numbering		Class of Precision Instrument				
B*	C**	0.001	0.0005	0.00025	0.0001	0.00005
0-50-0	0-100	0.250				
0-25-0	0-50	0.125	0.125			
0-20-0	0-40	0.100	0.100			
0-15-0	0-30	0.075	0.075	0.075		
0-10-0	0-20	0.050	0.050	0.050		
0-5-0	0-10			0.025	0.025	
0-4-0	0-8				0.020	
0-2-0	0-4				0.010	

*Balanced   **Continuous

Notes:
1. Based on 2 1/2 turns range.
2. Same for all dial sizes.
3. Same for counterclockwise rotation.
4. See manufacturers' catalogs for ranges of long travel types.

Fig. 10-36 The range, as well as the sensitivity, must be considered when selecting an indicator.

DIAL INDICATOR SELECTION

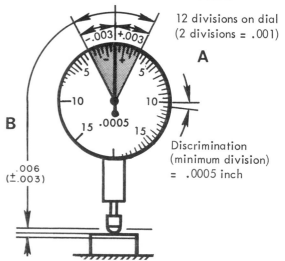

12 divisions on dial (2 divisions = .001)

A. Discrimination — approximately 10% of tolerance (total tolerance should cover about 10 divisions).

B. Range — total tolerance more than 10% of dial and less than 25% of dial.

C. Readability — select largest dial that satisfies A and B and physically meets the requirements.

Fig. 10-37 These steps assure a practical selection for most measurement requirements.

Fig. 10-38 shows why this is particularly important when measuring production parts. Tolerances establish limits. Indicator error at these limits will cause good parts to be rejected and bad parts to be passed. The measurement must be confined to a portion of the dial range that will minimize these errors.

When there is a choice of several indicators, all of which meet the measurement requirements, the final decision is based on readability. This involves more than size alone. A very large indicator may have to be placed at an angle in order to fit into the required space. This could introduce parallax error. Two indicators of the same size may differ in the legibility of their numerals or glare characteristics.

USE OF DIAL INDICATORS

Dial indicators are frequently built into other measuring instruments as the read-out portion, Fig. 10-39. These instruments are more akin to attribute gages than to measur-

EFFECT OF INDICATOR ERROR

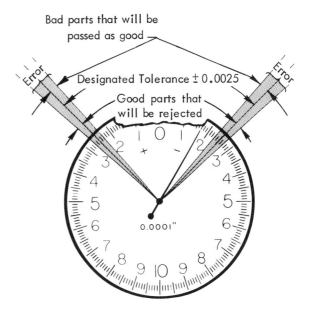

Fig. 10-38   This exaggerated drawing shows that indicator error causes bad parts to be accepted and good parts to be rejected when used for production inspection.

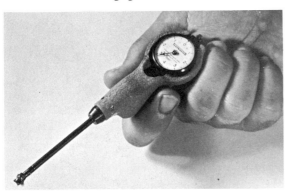

Dial Bore Gage

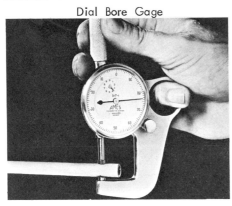

Dial Micrometer

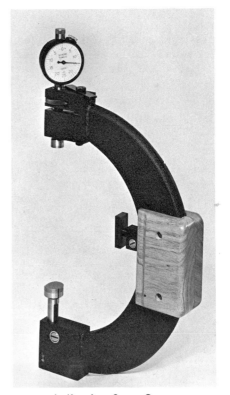

Indicating Snap Gage

Fig. 10-39   Dial indicators are often used as the read-out portion of gages. They are reliable when treated as special forms of comparators rather than as direct-measurement instruments.

ing instruments, and will not be discussed here in any detail. They are most reliable when treated as comparators; that is, when they are set with gage blocks or setting masters to the basic size and readings taken with minimum travel from that size. It is highly important that these gages be frequently calibrated.

The principal concern in this chapter is the use of dial indicators for comparison measurement. Among the principal uses of dial indicators are surface plate measurement and inspection. In later chapters much more will be said about the positional relationships of comparator instruments.

The remarks about cleanliness apply even more emphatically to dial indicators than to the previously discussed instruments. Whenever the amplification of an instrument is increased, the effect of unwanted disturbances such as dirt and vibration is correspondingly increased. The dial indicator is a rugged instrument, but its maze of moving parts are all the more reason to keep the instrument clean, away from abrasive dust and cutting fluid.

Care must be taken to prevent abuse. The slender spindle can be easily damaged. Avoid sharp blows, side pressure and overtighten-

ing of the contact points. The latter actually bulges the spindle sufficiently to interfere with its free travel in the spindle bearing. Clamping on the indicator stem may similarly cause binding. When an indicator is to be stem mounted it should be equipped with a threaded stem that screws into a hardened steel adapter. These are standard parts.

Among the most important considerations is rigidity. It is difficult to realize the rubbery action of steel until you begin measuring to a *tenth mil* or closer. Fig. 10-40 shows some precautions to observe when making comparison measurements. A vivid demonstration can be made easily using a standard indicator stand and accessories, Fig. 10-41. Everyone should witness this demonstration before being asked to do any measurement involving *tenth-mil* increments.

COMPLETE INSTRUCTIONS FOR REPAIR OF DIAL INDICATORS

Carefully pack and return to manufacturer.

CALIBRATION OF DIAL INDICATORS

Calibration of dial indicators is much the same as the calibration of the previously dis-

DO'S AND DON'TS FOR INDICATOR USE

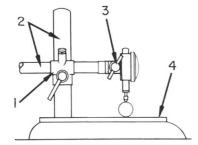

DO'S

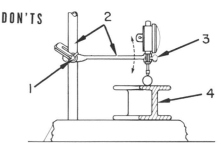

DON'TS

1. Properly designed clamps are available. Use them. Check frequently for play.
2. Use ample supports and keep them as short as possible.
3. Use reference surface whose features are known and in calibration.
4. Clamp to indicator back whenever possible.

1. Makeshift clamping causes shifts that destroy accuracy and make setups difficult.
2. Slim posts and arms do not provide sufficient support.
3. Do not use odds and ends for supporting reference surfaces. If you do, don't blame the indicator.
4. Do not clamp to indicator stem unless adaptor is used.

Fig. 10-40   These general precautions will add reliability to comparison measurement. *(Federal Products Corp.)*

cussed instruments. A series of measure-ments are made with standards and the read-ings compared with the true values. The most practical method requires gage blocks and will be discussed later.

There are two points that should be men-tioned now, however. First, *when using comparison instruments you must know that both the instrument and the standard by which it is set are in calibration.* Neither one alone is enough for reliable measurement. Second, the conditions under which the measuring is to take place should be duplicated as closely as possible for calibration. To remove the indicator from the setup, return it to the gage department for calibration and then replace it, is a good start--but it is not a calibration of the complete setup, nor is it a calibration under the conditions of use.

ACCESSORIES AND ATTACHMENTS

It is not the purpose of this book to go into the modifications, attachments and acces-sories for the instruments discussed beyond that which is necessary for an understanding of the principles. Manufacturers' literature and operating instructions are available for all instruments, and many books go into min-ute detail about their design and use.

Dial indicators are versatile instruments because their mountings adapt them to many methods of support, and the interchangeable contact points adapt them to many measure-ment situations.

Contact points are available in a variety of hard, wear-resisting materials including boron carbide, sapphire and diamond. In general, it is assumed that a chromium-plated point lasts ten times as long as a hard-ened steel point, and that a carbide or harder point will last 100 to 1,000 times as long.

Both flat and rounded points are available. Figure 10-42 gives the nomenclature of the standard ones. The pros and cons of flat points are shown in Fig. 10-43. Spherical contacts usually are preferred because they present point contact to the mating surface whether it is flat or cylindrical. Some precautions are necessary, of course, Fig. 10-44. Ex-cessive gaging pressure may indent the part. When cylindrical parts or standards are used, care must be taken to pass them through the center line of the spindle. The highest read-ing will be the diameter. All other readings are chords.

The use of contact points on spherical sur-faces presents special problems, Fig. 10-45.

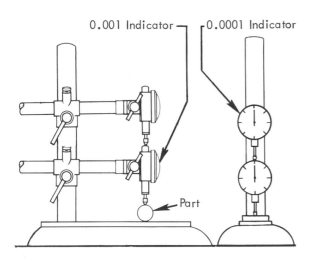

Fig. 10-41 This simple demonstration will show viv-idly the deflection in gaging setups. First, zero the thousandths indicator on the part. Remove the part. Zero the tenths indicator on the thousandths indi-cator. Replace part and notice the deflection on the tenths indicator. That is only part of the deflection because the tenths indicator is also being deflected.

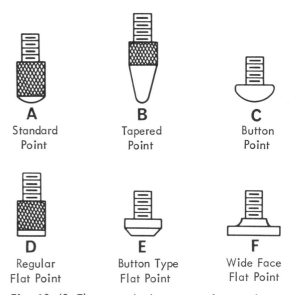

Fig. 10-42 These standard contact points are inter-changeable and available in a variety of wear-re-resisting materials.

FLAT CONTACT POINTS

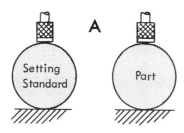

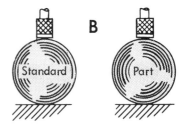

MOST
RELIABLE When both the setting standard and the part are cylindrical or spherical .

RELIABLE when either the standard and/or the part are spherical, providing the con-tact areas are square to the spindles.

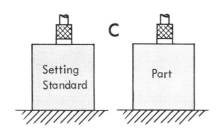

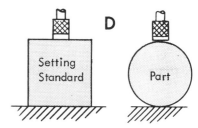

RELIABLE for most measurement. Not re-liable closer than .0001 inch because of varying air and oil film.

LEAST
RELIABLE when zeroed on flat surface be-cause of air film on flat and absence of film on cylinder.

Fig. 10-43 Flat contact points have an important role but can be a serious cause of error when improperly used.

SPHERICAL CONTACT POINTS

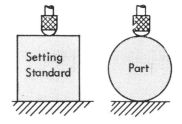

RELIABLE providing both standard and part are passed completely under contact point and readings taken at high points.

RELIABLE Point contact of spherical con-tact should rupture air and oil film on flat surface. Same with flat standard and flat part.

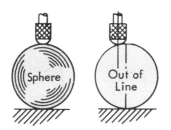

NOT RELIABLE Sphere to sphere contact makes high point difficult to find. Out of line measurement from failure to pass cyl-inder under contact measures chord instead of diameter.

Fig. 10-44 Excluding the measuring of balls, spherical contacts usually provide greater reliability than flat contacts. At very-high amplifications, or with heavy gaging force, they present problems.

SPHERICAL MEASUREMENT

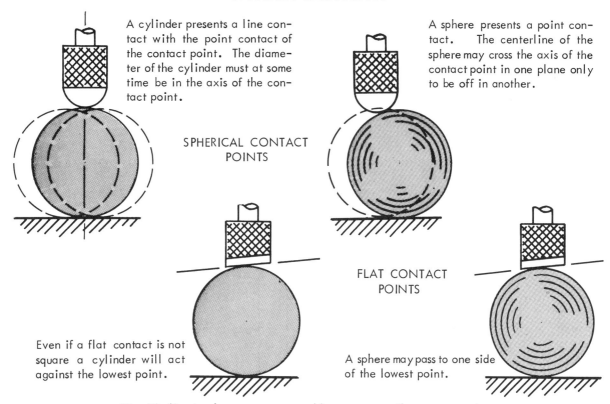

A cylinder presents a line contact with the point contact of the contact point. The diameter of the cylinder must at some time be in the axis of the contact point.

A sphere presents a point contact. The centerline of the sphere may cross the axis of the contact point in one plane only to be off in another.

SPHERICAL CONTACT POINTS

FLAT CONTACT POINTS

Even if a flat contact is not square a cylinder will act against the lowest point.

A sphere may pass to one side of the lowest point.

Fig. 10-45 A sphere presents a problem even to a flat contact point.

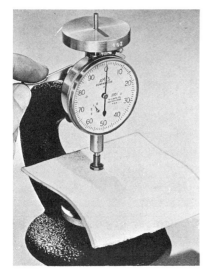

(B. C. Ames Co.)

Fig. 10-46 Weights are used instead of pull-back springs, when gaging pressure must be constant and of a stipulated amount. This is required when measuring compressible materials. A wide, flat contact point is usually required as well.

(B. C. Ames Co.)

Fig. 10-47 There are several types of tolerance hands that can be set to the high and low limits of the measurement. The type shown in Fig. 10-48 can be sealed to prevent tampering and has virtually no parallax error. The type shown above can be quickly set from the knurled knobs in the center of the crystal. Both turn with the dial.

(Federal Products Corp.)

Fig. 10-48 The maximum reading hand is a convenience for indicating the maximum value of a fluctuating variable such as checking eccentricity or runout. Obviously, it could be a serious cause for error if the total travel required is more than one revolution.

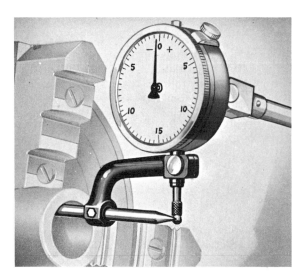

Fig. 10-49   The hole attachment is used primarily for positional checking, such as concentricity, rather than comparison measurement. *(Federal Products Corp.)*

Only a flat point is suitable unless the sphere is large and you have lots of patience to find the highest reading.   Even a flat point is unreliable unless it is accurately square with the indicator spindle and parallel to the support surface.   This is extremely difficult to attain except with comparator stands built for that purpose.   Figures 10-46 to 10-50 show other dial indicator accessories that you will frequently encounter.   Figure 10-51 is a popular indicator modification.

INDICATOR STANDS

As important as it is, the dial indicator is only one part of the measurement system. It is nothing more than a calibrated measured point.   The accuracy of the measurements depends every bit as much on the positional relationships as upon the indicator.   For this reason we must investigate indicator stands.

These are roughly of two types which the author terms *loose joint* and *fixed travel.* When an indicator stand is used on a reference surface it is termed a *test stand*, Fig. 10-52.   When a reference surface is attached it is termed a *comparator.*

Test stands are used on machine tools, on surface plates, on the parts being inspected and even within the interior of mechanisms. These stands need the versatility of the loose-joint construction.

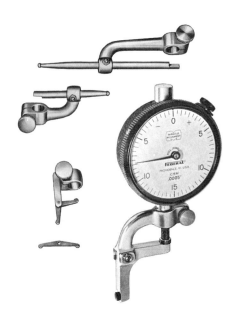

Fig. 10-50   Right-angle attachments permit the indicator to be located for best viewing position. Various types are shown at the top and bottom left. An indicator with a right angle attachment is on the right.   For permanent setups the perpendicular indicators are preferred. *(Federal Products Corp.)*

Fig. 10-51 Perpendicular indicators have the spindle at a right angle to the face. They are mounted by a rod on the side. *(Federal Products Corp.)*

There is a branch of mechanics that deals specifically with movement and position considerations.   It has a formidable name, *kinematic mechanics*,   or   simply *kinematics.* * Although its terms are not in everyday shop use, they furnish the simplest, clearest and

_____

* Wolowicz, C. S., Rindone, C. P. and Wilcox, R. J., "Axioms of Drafting Technology - VII," American Machinist/Metalworking Manufacturing, New York, McGraw-Hill, September 3, 1962, p. 101

LOOSE JOINT INDICATOR STANDS

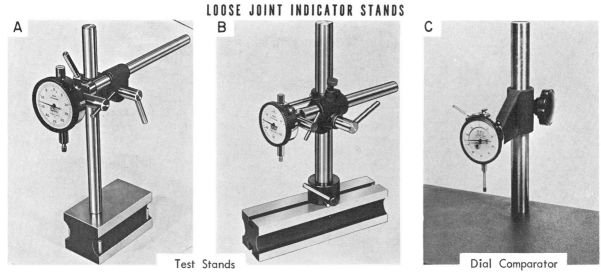

Test Stands   Dial Comparator

Fig. 10-52 These indicator stands permit adjustment in all three axes. This gives them both versatility and un-reliable positioning. A and B are called test stands because they are intended for use on a reference surface. C is called a dial comparator because it furnishes its own reference surface. *(Chicago Dial Indicator Co.)*

most concise method for discussing positional relationships in gaging.

Referring to Fig. 10-53 you will see that three axes have been constructed through a test stand. The left-to-right axis is conventionally called x; the front-to-back axis, y; and the up-and-down axis, z. Every object potentially can rotate about these axes and move along the axes; potentially, but not if constrained from doing so. The adjustments of the test stand in Fig. 10-53 allow the indicator to have all six of these *degrees of freedom.*

This means that the critical point, the measured point, also has six degrees of freedom. In order to make an accurate measurement, the part and the indicator must be po-

MOVEMENTS IN MEASUREMENTS

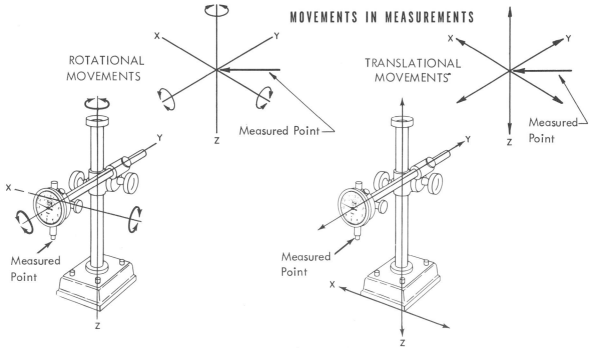

Fig. 10-53 The loose-jointed test stand shows the six separate movements that the measured point may have.

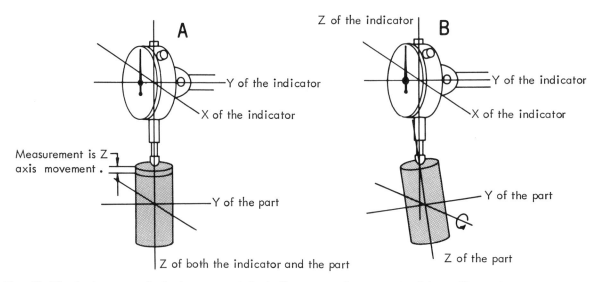

Fig. 10-54 During setup, both the part and the indicator must be constrained from all x and y movement and their z axes brought in line. The indicator then measures movement along the z axis. If z of the part is not z of the indicator, the measurement will not be accurate, as shown on the right.

ALL ELEMENTS MUST BE CONSIDERED

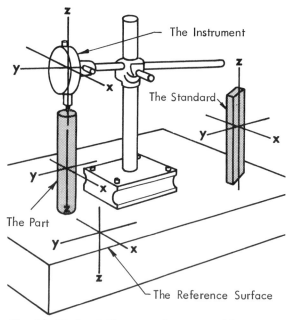

Fig. 10-55 A rotation around any one of four separate x and/or y axes will destroy the accuracy of the measurement.

So far we have only considered two of the elements; now add the standard and the reference surface and you have the complete picture. Fig. 10-55 shows that a change in any of four elements will cause errors.

One way to eliminate possible misalignment is to design them out. By that method the test stand becomes the comparator. In the comparator the reference surface is attached to the stand and thereby becomes part of it, as at C in Fig. 10-57. Presuming that the manufacture has been satisfactory (a very bad presumption, but necessary here), the positional relationship of the reference surface is no longer of concern.

The loose-joint-support arrangement for the dial indicator still permits all of the movement shown in Fig. 10-53. If the instrument is used only for comparison measurements some of these movements are not needed. *Any unneeded movement, even if provided with a lock, is a potential source of error.*

Instruments of this type are shown in Fig. 10-56. The kinematic considerations are shown in Fig. 10-57. Note that all movement within the instruments and stand is constrained except that in the z axis. The head can be raised or lowered in that axis for measuring parts of various height and the actual measurement movement takes place in that axis.

sitioned and constrained so that there is only one degree of freedom left--and that is the one we measure. This is shown by A in Fig. 10-54. B shows what happens when either the indicator or comparator is not sufficiently constrained and shifts during measurement.

(B. C. Ames Co.)

Fig. 10-56 Dial comparators are indicators with stands that incorporate their own reference surfaces. Note that all the indicators shown have spindle-lifting levers.

DIAL COMPARATORS

FUNCTIONAL
FEATURES

METROLOGICAL
FEATURES

Spindle
Lifter

Arm Clamp

Indicator

Head

Arm

Table

Column

Base

z

Line of
Measurement

Measured
Point

X   Y

Reference surface square to line
of measurement in X and Y axes.

Fig. 10-57 The dial comparator has the general C shape of all of the outside diameter instruments.

The platen, anvil or table must be precisely square to the z axis, which is the line of measurement. Now, those who are trigonometry whizzes will work this out and challenge the seriousness of the error caused by lack of squareness. For those who are not, this has been done in Fig. 10-58.

In this example, the axis of the indicator is one degree out of alignment with the line of measurement of the part. The indicator is

COSINE ERROR

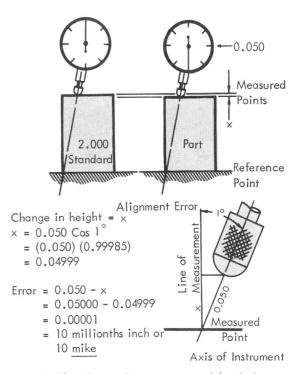

-0.050

Measured
Points

x

2.000
Standard

Part

Reference
Point

Change in height = x
x = 0.050 Cos 1°
  = (0.050) (0.99985)
  = 0.04999

Error = 0.050 − x
     = 0.05000 − 0.04999
     = 0.00001
     = 10 millionths inch or
       10 <u>mike</u>

Alignment Error   1°

Line of Measurement

0.050

x

Measured
Point

Axis of Instrument

Fig. 10-58 The cosine error caused by 1 degree applies only to the travel of the indicator and figures out to an error in length of only 10 millionths of an inch. Note that the same misalignment with a vernier height gage creates an error based on the full length and would be <u>3 tenths mil</u> (0.0003 in.).

zero set with a 2-in. standard and then the part placed under it. A *fifty-mil* (0.050-in.) difference is read on the indicator. Multiplying this by the cosine of the angle shows that the actual change in length is *ten mike* (0.00001 in.). This is not a serious error for most measurement but could be cataclysmic in gage calibration.

Although cosine error gets most of the publicity, it is the least of the errors caused by misalignment. In this same example, consider what would happen if a flat contact had been used. The error would be a function of the radius of the contact surface. As shown in Fig. 10-59, this example would result in an error of over *two and two-tenth mil* (0.0022 in.).

This example shows why flat contact points are avoided whenever possible. Using a regular point which has a spherical end eliminates part of this error but not all of it. Because both the point and the material which it contacts are compressible, there is area contact. Where area contact accompanies positional error, a sine error will be introduced to the reading.

Suggestions for reliable use of dial indicators are summarized in Fig. 10-62. The general metrological information about dial indicators is summarized in Fig. 10-63.

PUTTING ERROR TO WORK

Error can be its own worst enemy. If you can isolate error you can always compensate for it. Sometimes you can even put it to work. The case in point is the use of sine and cosine errors to square a comparator table with the axis of measurement. The error caused by misalignment can be used to show the direction, magnitude and method for correcting the misalignment.

Let's assume that either the indicator head or the base is adjustable, and that you desire to square them before beginning to use the setup. First, note A and B in Fig. 10-60. If a gage block is pushed uphill the indicator registers an increase in height (sine error). If the block is pushed downhill (C and D) the reading does not change. This tells you both the direction of the slope and its sign (plus or minus, up or down). By repeating this from various paths beneath the spindle, you can find the axis in which the deviation is greatest. Then you know what to adjust.

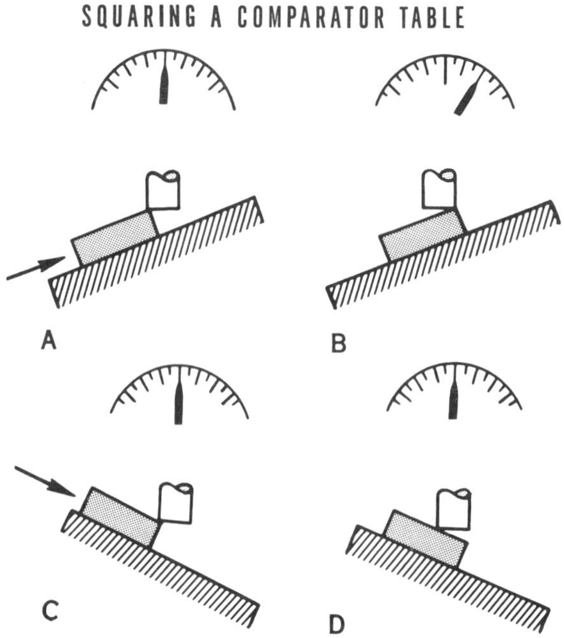

SQUARING A COMPARATOR TABLE

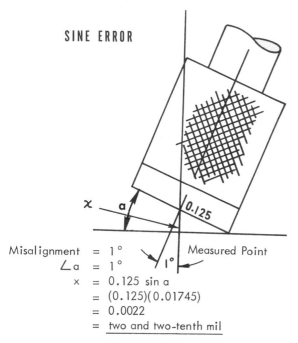

SINE ERROR

Misalignment = 1°
∠a = 1°
x = 0.125 sin a
  = (0.125)(0.01745)
  = 0.0022
  = two and two-tenth mil

Fig. 10-59   The sine angle caused by misalignment with a flat contact is a much more serious cause for error than the cosine error. Note: A misaligned micrometer or snap gage would have this error at each contact surface.

Fig. 10-60 By passing a gage block between spindle and anvil, along several paths, the misalignment can be detected and corrected.

These things must be kept in mind. A stationary reading might mean (1) that squareness has been attained, (2) that the movement is from high to low, or (3) that the axis of the movement might be along the one square line that always exists. We will not argue the last point now. It is taken up in the discussion of levels in Chapter 15. All of these

can be checked out by repeating from other positions.

Adjustments are anathema to precise measurement. In this case, and in all others, repeat the test after the final clamping. All too often you will find that it was not final after all and additional adjustment is required.

CALIBRATION CHART

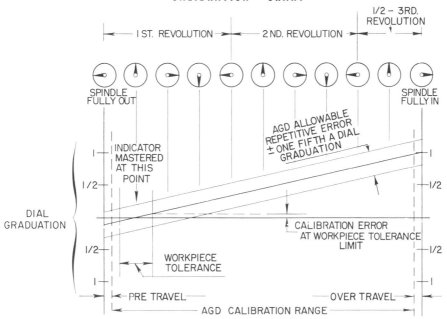

Fig. 10-61 The relationship between metrological and functional considerations is shown in this graph. It is based on AGD (American Gage Design Committee) Specifications. These specifications permit a maximum repeatability error (precision) of one-fifth the smallest dial graduation at any point on the 2 1/2 turns of useful range. The calibration error (accuracy) is limited to one graduation in the total useful range. Dial indicators should be "mastered" (zeroed) when the pointer is at the 12-o'clock position on the first turn. In this graph a plus calibration error is shown but it could be either plus or minus. (Gettelmen, K. M. "When You Select A Dial Indicator," Modern Machine Shop)

Summary — Milestones in Measurement

● This chapter marked an important milestone in your progress – measurement by comparison. Hitherto the standard of length was built into the instrument. Here it is separate. The instrument is nothing more than a means for comparison of a known with an unknown. By separating the standard from the means of comparison we have broken the bonds on amplification. Measurement involves a change of length. With comparison instruments this change can be magnified many times, thus making very small changes discernible.

● Dial indicators (calibrated measured points) are by far the most important comparison instruments. They achieve their amplification by a rack and gearing plus their pointer length. They are generally available with discriminations as fine as *one-tenth mil* (0.0001 in.). In themselves they provide no reference point. They must always be used in conjunction with a stand or holder. This has a C-frame configuration for external measurement and a straight frame for internal measurement. Both conform with Abbe's law and are inherently accurate.

● Because a dial indicator is loaded by the measurement movement, the energy to operate the instrument is limited. Amplification beyond a certain point does not increase sensitivity and is without value. To use fully an instrument's sensitivity it must have good resolution, which is part of its readability. This can be tested by repeated measurements. The dispersion of readings is the precision, not the accuracy.

● The accuracy is the difference between the readings and the true values. All comparators have some error. The amount that gets into measurements is proportional to the travel. Thus the dial indicator has its greatest accuracy at the place it is zero set. That can be anywhere in the working range. At that point accuracy is reduced only by sensitivity. All comparison should be as close to zero as possible.

● It is desirable always to get a usable reading directly without computational steps. Balanced and continuous dials conform with bilateral and unilateral tolerances respectively. Clockwise and counterclockwise are available for the same reason.

● Long travel indicators have revolution counters. They are guaranteed trouble makers when used to read the distance traveled from a reference. They are highly reliable when used to show that a reference has been reached. Useful measuring range for all indicators starts after the starting tension and ends before the overtravel. This range is independent of size or discrimination. Both discrimination and range are involved when selecting an indicator. The discrimination should spread the tolerance over 10 dial divisions. Select the range so that the total tolerance occupies at least 1/10 but nor more than 1/4 of the total dial.

RELIABLE COMPARISON MEASUREMENT WITH DIAL INDICATORS

1. Determine that this is best way to make the measurement.

2. Select a dial indicator with sufficient precision to spread the tolerance over about 10 divisions and with sufficient range to contain the tolerance in not more than 1/4 of the dial.

3. Select standards of suitable precision.

4. Check calibration data for indicator and standards.

5. Select contact point, indicator stand and reference surface that will assure correct alignment.

6. Thoroughly clean part, instrument and all other components of the setup.

7. Make setup. Check alignment. Check security and rigidity of all joints.

8. Zero set indicator on standard. Repeat until satisfied.

9. Measure part with indicator and adjust with the value of the standard to provide measurement. Repeat.

10. Recheck all above steps.

11. Consciously inquire into possible bias.

12. If it is a critical measurement have someone else make measurement. Any discrepancy greater than 10% of tolerance is warning that some step is inadequate (probably #1).

Fig. 10-62   By now you may have noted a strong similarity among these check lists. This emphasizes their importance.

METROLOGICAL DATA FOR DIAL INDICATORS

INSTRUMENT	TYPE OF MEASUREMENT	NORMAL RANGE	DESIGNATED PRECISION	DISCRIMINATION	SENSITIVITY	LINEARITY	RELIABILITY PRACTICAL TOLERANCE FOR SKILLED MEASUREMENT	RELIABILITY PRACTICAL MANUFACTURING TOLERANCE
Test indicators	Comparison	0.030 in.	0.001 in.	0.001 in.	0.0005 in.	2%	Not for measurement	Not for measurement
0.001 indicators: on height gage stands	Comparison	0.250 in.	0.001 in.	0.001 in.	0.0005 in.	2%	0.001 in.	0.010 in.
on comparator stands	Comparison	0.250 in.	0.001 in.	0.001 in.	0.0005 in.	2%	0.0005 in.	0.005 in.
0.0001 indicators: on height gage stands	Comparison	0.050 in.	0.0001 in.	0.0001 in.	0.0001 in.	2%	0.0001 in.	0.001 in.
on comparator stands	Comparison	0.050 in.	0.0001 in.	0.0001 in.	0.00005 in.	2%	0.00005 in.	0.0005 in.
0.00005 indicators: on height gage stands	Comparison	0.010 in.	0.00005 in.	0.00005 in.	0.0001 in.	2%	0.0001 in.	0.001 in.
on comparator stands	Comparison	0.010 in.	0.00005 in.	0.00005 in.	0.00005 in.	2%	0.00003 in.	0.0003 in.

Fig. 10-63   A practical comparison of dial indicators cannot be made without considering the means of supporting the indicator.

● Treat an indicator as you would your spouse. Give it lots of tender care. If an operation is necessary see a specialist, and a check up now and then is essential.

● Indicator versatility derives from the large number of contact points and mountings available. All of the latter are to be mistrusted. For every joint or adjustment in an indicator setup, double your distrust. Fortunately, the errors in stands can be analyzed systemati-cally which permits some corrections and compensations. Figure 10-63 will help in the selection of indicators and Fig. 10-62 with their use. This is graphed in Fig. 10-61.

● The pattern of this book has been to move constantly to instruments of higher amplification. That is why this chapter stopped at dial indicators. They certainly are not the only comparator instruments. The next chapter takes up the high-amplification types.

TERMINOLOGY

Direct measurement	Measurement with an instrument that incorporates its own standard of length.
Comparison measurement	Measurement by comparing the unknown length with a known length or standard.
Amplification	An increase in the output of a system as compared to the input.
Power	The ratio of the output to the input.
Measurement movement	The distance traveled to compare an unknown with a known length.
Comparator	Any instrument that provides the amplified difference in size between two lengths.
Test indicator	A mechanical comparator providing only a two-value signal (too large and OK, yes and no, on size and out of size).
Mechanical comparator	A mechanical comparator providing a multi-value output signal. It gives an approximation of the size difference.
Dial indicator	Familiar form of mechanical comparator.
Sensitivity	Minimum input that produces a discernible output.
Resolution	The ratio of the width of one output unit to the width of the read-out element.
Limit	Maximum or minimum stated value.
Zero setting	Bringing the zero position of the instrument into correspondence with a reference of the measurement.
Tolerance	Permissible variation in part size.
Unilateral tolerance	Tolerance entirely on one side of the basic dimension.

Bilateral tolerance	Tolerance partially on both sides of the basic dimension.
Allowance	Space between mating parts.
Indicator range	Useful portion of total travel.
Indicator stand	Self-supporting contrivance providing a reference point for indicator.
Loose-joint stand	Stand adjustable in several ways.
Fixed-travel stand	Adjustable only in direction of measurement movement.
Kinematics	Branch of mechanics dealing with movements.
Degree of freedom	Freedom to move along a line or to rotate.
Constraint	Elimination of a degree of freedom.

Fig. 10-64 The new words in this chapter enable you to discuss measurement with greater precision -- but the accuracy is up to you.

QUESTIONS

1. What are the two general types of measurement?
2. Upon what four items does the reliability of direct measurement depend?
3. Comparison measurement depends upon three points. What are they?
4. All measurement is made up of measurement of time, mass, or length. In each of these cases name the three elements involved.
5. How is precision increased in most direct-measurement instruments discussed thus far?
6. Are there disadvantages involved when an instrument's amplification is increased?
7. What are some of these disadvantages?
8. When an instrument incorporates its own standard, what limits its discrimination?
9. What advantage is achieved by separating the standard from the instrument?
10. What disadvantage is encountered when simple levers are used for amplification?
11. How can we achieve the same mechanical advantage as levers but not limit scale range?
12. Do dial indicators measure the length of parts?
13. Does a dial indicator have a reference point?
14. What is the basic difference between a test indicator and a dial indicator?
15. How does the term "comparator" differ from the familiar term dial "indicator"?
16. How does the movement in direct measurement differ from comparative measurement?
17. When amplification is increased what happens to the precision and the sensitivity of the instrument?

18. What are some of the factors that influence readability?
19. What determines the number of graduations that can be put on an indicator dial?
20. Is precision a measure of accuracy?
21. Is the accuracy of a dial indicator consistent throughout its range?
22. Standard indicator dials have four different discriminations. What are they?
23. What two common errors does the zero setting adjustment eliminate?
24. What are the two types of indicator dials?
25. For a specific measurement, why should the shortest range be selected that will contain the tolerance?
26. Why are there many types of contact points available for use with dial indicators?
27. How many different movements are possible with a loose-jointed indicator stand?

DISCUSSION TOPICS

1. Explain several methods that can be applied to increase amplification of measuring instruments.
2. Explain how gears can achieve high amplification.
3. Discuss the basic features of dial indicators.
4. Explain the terms "sensitivity" and "resolution" as related to comparison instruments.
5. Of what value are revolution counters on dial indicators?
6. Explain the basic rules for applying clockwise and counterclockwise indicator dials.
7. Explain the rules for carefully selecting the correct indicator for the specific measurement job.
8. Discuss the "do's" and "dont's" for dial indicator use.
9. Explain the application of various dial indicator accessories and attachments.

Man is a tool-using animal. Weak in himself and of small stature, he stands on a basis of some half-square foot, has to straddle out his legs lest the very winds supplant him. . . . Nevertheless, he can use tools, can devise tools; with these the granite mountain melts into light dust before him; seas are his smooth highway, wind and fire his unwearying steeds. Nowhere do you find him without tools. Without tools he is nothing; with tools he is all.

Thomas Carlyle

High Amplification Comparators
Sensing Millionths

The discussion of comparison measurement so far has centered around dial indicators. They are inexpensive, are found everywhere, and are reliable. Reliable, that is, if you do not exceed their capability nor overlook their inherent shortcomings.

Their reliability is good when they are used as comparators, poor for linear measurement. In other words measurement should be made as close to the zero setting as possible. Their inherent shortcoming results from their chief method of amplification, gear trains. Theoretically, there is no limit to the amplification possible. Practically, there is a very real limit because of gear errors, friction and inertia, not to mention hysteresis and even more subtle factors. Although there are dial indicators that have discrimination of *50 mike* (0.00005 in.) these are unusual. The practical limit is *tenth mil* (0.0001 in.). More sensitivity is lost when this is exceeded than there is improvement in discrimination to compensate.

For these reasons, better means of amplification have been devised for comparison measurement. The means fall into four classes, other mechanical actions, optical, pneumatic and electronic, Fig. 11-1, plus some unusual ones such as the level comparator discussed in Chapter 15.

GAGES AND MEASUREMENT INSTRUMENTS

Up to now we have not distinguished between gages and measurement instruments. In popular use they are almost interchangeable. Air gages, for example, are high amplification instruments. Yet they are omitted from this chapter. To explain this we must distinguish between gages and other measurement instruments.

The various departments of the federal government that purchase material have done much to advance standardization. The definitions they have agreed upon are condensed in Fig. 11-2. Note that gages are used primarily to check or inspect. Someone had to do considerable measurement to design and build the gage. The user of the gage, however, is concerned only with the reaction of the gage to the part feature being checked. Often this

Fig. 11-1 Practical comparators with extremely high amplifications utilize pneumatic, mechanical and electronic means.

```
┌─────────────────────────────────────────────────────────────────────────────┐
│                    GAGES AND MEASURING EQUIPMENT                              │
│                                                                              │
│ 1.1.3.1   Length Standards.  Standards of length and angle from which all measurements of gages are derived. │
│ 1.1.3.2   Master Gages.  Master gages are used for checking and setting inspection of manufacturers' gages.  │
│ 1.1.3.3   Inspection Gages.  Inspection gages are used to inspect products for acceptance. │
│ 1.1.3.4   Manufacturers' Gages.  Manufacturers' gages are used for inspection of parts during production. │
│ 1.1.3.5   Nonprecision Measuring Equipment.  Simple tools used to measure by means of line graduations. │
│ 1.1.3.6   Precision Measuring Equipment.  Tools used to measure in thousandths of an inch or finer. │
│ 1.1.3.7   Comparators.  Comparators are precision measuring equipment used for comparative measurements │
│           between the work and a contact standard such as a gage or gage blocks. │
│ 1.1.3.8   Optical Comparators and Gages.  Optical comparators and gages are those which apply optical │
│           methods of magnification exclusively. │
└─────────────────────────────────────────────────────────────────────────────┘
```

Fig. 11-2   These definitions are condensed from Military Standard, Gage Inspection, MIL-STD-120. This was prepared by the Munition Board Standard Agency of the Department of Defense and is used by the armed forces and other government departments.

is simply a right or wrong (usually termed *go* and *not-go* in gaging) decision. Even when numbers are involved in the *read out*, they are arbitrary units. They are useful to the quality control department but have little direct meaning to the person checking. Measurement instruments (or measuring equipment as they are called in Fig. 11-2) provide numerical information which can be of immediate use. Often measurement instruments are used for gaging, but rarely are gages used for measurement. That is due to the fact that the former are versatile but the latter are highly specialized.

The *air gages* (or *pneumatic comparators* as they are properly termed) are in the latter class. They compare parts with the masters. They are an extremely valuable and important part of modern quality control, particularly for the troublesome task of inspecting small hole diameters. They are not generally used as measurement instruments and therefore are omitted from this chapter. It must be added that recent developments are rapidly advancing air gages into the versatile measurement instrument class.

Chapter 12 is devoted to these important instruments. In it the distinction between the terms discussed here is made to greater precision.

ANALOG AND DIGITAL INSTRUMENTS

Thanks to the publicity for electronic computers these words are almost household expressions. They are by no means restricted to computers, however. They are very valuable for describing measurement activities, particularly for some of the more complex instruments. Up to this point we have only been concerned with analog instruments. Gages, in many cases, fall into the digital class.

Analog instruments use a physical quantity to correspond to numbers. This quantity may be almost anything. If a block of wood stands for one inch, or one mile or one dozen roses, then two blocks represent two inches, two miles, etc. Numbers are represented by measurable quantities. In the case of the dial indicator, the amount of the dial swept by the pointer is the quantity. In electronic instruments the electrical signal is the quantity; and in pneumatic comparators, an air stream is the quantity.

In contrast a digital instrument deals with numbers directly. For example, a block of wood when placed in a certain slot might represent ten, regardless of its length. In short: *An analog instrument measures. A digital instrument counts.* The gas gage on your car is an analog instrument. The mileage is read on a digital odometer. The instruments met in this chapter are analog. Their first stage consists of translating a change of length on the part to a physical quantity. This quantity is amplified by the intermediate stage. The final stage is read by measuring that amplified quantity.

NO CEILING ON AMPLIFICATION

By the ingenious use of mechanical, optical and electronic devices, comparators with as high as one-million-to-one amplification are actually in use today. As we have demonstrated with dial indicators, the working range reduces as the amplification increases. Most of the practical instruments have much lower amplification, 1,000 to one to 20,000 to one, in order to have a usable range. This is still enormous compared with dial indicators.

Now the newly developed laser promises to break through the bonds that have tied range to amplification. It is entirely possible that soon many of you will be measuring with instruments that are inconceivable today.

Dial indicators are all rather similar, although many of them have exceptional features. Not so with the high-amplification instruments. They vary as much as the separate manufacturers producing them. This being a book on principles instead of specific instruments no attempt is made to discuss them all. Most manufacturers of these instruments have extensive sales literature. Some have adequate operation manuals.

The mechanical instruments are discussed first because they show the natural evolution from dial indicators to more sophisticated instruments. They also will give you a preview to the advantages of optics in measurement. Following these, the electronic instruments will be discussed.

Electronic instruments have a unique feature, power amplification, not possessed by the other instruments. They are the fastest developing type of high-precision measurement and gaging instrumentation. Unlike all previously discussed instruments, many can provide two or more scale ranges. This versatility plus their relatively low cost (don't plan to buy one for your personal tool kit, however) explains their almost universal use for everything from *line inspection* of production parts to *master gage calibration*.

The general design of these gages will be discussed first. We will then get into the important part – *reliable measurement to tenth-mil increments* (0.0001 in.) and finer. To make sure that we are all talking about the same values, Fig. 11-3 reviews measurement terminology as it applies to very small increments.

SMALL INCREMENTS

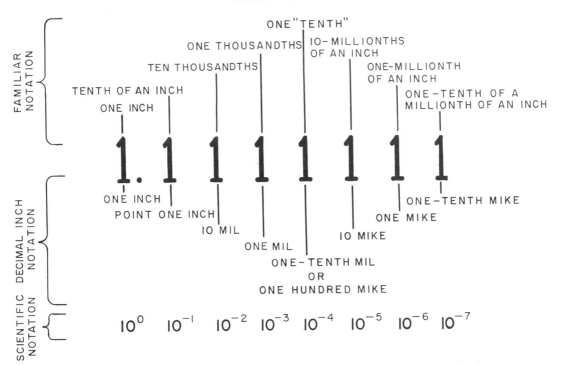

Fig. 11-3   At high amplifications, it is easy to misplace a decimal point.

HIGH-AMPLIFICATION
MECHANICAL COMPARATORS

As you might expect, in the race to achieve high amplification the dial indicator has been carried to its ultimate refinement. The resulting instruments are shown in Figs. 11-4 and 11-5.

A dial indicator is a mechanical system involving a *cascade* of stages, each one at a higher amplification. An error in the gearing at the final stage has a relatively small effect on the accuracy. An error in a stage close to the input has a tremendous effect upon accuracy. Therefore the weakest stage in a dial indicator is the rack and pinion gear. In the instrument shown in Fig. 11-4, this has been replaced with a precision steel ball rolling along a finely lapped sapphire surface. Extremely minute changes in position of the spindle are directly translated to the gear train. In this design the loading of the mechanical parts is independent of the contact pressure. This minimizes windup and other distortion. Note that in the first two gear stages, the movements are so slight that only segments of gears are required.

These instruments are available with discrimination to *20-mike* (0.00002-in.) increments. Attempts for further advances in completely mechanical amplification have shown considerable ingenuity. The example in Fig. 11-6 has eliminated all friction except intermolecular. That means that the

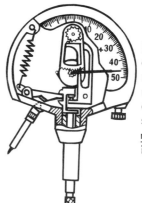

Fig. 11-4   By substituting a more accurate linear-to-rotary transducer than the rack and pinion of dial indicators, an instrument resulted that is capable of 20 mike (0.00002 in.) discrimination. *(Mahr Gage Co., Inc.)*

only friction consists of molecular particles sliding against each other when the twisted-band heart of the mechanism is stretched.

The twisted band is shown in Fig. 11-6. It is fixed at one end and connected to the spindle at the other. In the center, the pointer is attached. The band is twisted clockwise on one side of the pointer and counterclockwise on the other side. When the band is stretched it attempts to straighten out, thus rotating the pointer. In keeping with this method of amplification, the suspension of the spindle and driving of the twisted band also is affected by spring suspension. There is no rubbing friction whatsoever. This not only results in high sensitivity and repeatability, it also assures low maintenance. A well-designed comparator incorporating this instrument is shown in Fig. 11-7.

Turning Z changes O thereby adjusting the amplification.

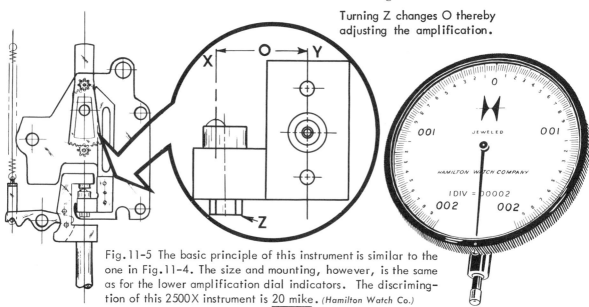

Fig.11-5 The basic principle of this instrument is similar to the one in Fig.11-4. The size and mounting, however, is the same as for the lower amplification dial indicators. The discrimination of this 2500X instrument is 20 mike. *(Hamilton Watch Co.)*

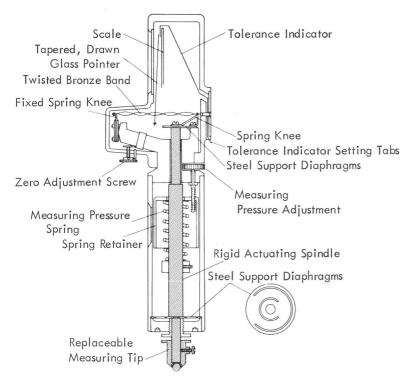

Scale
Tapered, Drawn
Glass Pointer
Twisted Bronze Band
Fixed Spring Knee
Tolerance Indicator
Spring Knee
Tolerance Indicator Setting Tabs
Steel Support Diaphragms
Zero Adjustment Screw
Measuring Pressure Adjustment
Measuring Pressure Spring
Spring Retainer
Rigid Actuating Spindle
Steel Support Diaphragms
Replaceable Measuring Tip

Fig. 11-6 This completely mechanical instrument has eliminated all friction except intermolecular to achieve amplifications up to 100,000X.

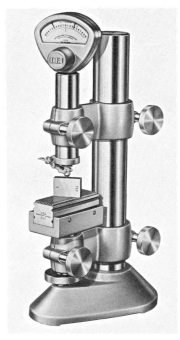

(The C. E. Johansson Gage Co.)

Fig. 11-7 Gage block standards are used to calibrate instruments of the accuracy and precision of this twisted-band comparator.

REED-TYPE COMPARATOR

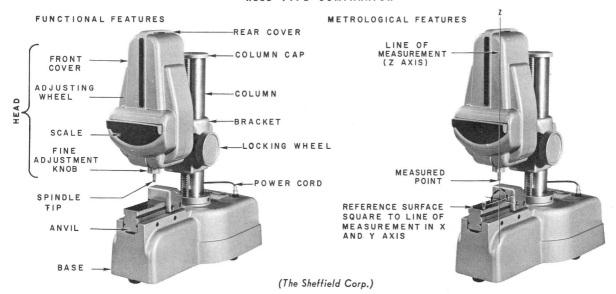

FUNCTIONAL FEATURES

HEAD

FRONT COVER
ADJUSTING WHEEL
SCALE
FINE ADJUSTMENT KNOB
SPINDLE TIP
ANVIL
BASE

REAR COVER
COLUMN CAP
COLUMN
BRACKET
LOCKING WHEEL
POWER CORD

METROLOGICAL FEATURES

Z

LINE OF MEASUREMENT (Z AXIS)

MEASURED POINT

REFERENCE SURFACE SQUARE TO LINE OF MEASUREMENT IN X AND Y AXIS

(The Sheffield Corp.)

Fig. 11-8   Metrological features of the reed-type comparators are the same as for the dial comparators.

THE REED-TYPE COMPARATOR

The reed-type comparator, Fig. 11-8, is similar to the previous instrument in several respects. It, too, requires only intermolecular friction to achieve its amplification and to support all of its moving parts. It is of particular interest because it achieves additional amplification by means of the optical lever.

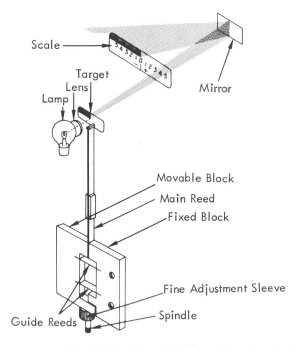

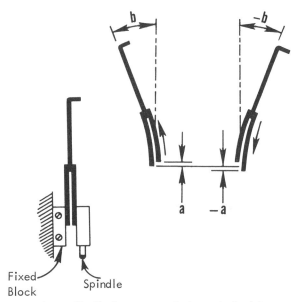

A small displacement of the spindle (a) results in a large swing of the pointer (b).

Fig. 11-9   An optical lever is used to further amplify the reed action. In the actual instrument, the optical path is longer. A prism is used instead of a mirror.

Figure 11-10 shows the principle of the reed mechanism.   Two blocks are attached to the reeds which are strips of spring steel. One block is fixed. The other is movable and attached to the spindle. The reeds join together to form the pointer. A small amount of spindle movement, (a), causes a large deflection of the pointer, (b). The amplification is the ratio of (b) to (a). The action is directional; if the direction of movement of (a) reverses, so does (b).

To further amplify this action an *optical lever* is used.   That principle also is shown in Fig. 11-9. The pointer moves an aperture (called the target) across a focused beam of light. The effect is exactly the same as a lever arm except that the light beam is weightless and frictionless.

Remember that, as shown in Chapter 10, the amplification of the hand or pointer multiplies the amplification of the mechanism. Where the hands of even the largest dial indicators are short, the effective hand of the reed-comparator is the length of the reed, followed by the optical lever. If the reed mechanism provided 50X, the hand leverage would only have to be 40 to one to increase the total amplification to 2,000X.

Note also in Fig. 11-9 that guide reeds are used to support the movable block. Unlike the bearings of dial comparators, this suspension is frictionless. The combination of low inertia (low weight of the moving parts due to the reed mechanism and optical lever) together with the freedom from bearing friction provides sensitivity to match the high amplifications of reed-type instruments.

The actual construction of the reed-type comparators is necessarily more complex than these illustrations.   There are adjustment screws in the fixed block to regulate both the tension and travel of the movable block. The knurled sleeve on the spindle turns a cam which provides the fine adjustment. Coarse adjustment is made by moving the entire head.

The scales for reed-type comparators are read in the same general way that dial indicators are read.   The important thing is to remember the discrimination so that the decimal point is not misplaced. This was mentioned in Fig. 2-4 and is reviewed in Fig. 11-3 with particular emphasis on *tenth-mil* and *mike* increments. Both metric and inch scales are available for reed-type comparators. Typical ones are shown in Fig. 11-10.

In addition to the improved sensitivity of the reed-type comparator as compared with the dial indicator, the entire scale is usable. No pretensioning is required as with a dial indicator. Like a dial indicator, however, the reed-type comparator has its greatest accuracy at zero. It should be mentioned again that this does not mean that one point in the movement has any greater accuracy than any other point. Any point within the range can be set to zero. Restated, it means that the accuracy *diminishes* as the difference between the standard and the part *increases.*

Considerable space is devoted to the electronic instruments. First of all the time has come to delve into the all important ten-to-one rule of measurement. Secondly, the speed with which electronic instruments are coming into widespread use, justifies an extensive treatment. Do not confuse space used for an instrument with its accuracy or even importance. Figure 11-11 shows that the advantages of electronic instruments are primarily functional. In many instances an inexpensive dial indicator may be preferred.

Their chief advantage is reduction of time lag. This becomes very long for most other types of high amplification instruments. Where the measurement rate might be 50 to 80 per minute with a mechanical instrument, rates over 500 per minute are possible electronically.* This particularly suits them for

---

* Judge, A. W., Engineering Precision Measurements, London, Chapman & Hall Ltd., 1957, 3rd Edition, p. 237.

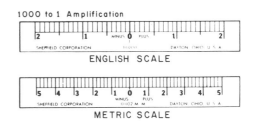

1000 to 1 Amplification

ENGLISH SCALE

METRIC SCALE

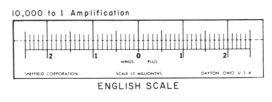

10,000 to 1 Amplification

ENGLISH SCALE

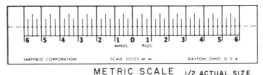

METRIC SCALE   1/2 ACTUAL SIZE

Fig. 11-10   Reed-type instruments are available in a range of amplifications with either metric or inch scales.

dynamic measurement. That is the measurement of a moving feature, or a changing value. The thickness of a strip passing through a rolling mill is an example of a moving feature. Monitoring of a deflection under varying load is an example of changing value. Some of their other features have created challenging opportunities; but, avowedly, gadget appeal also has had something to do with their popularity.

ELECTRONIC MEASUREMENT

Functional Features

1. Rapid operation even at high amplifications.

2. Multiple amplification ranges in same instrument.

3. Remote operation and multiple input operation.

4. Limited self checking (one scale against another).

5. Convenience (most are portable, some entirely self contained, controls are easy to understand).

6. Versatility (large number of measurement situations can be handled with standard components).

Metrological Features

1. High sensitivity in all ranges.

2. Favorable instrument accuracy compared to other instruments.

3. Signals can be combined electronically for added, subtracted and differential measurements.

Fig. 11-11   These points are generalities. Any one of them might be sufficient reason to accept or reject an electronic instrument for a particular measurement.

POWER AMPLIFICATION FOR MEASUREMENT

The discussion of measurement instruments, thus far, must have made it clear that inertia and friction are severe limiting factors in mechanical instruments. Electrical systems do not have these problems. Even more important, however, is their *power amplification*. No other measuring system that we have discussed possesses this extraordinary ability.

In our progression from low to high-amplification instruments, only the measurement movement has been amplified in every case. All of the energy required to operate the instrument had to be generated by the action of the part on the instrument. This is called *loading*.* The power generated by this action went steadily downhill between the input stage of the instrument and the read-out stage. This was caused by the losses in the instrument, mostly frictional. A point is reached in the design of dial indicators that so much of the loading is used to wind up

___

* Beckwith, Thomas G., Buck, N. Lewis, Mechanical Measurements, Reading, Mass., Addison-Wesley Publishing Co., Inc., 1961, p.78.

shafts, overcome bearing friction and cause gear teeth to slide over each other that not enough remains to move the read-out element. That is why high amplification is often coupled with poor sensitivity.

Great ingenuity has been used to conserve the loading, the reed-type comparators being an example. No matter how successful, however, a limit is reached as the amplification goes up. In contrast, electronic instruments use power from an outside source. This opens up entirely new opportunities. It makes possible measurement instruments that cannot only leave records of the measurements, but actually can use the read-out signal to control machines or processes. Without this there could be no automation or numerical control of machines.

ELECTRONIC MEASUREMENT INSTRUMENTS

All measurement instruments are described by the generalized system** in Fig. 11-12. This includes those that measure mass

___

** Beckwith, Thomas G., Buck, N. Lewis, Mechanical Measurements, p. 9.

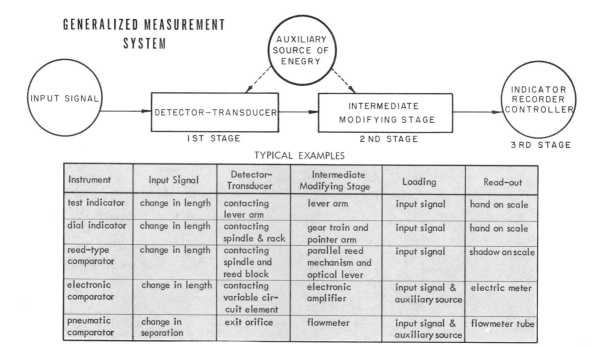

GENERALIZED MEASUREMENT SYSTEM

TYPICAL EXAMPLES

Instrument	Input Signal	Detector-Transducer	Intermediate Modifying Stage	Loading	Read-out
test indicator	change in length	contacting lever arm	lever arm	input signal	hand on scale
dial indicator	change in length	contacting spindle & rack	gear train and pointer arm	input signal	hand on scale
reed-type comparator	change in length	contacting spindle and reed block	parallel reed mechanism and optical lever	input signal	shadow on scale
electronic comparator	change in length	contacting variable circuit element	electronic amplifier	input signal & auxiliary source	electric meter
pneumatic comparator	change in separation	exit orifice	flowmeter	input signal & auxiliary source	flowmeter tube

Fig. 11-12 All measurement instruments, including those that measure mass and time, as well as length measurement, fall into the generalized system.

and time as well as those that measure length. In the systems described thus far the stages were not easily separable. With electronic instruments they may be considered as individual elements working together.

Virtually every electrical property, resistance, capacitance, inductance and simple on-off switching has been used for linear measurement together with various amplifier circuits. The circuit generally used is the balanced bridge.

You probably studied the Wheatstone Bridge in physics and Fig. 11-13 will look familiar. In this typical resistance bridge, if the value of each of the four resistances is equal, there will be no voltage drop across the meter. If, however, the value of any one of the resistances is changed, the meter hand will move a proportional amount. A mechanical arrangement could be made so that a change in length would change the resistance, such as the volume control on a radio.

The meter reading in such a simplified resistance bridge also would be proportional to the voltage from the battery. This would result in an unstable measurement instrument. To correct this, impedance bridges are used. An impedance bridge is sensitive to impedance change. It is energized by an oscillating alternating current. This can be held very constant.

Instead of a variable resistance for the detector-transducer element it uses an impedance device, a transformer. This is a very special transformer in which the impedance change is very nearly proportional to a change in the core or armature position. It is called an LVDT because its real name, *linear-variable-differential transformer*, is quite a mouthful. The LVDT, when combined with a suitable contact point in a housing, is called the *pickup head, gage head* or simply the *head*, navy men notwithstanding.

The unbalance of the bridge is amplified electronically and then connected to a sensitive meter that is calibrated in units of length. A means to change amplifications and to electrically zero set complete the instrument.

It is not necessary to describe the circuitry in greater detail because there is no field

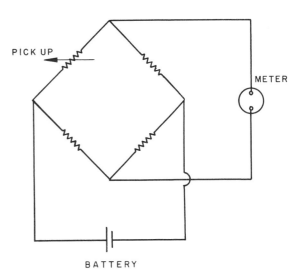

Fig. 11-13 This simplified bridge circuit is similar to those used in electronic comparators.

maintenance on these instruments. If they malfunction they are returned to the manufacturer. Of course, the absence of mechanical parts makes the electronic comparators exceptionally rugged for instruments of such high precision.

A large number of balanced bridge-type electronic comparators are available; they vary considerably in appearance. There are also other electronic measurement instruments. The basic principles of operation are the same. The one selected for discussion embodies most of the features popular in these instruments.

For all previous instruments we have furnished a drawing that showed the functional and metrological features separately. This is no longer necessary, nor even possible. Because the first stage of the measurement system can be separated from stages two and three, great flexibility is possible. The electronic measurement head can be incorporated into caliper-type instruments, height gage setups and comparators equally well. This chapter is concerned primarily with the comparator uses because they utilize the electronic components most completely.

THE ELECTRONIC AMPLIFIER AND METER

The instrument in Fig. 11-14 consists of the intermediate modifying and indicator stages of the generalized measurement system. Stage one, the detector-transducer or gage head as it is usually called, connects to it by means of a cord and a plug on the back of the case.

This particular instrument has five separate amplification ranges, Fig. 11-15. It can be operated either by battery or as a line-powered unit. The line cord detaches when not in use, making it a self-contained portable instrument.

The knob on the right has eight positions. Off is at the bottom. When placed at BAT the pointer will move to BAT on the scale if the battery is charged. When the knob is placed at CHG and the line cord plugged into a 115-volt, electric outlet the battery is recharged.

The remaining positions select scales. They are color coded alternately red and black to correspond with the red and black scales of the meter. For example, when placed at 0.001 in. the meter is read on the top scale which is black. The numbered divisions then represent *ten-mil* increments (0.010 in.) and each small division is *one mil* (0.001 in.). If the knob is turned two positions to the *tenth-mil* (0.0001 in.) position, the top scale is used again but the numbered divisions are *one-tenth* their former value, or *one mil* (0.001 in.), and the small divisions represent *one-tenth mil* (0.0001 in.) each. Two further positions bring us to the finest scale. Here the numbered divisions on the same scale we have been considering are *one-tenth mil* (0.0001 in.) and the small divisions are *ten-mike* increments (0.00001 in.) or ten millionths of an inch each.

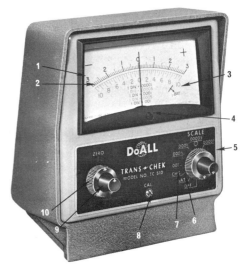

1 – BLACK SCALE	6 – BATTERY CHECKING POSITION
2 – RED SCALE	7 – BATTERY CHARGING POSITION
3 – BATTERY CHECKING MARK	8 – CALIBRATION ADJUSTMENT
4 – POINTER ADJUSTING SCREW	9 – FINE ADJUSTMENT ZEROING
5 – SCALE SELECTOR-COLOR CODED TO SCALE	10 – COARSE ADJUSTMENT ZEROING

Fig. 11-14 Stages 2 and 3 of the generalized measurement system are contained in one compact case. Stage 1, the gage head, is attached by a cord which plugs into the back.

The red, coded positions of the knob similarly correspond to the lower scale of the meter which is red. The smallest divisions for the two red positions have values of *five-tenth mil* (0.0005 in.) and *fifty mike* (0.00005 in.) respectively.

The left knob is for zero setting. Because of the great range of the instrument, this is a coaxial control providing both coarser and fine adjustment. The zero setting of the 5 scales are all within one-scale division of each other. If the finest scale is zero set, the lower scales are simultaneously zero set. The instruction manual for this instrument provides full information for its use and care.

TYPICAL ELECTRONIC COMPARATORS
POWER, PRECISION, AND RANGE RELATIONSHIPS

Scale	Color Code	Amplification	Discrimination	Range
mil	black	60X	0.001	0.030
five-tenth mil	red	170X	0.0005	0.010
one-tenth mil	black	600X	0.0001	0.003
50 mike	red	1700X	0.00005	0.001
10 mike	black	6000X	0.00001	0.0003

Fig. 11-15 The instrument in Fig. 11-14 provides a wide selection of discriminations and ranges.

TYPICAL SCALE SELECTIONS

SCALE	COMPARATOR APPLICATIONS	HEIGHT GAGE APPLICATIONS
0.001 in.	Production Inspection: appliance parts, agricultural machinery parts, builders' hardware parts, plastic parts, small castings and small forgings. Tool and Gage Inspection: not recommended.	Plate Inspection: heavy equipment parts, marine equipment parts, heavy engine parts, forgings and castings. Tool and Gage Inspection: assembly jigs and fixtures, patterns and templates. Setup Measurement: large planer and boring mill work.
0.0005 in.	Production Inspection: truck and automotive parts, motor shafts, machinery shafts, bushings, bearings (not anti-friction), small gears and precision hardware. Tool and Gage Inspection: wide tolerance production gages and small cutting tools.	Plate Inspection: machinery and machine tool housings and parts, engine parts, motor and generator parts. Tool and Gage Inspection: machining and inspection jigs and fixtures, precision-assembly jigs and fixtures, templates, cams and large cutting tools. Setup Measurement: milling machine and boring mill work, table positioning and rough surface grinding.
0.0001 in.	Production Inspection: high-speed engine parts, pump parts, small precision gears and firearm parts. Tool and Gage Inspection: production gages and precision cutting tools.	Plate Inspection: plastic and injection molds, dies, precision machining and inspection jigs and fixtures, aircraft parts and large instruments. Plate Inspection: plastic and injection molds, dies, precision machining and inspection jigs and fixtures, aircraft parts and large instruments.
0.00005 in.	Production Inspection: instrument and control parts, electronic components and antifriction bearings. Tool and Gage Inspection: close tolerance production gages and master gages.	Plate Inspection: precision aircraft and missile parts. Tool and Gage Inspection: precision inspection fixtures and instruments. Setup Measurement: jig borer, measuring machine and precision boring machine work.
0.00001 in.	Production Inspection: high-precision hydraulic and electric parts. Tool and Gage Inspection: highest accuracy master gages.	Plate Inspection: Not generally applicable Tool and Gage Inspection: '' '' '' Setup Measurement: '' '' ''

Fig. 11-16 Multiple scale selection provides the best amplification for each role in which the instrument is used.

PURPOSE OF MULTIPLE SCALES

An obvious advantage of an instrument which can be set for several scales is its versatility. It can eliminate the need for several separate instruments. This is shown vividly in Fig. 11-16. The table divides each scale into two parts, comparator applications, Fig. 11-17, and height gage applications, Fig. 11-18. The comparator applications are separated into production inspection and tool-and-gage

inspection. Following each of these are listed the recommended applications. In some cases none is recommended.

The height gage applications are setups in which either the part itself or a surface plate furnished the reference surface. Because of the considerations shown in Fig. 10-55, these applications cannot be made with the same reliability as comparator applications.

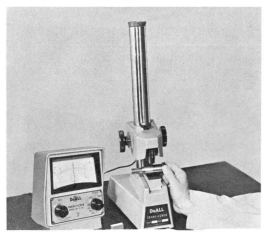

Fig. 11-17 Comparator applications utilize a stand that provides an accurately flat reference surface square to the gage head.

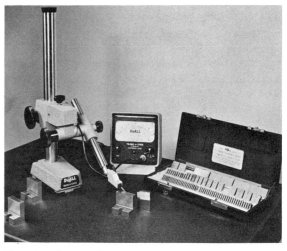

Fig. 11-18 Height gage setups use a surface plate or the part itself as a reference surface.

The multiple-scale feature has two roles far more important than simply saving the cost of purchasing additional instruments. First, it simplifies the tedious task of getting into scale range when using a high-amplification instrument. Second, it permits the best scale to be selected for every measurement.

It was mentioned in Chapter 10 that dial indicators are available with a discrimination of 50 millionths but that they are not considered to be general application instruments. One of the reasons is the difficulty of zero setting. Even with a fine adjustment the entire scale range can be passed so quickly that it

is impossible to tell whether the hand has turned part of a turn, one turn, two turns or more. The example in Fig. 11-19 is funny only when it is not happening to you.

Consider how much easier this is with a multiscale instrument, Fig. 11-20. The lower scales get you in the ball park before you begin to pitch.

The importance of always having a scale that best suits the measurement required cannot be overemphasized. In explaining the reason you will understand the remarks that have crept in through the earlier chapters about the importance of a ten-to-one ratio.

ZERO SETTING SINGLE SCALE INSTRUMENTS

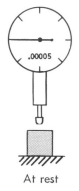 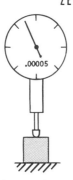 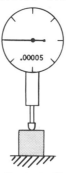 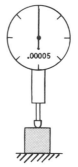 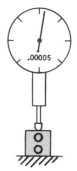 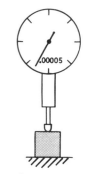

| At rest | Lowered onto part, but has hand made 0 full turn, or 1 or 2? | Backed off to at rest, preparatory to another attempt. | Lowered again and dial zeroed | Part inserted. Looks good but is it or was zero setting too close to end of range? | Replace standard and start all over again. |

Fig. 11-19 The trials and tribulations of zero setting a high amplification computer, that has only one scale, are often worse than this example.

ZERO SETTING MULTIPLE SCALE INSTRUMENTS
(comparator stand not shown)

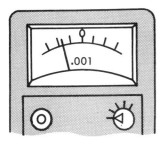

Set on 0.001 scale. Lower head onto standard until close to zero. Zero set with coarse adjustment. Then switch to 0.0005 scale.

Use the stand adjustment until hand is close to zero. Zero set with coarse adjustment. Then switch to 0.0001 scale.

Continue until on desired scale. Make final zero set with fine adjustment.

Fig. 11-20 Multiple-scale selection provides the best amplification for each role in which the instrument is used.

THE RULE OF TEN TO ONE

Over and over again, wherever people are concerned with practical measurement, you will hear it stated that the measuring instrument should be 10 times as accurate as the part. If you ask why enough times, eventually someone will tell you that it is done to limit the amount of instrument error that can creep into the measurement to one per cent. This is an example of the language problem discussed in Chapter 2. What does the rule really mean? What is the accuracy of the part? Its tolerance? The number of decimal places in the dimension? The "true value" of the part? The standard by which it is judged?

Restated the rule is simply: *The instrument should divide the tolerance into ten parts.* You will also hear this rule called the "one-to-ten rule" and "the rule of ten". All refer to the same thing.

In Fig. 11-21 we have greatly enlarged a portion of the scale of a comparator with a

discrimination of *one-tenth mil* (0.0001 in.). It is being used to measure a part with a tolerance of *two-tenth mil* (0.0002 in.). In other words, the instrument can only divide the tolerance into two parts.

The hand is zero set. The gray area represents the *uncertainty* of the instrument. We have not used that term before. It means *repeat error* or dispersion of repeated readings. It is based on *sigma 3*. When you study statistical quality control, you will find that this means that over 99% of all readings will fall into the zone. There may be some stray ones which we will ignore.

In an actual instrument the zone of uncertainty should not be more than one-fourth of a division. For clarity it has been made wider.

The zone of uncertainty does not create any problem when it is entirely within the tolerance. At B, for example, the part could be right on the low tolerance or on the basic dimension. In any case it is within tolerance.

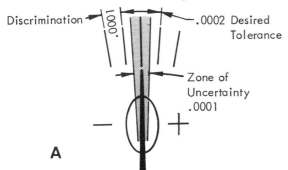

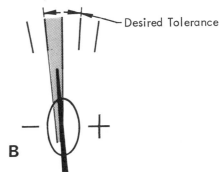

Fig. 11-21 The dispersion of readings does not cause a problem as long as it is entirely within the tolerance.

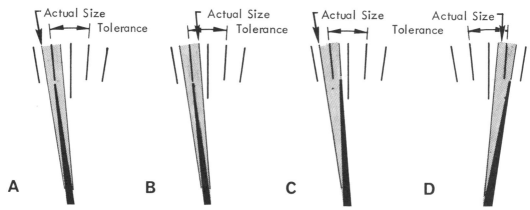

Fig. 11-22   These examples show that the apparent reading may be very different from the actual size as a result of the instrument uncertainty.

Figure 11-22 shows what happens when the tolerance as read on the scale is used to evaluate the parts. At A the hand is on the low limit of the tolerance. The actual part size is well below that limit as shown by the arrow. This would have been a bad part passed as good. At B the hand is in the same position, yet the part is well within tolerance. C shows the extreme out-of-tolerance situation being passed as well within tolerance. These cases have been at the low limit. They could all be repeated at the high limit. D, for instance, shows a good part being rejected as being past the high-tolerance limit.

In order to achieve reliable measurement, compensation must be made for the zone of uncertainty. The objective may be not to reject any good parts or not to accept any bad parts. The results are shown in Fig. 11-23. The reading must be double the tolerance (A) to get the zone of uncertainty out of the tolerance range, if no good parts are to be rejected. If no chances are to be taken of accepting a bad part only the space between the zones of uncertainty provide safe ground. Unfortunately in this case (B) there is none left. So let's go to a finer discrimination.

DISCRIMINATION DIVIDES TOLERANCE IN 2 PARTS

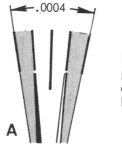

Range that passes all good parts. 50% additional range at each limit thereby passes bad parts.

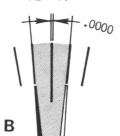

Range that rejects all bad parts. 50% of tolerance range thereby removed from each limit.

Fig. 11-23   An allowance must be made for the zones of uncertainty.

DISCRIMINATION DIVIDES TOLERANCE IN 4 PARTS

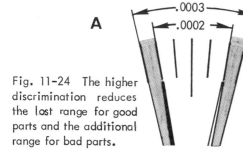

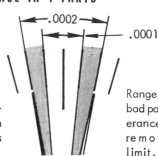

Fig. 11-24   The higher discrimination reduces the lost range for good parts and the additional range for bad parts.

Range that passes all good parts. 25% additional range at each limit thereby passes bad parts.

Range that rejects all bad parts. 25% of tolerance range thereby removed from each limit.

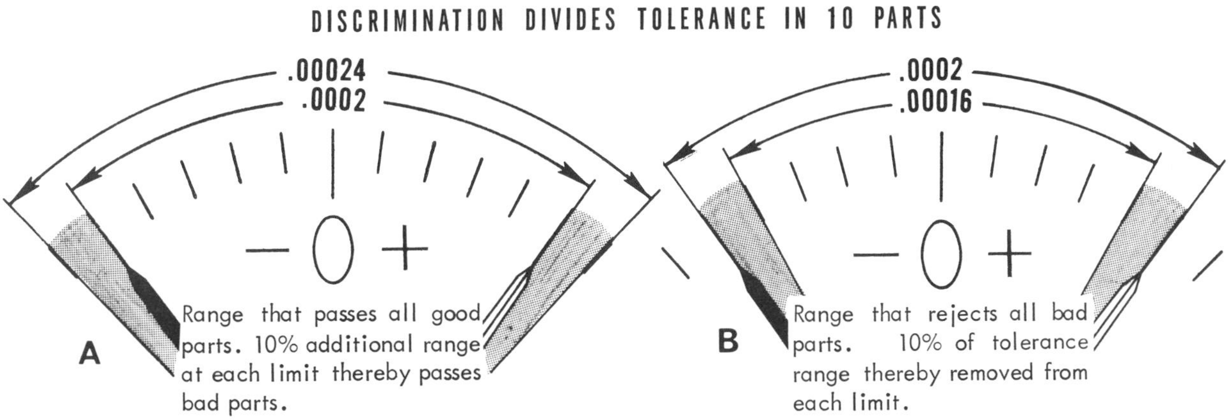

DISCRIMINATION DIVIDES TOLERANCE IN 10 PARTS

A Range that passes all good parts. 10% additional range at each limit thereby passes bad parts.

B Range that rejects all bad parts. 10% of tolerance range thereby removed from each limit.

Fig. 11-25 When the instrument has sufficient discrimination to divide the tolerance into ten parts, only 10% of range gives trouble at each limit.

In these examples the tolerance remains the same, 0.0002 in. The zone of uncertainty remains the same, one scale division. The discrimination, however, now divides the tolerance into four divisions instead of two as in the previous examples. In Fig. 11-24 it will be seen that this cut in half the range in which good parts are lost in B, and the range in which bad parts are accepted in A.

Figure 11-25 carries the matter to the recommended discrimination. The instrument now divides the tolerance into ten parts. As expected there is only 10% shrinkage of the scale for good parts and 10% expanding it for bad part admittance at each limit.

If you have followed this carefully, you might think you caught me on one. In the beginning of the ten-to-one rule discussion, I said that the reason for the rule was to assure that only 1% of the instrument error got into the actual measurement of the part. Yet Fig. 11-25 shows that there is 10% of the range at either limit that can pass bad parts or reject good ones. That adds up to 20% not 1%.

First, for clarity, we drew the zone of uncertainty four to five times larger than would normally be expected. That drops the percentage to four or five. Second, and by far the most important, the number of parts passed or rejected is almost never proportional to the scale length. Figure 11-26 shows that for a controlled operation most of the parts are grouped in one-third of the tolerance, and that as you get to limits of the tolerance there is just an occasional straggler.

Unfortunately this is not always the case. Keeping a manufacturing operation perform-

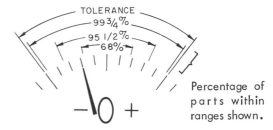

Fig. 11-26 In a typical manufacturing operation, most of the parts will be grouped close to an average size. The purpose of quality control is to assure that the average is close to the basic desired size.

ing this way is the function of the quality control department. They apply statistics to the data received by measurement in order to devise the most efficient control procedures. Obviously, their effectiveness depends upon the value of the measurement data they receive.

When you study statistical quality control, you will find that nothing is sacred to the statistician, not even the time-honored, ten to one rule. The one percent instrument error permitted in the measurement is questioned from all sides. Is it really one percent? Is that enough? Not enough? Too costly to maintain? Too costly not to maintain? You will find that there no easy answers and each problem must be evaluated on its own. Most of the questioning centers around whether or not the rule is economically justifiable.* No one questions its desirability in all cases except the most extreme

---

* Juran, J. M., *Quality Control Handbook,* New York, McGraw-Hill Book Co., Inc., 2nd Edition, Section 9, p. 11.

THE TEN-TO-ONE RULE

Rule:
The instrument * must be capable of dividing the tolerance into ten parts.

The Purpose:
To eliminate 99% of the instrumentation errors of previous steps in a measurement.

When Applied:
To every step in the measurement sequence until the limit of the available instrumentation is reached.

The Results:
Fewer bad parts accepted and good parts rejected.

* Instrument includes standards.

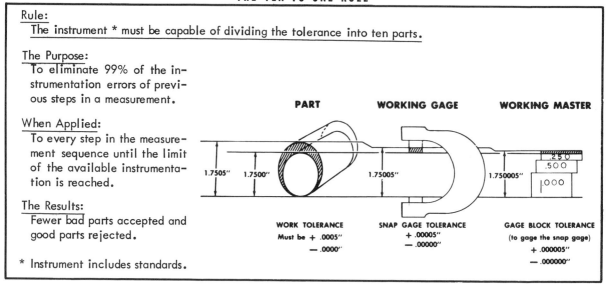

PART   **WORKING GAGE**   **WORKING MASTER**

1.7505"   1.7500"   1.75005"   1.750005"

WORK TOLERANCE
Must be + .0005"
        — .0000"

SNAP GAGE TOLERANCE
    + .00005"
    — .00000"

GAGE BLOCK TOLERANCE
(to gage the snap gage)
    + .000005"
    — .000000"

Fig. 11-27 This is a pragmatic rule, not a law. It is based on practical results. Often, practical considerations force us to deviate from it.

precision applications. It is summed up for you in Fig. 11-27.

You will notice from the examples in the figure that there is a tremendous pyramiding of precision. The standard may require 1,000 times the precision of the actual part. Figure 11-28 shows that this easily can exceed present day measurement ability.

You will note, too, that as stated the rule deals only with precision, accuracy is not mentioned. That is because *the accuracy is derived from the standard.* If it is not accurate, all is lost even though the rule is followed carefully, Fig. 11-29. That is why the next chapter on gage blocks is of such tremendous importance to anyone who expects to make reliable measurements.

With this knowledge of the rule of ten to one it will be easy to understand the major stride forward that multiple-scale instruments provided to practical measurement.

SELECTING THE BEST SCALE FOR THE MEASUREMENT

Theoretically the rule for scale selection is: *Always select the most sensitive scale that will contain the tolerance of the dimension to be measured entirely.*

LIMITATION OF TEN-TO-ONE RULE

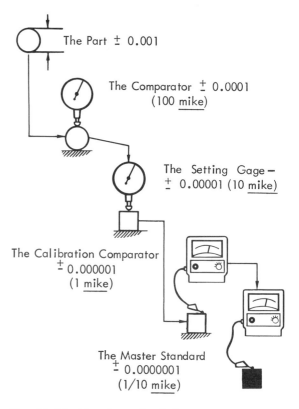

The Part ± 0.001

The Comparator ± 0.0001
(100 mike)

The Setting Gage —
± 0.00001 (10 mike)

The Calibration Comparator
± 0.000001
(1 mike)

The Master Standard
± 0.0000001
(1/10 mike)

Fig. 11-28 The demands of the ten-to-one rule quickly exceed present measuring instrumentation. The Bureau of Standards is currently investigating 1/10 millionth measurement.

TRACEABILITY AND THE TEN-TO-ONE RULE

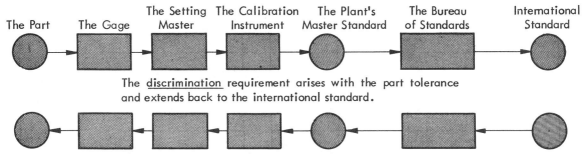

| The Part | The Gage | The Setting Master | The Calibration Instrument | The Plant's Master Standard | The Bureau of Standards | International Standard |

The <u>discrimination</u> requirement arises with the part tolerance and extends back to the international standard.

The <u>accuracy</u> arises at the international standard and progresses through each calibration step to the part.

<u>Traceability</u> is the procedure for determining that none of the connecting links between the part and the Bureau of Standards has been severed. It is required by law on government contracts and by sound economics for all other measurements.

MIL-1-45208A Inspection System Requirements, MIL-Q-9858A Quality Control Requirements, and MIL-C-45662A Calibration System Requirements are the applicable government standards.

Fig. 11-29 The ten-to-one rule deals with the precision of the measuring instruments but carried to its limit it reaches the international standard. By definition, this is the ultimate existing accuracy. If any link along the way is broken, the accuracy is lost, hence the need for <u>traceability</u>.

In the last chapter you were told why you could not follow this rule with dial indicators. You can with the electronic comparators. For the particular instrument we are using as an example, the five scales provide ranges from plus or minus *30 mil* to plus or minus *three-tenth mil* (±0.030 to ±0.0003 in.). This covers nearly any tolerance likely to occur in practical measurement.

For the few that are missed at the extreme high end, the manufacturer furnishes a companion instrument which is the same in most respects but has scales with discrimination up to one millionth of an inch. Because the highest scale must be used under laboratory conditions this instrument is used primarily for gage calibration.

It is not necessary to allow overtravel for the occasional part that is out of tolerance because it is easy to switch to a lower scale for such a part. The reverse situation also is encountered in repetitive measurement with a comparator. One or more parts is met that do not seem to move the hand at all. With a dial indicator and many other instruments you wonder if something has gone wrong, and usually you take time to recalibrate rather than to take a chance. With the multiple-scale electronic comparator you simply switch to a higher scale and immediately know whether

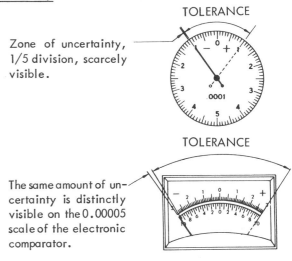

TOLERANCE

Zone of uncertainty, 1/5 division, scarcely visible.

TOLERANCE

The same amount of uncertainty is distinctly visible on the 0.00005 scale of the electronic comparator.

Fig. 11-30 Both of these instruments fulfill the requirements of the ten-to-one rule. The electronic gage provides readings in the uncertainty zone of the dial indicator.

the parts are exceptional or that something has gone awry.

As you were shown in the ten to one rule discussion, the greater the precision of the instrument in respect to the tolerance, the smaller will be the zones of uncertainty at each of the limits. The multiple scale feature enables full advantage to be taken of this, Fig. 11-30.

There is one situation in which the highest precision that will provide the necessary range may not be desired. If routine production inspection is being performed by unskilled personnel, the great sensitivity of the high-amplification scales may create the impression that the instrument "has gone crazy" and invite cramping, attempts at adjustment or worse. Of course, proper training is the best way to correct this. As a temporary measure, a lower scale will sometimes suffice.

ELIMINATION OF FRICTION AND INERTIA

The principal advantage of the electronic type of measurement system is the nearly complete absence of mechanical parts with their unavoidable vagaries. In the generalized measurement system, even with electronics, there are two places where mechanical action is hard to avoid. They are in the detector transducer and the final read-out stage. Ingenuity, however, has reduced the friction and inertia to a surprising minimum.

The read-out is a sensitive meter. The moving element is a tiny armature connected to the long, slender hand. Traditionally such meters have used jeweled bearings as in watch construction. To reduce the friction even further, the meter used in the instrument chosen for an example has a torque band movement.

The torque band is a taut, steel band held between end blocks. The armature and hand assembly is fastened to the band. The amount of twist to the band is proportional to the magnitude of the electrical signal acting on the armature. The band serves several functions – all good. It eliminates the friction of a jeweled movement. It substitutes a thin band for the shaft thus reducing the inertia of the moving parts. It eliminates the hairspring and its attachment arrangements. Best of all, it provides a resilient support that is nearly insensitive to shock. Even though utmost care is given to precision instruments, in practical industrial use they will receive more abuse than encountered in the laboratory.

IT ALL BEGINS IN THE HEAD

The need for low inertia and low friction is just as necessary in the other end of the system, the detector-transducer. In elec-

Fig. 11-31 Some electronic instruments are highly portable. This typical example operates on either a rechargeable battery or on line power. The finest of the three scales discriminates to 20 mike (0.00002 in.). *(Cleveland Instrument Co.)*

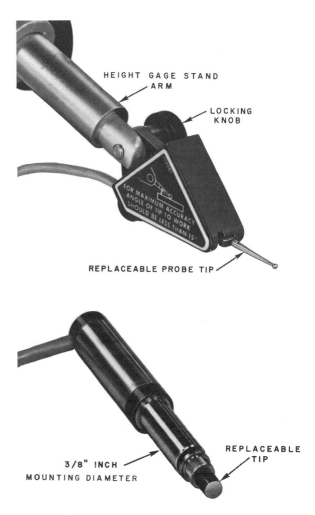

Fig. 11-32 The probe-type head (top) and cylindrical head (bottom) fulfill nearly all requirements.

tronic comparators the devices for this are
called *gage heads*. Agreed that this is contrary to the usual definition for gages, but that
is the way it is. There have been many types
of gage heads but the frictionless type is
clearly preferred from the reliability standpoint. These, like the meters we have just
discussed, make use of intermolecular friction to replace mechanical friction.

Two general constructions of gage heads
fulfill nearly all electronic gaging requirements, *probe-type* and *cylindrical*, Fig. 11-32.
The probe-type head usually is called simply
probe head or *indicator head*. It is attached
to an arm and extensively used for height-
gage measurements. It sacrifices some reliability to gain versatility. The cylindrical
type makes no compromises and therefore is
used when the highest reliability is needed,
such as in comparator setups. Its compact
design recommends it for use in gaging instruments which require a built-in head.

PROBE-TYPE GAGE HEADS

The probe head, or indicator head, has the
largest range of applications of any type of
gage head. The one chosen for our discussion
has an adjustable gaging force range of 5 to
100 grams, measured at the tip. It is used
when it is desired to have a pickup on the end
of an arm that can reach over obstructions
and into otherwise inaccessible places. The
probe head consists essentially of a housing,
transducer probe tip and lever assembly. The
transducer is an assembly of an LVDT (linear-
variable-differential transformer) with matched
coils and a core in a compact cartridge,
Fig. 11-33.

The transducer core is mounted on frictionless, stainless steel membrane springs.
The transducer head is more sensitive to fine
measurements than mechanical devices. The
friction-clutch mounting of the probe allows
a wide choice of gage positions. The probe
tip measuring range is *60 mil* (0.060 in.) plus
two-mil (0.002-in.) overtravel.

The probe tip and lever are pivoted on
a frictionless, double-leaf suspension. The
contact tips are 0.090-in. diameter, tungsten
carbide balls. The probe can be replaced
easily when worn, by simply unscrewing it.
It is important, however, not to replace with

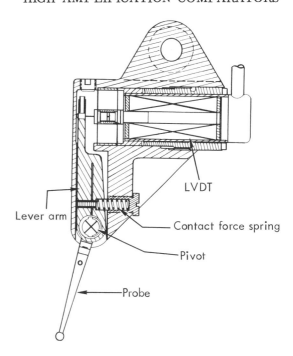

Fig. 11-33   The probe head is simply an adjustable
probe that acts upon a transducer cartridge in a
housing for convenient mounting.

a probe of different length. This would
change the geometry of the head and the meter would no longer read probe displacement
directly.

The mounting for probe heads has been
standardized, and is covered by an AGD specification. AGD specifications were mentioned
in the discussion about dial indicators.

The probe head may be used in a wide
variety of positions, as shown in Fig. 11-34.

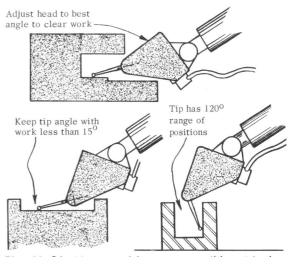

Fig. 11-34 Many positions are possible with the
probe head.

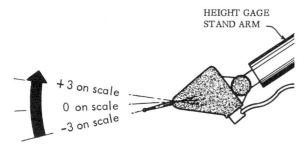

Fig. 11-35 Some probe heads have reversible action. Most do not. The direction of movement must be remembered. To operate in opposite direction, simply turn head 180 degrees.

There are probe heads that are reversible, however, the majority measure from one direction only, Fig. 11-35. This must be remembered in using the head.

COSINE ERROR AGAIN

We first met cosine error in the discussion of indicator stands, Fig. 10-58. Anyone concerned with reliable measurement necessarily becomes well acquainted with this source of trouble.

The reason they affect the probe head is shown in Fig. 11-36. Note that as the angle increases, the amount of vertical movement that is required to encompass the same arc increases. If the signal generated by the head is considered to be proportional to the angle, the effect of the angle of the probe can be appreciated. It must not be overlooked.

The amount of the cosine error is shown in Fig. 11-37. Note, in Fig. 11-38, that a *plus error in the instrument results in a plus error in the reading but a minus error in the actual part size.* The reverse is true also.

Cosine errors give us an opportunity to drive home again a point that has been made before. *A comparator has its greatest accuracy at zero.* As nearly as possible, use it as a comparator – not as a direct-measurement instrument. There are circumstances that require an angle of the probe that would produce a large cosine error. When the comparator is zeroed close to the actual part size, the cosine error in the reading is minimized. The amount of cosine error in the reading can be reduced even further by using the ten-to-one rule. Both methods are illustrated in Fig. 11-39.

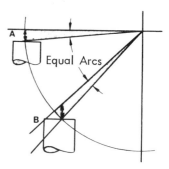

Fig. 11-36 It takes considerably greater travel of the contact in B to produce the same arc that was produced by the travel of the contact at A.

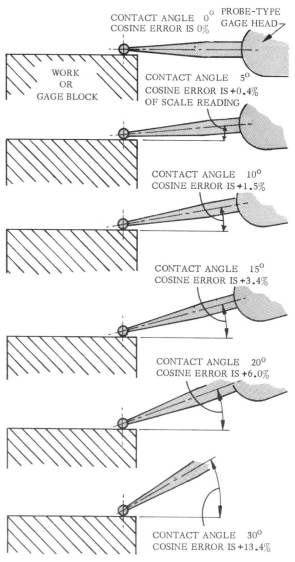

Fig. 11-37 The cosine error is dependent upon the probe position.

EFFECT OF INSTRUMENT ERROR

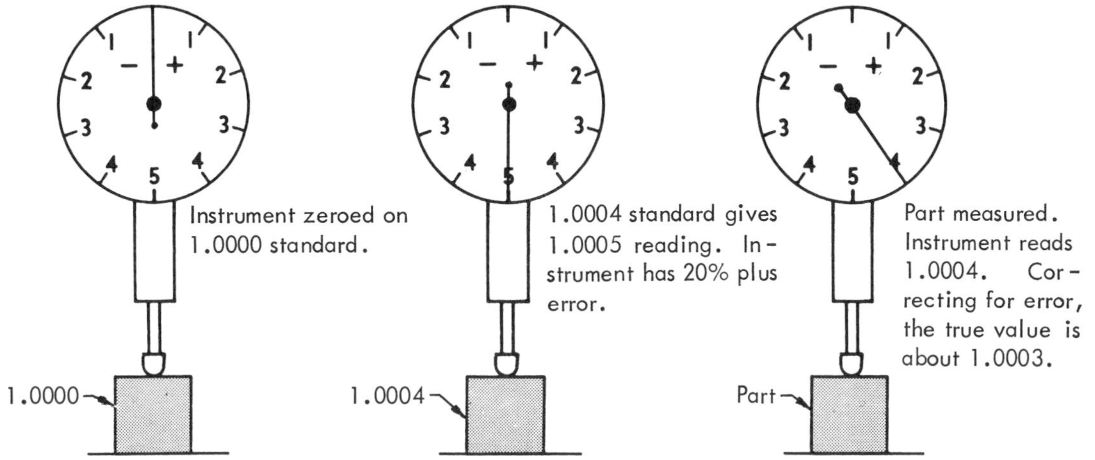

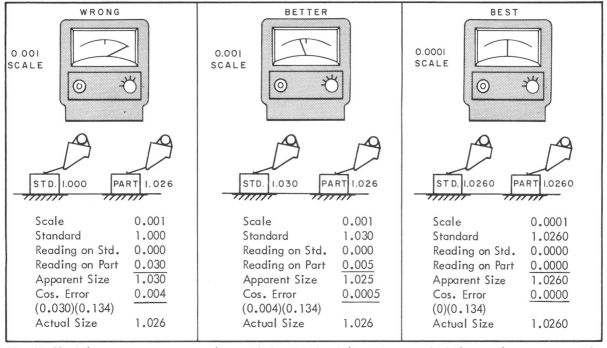

Fig. 11-38   Effect of instrument error is reversed on part. Plus instrument error decreases actual size. Minus instrument error increases actual size.

Fig. 11-39   When cosine error cannot be avoided, minimize it by using a standard close to the part size and by using rule of ten-to-one to select proper scale.

The steps recommended for minimizing cosine error are not restricted to that particular error. They are simply good practice and will, at the same time, minimize many other measurement errors. As a general rule the probe angle should be as close to zero as possible and never greater than 15 degrees, either direction.

GAGING FORCE

First we should review the distinction between *force* and *pressure*. Everyone speaks about "gaging pressure." Sometimes they mean it — more often they mean gaging force. The distinction was made in the discussion about micrometer ratchet stops, see Fig. 7-26.

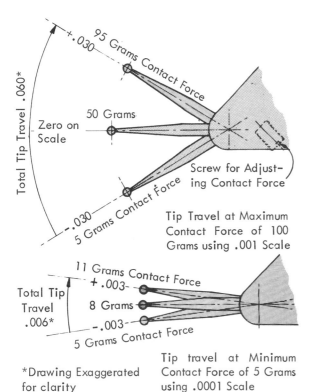

Tip Travel at Maximum Contact Force of 100 Grams using .001 Scale

Tip travel at Minimum Contact Force of 5 Grams using .0001 Scale

*Drawing Exaggerated for clarity

Fig. 11-40 The probe head may either be set for maximum travel or minimum gaging force.

It has been emphasized that the gaging force must be kept low. When measurements are made to *tenth mil* (0.0001 in.) or finer, even the addition of a few grams of force will cause measureable distortion, bending and compression. On the other hand, there must be sufficient force to assure positive contact between the instrument contact and the part. For these reasons the force of the tip of the probe head is adjustable. The force is proportional to the displacement. When it is decreased the displacement is also decreased. It may be set either for minimum force or maximum displacement. This is shown in Fig. 11-40.

The head is normally factory-adjusted for maximum displacement. In order to adjust for minimum force, first set the comparator on the desired scale. With the probe not deflected, back off the adjusting screw until the hand is just off the scale on the minus side.

THE CYLINDRICAL GAGE HEAD

The cylindrical head is nearly as versatile as the probe head, but in different ways.

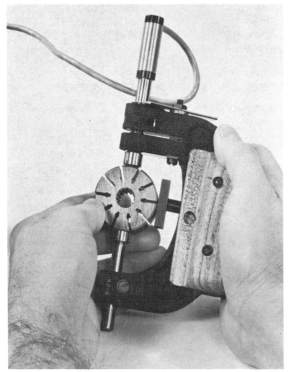

Fig. 11-41 The AGD mounting stem of the cylindrical head makes it interchangeable with dial indicators for use in fixed gages. (*DoALL Co.*)

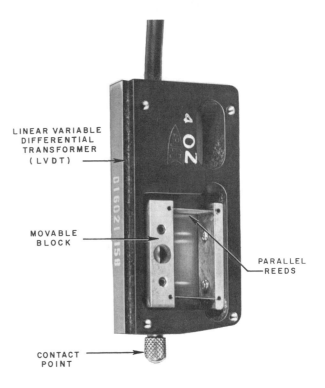

Fig. 11-42 The frictionless head uses a reed-spring suspension to eliminate sliding members. (*DoALL Co.*)

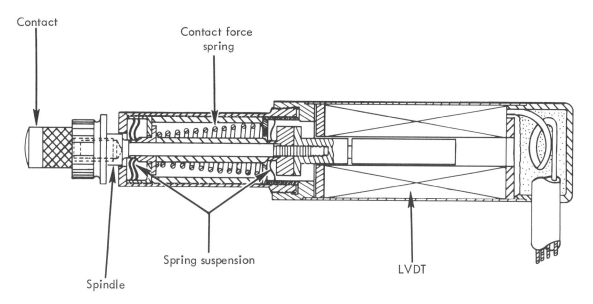

Fig. 11-43   The frictionless cylindrical gage head may be used in comparator setups or incorporated into fixed gages.

It has the same diameter as the AGD standardized stem of dial indicators. They, thereby, can be substituted for dial indicators in the tremendous variety of gages for which dial indicators were originally intended, Fig. 11-41 is an example.

Originally cylindrical heads used a sliding action. The contact attached to the spindle moved the core in the LVDT. A spring returned the moving parts. It was intended only as an improved replacement for dial indicators. When comparator work was to be performed to the highest precision a frictionless head was used, Fig. 11-42. Note particularly that these heads are in complete conformity with Abbe's Law, and owe much of their reliability to it.

Although both the sliding-type cylindrical head and the reed-suspension type frictionless head are still in wide use, their advantages have been combined. For that reason only one need be discussed here. This has been accomplished by use of the Bellville-type of springs for suspension, Fig. 11-43.

The gaging force is 6 ounces or 170 grams. While greater than the probe head, it is nearly uniform throughout the travel. The travel is *65 mil* (0.065 in.). The transducer components in the probe and cylindrical heads are the same and the heads, therefore, are interchangeable.

STANDS FOR ELECTRONIC COMPARATORS

For most high-amplification measurement instruments all three elements of the generalized measurement system are combined into one unit. This necessarily restricts them to compartor-type measurements. In the electronic systems the detector-transducer is independent of the remainder of the system. This permits three modes of operation instead of one. They are used in fixed production gages, Fig. 11-41, and for both comparator and height gage applications. The height gage applications are discussed along with surface plate measurements.

The role in which the electronic instruments can make the most complete use of their high amplification is as comparators. We have now run up against another ambiguous term in gaging. All of the instruments we have discussed from dial indicators on are used for comparison measurement, and are therefore, in the literal sense, comparators. The term *comparator*, however, is usually used to mean a rigid-frame instrument that furnishes both the reference and measured points. When an electronic comparator has its head installed in a loosely jointed stand that rides on a reference surface, it is known as a height-gage setup, or height-gage instrument.

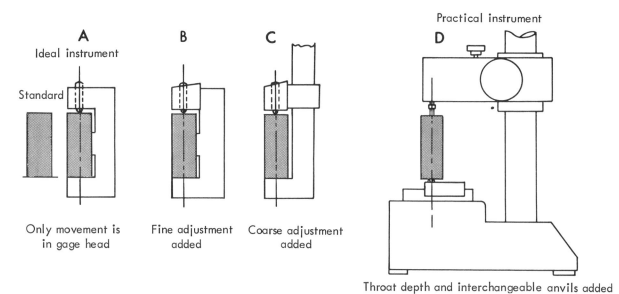

Fig. 11-44 The theoretically ideal instrument gives up reliability to gain range and versatility.

The higher the amplification, the more critical the positional relationships become. Consider an ideal situation, A in Fig. 11-44. In it the standard would not only be very nearly the same length as the part, but it also would have the same shape and be constructed of the same material. The instrument would hold the standard and the part in the same position and would have no movable members. The only movement possible would be displacement of the LVDT which would be along the line of measurement, or Z axis.

This would be of very limited use and difficult to set up. The first compromise with the ideal would be a fine adjustment of the head B in the figure. That is an extension of the Z-axis movement. Immediately a chance for positional and other errors is introduced which lowers the reliability. The next logical compromise would be to arrange the head so that the instrument could have greater range of measurement, as at C. While this greatly extends the Z axis it also greatly reduces the reliability. For example, the stop on the frame that locates the parts and standard in A and B could no longer be used when the parts differ widely in height.

The remaining compromise recognizes that the standard and the parts will rarely be the same shape and that no one arrangement for the part support will give adequate versatility. Therefore, at D, the much compromised ideal instrument has evolved into a practical measurement instrument with adequate work range both in height and throat capacity.

The value of the practical instrument is its ability to duplicate the ideal instrument during the actual act of measurement. When the setup is completed, the head is locked, the fine adjustment made, and the meter zeroed, there should be no movement possible in any axis except the Z-axis movement that displaces the LVDT.

That is the reason that a height gage application, Fig. 10-55, using a jointed stand with long arms, relying on a separate reference surface and whose elements must be aligned by eye cannot provide the reliability of a comparator setup. In a comparator, unlike a height gage, the relationship of the Z axis of the LVDT to the reference surface is controlled by the design and manufacture of the instrument.

These points were made in the previous chapter but assume new and greater importance with every increase in amplification. The metrological features of a comparator stand for high-amplification measurement are the same as for a dial comparator, Fig. 10-57.

The functional features of comparator stands are shown in Fig. 11-45. Although the constructional details may vary they must combine three qualities. First, the stand must provide positional relationships sufficiently

accurate for the measurement to be performed. Second, it must be sufficiently rigid
that the ideal instrument is approximated during measurement. And third, it must be
readily adjustable so that its part size may be
changed quickly and without danger of damage.

These requirements are not easily met.
At high amplification, for instance, on a *ten
mike* (0.00001-in.) scale, it is not uncommon
for the instrument to be swung out-of-scale
range when the head clamp is tightened. Naturally the instruments developed to overcome these difficulties vary considerably in
design, Fig. 11-46.

REFERENCE SURFACES

While the dial comparators often use a
simple, flat platen as the reference surface,
Fig. 10-56, these are not suitable for the
high-amplification comparators. Nearly all
use separate anvils that are attached to the
stands.

A typical anvil is shown in Fig. 11-45. It
is a hardened, ground and precisely lapped
steel block, 1 1/4 in. wide x 1 1/2 in. high x
4 1/4 in. long. It is held in place on the stand
by set screws that bear against the V-slots
along its sides. This forces the bottom of the
anvil into firm contact with the stand.

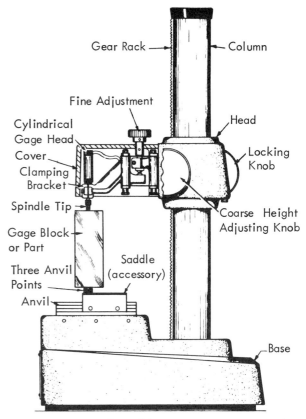

FUNCTIONAL FEATURES OF COMPARATOR STANDS

Fig. 11-45   High amplification comparators require
stands that combine rigidity with ease of operation.

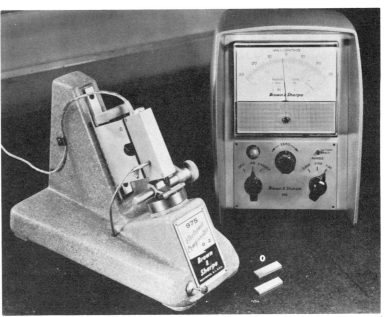

(Federal Products Corp.)

(Brown & Sharpe Mfg. Co.)

Fig. 11-46   High amplification comparators take very different forms as these two examples show.

The top and bottom surfaces are parallel within *ten mike* (±0.00001 in.) per in. of length and width. One surface is flat, the other is serrated. Either may be the reference surface. *The serrated side is used for flat parts* because of air film. This will be discussed in the chapter on calibration. The flat side is used for cylindrical, spherical or other parts that present line or point contact with the reference surface.

The second set of V-slots along the side is for the attachment of accessories. One of the most important of these is the 3-point saddle, Fig. 11-47. This device is essential for accurate measurement of parallel surfaces when the highest scales are employed. Its use is discussed on pages 288 and 289. Other important accessories that attach to the anvil are shown in Fig. 11-48.

An important accessory that does not attach to the anvil is the spindle lifter. It may be seen in Fig. 11-47. It is used to raise the contact point so that parts or standards may be placed in the comparator or removed without disturbing the setup or marring the fine surfaces.

Pressing the thumb button withdraws the contact point *60 mil* (0.060 in.) and holds it there until released by a release disk. Fig. 11-49 shows the details.

FOR RELIABLE MEASUREMENT, KEEP COOL

So far two points have occurred over and over. They are axioms for reliable measurement. First, is the necessity for extreme cleanliness; and, second, the necessity for correct positional relationships. Without these you cannot achieve reliable measurement. A third axiom has recently been made – the instrument must be of much higher precision than the tolerance of the part, preferably ten times higher.

Now a fourth axiom is required. In brief it is "keep cool." Even in the slang sense of not being panicked into snap judgments it is important in measurement. It is intended here in a more literal sense. Nearly all engineering materials that you will encounter, expand when their temperature rises. To complicate matters they expand at very different rates.

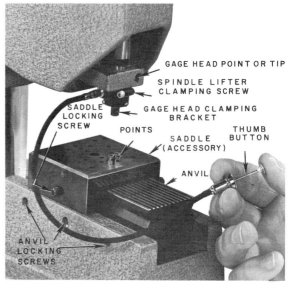

Fig. 11-47 Two important accessories are shown, the spindle lifter and the 3-point saddle. *(DoALL Co.)*

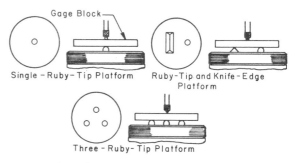

Single – Ruby– Tip Platform　　Ruby-Tip and Knife-Edge Platform

Three – Ruby- Tip Platform

Fig. 11-48 The work support platforms that attach to comparator anvils, often have support points of precious stones such as ruby or sapphire. The small contact areas would accelerate wear if extremely hard and abrasion-resistant materials were not used. (Dana, R.A.,"When and How to Check Dial Indicators,"Tool and Manufacturing Engineer, Nov. 1963.)

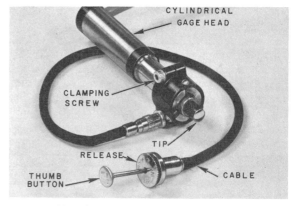

Fig. 11-49 The spindle lifter will either provide momentary retraction or will hold the spindle in the retracted position until released. *(DoALL Co.)*

These are known as their coefficients of expansion. They are usually expressed as *mike* per degree (0.000001 in. per degree) of temperature increase for one inch of material length.

For this reason, the international standards are established for one temperature, 68 degrees Fahrenheit (68 deg. F.) or 20 deg. Centigrade (20 deg. C.). Measurement of the highest precision is performed in laboratories maintained at those temperatures.

Fortunately, the majority of materials that are involved in most industrial use of measurement at high amplifications are steels with nearly the same coefficients. The only precaution that must be taken is to assure that all elements of the measurement are at the same temperature. When you are dealing with *tenth mil* precision (0.0001 in.) that begins to be a problem. From *50 mike* (0.00005 in.) and finer it is a very real problem. Factors that we never consider in ordinary life can become perplexing sources of thermal error.

There are three ways that heat is transmitted; by convection, radiation and conduction. All are involved in measurement and you must be constantly on the watch for them.

Convection is the transfer of heat by air currents. Drafts can have a very troublesome effect on precision measurement. Fortunately, the instrument and setup usually can be shielded from drafts when you become conscious of this factor.

Radiation is the transference of heat by rays such as light. Every cool body receives radiant heat from the adjacent warmer bodies. When measuring to *mike* (0.000001 in.) increments, the radiating person using the instrument can seriously change the readings. Even when measuring to lower precision you should be watchful for excessively close lights that can be very troublesome. Reflected light is preferred to direct.

The most common source of heat disturbance is conduction. Parts may have been held in a colder or warmer area than that in which the measurement is to be made. While there they will have acquired that temperature. Part of this will be due to convection from the air and part from conduction with other parts, machines, floor, etc. The most troublesome conduction, however, is from the hands of the inspector during measurement. When much handling is required to properly align the part and the instrument, it will be necessary to let it cool before taking the reading. Usually this involves, not 30 seconds of cooling, but an hour or more.

It takes much longer for a part to cool than it does to warm. This can be demonstrated easily as shown in Fig. 11-50. Using a high scale, the comparator is zeroed on a part, preferably rather long. The part is clasped in the hand and rather quickly the meter will begin showing the thermal expansion. Note the length of time required for a given expansion, say *10 mike* (0.00001 in.). Then release and notice the length of time required

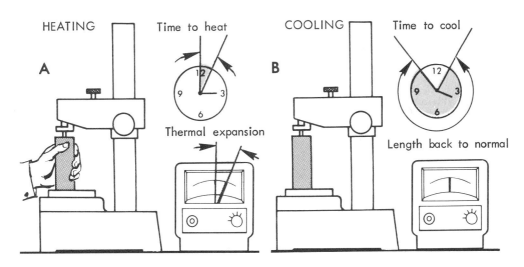

Fig. 11-50   It takes much longer for a part to cool than to heat.

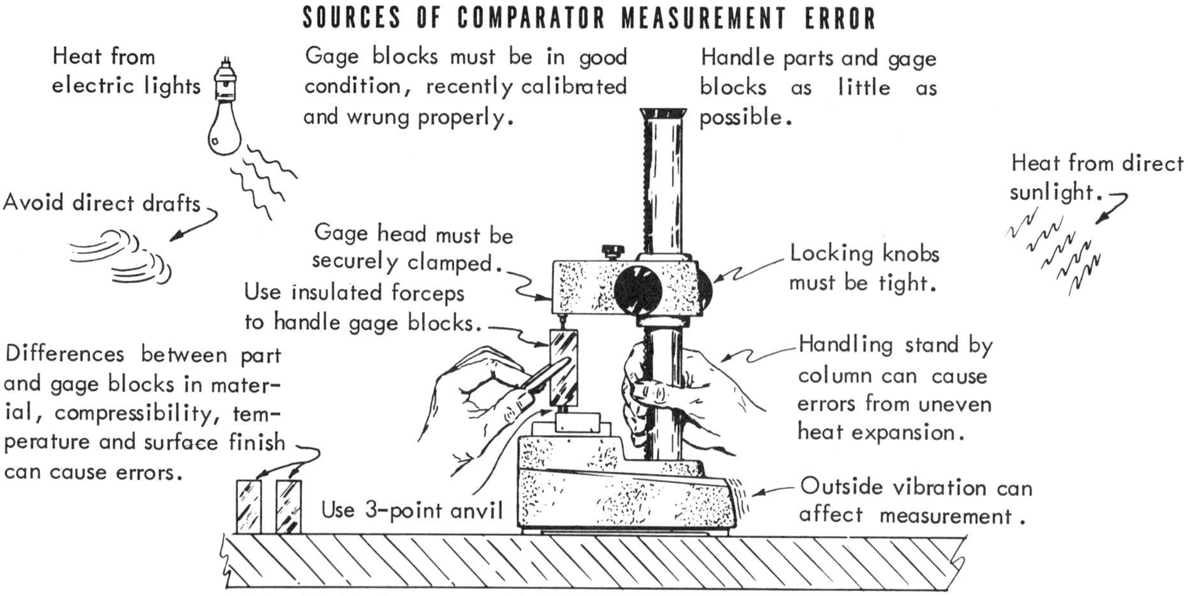

SOURCES OF COMPARATOR MEASUREMENT ERROR

Heat from electric lights

Gage blocks must be in good condition, recently calibrated and wrung properly.

Handle parts and gage blocks as little as possible.

Heat from direct sunlight.

Avoid direct drafts

Gage head must be securely clamped.

Use insulated forceps to handle gage blocks.

Locking knobs must be tight.

Differences between part and gage blocks in material, compressibility, temperature and surface finish can cause errors.

Handling stand by column can cause errors from uneven heat expansion.

Use 3-point anvil

Outside vibration can affect measurement.

Fig. 11-51 All sources of potential errors should be consciously investigated whenever a high-precision measurement is made.

for the part to *normalize*. That term means "to bring to the ambient temperature". In other words, it means that the part is brought to the same temperature as the instrument and the standard. The higher the scale, the quicker this demonstration can be made. However, it can be performed even with a *tenth-mil* (0.0001-in.) indicator, but requires more patience.

All of these heat problems are greatly increased when materials with differing coefficients of expansion are involved. For example, a steel part one-inch long will change *60 mike* (0.00006 in.) with a 10-degree temperature change. A one-inch brass part will change *90 mike* (0.00009 in.) under the same condition; and if aluminum the change would be *130 mike* (0.00013 in.).

A heat "sink" is one of the best methods to speed normalizing. It is simply a relatively massive body with high rate of heat transfer and a good surface finish. It is located alongside the comparator and parts, as well as the standards, are placed on it well in advance of the time that they will be measured.

The heat considerations are shown in Fig. 11-51 along with some other important sources of measurement error. When a problem arises investigate all possibilities. Sometimes errors that should not happen under any circumstance do happen. Misplac-

ing a decimal place is almost excusable compared with troubles caused by the tinkerer who cannot keep his hands off anything with knobs.

DIFFERENTIAL MEASUREMENT

The flexibility of electronics makes possible adaptations of the electronic comparator that would be nearly impossible in mechanical instruments. An outstanding example is the differential comparator. This is a comparator that provides measurement information by the algebraic sum of the signal from two pick up heads. Typical applications are the measurement of concentricity, roundness, parallelism and other part features. The advantage is their relatively high precision without expensive special fixtures.

The amplifier for differential measurement is similar to the ones previously discussed. It has, however, provisions for two input channels and two additional switches to select the mode of operation. In a typical differential instrument one of these switches is marked A and A-B. In the A position only one channel is used. In the A-B position the second channel is included. The second switch controls the way the two channels are combined. When set at plus (+) the channels are combined as +A-B. When at minus (-), they combine as -(A+B) or -A-B. Remember, that plus and minus indicate the direction of the pointer movement for a given movement of the contact in the gage head.

DIFFERENCE MEASUREMENTS

The general case for *difference measurements* is shown in Fig. 11-52. The two gage heads, A and B, are aligned along parallel lines of measurements on the same side of the part. With the reversal switch at plus, the selector switch is placed in A position and a standard placed between head A and the reference surface. The instrument is zeroed electronically. The selector switch is then turned

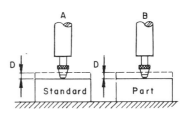

Fig. 11-52   This is the general setup for difference measurement.

ELIMINATION OF VARIABLE BY DIFFERENCE MEASUREMENT

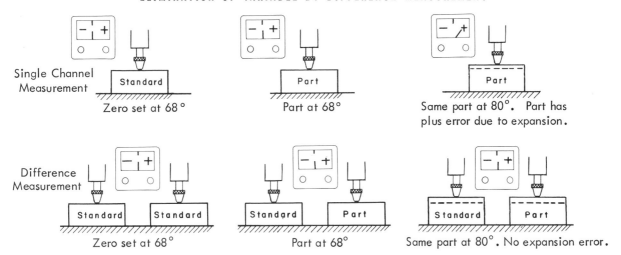

Single Channel Measurement

Zero set at 68°   Part at 68°   Same part at 80°. Part has plus error due to expansion.

Difference Measurement

Zero set at 68°   Part at 68°   Same part at 80°. No expansion error.

Fig. 11-53   Thermal expansion provides an example of the elimination of a variable by means of difference measurement.

to A-B, and the head B set to a standard. In this case the instrument is zeroed mechanically so that the adjustment of head A is not disturbed. The settings should be rechecked before measurement begins. Either of the standards can now be replaced with the part to be measured. The reading will be the difference between the two heads, or the amount that the part differs from the standard.

Why all this fuss when we can achieve the same thing with an ordinary comparator set up? Figure 11-53 shows the basic reason based on temperature change. In this case we have eliminated (or at least minimized) the variable caused by thermal expansion and contraction. This method is by no means limited to minimizing temperature errors. In Fig. 11-54, a setup is shown in which concentricity is measured between two surfaces without regard to the many variables that could affect this measurement if only one channel were used. Figure 11-55 is a similar case in which a taper is checked without regard to diameter.

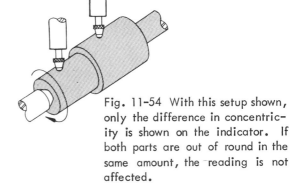

Fig. 11-54   With this setup shown, only the difference in concentricity is shown on the indicator. If both parts are out of round in the same amount, the reading is not affected.

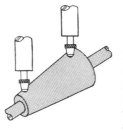

Fig. 11-55   By placing two gage heads parallel on a tapered part, it is possible to check the degree of taper without regard to its diameter. The taper must be set to a master, of course.

SUM MEASUREMENTS

The general case for *sum measurements* is shown in Fig. 11-56. The two gage heads A and B are aligned along the *same* line of measurement but on opposite sides of the part. The selector switch is placed in the A position as in the previous case, but with the reversal switch at minus. As before, head A is zeroed electronically on a standard. The selector switch is turned to A-B. Leaving the standard in place the instrument is me-

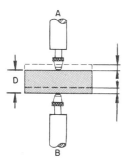

Fig. 11-56   This is an example of sum measurement using opposing bands.

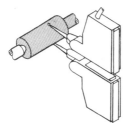

Fig. 11-57   Sum measurement in this example provides a measurement of the diameter even if the part is not exactly on center.

chanically zeroed again. After rechecking these settings, the part is substituted for the standard. The pointer now reads the *sum* of gage heads A and B.

This results in a setup that permits the part to be moved perpendicular to the line of measurement without changing the measurement. This is seen in Fig. 11-56. The part feature is D and that reads zero on the instrument. If D moves up a distance x then a +x is added to head A and a -x to head B. The net result is no change of reading. The example in Fig. 11-57 shows the elimination of the centering variable when measuring the part diameter. Figure 11-56 shows a *sum measurement* setup for comparing gage blocks.

Obviously, in all of these examples differential measurement is extremely helpful but should not be considered a cure all. In the last example, centering error could exceed the scale range. Furthermore, if the diameter is very small the centering error could displace the line of measurement sufficiently far from the true diameter to cause a minus error. The general precautions for reliable comparison measurement are reviewed in Fig. 11-58. Because of the extra steps required in differential measurement, additional precaution must be exercised in following these precautions.

RELIABILITY CHECK LIST FOR RELIABLE COMPARISON MEASUREMENT

1. Is there a better way to make the measurement?
2. Can the available instrument divide the tolerance into ten parts? Is there a record to support this? If not, is lower precision adequate?
3. Are the length standards in calibration (i.e.: is their accuracy traceable to The Bureau of Standards by up-to-date calibration records)?
4. Has the most reliable support been selected? Don't use a height gage if a comparator can be used. Don't use a comparator if a fixed gage can be used.
5. Are the instrument, parts and standards scrupulously clean?
6. Are all parts of the setup locked and secured to eliminate all movement except displacement in the gage head?
7. Has the environment been checked for drafts, direct light, vibration and other error causing disturbances?
8. Have the instrument, parts, standards, workholders and reference surfaces (whichever apply) been fully normalized?
9. Has the best scale been selected (electronic instruments only) and are the line values for reading that scale understood?
10. Has the measurement been repeated as a check?
11. If critical, has the measurement been repeated by someone else and the results compared?

Fig. 11-58   Temperature considerations have been added to this check list because of their importance in high amplification measurement.

METROLOGICAL DATA FOR HIGH AMPLIFICATION COMPARATORS

Instrument	Type of Measurement	Normal Range	Designated Precision	Discrimination	Sensitivity	Linearity	Reliability	
							Practical Tolerance for Skilled Measurement	Practical Manufacturing Tolerance
Mechanical (5000X)	Comparison	0.001 in.	0.000025 in.	0.000025 in.	0.00001 in.	2%	± 0.00002 in.	± 0.00025 in.
Electronic 0.0001 scale	Comparison	±.0024 in.	0.0001 in.	0.0001 in.	0.00005 in.	2%	± 0.00005 in.	± 0.0001 in.
0.00005 scale	Comparison	±.0016 in.	0.00005 in.	0.00005 in.	0.00002 in.	2%	± 0.00004 in.	± 0.0005 in.
0.00001 scale	Comparison	±.00024 in.	0.00001 in.	0.00001 in.	0.000005 in.	2%	± 0.000008 in.	± 0.0001 in.

Fig. 11-59   At very high amplifications the many variables to measurement accuracy are even more important than the inherent capability of the instrument itself, hence these conservative data.

Summary — Never Use High Amplification Measurement . . . .

● ...unless you have to. Unfortunately, you frequently have to and that is why it might pay to sum up this chapter before moving on. Dial indicators are inexpensive and reliable when properly applied. They are generally limited to *tenth-mil* (0.0001-in.) discrimination.

● Other means are used for finer discriminations. High amplification mechanical instruments achieve fine discrimination together with excellent reliability and low cost. They give up range and versatility as the penalty for power. Another large class of high amplification instruments are the pneumatic ones. They rate a chapter of their own.

● Of next greatest importance are the electronic instruments. Study of these requires recognition of the generalized measurement system which exists for all measurement. For electronic instruments the intermediate-modifying stage is different than for the previous instruments. They have power amplification. They are not limited by loading during the act of measurement. This gives them both their versatility and their high sensitivity.

● Electronic instruments make measurement in *mike* increments (millionths of an inch) deceptively easy. Therefore we must decide when to use such high precision. The ignorant who are careless simply guess. The ignorant who are careful base it all on the rule of ten to one. The best discrimination cannot be determined by any one simple rule. The rule states that the instrument must divide the tolerance of the measurement into ten parts or more. It is fine as far as it goes. The trouble is that you run out of rule and end up requiring instruments that have not yet been invented. On the other hand, such as making 50,000 functioning parts in a missile system all work together, the rule may be hopelessly inadequate. Anyway, traceability requirements often may do the deciding for you. Because of the human factors involved, lower sensitivity instruments may sometimes produce better results than high sensitivity.

● There are two general styles of gage heads used with electronic gaging, probe and cylindrical. The probe is used in height gage setups. The cylindrical is used in C-frame instruments of all types from hand-held snap gages to ultraprecise comparators. The probe head does not obey Abbey's law and therefore must be used with care. The most reliable heads of either style are the frictionless type. These use flexing members instead of rotating shafts on sliding surfaces.

● Stands for high amplification comparators must be selected carefully. Versatility is achieved by sacrifices in reliability. All unnecessary adjustments should be eliminated. Thermal considerations become a problem at any discrimination finer than *one-tenth mil* (0.0001 in.) and are nearly insurmountable except under laboratory conditions when it reaches *ten mike* (0.000010 in.). Cleanliness and contact between parts similarly takes on new importance at high amplifications.

● A distinct advantage of electronic measurement is the ease with which an electrical signal can be manipulated. This makes possible differential measurements. These require two heads whose outputs either add or subtract. Both are used to eliminate un-wanted variables from measurement. They find good applications in checking roundness, concentricity and parallelism.

● Reliable high amplification measurement requires careful attention to details. Every part of the measurement system including the observer must be checked for suitability, precision and accuracy. Both knowledge and skill are needed. Intelligent application, patience and complete objectivity are prerequisites. The data in Fig. 11-59, although general, will help. The terminology of Fig. 11-60 should keep you from being misunderstood — but probably won't.

TERMINOLOGY

Go	The interaction of gage and part that indicates within tolerance.
Not-Go	Interaction indicating out-of-tolerance condition. Often spelled "no-go."
Read-out	The data that the measurement system presents to the observer.
Analog Instrument	An instrument in which the sizes of physical quantities correspond to numerical values.
Digital Instrument	An instrument that uses counting methods.
Reed-Type Instrument	One in which spring suspension is substituted for shaft bearings, knife edges and other mechanical friction-producing suspensions.
Optical Lever	A weightless, frictionless beam of light used in the manner of a mechanical lever arm.
Loading	Imparting the input signal to measurement system by the energy it contains.
Power Amplification	The amplification of the loading as well as the amplitude of the input signal.
Electronic Measurement Instruments	Those in which the amplification is obtained by electronic means. (In most instruments the initial pickup is mechanical.)
Bridge Circuit	A closed network usually with fan branches in which a change in any branch alters the potential across the bridge.
Amplifier	The intermediate modifying stage of an electronic measurement system. May incorporate other functions such as filtering of the signal and comparison of signals.
Multiple Scale Instruments	Those instruments that can be switched from one amplification to another.

Ten-to-One Rule (rule of ten)	General rule that instrument should be capable of dividing part tolerance into ten parts.
Traceability	Documentation to establish that standards are known in relation to successively higher standards until the National Bureau of Standards is reached.
Pickup Head Gage Head Head	The detector-transducer stage of an electronic measurement system that converts a length change into its electrical analog.
LVDT	Abbreviation used in place of linear-variable-differential transformer, the most common pickup element in electronic instruments.
Probe Head	A pickup head whose moving member is a lever arm that can be inserted in openings too small for the head itself.
Cylindrical Head	A standardized head whose moving member travels along the axis of the cylindrical body.
Frictionless Head	A head using reed suspension of the moving members so that mechanical friction is eliminated. (Note that intermolecular friction remains.)
AGD Specifications	Standardized designs developed by the gage design committee of the American Standards Association.
Anvil	Common name for comparator reference surface.
Platen	Term frequently used for anvil or reference surface.
Serrated Anvil	Anvil having large number of closely spaced grooves to reduce surface area. Used to minimize effect of air film.
Air Film	The minute layer of air that separates two finely finished surfaces when wrung together.
Convection, Radiation and Conduction	The three means by which heat passes from one body to another.
Differential Measurement	The algebraic combination of input signals within the measurement instrument. Used to obtain concentricity, roundness and other measurements directly.
Difference Measurement	Differential measurement in which one input signal is subtracted from another one.
Sum Measurement	Differential measurement in which one input signal is added to another one.
Dynamic Measurement	Measurement of a moving feature or changing value.

Fig. 11-60 Many of these terms are not used in discussions of practical measurements. They will help you understand what is taking place within the instrument.

QUESTIONS

1. The practical discrimination limit for dial indicators is tenth mil (0.0001 in.). What better means of amplification have been devised?
2. How does the optical lever, in the reed-type comparator, further amplify the reed action?
3. How does the amplification of the pointer (hand) affect the amplification of the reed mechanism?
4. How are tension and travel adjusted for the movable block?
5. In our progression from low to high amplification instruments, what has been amplified in every case?
6. What energy is required to operate most indicators and mechanical comparators?
7. What is an electronic gage?
8. What type of circuit is most frequently used in electronic comparators for linear measurement?
9. How is this circuit used to provide accurate linear movement?
10. Why is greater flexibility possible with an electronic comparator?
11. Of what importance is the choice of various scales or the ability to change amplification on electronic comparators?
12. Of what importance is "zero setting" at higher amplifications?
13. Are there any situations in which dial indicators are superior to electronic gages?
14. What are the main reasons for using electronic gages?
15. What is the rule of "ten to one"?
16. Why is this rule needed?
17. What is usually meant by the "zone of uncertainty" in a measuring instrument?
18. How wide can this "zone of uncertainty" be?
19. Why do the demands of the ten-to-one rule exceed present measuring equipment?
20. Has the ten-to-one rule lost its significance in the present era of high precision electronic comparators?
21. What can be done if the ten-to-one rule cannot be followed?
22. What is the basic rule for proper scale selection when using electronic amplifiers?
23. What are the two basic types of gage heads that are commonly used with electronic comparators?
24. Name three important constructional features that a comparator stand should contain.

DISCUSSION TOPICS

1. Discuss the features of the reed-type comparator.
2. Explain how reed-type instruments provide high sensitivity.
3. Explain the principal advantages of electronic comparators over mechanical comparators.
4. Explain the principle of the simplified bridge circuit that is found in most electronic comparators.
5. Define and discuss the term "traceability."
6. Explain the features of an electronic comparator stand.
7. Discuss the principal sources and reasons for measurement error on electronic comparator assemblies.

Pneumatic Measurement
Amplification By Air

A remarkable means for measurement uses air. Changes in a calibrated flow respond to changes in the part feature. This is achieved several ways, all known as *pneumatic gaging*, *air gaging*, or more appropriately, *pneumatic metrology*. This entire book is concerned with principles and practices rather than nuts and bolts. That is particularly apparent in this chapter. Pneumatic metrology equipment varies considerably in operation and in metrological characteristics. Actual selection and use of equipment should be based on the ample material available from the manufacturers. The data included here are formulated from widely used equipment that has been important in the rapid spread of pneumatic metrology.

Pneumatic metrology owes its success to two basic features helped along by a third one. It *eliminates metal-to-metal contact* and it *provides power amplification* along with linear amplification. Its *low cost*, compared to other methods with similar features, assured it the attention of industry. Perhaps because of its newness, it has not found, as yet, the role it deserves in science. It is hoped that this text will help spread the gospel.

Pneumatic comparators share with electronic comparators the feature of power amplification. Not only is the change-in-length signal amplified, but the power to operate the measurement system comes from an external source. They are not dependent upon the energy imparted to the pick-up element by contact with the part. Therefore high amplification can be obtained together with high sensitivity.

Because there is no metal-to-metal contact, pneumatic comparators provide advantages not presently obtainable with any other practical instruments. And to show their versatility, pneumatic comparators do not limit themselves to air sensing. When required, contact-type pick-up elements can be used as well.

FROM HISTORY TO SEMANTICS

Most of the developments in pneumatic metrology have occurred since World War II.* The principles can be traced back to about 1917, however. Many investigators throughout the world developed methods for using fluid flow to solve particular measurement problems. These ranged from means for regulating paper manufacture to gaging rifle barrels. The first publicized use of pneumatic measurement was in France to control the size of automotive carburetor jets. It received great impetus during World War II.

These original uses were all production inspection applications, hence the term "gaging." It has only been in recent years that the basic principle has been expanded into versatile measurement instruments. Because it lends itself so well to the gaging of several features at once, it remains one of the most important means of production inspection, in use today, Fig. 12-1.

* Polk, L., A Review of Pneumatic Dimensional Gages, Symp. on Engineering Dimensional Metrology, London, 1955, pp. 205-224.

Fig. 12-1   This inspection instrument checks 47 dimensions and relationships on three sizes of diesel engine crankshafts by means of pneumatic gaging.

Pneumatic metrology is also important in the development of this text because it forces us to give somewhat more precision to three terms that so far are defined only loosely: *measurement, gaging,* and *inspection.*

The least precise thing about measurement is the language. It is ironic that, even though we readily have available instrumentation to measure to *mikes* (millionths of an inch), when you say "gage" it means entirely different things to the quality control man, the machine operator, the plant manager and the research scientist.

In the *Gages and Measurement Instruments* section of the last chapter the general distinctions between these two major classes was made. A portion of MIL-STD-120 in Fig. 11-2 nailed them down. In pneumatic metrology popular use made gaging and measurement almost synonymous. This is sloppy speech and can lead to sloppy thinking; therefore, this book uses these terms in their more precise sense.

BACK PRESSURE INSTRUMENTS

Pneumatic comparators, like the previous instruments, are analog devices. The quantity whose changes correspond to part feature changes is a flow of a very abundant liquid, air.* All pneumatic comparators are based on two of the several possible changes, pressure and flow rate. Of these, pressure was the first to be developed.

Figure 12-2 shows the double orifice system that is the basis of these pneumatic comparators. The intermediate pressure is dependent upon the source pressure and the pressure drops across the restrictor orifice and the nozzle orifice.** Pressure drop is the loss in pressure between any two points in a fluid system. It is largely due to eddies in the flow. The restriction caused by the nozzle orifice varies with the clearance distance, x. When x is changed the intermediate pressure changes. A measure of this change is an analog of the change in clear-

---

* Precision - A Measure of Progress, Detroit, Michigan, General Motors Corp., 1952, p. 45.

** Beckwith, T. G., and Buck, N. L., Mechanical Measurements, Reading, Mass., Addison-Wesley Publishing Co., Inc., 1961, p. 215.

ance. If the plate in the illustration is moved closer to the orifice, the flow of air is restricted and as a result less escapes to the atmosphere. This increases the *back pressure* and is read on the pressure gage – hence the term *back-pressure gage.*

With these instruments you might expect the constant pressure from the pressure regulator to result in a constant flow through the nozzle orifice. Of course, it does not – and here lies the limitation of the back-pressure air gage. The back pressure changes the rate of flow.

The amount of air that enters the system must equal the amount that escapes through the orifice. The pressure in the conduit must be that which permits the same amount to exit through the orifice as is entering through the pressure regulator. The balancing, although automatic, is far from instantaneous. The time it requires is called the *lag.* It results from the compressibility of air. The amount of lag depends upon the relationship between the volume of the conduit and the size of the orifice. It decreases the sensitivity.

To demonstrate, consider first a pneumatic comparator with a small volume of conduit and a large orifice. If the orifice is nearly restricted by a small distance, Fig. 12-3, the lag will be short. That is good, but if the x distance is large we have a bad situation. Then the back pressure caused by the conduit affects the pressure gage more than the changes in the orifice restriction (x distance). This air gage has high sensitivity but a short range.

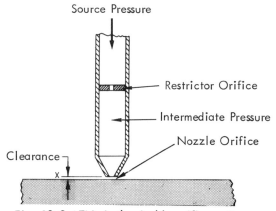

Fig. 12-2 This is the double orifice system upon which back-pressure pneumatic comparators are based.

BACK-PRESSURE AIR GAGE

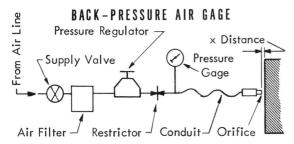

Fig. 12-3   In the basic back-pressure air gage, a change in x alters the conduit pressure and is read on the pressure gage.

In the reverse case we have a large conduit and a small orifice. Now the changes of restriction of the orifice (changes in x distance) have the predominant effect on the pressure gage. The range can be extended; but, unfortunately, the lag is extended also. Thus the back-pressure air gages are limited, both in sensitivity and range.

(Pratt & Whitney Co., Inc.)

Fig. 12-4   This compact air gage is a back-pressure type using a bourdon-tube pressure gage as the indicating element.

The type of back-pressure pneumatic comparator in Fig. 12-3 is built with two means of read out. The example in Fig. 12-4 uses the familiar bourdon-tube pressure gage. Its best amplification range is from 1,000X to 5,000X. Some are higher. Although a rugged and practical instrument, it has the problem of wear of the moving parts in final stage. A variation eliminates these problems by the use of a water column. This has not proved popular for general use, however.

Refinements of design have extended the range and sensitivity of the back-pressure pneumatic comparator. But, the problem of lag limits the basic design to relatively short conduit lengths. These limitations have been removed by two very different departures. One is the balanced-type system; the other is the rate-of-flow system.

BALANCED SYSTEMS

The balanced system, Fig. 12-5, starts off much the same as the basic back-pressure system, Fig. 12-3. The flow of air is divided into two channels; both of these have restrictors. The reference channel then exits through another restrictor, the zero setting valve. The measuring channel continues through the air hose to the gaging spindle. Between the channels is a differential pressure gage. This arrangement is analogous to the bridge circuit used in electronic comparators.

BALANCED SYSTEM

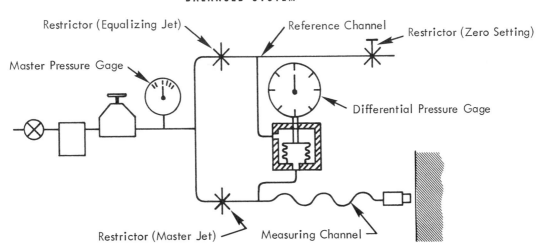

Fig. 12-5   The balanced system has fixed amplification. The only adjustment is zero setting.

The basic back-pressure system in Fig. 12-3 shows changes in the gaging pressure in respect to atmospheric pressure. There is no control over the latter. In the balanced system of Fig. 12-5 and Fig. 12-6, the changes in the measuring channel pressure are shown in respect to the reference channel pressure. Both channels are subject to careful control by the restrictors.

By these means some of the difficulties with back-pressure systems are eliminated and some desirable features obtained. Unlike the other systems, this one requires only one master and one adjustment, zero setting. It is claimed further that wear of the spindle does not change the amplification of the system. The relation between amplification, discrimination and measurement range for one line of these instruments is shown in Fig. 12-7. A unique, although very practical, application is shown in Fig. 12-8.

BALANCED SYSTEMS DATA		
Amplification	Discrimination	Measuring Range
1250 : 1	0.0001"	0.006"
2500 : 1	0.00005"	0.003"
5000 : 1	0.00002"	0.0015"
10,000 : 1	0.00001"	0.0006"
20,000 : 1	0.000005"	0.0003"

Fig. 12-7 The balanced systems are available in a wide selection of amplifications and with relatively wide measuring ranges.

Fig. 12-6 The balanced system results in a compact instrument that only requires one set-ting master for each gaging spindle.

RATE-OF-FLOW AIR-GAGE INSTRUMENTS

The *constant-pressure, rate-of-flow* or *velocity* pneumatic comparator was developed to eliminate the limitations of the back-pressure type. In this type, the pressure is constant but the change in flow rate is used to show changes in the x distance. The basic circuitry is shown in Fig. 12-9. It differs

Fig. 12-8 This flatness gage uses a balanced system. The part is placed on the surface plate and passed over the air jet. Deviation from flatness is read directly from the instrument dial.

from the back-pressure type in that the flow through the conduit is not deliberately restricted. The air can pass as rapidly through the conduit as the restriction at the orifice (x distance) allows. The rate at which it flows is shown by the level of the float in the internally tapered flow meter tube.

In this system a much higher flow of air can be used. Consequently the working range can be extended considerably longer than the back-pressure type. And because the restriction caused by the system itself is slight, the lag is also slight. Anything that interrupts the escaping air slows the rate of flow throughout the entire system almost instantaneously.

The general response of the pneumatic comparator is shown in Fig. 12-10. The height of the curve shows the volume of air flowing at various amounts of restriction

CONSTANT-PRESSURE RATE-OF-FLOW AIR GAGE

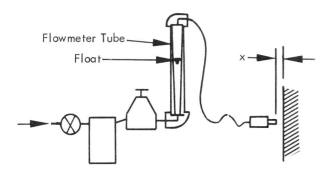

Flowmeter Tube

Float

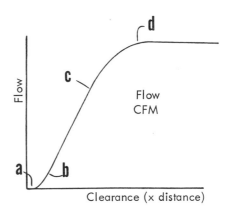

Fig. 12-9   In the constant-pressure rate-of-flow air gage the conduit pressure remains constant. Changes in x alter the rate of flow. This is read by the height of a float in a tapered flowmeter tube.

Fig. 12-10   The flow changes when the clearance is changed.   A portion of the response curve (from b to c) is nearly straight and suitable for measurement.

starting with x distance equal to zero. The first small increase in the clearance shows no flow (a). Then as the clearance increases, the amount of flow change per unit-of-clearance change becomes greater until the maximum slope is reached (b). The flow-to-clearance ratio then remains constant for a major portion of the curve. This is the portion that is usable for measurement. When the clearance becomes sufficiently large, changes in it no longer cause equal changes in the flow (c). The flow increases, then gradually diminishes until a point is reached (d), after which increases produce no further change.

The column-type flow circuit pneumatic comparator has many desirable features, Fig. 12-11. It directly provides amplifications in excess of 50,000 X. With auxiliary equipment this may be raised to 100,000 power.   It is completely free of the wear and the hysteresis of mechanical systems. The response is immediate.   Calibration is simple. And it shows, with the other pneumatic comparators, an application versatility that enables many difficult measurements to be performed easily and reliably.   This versatility derives from the unique gage heads that can be used with pneumatic gaging.

FEATURES OF PNEUMATIC METROLOGY

Functional Features:

1. No wearing parts

2. Rapid response

3. Remote positioning of gage heads

4. Self cleansing of heads and parts

5. No hysteresis

6. Adaptability to diverse part features

7. Small size of gage head

Metrological Features:

1. No direct contact

2. Minimum gaging force

3. Either variables inspection (measurement of size) or attribute inspection (go and not go gaging of limits)

4. Range of amplification

5. Adaptability to several modes of measurement (length, position, surface topography)

Fig. 12-11   These features recommend air gaging for small holes, highly polished parts, fragile or easily determined parts, remote measurements and the inspection of multiple part features.   This is based primarily on the rate-of-flow type of air gage with flowmeter columns, but many of the features apply to other air gages as well.

THE GAGING ELEMENT

In pneumatic metrology the first step of the generalized measurement system, Fig. 11-12, is called a *gaging element*. It corresponds to the *pickup* and *gage head* of previously discussed instruments. There are three general types of elements. In type 1, the hole being measured is the exit nozzle of the gaging element. In type 2, an air jet not in contact with the part is the gaging element. In type 3, the air jet is mechanically actuated by contact with the part.

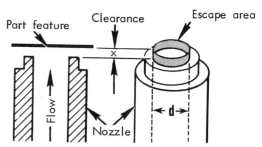

Fig. 12-12 The airflow through the gaging element is proportional to the exit nozzle cross section and the clearance or x distance.

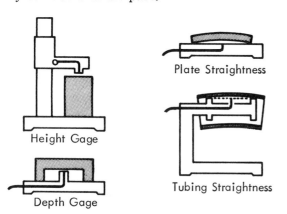

Height Gage

Depth Gage

Plate Straightness

Tubing Straightness

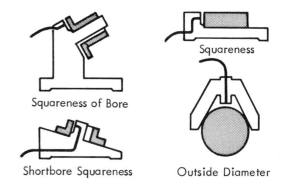

Squareness of Bore

Shortbore Squareness

Squareness

Outside Diameter

Fig. 12-13 Single jet nozzles form the basic gaging element as in these examples.

Type 1 is only suitable for inside measurement. It is used when the cross-sectional area is to be controlled rather than the shape. This method has limited but important applications. Carburetor jets and nozzles for the production of synthetic fibers are examples of its use.

The gaging element in Figs. 12-3 and 12-9 are type 2 of the simplest form. The rate of flow depends upon the cross-section area of the nozzle and the clearance between the nozzle and the part feature, Fig. 12-12. Essentially this is an air jet placed close to the part as shown in the examples in Fig. 12-13. Obviously each of these must be calibrated to standards. This is discussed later.

In the discussion of electronic gaging the advantages of differential measurement were pointed out. This consisted of electronically combining the output of two gage heads, and made possible measurements that would be difficult with a single head at the same level of reliability. A unique feature of pneumatic gaging is the ease with which this same principle can be applied. In fact, unlike electronic

gaging, it can be done entirely in the gage head without any alterations in the rest of the system. Furthermore, it is not limited to two inputs. The combinations that can be used are limited only by the understanding of the principle and ingenuity in its application.

It is for this reason that pneumatic gaging is perhaps the most important method for the inspection of holes. The gaging elements, or *spindles* as they frequently are called, can be adapted to measure nearly any feature of the hole including diameter, roundness and straightness. Consider a spindle with only one jet, located on one side, Fig. 12-14. The float height in the flow meter tube will rise

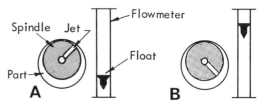

Fig. 12-14 Rotation of the single jet spindle changes the height of the float.

and fall as the spindle rotation alternately opens and then closes off the one jet. If, however, there were two opposed jets we then have a differential measurement. This is shown in Fig. 12-15. Clearance A added to clearance B always results in the same equivalent x distance.

A cross-section of a spindle of this type is shown in Fig. 12-16. This spindle would be used for determining roundness as well as diameter. By rotation it will measure any changes in the diameter. If only the average diameter is to be measured, three or four jets could be used, Fig. 12-17.

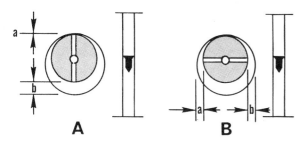

Fig. 12-15 With opposite jets, rotation of spindle does not change float height because the sum a + b equals x for any position.

SINGLE-JET PLUGS - check concentricity, location, squareness, flatness, straightness, length, depth.

TWO-JET PLUGS - check inside diameters, out-of-round, bell-mouth, taper.

THREE-JET PLUGS - for checking triangular out-of-round.

FOUR JETS - are used to furnish average diameter readings.

SIX JETS - will show average determinations for both two-jet and three-jet conditions.

Fig. 12-17 By simply changing the spindle, a variety of characteristics may be checked.

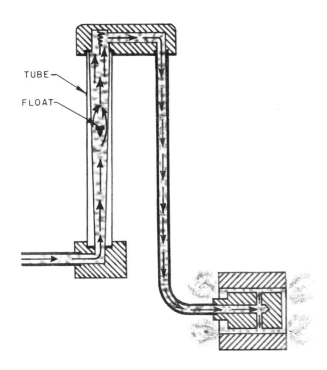

Fig. 12-16 Cross section of typical spindle for the measurement of inside diameters.

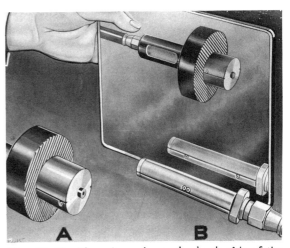

Fig. 12-18 The mirror shows the back side of the spindle for checking squareness (A) and another for checking straightness (B).

Further modifications of the basic spindle adapt it for the measurement of still other characteristics of the hole. In Fig. 12-18, a mirror is used to show both front and back of two spindles. The one at A measures the squareness between a face and the axis of bore.

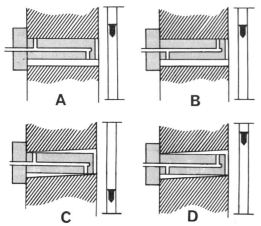

Fig. 12-19   If the axis of the hole is square to the shoulder , rotating the spindle does not alter the height of the float as in A and B. Lack of squareness is shown by a fluctuating float height as in C and D.

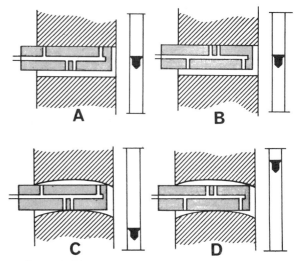

Fig. 12-20   This spindle checks straightness. Rotating the spindle from A to B does not change the combined clearances if the hole is straight. If it is not, a maximum and minimum reading results.

It is explained in Fig. 12-19   Note that two jets are used. Both measure clearance along a radius but one is located close to the shoulder, the other at the far end of the spindle. If the hole is square, rotating the part will not alter the reading. If the hole is not square, rotating the part will cause a fluctuating reading.   The other spindle (B) in Fig. 12-18 is designed particularly for checking hole straightness. Figure 12-20 shows the way it operates.

The ease of differential measurement by pneumatic metrology makes possible computing measurement instruments. The basis for this is shown in Fig. 12-21. It amounts to separate pneumatic comparators for the large and small diameters plus an interconnection that directly reads in taper. A simplification of this design is shown in Fig. 12-22. In this case the interconnection that provides the taper information consists of a slidable scale. The same principle can be used to assure the proper fit of mating parts. In that case a three-column instrument is used to show o.d., i.d. and clearance.

Type 3 gaging elements are contact type. They resemble the cartridge-type of electronic pickup head in both appearance and use, Fig. 12-23. Instead of an LVDT transducer, they incorporate an air valve which changes the air flow in proportion to the linear change.

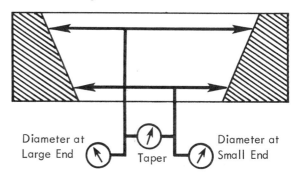

Fig. 12-21   Computing gages are readily constructed with pneumatic gaging components.

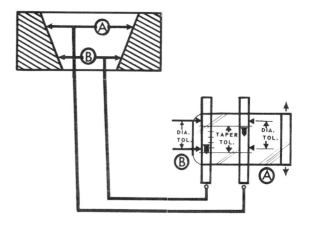

Fig. 12-22 This simplification of the computing gage works on the principle of the nomograph.

SURFACE FINISH CONSIDERATIONS

Your question is obvious, "If absence of metal-to-metal contact is such an important feature of pneumatic comparators, how do you justify going back to contact-type gaging"? Pneumatic metrology obeys one of the fundamental laws of measurement that has been repeated many times in this book: Whenever you gain something in measurement you must give up something. A pneumatic comparator with the type 2 element provides high amplification, high sensitivity, fast response, good accuracy, and assorted other benefits. To do this it gives up range and becomes as sensitive to surface condition as a Ferrari on a cow path. The air gages have about 30 per cent less range than the better electronic instruments. They more then compensate for this with their superior *resolution* which makes for exceptional *readability*. More about this later.

Their response to surface finish is another matter. The measured point for the type 2 gaging element is the *pitch line* of surface roughness. You will hear more about that term when you investigate screw threads. For now you can consider this to be the *mean* value between the highest and lowest points of the surface finish, Fig. 12-24. This is not the same measured point that would result if contact-type instruments were used. Therefore, the difference is a source of error.

Consider a surface finish of 100 microinches* (0.0001 in.) and a pneumatic comparator with a range suitable for a *three-mil* (0.003-in.) tolerance. For ordinary levels

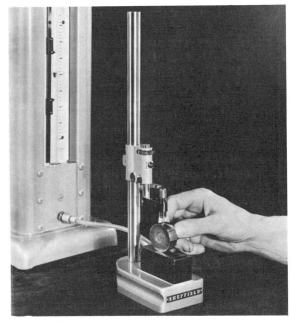

Fig. 12-23   Contact-type air gaging cartridges extend the role of air gaging. They can often be used interchangeably with electronic gage heads.

of quality assurance the maximum permissible uncertainty in the measurement is *three-tenth mil* (0.0003 in.). The surface finish alone could use up one third of this, not leaving enough for the frailties of the instruments and the inspector. Generally, as a rule of thumb, the type 2 gaging element is not used if the surface finish exceeds 60 to 65 microinches.

---

* The use of microinch for surface finish measurement is so ingrained that decimal-inch terminology has not been substituted.

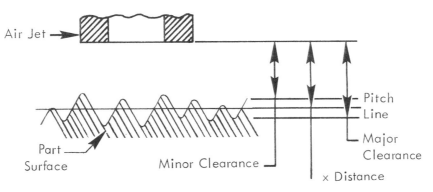

Fig. 12-24   With the type 2 gaging element, the reference line is the pitch line between the crests and valleys of the surface.

SELECTION OF GAGING ELEMENTS

Element	Operation	Measurement
Type 1	Part forms own exit nozzle	Cross sectional area
Type 2	Clearance between part and nozzle regulates flow	Tolerances not greater than 0.003 in. on parts with surface finishes 65 microinch or finer
Type 3	Mechanical contact with parts regulates flow	Tolerances greater than 0.003 in., or parts with surface finishes rougher than 65 microinch, or parts whose surface finish differs from the master by 50 microinch or more.

Fig. 12-25 One of these three gaging elements will reliably accommodate nearly every measurement requirement.

Surface finish must be remembered when setting a type 2 gaging element to its masters. If the master and the part have the same surface finish there is no problem. Most masters, whether ring or cylindrical, are lapped to a finish of 5 microinches or better. If an instrument is set to a 5 microinch master and then used to gage a 200 microinch part feature, a large error is possible. For most applications there should not be more than a 50 microinch difference in surface finish.* Even this is excessive if the tolerance is *one*

---

* Kennedy, Clifford W., <u>Inspection and Gaging</u>, New York, Industrial Press, 1951, p.293.

mil (0.001 in.) or closer. This then gives us the third justification for the use of type 3 gaging elements. These are summarized in Fig. 12-25.

There are still other places in which the type 3 gaging element is preferred, Fig. 12-26. For example, when the surface width is less than *one hundred mil* (0.100 in.) the type 2 rarely can be used. Serrations or grooves in a surface also may disturb the reliability of type 2 gaging elements. In these cases the type 3 can be substituted. They also can be used in the multiple inspection setups for which pneumatic metrology is renowned. Figure 12-27 is an example.

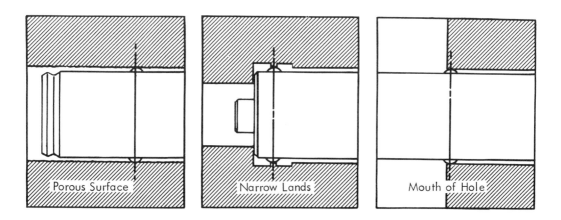

Porous Surface   Narrow Lands   Mouth of Hole

Fig. 12-26 For these situations the type 3 gaging element is preferred to the air sensing of type 2.

"MASTERING" THE PNEUMATIC COMPARATOR

Although the instruments differ greatly they are set up similarly for measurement. Three distinct steps are required. The instrument is set to a master at the high or the low limit. The amplification is adjusted, and the instrument is set to a master at the second limit. Two things are different from previous comparators. First of all, the amplification is subject not only to selection but also to adjustment. Secondly, because of this, both limits are set to masters.

Opponents of pneumatic metrology cite the latter as a disadvantage. The use of two masters is a distinct advantage and materially reduces the uncertainty in setting the pneumatic comparator. All too often other comparators are used for direct measurement. This is nearly impossible with pneumatic comparators. The pneumatic comparator is a comparator and behaves like one!

Gaging elements for inspecting small holes are set to ring gages, Fig. 12-28. Those used for inspecting large inside measurements as well as outside measurements usually are set to gage block calibration fixtures. Two of these are shown in the next chapter on calibration.

When measurements to a precision of *one-tenth mil* (0.0001 in.) or finer are required, some precautions are required. The instrument should be *mastered* in as nearly the same manner as it is used for measurement. If the jets of the gaging element, for example, are in a vertical line when checking parts, they should also be vertical when the instrument is set to its master. And as mentioned before, the masters and the part should have as nearly the same surface finishes as possible.

METROLOGICAL ADVANTAGES OF PNEUMATIC COMPARATORS

The next questions that logically arise concern the metrological features. What are the amplification and precision relationships that govern pneumatic comparators? As mentioned in the beginning of the chapter, there is a large selection in pneumatic metrology equipment. The following data are based on the instrument in Fig. 12-28 which is a familiar example. The functional parts of this instrument are shown in Fig. 12-29.

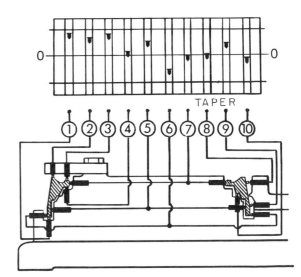

Fig. 12-27 The vertical flowtubes of pneumatic comparators particularly suit them for multiple inspection setups. Either type 2 or 3 gaging elements may be used. This example shows length, width, thickness and taper dimensions being checked simultaneously with ten type 3 gaging elements.

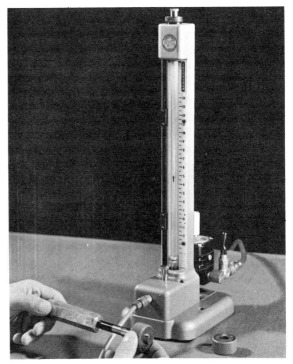

(The Sheffield Corp.)

Fig. 12-28 This is a popular form of pneumatic comparator. It is available in a wide range of amplifications and may be used with types 1, 2 or 3 gaging elements. The gaging element shown here is being calibrated with a master setting ring.

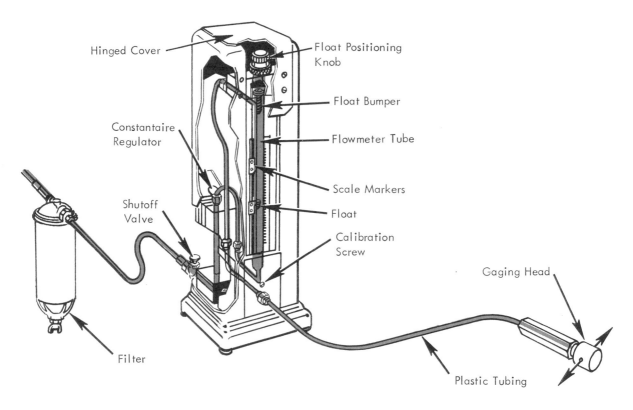

Fig. 12-29 This flow diagram shows the simple circuit and adjustments required for a typical rate-of-flow type pneumatic comparator.

POWER AND DISCRIMINATION — PNEUMATIC COMPARATORS

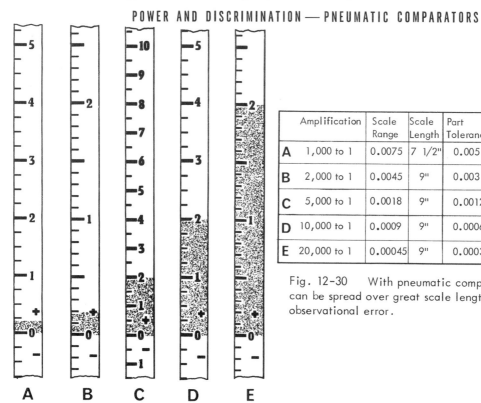

	Amplification	Scale Range	Scale Length	Part Tolerance	Tolerance Spread	Discrim- ination
A	1,000 to 1	0.0075	7 1/2"	0.005	5"	0.0002
B	2,000 to 1	0.0045	9"	0.003	6"	0.0001
C	5,000 to 1	0.0018	9"	0.0012	6"	0.00005
D	10,000 to 1	0.0009	9"	0.0006	6"	0.000020
E	20,000 to 1	0.00045	9"	0.0003	6"	0.000010

Fig. 12-30 With pneumatic comparators, tolerances can be spread over great scale lengths, thus minimizing observational error.

In Fig. 12-30 the discrimination and range for several amplifications is shown. Note that the scale lengths are substantially longer than those for other forms of comparators with similar amplifications. This is one of the important reasons that the pneumatic comparator provides exceptional reliability.

Before continuing, here is a quick review of the metrological points made in previous chapters. The *accuracy* of an instrument is its adherence to some standard. The *precision* is the fineness that it subdivides the standard. While its fidelity, or truthfulness, is measured by its accuracy, its usefulness is measured by its precision. This is limited by the *sensitivity* which is the smallest input signal that produces a discernible output signal. Precision is increased by *amplification*. This is any means for stretching out the scale so that the output signal will be greater in magnitude than the input signal. However, precision that cannot be read is not useful. One of the important factors that determines *readability* is the *resolution*. This is the ratio of the width of a scale division to the width of the read-out element.

Now let's see how the pneumatic comparator measures up along these lines. In Chapter 10 it was shown that in mechanical instruments an increase in amplification does not produce an equivalent increase in sensitivity. Thus the scale can only be stretched out so far and that is all. This is true in pneumatic comparators too, but not nearly to the same extent. Therefore, the scale can be very long. The long scale permits wide graduations. The high sensitivity theoretically makes these wide graduations useful and the high resolution makes them useful from the practical consideration of readability.

READING THE PNEUMATIC COMPARATOR

The basic instrument in Fig. 12-28 will accommodate a wide assortment of scales. These vary in both style and amplification. All of those in Fig. 12-31 are the same amplification, 2,000X; and have the same discrimination of *one-tenth mil* (0.0001 in.). Right hand and left hand refer to the side of the flowmeter tube along which they are placed.

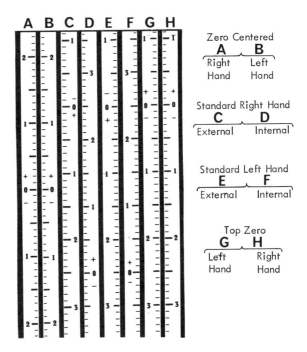

Fig. 12-31 These are all 2000X scales. Similar ones are available for higher and lower amplifications. They are printed on both sides and easily changed.

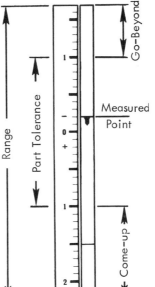

Fig. 12-32 "Come-up" and "Go-beyond" are terms peculiar to pneumatic comparators.

Before considering the purposes of the other scale styles, it is time to introduce some of the terminology invented by users of pneumatic comparators, Fig. 12-32. For once, the words mean what they sound like. The float does "come up" as the element approaches the part tolerance, and it does "go beyond" when the part tolerance is exceeded.

Internal Measurement   External Measurement

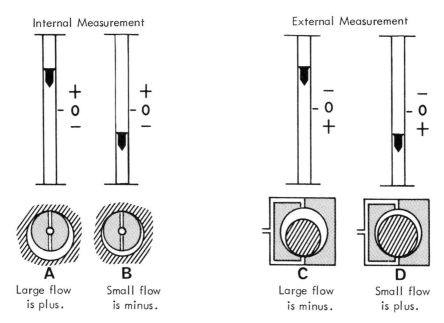

Fig. 12-33 Separate scales are needed for internal and external measurement because the relationship of flow rate and the plus-minus sign is reversed.

Plus and minus are reversed for inside and outside measurements. This is shown in Fig. 12-33. At A, the gaging element is in a large hole. The escape of air through the jets in the gaging element is only slightly restricted. It races through the system carrying the float high on the scale. In B, the same element is closely confined by the smaller hole. The flow is retarded and the float stops low on the scale. In A the hole is large, hence the plus reading. In B it is small for which a minus reading is expected.

In C and D the matter is reversed, the part feature is a diameter and the jets are in a ring around it. Now the flow is rapid when the feature is small — the reverse of the previous case. This is shown in C. In D the larger part restricts the flow. Thus for external measurement, minus represents a large flow and plus, a small flow.

This difference between inside and outside measurement exists for all measurement. It is only the pneumatic comparators that permit quick scale changes in recognition of it. The change of sign is taken care of in set up (by the scale change), not computationally for each measurement. This eliminates an important variable of uncertainty. The reason for the other scales will be seen when we consider scale selection.

The actual reading of the scales is very simple. They are all marked with amplification, discrimination, and range. The float's flat top provides a clear reference. Because higher amplifications are easily selected, there is no excuse for interpolation.

SCALE SELECTION

In order to select the proper scale you must make two sets of decisions. One determines the magnification. The other determines the style of scale.

The general rule for scale selection is: Always select the most sensitive scale that will contain the tolerance of the dimension to be measured entirely. In Chapter 10 you were told that you could not always follow this rule because of the limitations of dial indicators. In Chapter 11 you were told that an advantage of multiple range electronic comparators is that they do enable you to follow the rule. Now you will take the next step and learn why it is not necessary to follow the rule under certain conditions. These conditions are made possible by the long scales of pneumatic comparators.

When space is limited, scale divisions must be crowded and readability is poor. You know what that means to reliability. Spreading out the scale obviously improves this situation.

Just as obviously there must be a limit to the amount of improvement possible. One chocolate eclair is good — but are two dozen of them 24 times better? The pneumatic comparator can spread out a tolerance well beyond the amount required for maximum readability, Fig. 12-32. The amount of spread actually required depends upon the reason for the measurement. Here are some "for instances."

If the *two-tenth mil* (0.0002-in.) tolerance in Fig. 12-34 is to be held on a part being cylindrically ground, the 5,000X scale would be best. It spreads the tolerance over one inch. This should give the grinder operator plenty of room in which carefully to control the operation. Yet by leaving most of the scale outside of the tolerance it helps him get into the critical area quickly and safely. By use of a *top zero* scale the amount of scale below the tolerance is increased. The long *come-up* range, Fig. 12-32, permits the operator to follow his progress and know when to slow down the rate of stock removal. To do this with a dial indicator would risk cutting right on past the low limit. To do it with a multiple-scale electronic instrument would risk forgetting to switch ranges at the proper times.

Agreed, the operator cannot discriminate as precisely with 5,000X as he could with, say, 20,000X. When he is right at the limits this could mean passing bad parts and rejecting good ones. This need not happen if the grinding operation is "in control" — and that is the only reason we have him checking it as he goes along with a measurement instrument.

This returns us to the statement in Chapter 1 that there are three states of measurement. In this case the purpose of the measurement is not to obtain a *precise description* about one part. It is to *control the manufacture* of many parts. If we wanted the former, then the higher the usable amplification the better. But for the latter, the purpose of the measurement is to keep the ground parts within their tolerance and away from the limits. The added *come-up* range helps achieve this better than additional discrimination.

Now we will change the conditions a bit and find out what that does for scale selection. Suppose that an automatic grinder is substituted for the manually operated one. Then the *come-up* range is of no importance. The scale may be used chiefly for containing the tolerance, so a 10,000X would be selected.

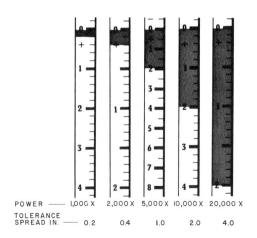

POWER	1,000 X	2,000 X	5,000 X	10,000 X	20,000 X
TOLERANCE SPREAD IN.	0.2	0.4	1.0	2.0	4.0

Fig. 12-34 The effect of amplification on tolerance spread is shown by this comparison of a two-tenth mil (0.0002 in.) tolerance on five scales.

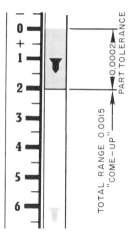

Fig. 12-35 The long come-up range speeds the grinding operation.

Is the purpose of the additional precision to give the inspection department a rougher time concerning borderline cases? While that cannot be ignored as long as humans are involved, there are better reasons. The best is that it permits a higher degree of control and the fullest ability of the automatic grinder can be utilized. Note that the immediate purpose of the inspector's measurement is descriptive, even though his function also is control of manufacture. Thus he is expected to take issue with production people on questionable parts. This is often minimized by furnishing both the inspector and the machine operator with instruments of the same power. All too often this creates a debating society and obscures the real issue — control of manufacture. The author is not unmindful of the exquisite tyranny of bird-brained inspectors armed with precision instruments. However, the inspector should have, whenever possible, a higher amplification instrument and he should have the last word. Both he and the operator should understand that their combined reason for being is to keep the operation in control.

"TOP ZERO" SPINDLE TECHNIQUE

These two examples show the respective roles for *zero centered* and *top zero* scales. Like so many desirable things, there is a catch to this. Consider how you raise the float when *mastering* (setting to standards) the comparator. You increase the flow rate. Remember that the instrument must operate in the straight line part of its performance curve, Fig. 12-10. That limits the amount that you can increase the flow and still have linearity. How else can the flow be increased? By increasing the clearance between the gaging element and the part. Therefore to take full advantage of *top zero* scales it is necessary to have spindles with greater than usual clearance.

In order to understand this we must make two things clear. The gaging element has a distinctly limited measuring range in pneumatic metrology. This is its most serious

limitation. And the gaging element wears. A big point has been made about the reduced wear because metal-to-metal contact is eliminated. That is true in so far as the sensing itself is concerned. The body of the gaging element is metal, and does wear, of course.

Unfortunately, the wear increases the clearance. This permits greater flow which results in a higher position of the float along its scale. That is why frequent *mastering* is required. At each *mastering* the float is moved slightly by reducing the flow. After this has been repeated many times, the clearance is beyond the measuring range. You will note as those for inside measurement are frequently called. By techniques known only to the trade, spindles are obtainable that will that in compensating for this wear the float is lowered each time. Clearly the way to make more of this adjustment possible is to start up higher on the scale in the first place. That requires special gaging elements, or *spindles* provide any of three conditions: (1) The maximum *come-up* range, (2) additional wear life, and (3) an optimum combination of both, Fig. 12-36.

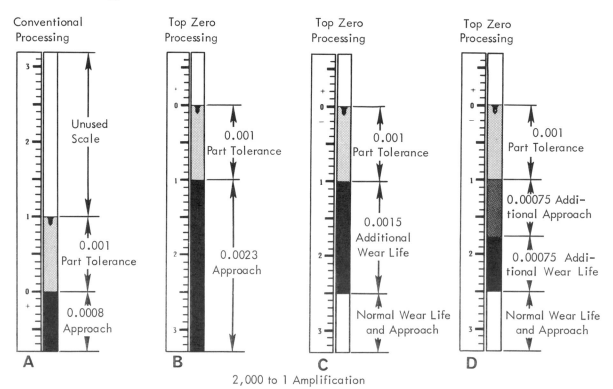

Fig. 12-36 Special gaging elements and scales are available to modify ordinary operation (A) in order to provide maximum come-up range (B), to provide additional wear life (C), or to provide an optimum combination of these (D).

Fig. 12-37 Dial-type pneumatic instruments are also available with the interchangeable scale feature.
(Moore Products Co.)

REASONS FOR PNEUMATIC INSPECTION OF SMALL HOLES

1. Pneumatic gaging elements can be very small. This permits precision measurement of smaller inside diameters than is possible by other means of measurement.

2. Deeper holes can be inspected because depth has little effect upon ease of measurement.

3. Greater accuracy is obtained because rocking and centralizing are unimportant.

4. Nominal gaging force permits thin-walled and yielding materials to be measured.

5. Absence of metal-to-metal contact protects finely-finished surfaces.

Fig. 12-38 In view of these advantages, it is small wonder that pneumatic gages are extensively employed for small hole measurement.

SUMMARY

● Most production gaging equipment developed from measuring instruments. Pneumatic comparators are the reverse. They are measuring instruments that developed from production gaging equipment, the field in which they still enjoy their greatest popularity.

● Pneumatic comparators are clearly the preferred method for the measurement of small holes (0.500 in. and smaller) when precision in *tenth-mil* increments is required, Fig. 12-38. Of course, they have many advantages for larger holes as well. They provide a remarkable combination of desirable features: high amplification without loss of sensitivity, easily selected scale ranges, differential measurement, excellent readability, no metal-to-metal contact of sensing element, freedom from critical gaging geometry, no mechanical parts, few opportunities for operator error, and very low cost for the basic instrument. They traditionally suffer from three serious limitations: short measurement range, sensitivity to surface finish, and need for expensive gaging elements and masters that offset the low instrument cost.

● Gaging elements with air jets may be considered suitable for amplifcations up to 100,000X, although 40,000X is generally the highest. This permits discrimination to *two mike* (0.000002 in.) under ideal conditions.

They may be used at as low an amplification as 1,000X which provides a *two-tenth mil* (0.0002-in.) discrimination and sufficient range for a part tolerance of *five mil* (0.005 in.). In the average case a *tenth mil* (0.0001 in.) tolerance can be measured adequately with a pneumatic gage having 5000 to 1 amplification. This provides *20 mike* discrimination and an operating range *one and a half mil* (0.0015 in.) operating range.* A surface finish of 65 microinches is required even for the low amplification ranges. For the high ranges it must be proportionately better, and must be nearly the same in the part and the masters.

● Many of the limitations are eliminated by the use of mechanical contact air gaging elements. These are used when the surface finish is poor, the area to be measured is small, and when greater range is required. They are available with ranges from *one mil* (0.001 in.) to *40 mil* (0.040 in.). The highest amplification generally used is 5,000X.

● This is summarized in the metrological data table, Fig. 12-39. When consulting it, the disclaimer with which this chapter began must be remembered. There is a diversity of equipment available. The values chosen are representative but by no means all inclusive. Nothing in this chapter can in any way substitute for a thorough study of the manual furnished with each pneumatic comparator that you have occasion to use. Those going into the metalworking industry will certainly meet many of them. It is hoped that some of the scientifically oriented students will remember the unique features of pneumatic metrology. They could simplify much lab work. An example is shown in Fig. 12-40. Whatever route you go, the pneumatic metrology terminology in Fig. 12-41 should help you to sound knowledgeable.

* Johnson, Herbert A., One Tenth to Nothing, Detroit, American Society of Tool Manufacturing Engineers, 1962, Paper No. 403, Vol. 62, Book 1, p. 3.

METROLOGICAL DATA FOR PNEUMATIC COMPARATORS

INSTRUMENT	TYPE OF MEASUREMENT	NORMAL RANGE	DESIGNATED PRECISION	DISCRIMINATION	SENSITIVITY	LINEARITY	RELIABILITY	
							SMALLEST PRACTICAL TOLERANCE FOR SKILLED MEASUREMENT	PRACTICAL MANUFAC- TURING TOLERANCE
Back pressure, dial type air gages. Similar to Fig. 10-3	Comparison							
62 1/2X type 3 gaging elements to 7500X type 2 or 3		0.060 in. 0.005	0.002 in. 0.0005	0.002 in. 0.00002	0.0004 in. 0.00002	0.001 in. 0.00003	0.005 in. 0.0001	0.010 in. 0.0003
Flow-rate type with type 2 gaging elements 1000X	Comparison	0.0075	0.0002	0.0002	0.00004	0.0001	0.001	0.003
40,000X		0.00022	0.000010	0.000005	0.000002	0.000005	0.00005	0.00015
100,000X		0.000090	0.000005	0.000002	0.000002	0.000002	0.00002	0.00005
Flow-rate type with type 3 gaging elements 62 1/2X	Comparison	0.080	0.003	0.003	0.0005	0.0015	0.008	0.016
5000X		0.0018	0.0001	0.0001	0.00005	0.00005	0.0002	0.0006

Fig. 12-39 These data are based on representative instruments. Some will vary from them considerably. The actual instrument specifications should be studied closely before selecting a pneumatic comparator.

Fig. 12-40   There are many opportunities for pneumatic comparators in scientific and laboratory work. As shown here, it has found extensive use in the textile field to determine fiber fineness. *(Sheffield Corp.)*

TERMINOLOGY

Pneumatic Metrology	Measurement in which amplification is achieved by a system using air or other gases.
Air Gaging Pneumatic Gaging	Measurement by means of pneumatic metrology.
Pneumatic Comparator	The instrument for comparing lengths, areas or finishes by means of pneumatic metrology.
Measurement	Determination of size according to a standard.
Gaging	Single-purpose measurement to determine if objects are between size limits or to sort out objects into size categories.
Inspection	Examination by measurement, gaging or other means to verify an object's compliance with predetermined standards.
Back-pressure Comparator	Pneumatic comparator utilizing pressure changes resulting from nozzle restriction.
Flow-rate Comparator	Pneumatic comparator utilizing flow rate changes resulting from nozzle restriction.
Velocity Comparator	Same as flow-rate comparator.
Bourdon-tube Gage	The most familiar form of pressure gage consisting of a bent tube that straightens with internal pressure.
Flow Tube	A vertical tube with air float or fluid column as read-out stage of a pneumatic comparator.
Gaging Element	The input stage of a pneumatic comparator.

Nozzle Orifice Jet	As applied to pneumatic comparators these are the exit openings of the gaging element.
Clearance	The distance between the nozzle and the part.
Spindle	Familiar name of a gaging element for holes, usually small holes.
Mastering	The use of setting standards for calibration of pneumatic comparators.
Differential Measurement	The combining of nozzles to permit direct measurements of features that would not show up if measured separately.
Computing Measurement	The combination of measurements to provide data directly about their mutual relationships.
Come-up	Travel of the float before it arrives at the part tolerance.
Go-beyond	Travel of the float beyond the part tolerance.
In Control	The variables that affect a process are known and are responding to sufficient regulation that the output is predictable.
Out of Control	Variables are either unknown or not regulated so that output of the process is not predictable.
Top Zero	Placement of the zero position high on the scale rather than in the center.

Fig. 12-41   Pneumatic metrology uses terms not generally applied to other measurement instruments but nonetheless useful.

QUESTIONS

1. What is meant by pneumatic metrology?
2. What other names are used for this?
3. What is the chief unique feature of pneumatic metrology?
4. In what respect is pneumatic gaging similar to electronic?
5. When is the term "gaging" properly used for pneumatic metrology?
6. Why is a pneumatic instrument an analog measuring device?
7. What two characteristics of the air are subject to calibrated change for measurement?
8. Which was used for measurement first and what are the instruments called?
9. What is "pressure drop"?
10. What is meant by "lag"?
11. Why was the balanced system developed?

12. What is the chief difference between back-pressure and balanced systems?
13. Why can the range of a rate-of-flow system be greater than that of the other two systems?
14. How are the back-pressure and balanced systems read?
15. How are the rate-of-flow systems read?
16. What features of pneumatic instruments in addition to no metal-to-metal contact are important in air gaging?
17. What is the name of the portion of a pneumatic system that corresponds to the detector-transducer of the generalized measurement system?
18. What are the three types of gaging elements?
19. What is type 1 used?
20. What are other familiar terms for gaging elements?
21. Upon what does the response of a type 2 gaging element depend?
22. What is the term for measurement by combining two or more nozzles?
23. What bearing does this have on pneumatic metrology?
24. What is the most frequent reason to use a type 3 gaging element in preference to a type 2?
25. What is the coarsest surface finish for type 2?
26. When else is a type 3 preferred to type 2?
27. What is meant by "mastering"?
28. What kind of master is used for measurement of holes?
29. What are meanings of "come up" and "go beyond"?
30. What is the highest amplification that can be used with pneumatic instruments?
31. What general precision are penumatic instruments?

DISCUSSION TOPICS

1. Explain significance of pneumatic measurement.
2. Explain significance of differential measurement.
3. Relate the terms "accuracy", "precision", "sensitivity", "readability" and "resolution" to pneumatic metrology.
4. Describe the reasons that pneumatic metrology is particularly suited for production work.

> *The only thing that makes the assembly line possible is our ability to make pieces so exactly alike that we can make any one of a thousand and drop it into place and have it fit.*
>
> C. F. "Boss" Kettering
> General Motors Corporation

Calibration
Measurement of Measurements

All too often an instrument is calibrated meticulously. The user then goes back to work smugly repeating the same errors that he was making before. He probably is worse off because now he enjoys a false sense of security that may mask even greater errors. This chapter, first, robs you of any sense of security where error is concerned. It then shows you how calibration, properly applied, can reduce your risk.

First, we must define the role of calibration. It frequently is considered a cure-all that can remedy any measurement problem. Instead, calibration determines only adherence to standards. We have emphasized that accuracy is adherence to a standard. Thus, calibration is the measurement of accuracy. With a fixed gage this is simply the determination of the amount that the gage is larger or smaller than its nominal size. With a measuring instrument it is more complex. It is the relationship between a change in input to a change in output.

In the strictest sense calibration of an instrument is the establishment of numerical values for the metrological features according to accepted standards. One of these features is the inherent accuracy. It can only be known in terms of a higher standard. Thus each calibration rests upon the calibration of a higher instrument or standard, until the international standard is reached. Figure 11-29 shows that there may be many steps that extend from a working measurement back to the international standard. Figure 13-2 shows that each calibration step depends upon many diverse factors.

The ability of the observer to measure is conditioned by the environment and the functional features, the things he can see, feel, and twist. These conditions merge with the design features of the instrument and result in its precision. All of them, aided by precision, interpret the accuracy. This must be repeated over and over, carrying the accuracy from the part directly back to the inter-

Fig. 13-1   High-quality equipment, qualified personnel, and good working conditions must be accompanied by systematic calibration procedures for reliable measurement. *(DoALL Co.)*

national standard. At each step all of the various factors contribute error that distorts the accuracy, yet only one of them, the inherent accuracy, carries the accuracy forward. Obviously this thin thread must be protected. When it is not, parts won't fit, ballyhoo replaces performance, the dollar buys less, and the moon becomes more distant. Protecting it is the job of calibration.

Figure 13-2 shows that effective calibration depends upon far more than impersonal instruments and standards. At each step individuals are involved: individuals making measurements, individuals transporting things, individuals who could not care less. The chance of accuracy getting through is remote — unless these individuals are organized.

Early in man's strident history the leader found that whenever he tried to plunder his neighbor he got clobbered unless he was better organized than that neighbor. His churches, governments, and ball clubs rapidly learned the same lesson and organization became the most distinguishing prerogative of man. For the first time in this book, we must consider the social aspects of measurement.

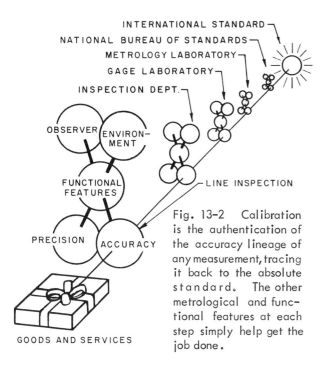

INTERNATIONAL STANDARD
NATIONAL BUREAU OF STANDARDS
METROLOGY LABORATORY
GAGE LABORATORY
INSPECTION DEPT.
OBSERVER
ENVIRON-MENT
FUNCTIONAL FEATURES
LINE INSPECTION
PRECISION
ACCURACY
GOODS AND SERVICES

Fig. 13-2 Calibration is the authentication of the accuracy lineage of any measurement, tracing it back to the absolute standard. The other metrological and functional features at each step simply help get the job done.

This does not mean taking extended coffee breaks to debate the length of the winning pass in Saturday's game. It is the recognition that measurement is something you cannot do alone -- not if it is to serve a very useful purpose. Whenever you measure you fit yourself into the vast technological upheaval that exploded out of the Industrial Revolution and has grown into the worldwide complex of scientific and industrial organizations. If your measurement fits into this only casually, the chances of its reliability are poor. If it is well integrated, the chances are good. Calibration is the organization of these relationships. It gives the instruments their pedigree. It provides you with a place to stand.

THE SCALAR PRINCIPLE OF METROLOGY

The organization of departments and laboratories is not within the scope of this text. *Fortunately, the principle of organization exactly parallels the principle of measurement.* Whether you find yourself in a physics lab measuring for your own research or in a giant corporation measuring to produce the world's goods, an understanding of these principles will help you understand your function.

In both organization structure and measurement, coordination is achieved by the *scalar principle.** As you might expect from the study of scaled instruments, this means that authority proceeds through an orderly series of steps from the highest to the lowest levels.

All social organizations, including business, government, church, military and fraternal, can be evaluated by their scalar organization. For simplicity we will compare measurement only to the military organization. The two fundamental aspects of measurement, accuracy, and precision, correspond to military *line* and *staff* functions.

First, consider the line function. All authority comes from the supreme commander (the international standard, Fig. 11-29). This authority is delegated through a series of subordinates down to the troops (the line inspectors). At each level some of the authority is retained. The portion that is passed on must clearly be an order directly traceable to the supreme commander. If, for any reason, this line of authority is broken, exaggerated, confused, or otherwise altered, a loss is certain to occur. At this point the similarity between measurement and military activity vanishes. Twenty thousand casualties at Iwo Jima at least glorified the traditions of "The Corps." But don't expect to get away with the purchase of an interferometer to check the thickness of bubble gum wrappers. In measurement, success must be weighed against expenditure.

In any organization the line function furnishes the authority and the staff furnishes the means for getting the job done. The latter consists of three related activities, *information, advice* and *supervision;* and is precisely what precision furnishes in measurement. Furthermore, it is precisely what the individuals who measure furnish to their organizations. If you head a research project you may be in a line position, but when you take a distometer reading you are a "staffer." Your information, advice and supervision determine the value of that reading. Its value lies entirely in its accuracy. At each step in the scalar line the accuracy could be impaired by error. So in keeping with our military analogy . . . . .

* Mooney, James D., The Principles of Organization, New York, Harper & Row, Inc., 1947, p. 14.

ERROR IS THE ENEMY

Line and staff functions, although different, must work toward common goals. This is true in measurement as well as in the military and in business. The enemy of the line function is *systematic error*. Of the staff function, it is *random error*.

Except for unusual cases we never know an absolutely *true value*. We must always assume that there is some error in our knowledge of the standard. When a measurement is repeated the results will disagree by an amount called the *discrepancy*.* The causes of discrepancy must also be the causes of differences between the readings and the true value. The differences are random *errors*. *The smaller the random errors, the greater the precision*. This agrees with our statement in Chapter 2 that precision is a measure of the dispersion of readings.

If all of the individual readings are the same it does not mean that there are no errors. They could all vary from the true value by a uniform error. This is called a *systematic* or *constant error*. A good example is a kink in a steel tape. All measurements that include the kink will contain the same amount of plus error – they will be too long. *The smaller the systematic errors, the higher the accuracy*. Most measurements contain both random and systematic errors.

If an error can be evaluated by any logical process it is a *determinate error*. If not, it is *indeterminate*. Random errors may be evaluated by a study of repeated readings and are, therefore, determinate. Some systematic errors cannot be determined by any experimentation and are only known to exist by inference. Fortunately, most of them can be evaluated by comparison to a higher authority. In measurement this is calibration. Therefore, whether a systematic error in linear measurement is determinate or indeterminate depends upon the availability of standards.

The correction of error also follows separate lines for random and systematic error. In the tape example, the systematic error caused by the kink could be determined by a calibration to a standard. When the amount is known the readings are simply corrected by subtracting the amount of the error. A typical random error would be the change in length of the tape with temperature change. This, too, is subject to calculation and correction. The difference in this case, however, is that the correction would not be constant but would have a different value for each temperature measured.

In addition to systematic and random errors there is one other large class of errors. These are the ones that should not happen, the *illegitimate errors*. The other two exist to some degree in all measurement. The illegitimate errors need not exist at all. Because they do, we must consider them. When we consider actual calibration situations you will find that systematic, random and illegitimate errors can all be subdivided into even more specific errors.

ERRORS IN CALIBRATION

If you recognize that all of the errors we have discussed apply to every step in the measurement act and to every person and object involved, you are ready to consider the particular ways that they apply to instrument calibration.

There are a number of ways that error can be classified. The one chosen** particularly fits calibration. Figure 13-3 organizes the information about measurement errors. Each of the general classes is divided into *types of errors* in the first column. The next column, *frequency*, correlates the occurrence of the error with some observable event, such as with change in measurement, change of observer, etc. The *location* column has been arbitrarily limited to instrument (including setup and reference surfaces, if separate), observer and technique. Under *cause* are the general categories of malfunctions. The *detection* column is based on the assumption that slipping quality control has already shown that a problem exists. It directs you to the general steps that may isolate the error. The remedy column is quite distinct from the *prevention* column, the latter being most important.

* Beers, Y., Introduction to the Theory of Error, Reading, Mass., Addison-Wesley Publishing Co., Inc., 1957, 2nd Edition, p. 4.

**Beckwith, T. G. and Buck, N. L., Mechanical Measurements, Reading, Mass., Addison-Wesley Publishing Co., Inc., 1961, p. 38.

CLASSIFICATION OF MEASUREMENT ERRORS

TYPES OF ERRORS	FREQUENCY	LOCATION	CAUSE	DETECTION	REMEDY	PREVENTION
1. Systematic or Fixed	Periodic	Instrument Observer or Technique	Recurring malfunction of one or more elements	Comparison or Substitution		
a. Calibration	Changes with measurement	Instrument	False elements, design and construction errors	Comparison to superior standard	Replace instrument	Instrument evaluation
b. Human	Changes with observers	Observer	Bias, physical peculiarities, behavioral traits	Substitute observer	Train observer	Training
c. Technique	Changes with measurements	Instrument or Observer	Use of the known method but in a situation for which it is not satisfactory	Substitute method	Change method	Education
d. Experimental	Changes with measurement	Instrument or Observer	Use of an unknown method in a situation for which it is not satisfactory	Substitute method	Change method	Education
2. Random or Accidental	Random	Instrument Observer Technique or Part	Erratic malfunction of one or more elements or part	Comparison or Substitution		
a. Judgment	Changes with observers	Observer	Lack of discipline or precise instructions	Substitute observer	Tighter controls and procedures	Training
b. Conditions	Changes without regard to system	Instrument or Part controlled	Disturbed element caused by external influences such as vibration or temperature change	Systematic substitution of elements	Replace troublesome element	Environmental controls, instrument standards
c. Definition	Changes with position of measurement	Part	Attempt to get more information out of the measurement system than was put in	Analysis of part	Change requirements of part or measurement	Education
3. Illegitimate	Random Periodic and Continuous	Instrument or Observer	Outside interference or other completely avoidable disturbances	Reappraisal of procedure		
a. Mistakes	Changes with observers	Instrument or Observer	Wrong decision in choice and/or use of measurement instruments	Analysis of measurement system and technique	Replace or retain faulty element	Education
b. Computational	Random or Continuous	Observer	Environmental factors, fatigue and poor instrumentation	Substitute computation	Change environment technique or instrument	Personnel and environment evaluation
c. Chaotic	Random	Observer or Instrument	Extreme external disturbance	Self detecting	Stop measurement until ended	Environmental analysis

Fig. 13-3   Any table such as this is necessarily general in nature. It is a guide rather than a rule.

Systematic errors can be detected by their regular, periodic recurrence. This also causes them to be overlooked. Things that are systematic tend to appear legitimate and desirable whether they are or not. *Calibration errors* do not refer to an incorrect setting of the instrument but to a disparity between the input signal and the reading. This is caused by false elements,* such as a scale spacing that does not match the amplification, or a probe that is too long (which results in the same thing). They are detected by calibration, usually to gage blocks. Sometimes they can be corrected by adjustment. If not, the instrument needs to be replaced – often to a lower precision application where it is satisfactory. Calibration errors are prevented by instrument evaluation.

Human errors are particularly difficult to detect when you are making them yourself. They include such common things as a tendency to read high, or low. A left-handed person may unwittingly cramp an instrument designed for a right-handed observer. Human errors generally change with change in observer. This is usually both the clue to their existence and the means for detection. Often they are remedied for the immediate situation as soon as they are disclosed. Personnel training is the best means of prevention.

Errors in technique and *experimental errors* are closely related. If a technique developed to calibrate vernier instruments is used to calibrate micrometers, the results will have high uncertainty because the technique is not appropriate. If, on the other hand, you do not know what technique to use, you might experiment and try one. If the unknown technique works out, it is no longer unknown, at least in that one situation. But if it fails to work out, it is an experimental error. Both can usually be detected by the substitution of another method, and remedied by change of method. The prevention is education – and that is why we are here.

Random errors are distinguished by their lack of consistency. The observer may "go all to pieces" on stormy days and his readings might then vary "all over the lot." These are *errors of judgment*. They are often caused

by lack of self-discipline and/or lack of adequate instructions. A change in observers is a test for this error, and training is its prevention in nearly all cases.

Variations in conditions are a common cause of random errors. These may occur in the technique, observer or instrument. When they occur in the instrument they usually are due to a *disturbed element*. The disturbed element may be anything from an improperly soldered connection to backlash, play, or temperature change in mechanical systems. Lack of constraint is frequently the culprit here as it was in the discussion of indicator stands in Chapter 10. The detection may be difficult, requiring the systematic substitution of elements. Obviously the troublesome element must be corrected or replaced. These errors can be prevented by environmental controls to eliminate one source, and adequate instrument specifications or standards to eliminate the other source.

Definition is an evaluation of consistency of the measured quantity. *It is an attempt to get more information out of the measurement system than is put in.* If an object has sides that are not smooth and parallel, one measurement can tell little about their separation regardless of the perfection of the measurement system. This is the uncertainty of the measured quantity itself. Definition error is detected by a full inspection and analysis of the part. Its remedy is procedural. Either the requirements of the part and/or the instructions for measurement must be changed. Its prevention is largely better understanding of measurement capabilities.

There should be no *illegitimate errors*. They are the result of mistakes and carelessness. The existence of illegitimate errors gives legitimacy to the use of repetition as a detection device in measurement. Detection of illegitimate errors is difficult because they may be of random occurrence, periodic or continuous. The latter ones often resist repetition as a means of detection.

Mistakes may be made at any level from the observer to the person calibrating the standards. Often the mistake is in the choice of the instrument or of the entire measurement system. Such mistakes might be noticed only when parts fail to assemble or function properly. Complete analysis of the measure-

---

* Whitehead, T. N., Instruments and Accurate Mechanisms, New York, Dover Publications, Inc., 1953, p. 54.

ment system may be required in order to remedy the faulty element. Better understanding is perhaps the only practical prevention for this error.

Computational errors may be random or continuous. An error once started has a tendency to establish itself. These errors are affected by environment, fatigue and instrumentation. The latter includes not only instruments that do not zero set but those that read in the reverse direction from the desired measurement. Computational errors can often be detected simply by having another person perform the computational steps. Their remedy and prevention are most difficult, requiring a deep understanding of the capabilities of observers in relation to changes of instruments, techniques and environment.

Chaotic errors are extreme disturbances that ruin or hide the measurement results. They may be vibration, shock, sudden glare, extreme noise, and even such rare events as flooding. Unlike all the other errors they call attention to themselves. The only remedy is to halt the measurement until the disturbance ends. A thorough analysis of environmental factors will usually disclose ways to prevent chaotic error.

INTERACTION OF ERRORS

It is not enough to simply detect errors. You also must determine what effect they have on each other, if any. They may be classified as independent errors, dependent errors and correlated errors, Fig. 13-4.

Consider, for example, a micrometer with an error. If it is used to measure the length and width of a rectangle, the resulting area calculation will contain both the length and the width errors of measurement. There is no way they can compensate or cancel each other. Errors such as these are *dependent errors*. They are total errors that cannot be separated from the separate errors that make them up.

On the other hand, let us assume that the micrometer used in the example was within calibration. This means that its total error is so small that it will not materially affect the reliability of the measurements for which it is used. As discussed in Chapter 11, this generally requires that the instrument error will not account for more than one per cent of the total measurement error. Suppose, however, that the person calculating the area of the same rectangle is careless in his measurements of length and width. These errors could compensate, or could cancel completely.

INTERACTION OF ERRORS

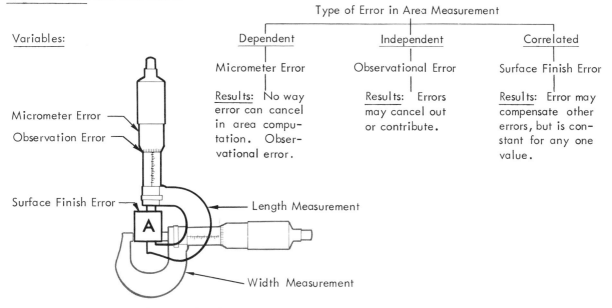

THE PROBLEM: determine area A

Variables:	Type of Error in Area Measurement		
	Dependent	Independent	Correlated
	Micrometer Error	Observational Error	Surface Finish Error
	Results: No way error can cancel in area computation. Observational error.	Results: Errors may cancel out or contribute.	Results: Error may compensate other errors, but is constant for any one value.

Micrometer Error

Observation Error

Surface Finish Error

A

Length Measurement

Width Measurement

Fig. 13-4 Even though errors can interact to produce accurate results, those results are not <u>reliable</u> unless the errors and their interaction are known.

A correct area computation could be made from totally inaccurate measurements. What is so bad about that? The matter of reliability. The person making the computation has no way of knowing when the results are correct and when they are not. This is an example of *independent errors*.

Correlated errors are errors formed by a systematic relationship between independent and dependent errors. If, in the example we are using, the surface finish of the rectangle is considered, we introduce a correlation. For each value of surface finish there will be an effect on the error contained in the area calculation. It may compensate for other measurement errors or increase them; but for any one surface finish, it will be constant.

All of the examples chosen, although possible, have been oversimplified for clarity. You can expect errors in real calibration situations to be far better concealed and much more mysterious about their private relationships than these.

In the discussion of errors you may have noticed that only a small part of them are metrological errors of the instrument. That emphasizes the importance we already have placed on cleanliness, temperature, condition of the reference surfaces and geometrical alignment considerations. Gage blocks cannot be effectively used to pinpoint calibration errors until those errors can be isolated in specific variables.

Even after these potentials errors have been disposed of, calibration is not a haphazard matter of checking a couple of readings. Each instrument and its applications must be analyzed separately. Fortunately, much of this is available in manufacturers' manuals, government publications and books on inspection.

BASIC CALIBRATION PROCEDURE

It is easier to discuss calibration if we turn the coin over and consider the reverse of accuracy, the *uncertainty* of the instrument. This is made up of all the composite errors from the many separate characteristics. Each possible source of error is known as a *variable*. Calibration, then, consists of three acts, all on the metrological side: 1. Determination of variables that might contribute to instrument uncertainty. 2. Measurement of the error contributed by each variable. And 3, determining the net effect of the interaction of the variables on the instrument's measurement capability.

Gage blocks provide a tool of almost unbelievable precision and accuracy for measuring the separate variables. Unfortunately, these variables are closely wedded to the functional characteristics of the instrument being calibrated. Most of the errors from the functional characteristics cannot be measured with gage blocks. Even worse, in many cases these functional errors prevent the gage blocks from measuring the metrological features with any degree of reliability.

If, for example, the jaws of a vernier caliper are badly worn they may never present the same contact twice to the gage blocks. The difference between the reading and the known value of the blocks will clearly be error. But is it functional error, metrological error or both? It was for this reason that the previous chapters emphasized the importance of the reference surfaces and geometrical relationships of the instruments discussed.

The general procedure for instrument calibration is covered by MIL-STD-124, by the manuals available with instruments, and by many books on industrial inspection. Therefore, we will review only some general steps that are common to most calibration.

The technican making the calibration determines that the calibration record, the instrument and specifications correspond. If the instrument is used for production work the part drawing is also required. The instrument is thoroughly cleaned. It is visually inspected for obvious damage such as burrs, scratches, cracks and bending. The precision surfaces are lightly stoned with a conditioning block or an Arkansas hard stone. Mechanical actions are checked for freedom and full travel. If it has recently been brought from a different area, it is given ample time to normalize. After these steps it is ready for calibration. Figure 13-5 is a check list.

PRECALIBRATION CHECK LIST

A. Identification

1. Does the information accompanying the instrument agree with the serial number and/or description of the instrument?
2. Do you have the record of past calibrations?
3. Do you have the manual for the instrument?
4. Do you have all applicable federal specifications?
5. Do you have a report from the user concerning real or imaginary troubles with the instrument?
6. Do you have instructions for disposing of the instrument after calibration?

B. Requirements

1. Is the environment for the calibration suitable for the precision expected? Have the following been checked and minimized: drafts, temperature change, vibration, interference?
2. Are suitable standards on hand and are they in calibration?
3. Are the necessary calibration instruments and accessories on hand and are they in calibration?
4. Is a heat sink of adequate capacity on hand?
5. Are the necessary supplies on hand?
6. Are paper and pencil on hand?
7. Are the necessary packaging materials on hand?

C. Preparation

1. Have the instrument, the standards and the calibration instruments been normalized?
2. Has the instrument been visually inspected?
3. Have the references and contact surfaces been inspected for damage, wear and alignment?

Fig. 13-5   These practical considerations will speed calibration without impairing reliability.

Measurement is only a means to an end. It means nothing if the instrument checks perfectly under the calibration conditions but produces large errors under actual measurement conditions. This is minimized by duplicating, as closely as possible, the actual measurement conditions. In an ideal situation the standard used to calibrate the instrument should have the same size and shape as the part the instrument is used to measure and should be made of the same material.

In a production gage that only measures one dimension, this can be approximated. In a measurement instrument that is expected to perform reliably throughout its range the problem is more complicated. Total error is made up of a number of separate errors. These may add up to one total for one reading and still another total for a different reading. Therefore, it is necessary to *distinguish the variables before you can check for them.* The differences in *feel* rate a discussion by themselves. This follows the section on micrometer calibration.

CALIBRATION OF A VERNIER CALIPER

A complete calibration would require a comparison to the standard of every point to which the instrument can discriminate. Consider a 24-in. vernier caliper. Its discrimination is *one mil* (0.001 inch) and its range is 24 inches. This would require 24,000 separate calibration measurements. If you are handy with gage blocks, and presuming that everything is already clean, you could probably complete this calibration in 10 weeks.

You would not be far into it, however, before you discovered a pattern to the errors. Perhaps the *observed values* get successively larger than the *true values*, or you may find a uniform fluctuation, or even a random change in magnitude of error. In all these cases except the last, once the pattern is observed you can skip many readings and still have a reasonably good change for an accurate calibration. This is done in calibration to eliminate an impractical number of measurements.

POSSIBLE SYSTEMATIC ERRORS

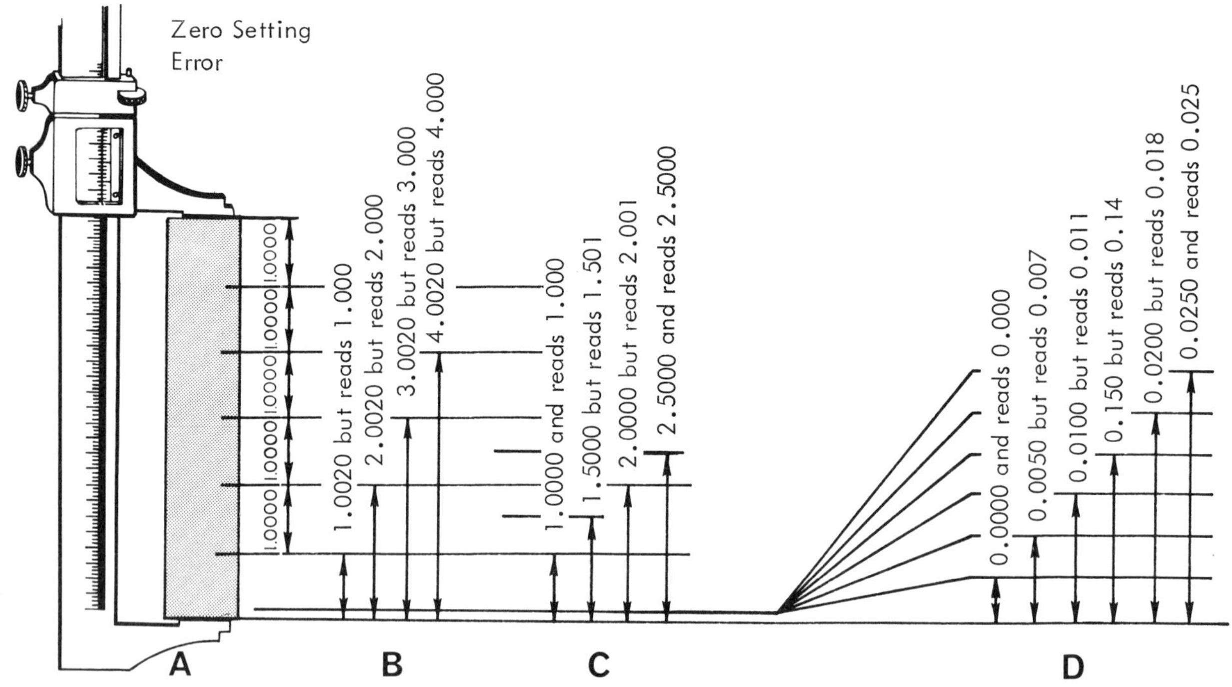

Zero Setting
Error

1.0000|1.0000|1.0000|1.0000

1.0020 but reads 1.000
2.0020 but reads 2.000
3.0020 but reads 3.000
4.0020 but reads 4.000

1.0000 and reads 1.000
1.5000 but reads 1.501
2.0000 but reads 2.001
2.5000 and reads 2.5000

0.0000 and reads 0.000
0.0050 but reads 0.007
0.0100 but reads 0.011
0.150 but reads 0.14
0.0200 but reads 0.018
0.0250 and reads 0.025

A   B   C   D

Fig. 13-6   Inch to inch calibration as at A could fail to disclose the systematic errors shown at B, C, and D. More than one could exist at once.

Suppose in the case of the vernier caliper you decided to save time by only checking from inch to inch, as in A of Fig. 13-6. This could check out "right on the nose" but fail to disclose serious errors. The most obvious error might be that, although the inch increments are uniform, they are also uniformly too high or too low in comparison with the true values. This is shown at B. In C another common form of error is shown. The readings depart from their true values but then return. They do the same thing in D but in this case it is restricted to a 0.025-in. range.

VISUALIZATION BY GRAPHING

The easiest way to visualize these errors is by graphing them. The graphs are shown in Fig. 13-7. In these graphs the true values are the vertical divisions, and the observed readings that correspond to the true values are shown by horizontal divisions. A perfect instrument would have a straight line as shown in graph A. Reading down or to the left results in the same number. Straight or not, the line is called a *curve*. If a *plus error* is encountered, as at 2 inches, the point is above the line. If a minus error as at 2 1/2 inches, the

point is below. Note that in these graphs the error has been exaggerated greatly. When you study statistical quality control you will learn better ways to graph variables, but this method quickly shows the characteristic curve of the instrument.

In graph B we have the condition shown by the B reading in Fig. 13-6. The curve is exactly parallel to the perfect instrument but too high at every calibration point. There has been a uniform plus *two-mil* (+0.002-in.) error. Carry the curve to the left and the reason is seen. The cause is a plus *two-mil* (+0.002-in.) error in zero setting which can be corrected by adjustment.

When the readings from C in Fig. 13-6 are plotted, their curve (C in Fig. 13-7) clearly shows a *topical error*. That refers to an error of one portion of a scale or surface. It is most commonly caused by wear. You might wonder why this is a systematic error rather than a random error. Its position on the scale of the instrument is certainly not systematic, but its effect on measurement is. Whenever a measurement falls into the worn area it will have a predictable error.

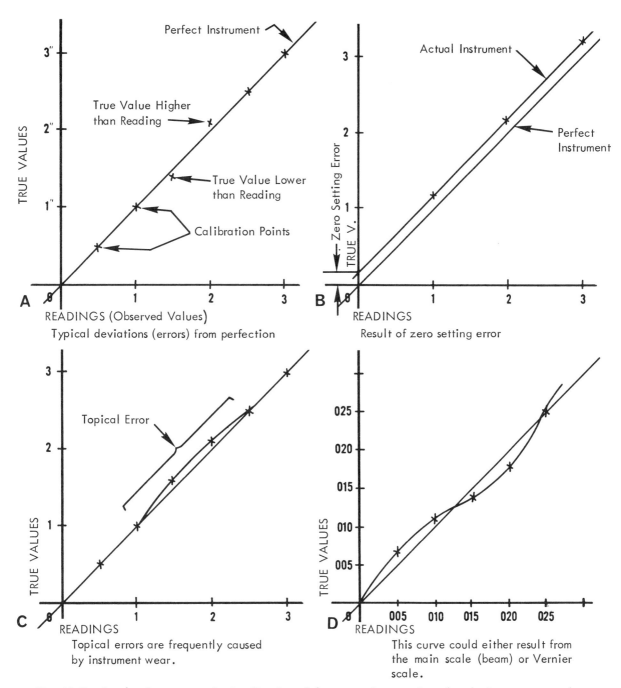

Fig. 13-7   Graphs of errors permit visualization of the types of uncertainty that the instrument may have.

The situation represented by D is similar in that the readings depart from their true values but return to them periodically. It is different in that the full cycle takes place between zero and *25 mil* (0.025 inch). This error could be wear of the beam of the instrument or it could be error in the vernier scale. If it repeats each *25 mil* (0.025 inch), it is the latter without question.

VARIABLES TO BE CALIBRATED

The problem becomes one of finding a minimum number of readings that will check each of the variables that may contain errors. Therefore, first determine the variables. In the case of the vernier caliper there are always at least four: the main scale, the vernier scale, the mechanical action and the con-

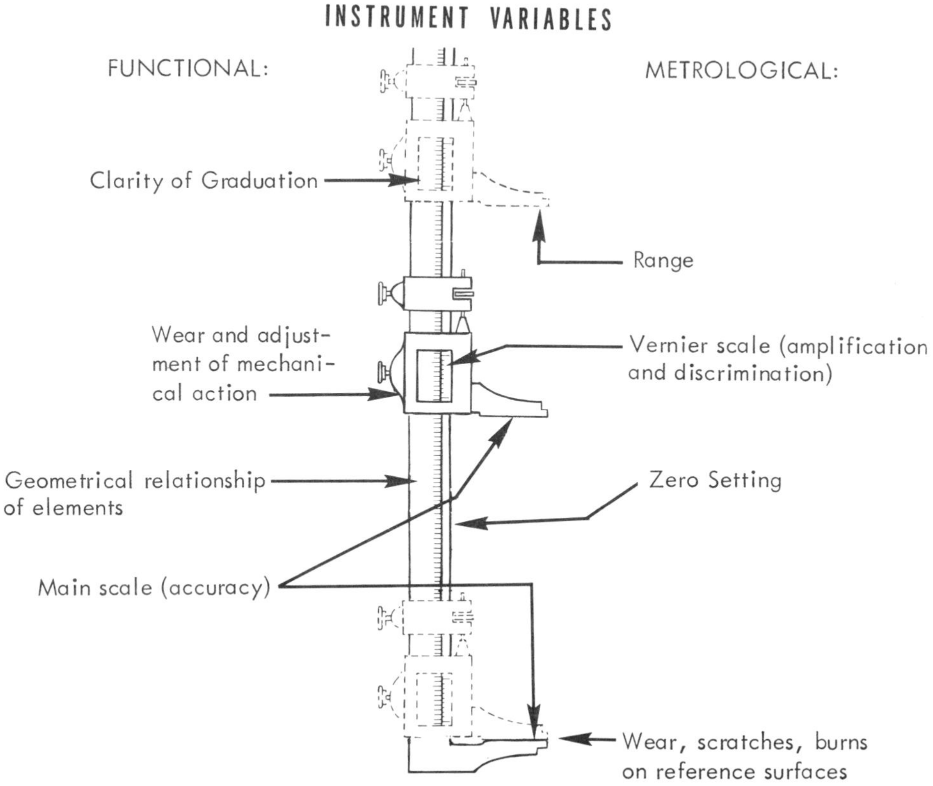

INSTRUMENT VARIABLES

FUNCTIONAL: METROLOGICAL:

Clarity of Graduation

Range

Wear and adjust-
ment of mechani-
cal action

Vernier scale (amplification
and discrimination)

Geometrical relationship
of elements

Zero Setting

Main scale (accuracy)

Wear, scratches, burns
on reference surfaces

Fig. 13-8   Although interrelated, the errors that can develop from functional variables are separate from those of a metrological nature.

dition of the contact faces, or jaws. The last two are functional, not metrological, Fig. 13-8. In this chapter we assume that once recognized they will be checked, and we can get on with the pursuit of metrological errors. They are the variables that might be caused by *use* of the instrument.

There are three variables of the main scale to be investigated. All are potential sources of systematic calibration errors although one might be illegitimate. They are errors of manufacture, wear and abuse. The first is highly unlikely except in poor-quality instruments. It may assume any of the characteristic curves that we have graphed. The second is wear. Its general form is shown in C of Fig. 13-7. If it occurs at one or more distinct places it is a topical error. If it occurs uniformly throughout the range it appears as reduced precision. The third is linearity error. That is a uniformly increasing or decreasing uncertainty. It is shown in Fig. 13-9. A possible cause was shown in Fig. 6-15. The reverse condition could also take place. If this were caused by anything other than normal service it would be an illegitimate error.

There are two variables of the vernier scale. Like the main scale, they are sources of systematic calibration errors. And like the main scale further, one is highly unlikely. The common vernier scale error is zero setting. This results in the characteristic curve shown in B of Fig. 13-7. The unlikely one is that there may be something wrong with the vernier scale itself. About the only way the author can envision this happening is for two vernier scales to become switched during calibration or repair. If it happened, the characteristic curve could be similar to D in Fig. 13-7. While the shape might be different, it would reoccur each $25 \, mil$ (0.025 inch). Need it be mentioned that this would be an illegitimate error?

These are the possible errors. Assume that you do not know that any of them exist and you do not know that they do *not* exist. You do know that you cannot take 10 weeks to check each *one-mil* (0.001-inch) reading, and by now you are sophisticated enough to know that a "spot check" could be very misleading. Your problem is to make a minimum number of calibration measurements provide a maxi-

LINEARITY, UNCERTAINTY AND ACCURACY

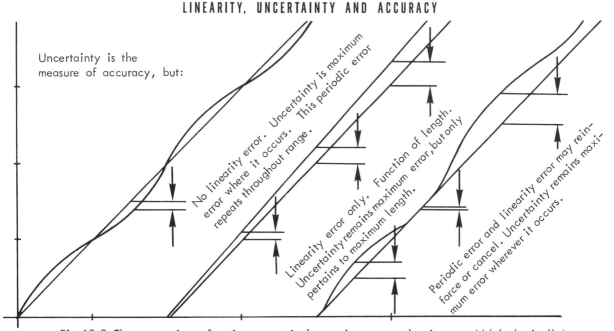

Uncertainty is the measure of accuracy, but:

No linearity error. Uncertainty is maximum error where it occurs. This periodic error repeats throughout range.

Linearity error only. Function of length. Uncertainty remains maximum error, but only pertains to maximum length.

Periodic error and linearity error may reinforce or cancel. Uncertainty remains maximum error wherever it occurs.

Fig. 13-9 The uncertainty of an instrument is the maximum error that it may add (algebraically) to a reading. It is the combination of all errors, only 2 of which are shown here.

mum amount of information about the instrument. Two techniques enable you to do that. One is to make each measurement relate to more than one variable. The other is to use the artillery technique of bridging the target and then "zeroing in" on it. If you know something about the previous use of an instrument, it is possible to get "on target" rapidly. To make it more interesting we will assume, in our example, that you have no clues to the errors you may find. Of course, you have taken the necessary preliminary steps and have paper and pencil on hand to record the calibration measurements.

Zero-setting error is common to both the linearity check and topical check, and thus should be made first. Next, go nearly to the extreme range of the instrument, say 20 in-

ches, and make a calibration measurement. This is done simply by carefully closing the caliper on a 20-in. gage block. The gage block should lie flat on a supporting surface and the caliper should be held the same as it would be if you were measuring an unknown part feature. Remember to *centralize* (see Chapter 6) carefully to duplicate the normal *feel*, Fig. 13-10. Read the vernier after carefully locking the movable jaw. Repeat operation until repeated measurements corroborate each other.

This repetition gives your built-in bias an excuse to run rampant unless you guard against it. The purpose of the corroboration is not necessarily to make all of the measurements read the same. It is to warn you of any illegitimate errors that may have occurred.

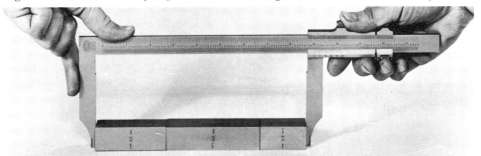

Fig. 13-10   When calibrating a vernier caliper to a long gage block, duplicate as nearly as possible the position and feel required to measure a part feature of the same length.

Suppose the measurement checks 20.000 inches which is as closely as you can read the vernier. And you had better use a magnifying glass at that. Does this prove that there is no linearity error in the instrument and no topical error at this point on the scale? No, because these are independent errors that could cancel out. If this reading had been in the range of the majority of use that could well have been the case. The long measurement was chosen because in most cases it would be beyond the range in which a topical error might compensate for a linearity error. In short, you do not yet have sufficient information to begin arriving at conclusions.

Take a measurement about midway, say 10 inches. If it also checks out to the *one-mil* place you have a reasonably good chance that the instrument is linear. If it does not you may have a topical error and/or linearity error at that point. Let us suppose that it does show an error. For illustration we will make it a *one-mil* (0.001-inch) error. You know that this is a clue that requires further investigation, but rather than run it down now, you continue to get the "big picture". That will require measurements on either side of the last one. Good choices would be 5 and 16 inches. Why 16 inches? Because there is an even-size gage block of that length. No point in wringing blocks when exact sizes are available for use.

Assume that the observed values (or "calibration measurements", or "readings", you will hear all these terms used) are 5.003 and 16.000 inches respectively. Jot these down on your pad. Now a pattern is beginning to emerge. It looks like this:

True values	Observed values
0.000	0.000
5.000	5.003
10.000	10.001
16.000	16.000
20.000	24.000

What do you know from this table? Very little for sure; but you can begin taking small bets. For instance, one "Coke" that the instrument is linear. The error at 5 inches now draws your attention away from the error at 10 inches. By coincidence could that be the greatest error? That is unlikely, so your next measurements bridge it to find which way it

changes. If the next measurement was at 3 inches and showed a *ten-mil* (0.010-inch) error you would, without further ado, assume that the instrument had been damaged and would send it back to the manufacturer for repair. If that reading came out 3.003 inches, however, you would try not to jump to conclusions and take a measurement at 7 inches (or thereabouts). If the 7.000 read 7.003 inches you would presume that there is a topical error with a maximum value, some place on one side or the other of 5 inches, and probably on the high side. You would jump to the conclusion because there was evidence of error at 10 inches. If a check at 6 inches shows *two-mil* (0.002-inch) error, the next check would be on the other side at 4 inches. If that reveals *two-mil* (0.002-inch) error, checks would be made in between until you had sufficient information.

How much is sufficient information? If the calibration was to determine whether or not the instrument could continue in use, sufficient information is the first verified out-of-tolerance reading. That would require that the instrument be withdrawn from service until repaired. If the purpose of the calibration was field adjustment of the instrument, sufficient information would be enough readings to know what variable required adjustment. In the caliper example, any time after the five-inch reading probably would be chosen to check the play of the movable jaw on the beam. If this was found to be loose it would be tightened and the same calibration measurements repeated. If this eliminated the error the calibration would only continue far enough to determine that it was now in calibration.

The calibration, thus far only has ascertained the condition of the main scale, and that the vernier scale has been zero set. Although highly unlikely, the vernier scale could be out of calibration. No measurement taken thus far would show this conclusively. Therefore a measurement should be made within the *25-mil* (0.025-inch) range of the vernier scale. If this disclosed an error the bridging and zeroing-in technique would be used within that *25-mil* (0.025-inch) range in the same manner that it was used in the 24 inches of the main scale calibration.

These examples are intended to show that arbitrarily choosing some nice, even interval is not the way to go about calibrating an instrument. The measurements should be made at intervals that will disclose the change in error that can be expected from the separate variables. Often a small number of measurements will provide more information than a large number. Of course, it is essential to recognize the variables in each particular case. The following example of micrometer calibration in contrast with the vernier caliper calibration will show this.

CALIBRATION OF MICROMETERS

The procedure for micrometer calibration was discussed in Chapter 7. The part omitted was the use of gage blocks to track down the variables peculiar to screw thread standards.

The *pitch* of a thread is the distance between one point on one thread and the same point on the next thread. It is usually expressed in terms of a given length, for example, threads per inch. The *lead* is the amount of movement between the male and female threads for one revolution. In a single-thread screw, the lead and the pitch are numerically the same. For convenience in discussing screws as length standards, pitch usually refers to advancement by full turns, while lead refers to advancement up to a full turn. These, then, are two variables in screw thread standards. They are independent variables that may be either plus or minus in each case. There are others, as you will learn when you study screw threads, but because of the short length of micrometer screws they can be omitted for now.

These two variables are roughly equivalent to the main scale and vernier scale variables of the vernier caliper. The same vernier scale variables are repeated over and over, while no two points of the screw thread are exactly the same. A defect in the lead of the thread caused either by manufacture or use would have a tendency to repeat itself. If, for example, slight eccentricity caused a thread to rub the mating thread unusually hard during a third of its revolution, this would also happen to the next thread, and so on. All the threads would probably have a wear pattern in the same general area, Fig. 13-11. This simplifies micrometer calibration because the error in the lead variable is usually a periodic type. In order to calibrate a micrometer screw, both of these variables must be

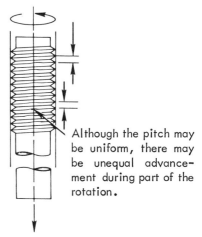

Although the pitch may be uniform, there may be unequal advancement during part of the rotation.

Actual advancement due to bad errors.

True advancement per degree of rotation.

Fig. 13-11 The axial and rotational thread errors are independent and affect micrometer accuracy.

checked. This is in marked contrast to what often passes as calibration.

Zero-setting check is undeniably the first step for calibration. It is a lead variable, but it provides very little additional information. There is no *feel* involved so it is not equivalent to anything encountered in actual measurement. The addition of a calibration check at 1 inch provides more information. Taken alone, however, this information provides little knowledge about the micrometer condition. The lead and pitch variables, being independent, could be compensating at the 1-inch position. Even one more calibration measurement helps determine whether that is the case. If, for example, the micrometer had a plus *two-tenth-mil* (0.0002-inch) error at 1 inch, a measurement at 1/2 inch that had the same error would, if nothing else, cause the zero setting to be rechecked. An error of more than plus *two-tenth-mil* (0.0002 inch) or any minus value would be ample indication that a thorough calibration was required before it could be trusted.

Often, in an attempt to be thorough, the 1/4-, 1/2- and 3/4-inch positions are chosen. Note that in each case the thimble rotation will be stopped in the same position. This has provided a comprehensive check of the pitch variable, but none at all of the lead variable. This problem is recognized in federal specifications. MIL-STD-120, which has

MICROMETER CALIBRATION POSITIONS

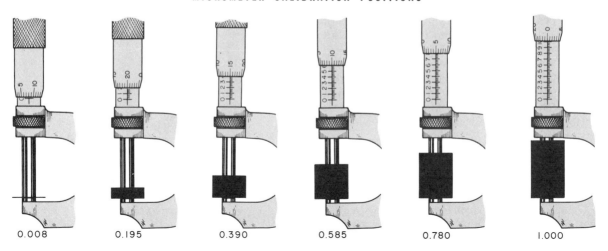

| 0.008 | 0.195 | 0.390 | 0.585 | 0.780 | 1.000 |

Fig. 13-12   This is one of many sequences of calibration positions that check both the pitch and lead variables.

been referred to before, recommends (section 8.4.2.1) that periodic lead error be checked by measuring at three successive *eight-mil* (0.008 inch) length-difference positions.   An authority* suggests that 0.195-, 0.390-, 0.585-, 0.780- and 1.000-inch intervals or similar series of intervals be chosen. Figure 13-12 shows that this checks both variables simultaneously.

THE OBSERVATIONAL VARIABLE IN CALIBRATION

It was mentioned earlier that the problem of feel in calibration was sufficiently important to be treated separately.   This problem is involved in all measurement.   Even in the most precise instruments in which the gaging force control is built in, the user has some effect by his handling of the instrument.   For instruments such as micrometers or gages such as snap gages, the problem is of direct concern during calibration.

Suppose, for example, that the user of a micrometer always applies three pounds of gaging force.   If the instrument was calibrated using three ounces instead, a minus error would exist in every measurement, probably a substantial one of *one mil* (0.001 inch) or more.   The logical conclusion seems to be that the instrument should be calibrated with the same feel as that used for measurement with the instrument.   But it's not necessarily so, as you will soon see.

---

* Kennedy, Clifford W., Inspection and Gaging, New York, The Industrial Press, 1951, p. 105.

The variable of feel introduces independent errors of considerable magnitude.   These errors may compensate for the instrument errors, but, even worse, they will probably dwarf the latter.   Furthermore, who is to say what feel should be used? If it could be stipulated, how could you guarantee it?

It is not in keeping with sound metrological practices to introduce unnecessary variables. The entire purpose of the precalibration check list, Fig. 13-4, is to eliminate errors so that the full value of the gage blocks can be applied to specific calibrations.   Therefore, the feel variable should be reduced as much as possible.   It is far more important that all measurements are made with as nearly the same observational uncertainty (which includes feel) as possible.   Fortunately, changes in feel have much the same effect as zero setting. Once the instrument has been calibrated, the feel of the individual users can be compensated for by the zero setting adjustment. *Reliability is not achieved unless the calibrated instrument is adjusted to the observer or the observer is trained to the calibrated instrument,* even though this is rarely done.

What can be done to assure minimum feel differences in the total observational variable?   The use of the ratchet stop will help. A better method is to apply the wringing technique.   As the micrometer is closed the flat micrometer contact surfaces are carefully slid and turned to achieve the effect of wringing.   Care must be taken that the sharp edges of the contact surface do not dig into and injure the precision surfaces of the setting standards or gage blocks. If these surfaces are carbide tipped particular care is required. Why are the edges not rounded, you might ask? A rounded edge has a tendency to ride up and over dust particles. A sharp edge instead pushes them ahead as it is slid along, but must be handled carefully.

Feel is not the only element of the observational variable, of course. The previously mentioned parallax error should not be minimized even with a micrometer, although modern micrometer design has minimized it, Fig. 7-10. Bias is part of the observational variable that must always be considered.

The error inherent in interpolation is perhaps the most serious observational variable. Fortunately, it is eliminated with vernier-equipped micrometers. A strong case was made against interpolation. Why is it required here? Because there is no other way to read *tenth-mil* increments with a plain micrometer, and there can be no calibration without reading to that precision. A magnifying glass and constant awareness of bias are the only reliable aids.

MINIMIZING OBSERVATIONAL ERRORS WITH HIGH-AMPLIFICATION COMPARATORS

The high-amplification electronic comparators can be advantageous for minimizing the feel variable in some calibration. The calibration of a vernier-height gage is an example. The calibration can be made by simply taking measurements of gage block stacks and comparing the observed values with the true values, A in Fig. 13-13. It is uncertain, however, whether you are reading instrument error, observational error, some combination of both, or even more serious errors such as shown in B of the figure.

If an electronic pickup head is fastened to the slide, the situation is improved, C. in Fig. 13-13. The pickup is easily zeroed against the reference surface. Then calibration consists of three steps. First, any combination of gage blocks is placed between the head and the reference surface. Second, the slide is set so that the vernier reads the same length as the gage blocks. Third, the error is read directly from the meter of the comparator.

This is an improved method although it requires an added element, the electronic comparator. Any added element must add its own errors to the measurements. If these errors are much smaller than the errors eliminated, this method is advocated. If the added errors approach the errors eliminated the method is not advocated. That is the reason the high-amplification comparator is recommended. This method of calibration is more commonly performed with dial indicators. If an indicator of sufficient discrimination and sensitivity is used the results may be favorable. You will remember from the section on sensitivity in Chapter 10 that this is not an easy combination to find. If Chapters 11 and 12 did their job, you already know that sensitivity does not suffer with increased discrimination in an electronic or pneumatic comparator. Hence, our preference is for the latter instruments whenever there is a choice.

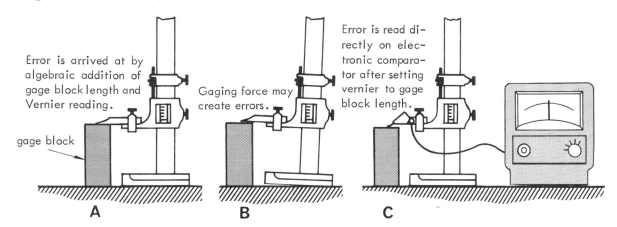

Error is arrived at by algebraic addition of gage block length and Vernier reading.

gage block

Gaging force may create errors.

Error is read directly on electronic comparator after setting vernier to gage block length.

A   **B**   **C**

Fig. 13-13 High amplification comparators can be used in some calibration to minimize error caused by feel and eliminate computational errors.

INTERNAL INSTRUMENT CALIBRATION

The principles involved for the calibration of internal measuring instruments are exactly the same as for external instruments. There are several additional variables that may add uncertainty and must, therefore, be considered. First of all, *gage blocks are male standards.* They are only interested in female measurements. They can be induced to male measurement only by the use of accessory equipment.

Standard gage block holders with caliper-bar end standards work out well for most situations. The outside caliper assembly in Fig. 13-14 shows the calibration of an inside micrometer. The controlled wringing interval provided by the gage block holder assists reliable calibration, but unfortunately takes time. This can be minimized by making up the outside caliper assembly for the maximum size, then wringing blocks to the inside of one of the extension-caliper bars for the various smaller increments desired.

Can the use of this inferior method be justified on time saving alone? Not if it introduces substantial errors. In this case the instrument being calibrated has a discrimination of *one mil* (0.001 inch). Even with experience the interpolation between graduations cannot be considered reliable finer than *two tenth-mil* (0.0002 inch). Assume that each wringing interval is *two mike* (0.000002 inch) smaller. It is unlikely that more than five intervals would be required. That would be *ten mike* or 1/20th of the finest reading (0.0002 inch). The amount of error introduced would not be great enough to seriously affect the reliability for most measurements.

When much internal calibration is required a specialized instrument is used. This instrument is a rigid frame in which two internal reference surfaces are spaced apart by the gage blocks. It has the advantage of rigidity and of spreading the wear over relatively large reference surfaces.

Very accurate measurement of inside diameters is very difficult at best. When sufficient precision is required to calibrate inside diameter gages or ring gages extreme care must be exercised. One method is to allow the ring gage to rest on a laterally floating table. The contacts for the reference and

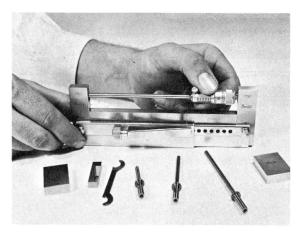

Fig. 13-14 Standard gage block holders with extension caliper bar and standards adapt gage blocks for calibration of inside-measurement instruments such as this inside micrometer. *(DoALL Co.)*

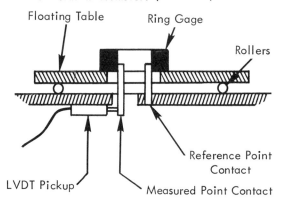

Fig. 13-15 The floating table of the internal comparator minimizes friction that would impair aligning the gage with the axis of the instrument.

measured points protrude through. The antifriction support of the table permits the line of measurement of the ring gage to be precisely aligned with the measurement axis of the instrument. This is shown in Fig. 13-15. When zero setting the instrument for a specific internal diameter, the gage blocks with end standards are substituted for the ring gage.

CALIBRATION OF DIAL INDICATORS

Dial indicators may be calibrated separately or as part of measurement setups. The interrelated discrimination, range and amplification of dial indicators are determined by design and are not subject to calibration. The sensitivity and accuracy are the two characteristics that are evaluated by calibration.

Fig. 13-16 This instrument is comparing an indicator with a large drum micrometer. It is fast but limited to the accuracy of the micrometer. *(Hamilton Watch Co.)*

Most dial indicator calibration is most haphazard, even when performed by manufacturers. It usually consists of running the indicator through its range by means of a micrometer head, Fig. 13-16. Any differences noted are the errors. This generally violates the rule of ten to one. Even if it did not, it is objectionable because of the great vulnerability to observational error. But it's quick and, therefore, may be performed frequently, Fig. 13-17. This discussion will pursue indicator errors somewhat more thoroughly.

In general the sensitivity is determined by a variable made up of friction, inertia and hysteresis. The accuracy is determined by a variable made up of the gear rack error and the errors in the various gears of the gear train. The individual errors in each of these variables are independent but the general action of the variables upon each other is correlated. For practical purposes the two can be calibrated as separate characteristics.

Sensitivity is checked by single-sample measurements. These are measurements identical in every respect except time. Single measurements never provide a real test of accuracy. That requires *multiple sample measurements*. Those are measurements made in as many ways possible, by many observers, at separate times and under changed conditions. Such tests provide a measure of accuracy but, of course, can only approximate it. *Absolute accuracy can never be achieved because the act of measurement itself changes both the measurement system and the object being measured.*[*]

FREQUENCY OF CALIBRATION

Dial graduations (inch)	0.0005	0.0001	0.00005	0.00002
	Dirty Shop			
Frequency of checking(mos.)	6-9	3-6	1-3	1-3
Checking method	A,B, C,D	B,C,D	C,D	D
	Clean Shop			
Frequency of checking(mos.)	9-12	4-9	2-4	2-4
Checking method	A,B, C,D	B,C,D	C,D	D
	Laboratory			
Frequency of checking(mos.)	12	12	3-6	3-6
Checking method	A,B, C,D	B,C,D	C,D	C,D

Method: Comparison to:
  A   Large-drum micrometer
  B   Wrung gage block
  C   Magnified large drum micrometer (left)
  D   Point contact gage block

Fig. 13-17 Frequency of calibration depends upon the discrimination and the use of the instrument, as well as the available equipment. (Robert A. Dana, "When and How to Check Dial Indicators," Tool and Manufacturing Engineer, Nov. 1963.)

Generally the results of single-sample measurements are called the *repeatability* of the instrument. The difficult task is making the measurements identical. Any difference between the measurements will show up as error in the instrument being calibrated. Because these are independent errors the results may either make the instrument appear better or worse than it actually is.

The sensitivity calibration is made by repeatedly measuring one gage block. To have value the readings must be beyond the starting tension point but less than the overtravel. To minimize observational error the indicator should be zeroed, *although that has nothing to do with the sensitivity.* The gage block is then passed repeatedly between the indicator and the reference surface - and that is where the problem lies. The difference between the contact of the gage block and reference surface with each successive pass may be greater than the error being measured.

* Beckwith, T. G. and Buck, N. L., Mechanical Measurements, Reading, Mass, Addison-Wesley Publishing Co., Inc., 1961, p. 36.

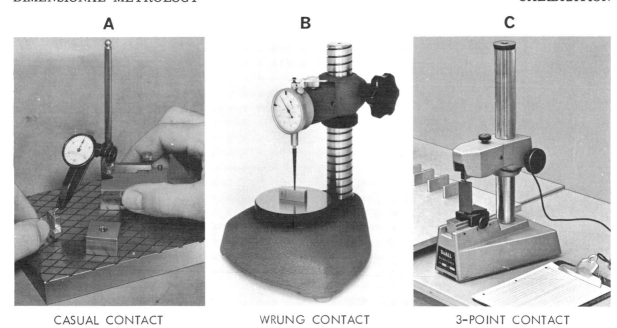

A   **B**   **C**

CASUAL CONTACT   WRUNG CONTACT   3-POINT CONTACT

Fig. 13-18   One of the variables is the contact between the reference surface and the gage blocks or part. Casual contact is not suitable for sensitivity calibration and only for the coarsest uncertainty calibration. Wrung contact is suitable for most calibration, but 3-point contact is required for highest precision.

There are three general contact methods, casual, wrung, and point, Fig. 13-18. All are determined by the type of reference plane available. The *casual contact* is formed by simply laying the clean part on the clean reference surface. It is only suitable for measurement with large tolerances, *one mil* (0.001 inch) or greater. When the indicator stand is not attached to the reference surface and is not very massive, as in A of the figure, this is the appropriate contact. It is not suitable for any sensitivity calibration, and is suitable only for the coarsest accuracy calibration.

The *wrung contact* is far more reliable than the casual contact. It requires that the reference surface and the standard both have sufficiently good surface finish to permit wringing. Furthermore, the support of the indicator must not introduce independent variables of sufficient magnitude to compensate for the sensitivity variable. Generally this requires that the indicator be held to a fixed stand, as shown in B of Fig. 13-18.

This method is suitable for checking the sensitivity of *mil*-reading dial indicators. The gage block is carefully wrung to the reference surface, the spindle is raised, the gage block slid under the spindle, and the spindle gently lowered on the block. The instrument is zero set. The spindle is then raised, the gage block slid somewhat, and the spindle lowered on it again. The reading is noted and the operation repeated. After four or five measurements it should be possible to detect any that contain illegitimate errors, such as the spindle landing on a particle of dust, and reject them. The balance shows the sensitivity. If the readings vary by more than one-quarter dial division, the instrument should be returned to the manufacturer for repair.

Because the sensitivity and uncertainty are correlated there may be circumstances under which their errors compensate. Therefore, it is a good precaution to repeat the sensitivity calibration at another position within the working range of the instrument.

The *three-point* anvil, C in Fig. 13-18, is the most common form of *point contact*. It is preferred for checking the sensitivity of tenth-mil (0.0001-in.)-reading indicators and all high-amplification comparators. It takes full advantage of the kinematic principles discussed in the section on indicator stands in Chapter 10. Only three points are needed to establish the reference plane, so all the other (redundant) points are removed. This greatly reduces the contact variable. The differences that can be caused by changing the wringing intervals are nearly eliminated, along with the effect of oil film and the chance of trapping dust between the surfaces.

With the three-point contact, sensitivity calibrations can be made reliably without the error introduced by the wringing interval variable. The small areas of the contact points develop high unit pressures even though the gaging force is low. This breaks through the air film that remains when surfaces are wrung together. It also restricts this method to small parts of low mass and to instruments with light gaging force.

A three-point anvil attachment for a high-amplification comparator is shown in Fig. 13-19. This attachment is called a *three-point saddle*. It is fastened to the standard anvil by set screws. Both the attachment and the anvil must be scrupulously clean and dust free before fastening them together. The contact points are tipped with hard, wear-resistant material such as boron carbide. They are precisely lapped to the same shoulder-to-crown heights. The several tapped holes in the top triangular plate provide a choice of spacings. When a larger working surface is required, the triangular plate can be removed as in Fig. 11-47.

The three points should be spaced as widely as possible without allowing the gage block or part to sag. The center line of the gage head contact point must fall within the triangle, Fig. 13-20. Preferably it should be close to the line connecting the two closest points, as at A. The gaging force then will cause the minimum bending of the block or part. The points close together, as at B, also minimize bending, but if the part or block is long, they may not provide a stable reference plane. Very thin blocks and parts can be measured with little bending by using only one reference point, shown at C. Of course, this requires that the block be held precisely perpendicular to the travel of the gage head. That is determined by spinning. There is no wobble when perpendicular.

In this discussion of the contact variable you may have noticed that the serrated anvil provides a compromise between the flat reference plane and the three-point contact. It decidedly has reduced wringing interval because of the higher contact pressure, and dust is more easily removed. Yet the serrated anvil provides substantially more support to the block or part than does the three-point contact. The serrated anvil, in fact, can be far more *accurate* than the three-point contact, which brings us up to our next topic.

Fig. 13-19 The 3-point saddle is used for instrument sensitivity calibration because it minimizes the effect of wringing interval variations.

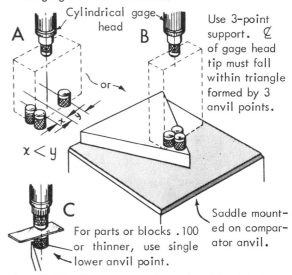

Fig. 13-20 The contact of the head should always be within the triangle formed by the points. When it is near to the line connecting the closest points, as at A, bending is minimized. For least bending only one reference point is used, as at C.

ACCURACY CALIBRATION

Sensitivity of an indicator-type instrument is just one of the variables that can be calibrated by gage blocks. Another one of equal, if not greater, importance is the *accuracy* of the instrument. Precisely what does this mean? Earlier you were told that *uncertainty* would be a more convenient term to use. The reason is that while accuracy is a measure of perfection, uncertainty is a measure of error. It is a lot easier to look for the gold in sea water than the sea water in gold. Accuracy is the degree to which an observed value agrees with the true value. Error is any deviation. Uncertainty is the maximum amount of error that can be expected under standard conditions. Standard conditions are simply the conditions under which a skilled person in normal circumstances would make the measurement. Sometimes these conditions are varied. When this happens the new conditions should be expressed. The newly stated conditions have, for the purpose of that particular measurement, become the standard conditions. For some measurement the standard conditions are specified by rules. These may be informal understandings within a plant, formal standards set up by trade or industry associations, or bylaws established by government bodies.

ACCURACY CALIBRATION OF A DIAL INDICATOR

Sensitivity should be calibrated first because accuracy expressed closer than sensitivity has little practical use. It does have design importance, however. If, for example, a steel rule has a discrimination of 1/64-inch, you would certainly not tolerate an accuracy error or uncertainty of 1/64-inch. On the other hand, a dial indicator that will repeat no closer than one-half division cannot be considered accurate any closer than one-half division. It may have less accuracy, and that is the reason for calibration.

The portion of the range to be calibrated, of course, is from the end of the tensioning travel to the beginning of the overtravel. An ordinary indicator with two turns of travel presents a simpler calibration problem than does a long-range indicator.

Each gear in the indicator may be the source of error. These will be periodic errors repeating their effect every time the gear is in the same position. Being dependent, they will compensate. The rack error will also compensate the others but will be a random error.

When calibrating the micrometer, uneven intervals were selected so that the periodic errors would not be encountered over and over again at the same magnitude of compensation. That seems logical for the dial indicator, but is not. To the observer making the calibration, there is no discernible relationship between the position of the internal elements and the pointer along the scale. Even intervals have as good a chance for random compensa-

tion of errors as do uneven ones. A point at the beginning of the range, one at the end and one or two in between are all that is necessary.

The indicator should be securely mounted to a comparator stand. That is a stand that has its own reference surface (the anvil) integral with it. A height gage stand, one that rests on a reference surface, should be used only in last resort because of the additional errors it adds.

A three-point anvil should be used, not because of its metrological features, but because it is faster than wringing. If it is not available the gage blocks should be wrung to the anvil. Even if the surface finish of the anvil is not sufficiently good for wringing you should go through the act of wringing to assure intimate contact.

The principle of bridging an error and then zeroing in on it should be practiced. If any of the readings contain greater error than the sensitivity, a reading should be made between them and the next reading. If that shows an error, a reading on the opposite side should be made. If this shows the same error, a reading between the two is indicated. This procedure is followed until the maximum error is located – or until an error greater than the permissible maximum is found.

In the latter case the instrument is returned for repair and/or replaced. If the maximum error is not sufficient to condemn the instrument it should be noted on the calibration record. This will serve a good purpose. It will speed the next calibration, and it may provide the user with a clue should problems arise in the use of the instrument.

If a long-range indicator is used as shown in Fig. 10-32, only the portion of the range producing usable readings needs to be calibrated for the accuracy. The sensitivity calibration should be made at the beginning of the travel end, as well as in the usable range, however. Otherwise a sensitivity error could be systematic and affect the entire range in much the same manner as a zero-setting error on a micrometer.

If the long-range indicator is used throughout its range it would be wise to check at several intervals, perhaps one inch, and zero-in on any errors disclosed.

CALIBRATION OF HIGH-AMPLIFICATION COMPARATORS

The procedure for calibration of high-amplification comparators is similar to that for dial indicators except that much greater care must be taken. The gage blocks must be of adequate accuracy for the instrument and they certainly must not have had severe use since their last calibration. Cleanliness, always important in measurement, now becomes a paramount consideration. Temperature considerations cannot be left to chance, and must be planned as carefully as any other part of the calibration procedure. And unless the positional relationships are (1) known, (2) within adequate tolerances and (3) secure, there is little reason to proceed at all.

If the calibration involves the reading of size differences from zero to *five mike* (0.000000 to 0.000005 inch), the calibration *cannot* be made under shop conditions.* The fact that the gage blocks have sufficient accuracy and the instrument has a discrimination of *one mike* (0.000001 inch) does not give license to call the readings accurate to millionths. These are readings *in* millionths, not *to* millionths. You might read millionths but you cannot answer the questions, *whose millionths?* They might be from any of the complex assortment of intermingled variables involved.

If the calibration involves readings in the 5- to 10-*mike* range, you may be able to make them under ordinary shop conditions − *but not reliably.* It will certainly be slow and tedious. For example, in the gage room to calibrate within this range, a heat sink may be adequate for the last hour of normalizing the gage blocks before the calibration begins. Under shop conditions only "togetherness" would assure that the gage blocks and the instrument are at the same temperature. This would involve making the setup and then waiting at least 20 minutes before making a reading, if a single block is used. If two or more blocks are wrung together, at least one hour will be necessary because of the heat pickup while wringing. Obviously the time required to take the large number of readings necessary for careful calibration would be prohibitively time consuming.

Note that, for the purpose of calibration, measurements do not have to be made at the standard temperature of 68°F providing the gage blocks and the instrument stand have nearly the same coefficient of expansion. If carbide blocks are used care must be taken to compensate for the different coefficients.

High-amplification comparators that will be used for measurement in the 10- to 50-*mike* (0.000010 to 0.000050 in.) range may be safely calibrated under shop conditions. It is essential, of course, that all extraneous controllable variables are either eliminated or at least identified. This requires the elimination of drafts by shields or guards or direct radiant heat from electric lights; sunlight or nearby hot equipment must be closed off. Vibration must either be eliminated or another site chosen for the calibration. If excessive dust is present in the air, this, too, requires screens or temporary enclosure of some kind.

As with dial indicators, the sensitivity should be checked first. You were cautioned to make this test within the working range of the instrument. Most high-amplification instruments cannot be read except in the working range. However, those electrically powered must be sufficiently warmed up before making the test. This simply involves turning on the power for at least as long as the period recommended by the manufacturer before performing the test.

If a three-point anvil is available the sensitivity test can be made quickly. The gage block is inserted several times, the spindle being carefully lowered onto it each time, and the readings recorded. If only a flat anvil is used, this part of the calibration is slow because the block must be wrung to the anvil for each reading. This itself takes time. The anvil and the block must be wiped free of dust each time. Furthermore a normalizing period must be allowed before each reading. If you are calibrating a 50-*mike* instrument, the normalizing time can be brief, perhaps three or four minutes. If a 10-*mike* instrument is being calibrated, 15 minutes will not be too short, even for previously normalized gage blocks. The temperature considerations were discussed in greater detail in Chapter 11.

---

*"Measuring in Millionths", <u>Machinery</u>, New York, N. Y., The Industrial Press, June 1962, p. 131.

As with dial indicators, a sensitivity of 1/4-scale division is usually considered satisfactory. If a wider deviation between readings is encountered do not immediately condemn the instrument. One by one, check the various variables that might add compensating errors that cause some of the readings to be larger than the instrument sensitivity and some to be smaller.

The accuracy calibration of high-amplification instruments is limited by their very short ranges. This was shown in Fig. 11-15. The total range in most cases will be less than one-scale division of a dial indicator. That means that the calibration will be made by successive intervals, all in *tenth-mil* increments. The 0.100025-, 0.100050- and 0.100075-inch gage blocks are helpful for high-amplification calibration.

It should be remembered that accuracy of high-amplification comparators is usually stated as a percentage of "full-scale deflection"; that means, as a percentage of the range. Figure 13-21 shows that the percentage must be evaluated in terms of magnitude of error. It also shows typical calibration points. Figure 13-22 shows how this varies for multiple-scale instruments.

ERROR — PERCENTAGE VS. MAGNITUDE

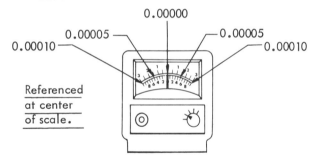

Referenced at center of scale.

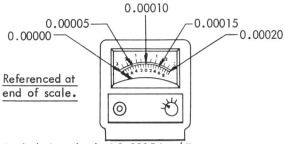

Referenced at end of scale.

Scale being checked 0.0005 in./div., instrument error 1%.

Fig. 13-21   The percentage of error remains the same but the magnitude varies with total distance traveled. In calibration it is desirable to stretch out travel to find the errors (bottom), while in use of the instrument the reverse is true (top).

ERROR — MAGNITUDE VS. AMPLIFICATION

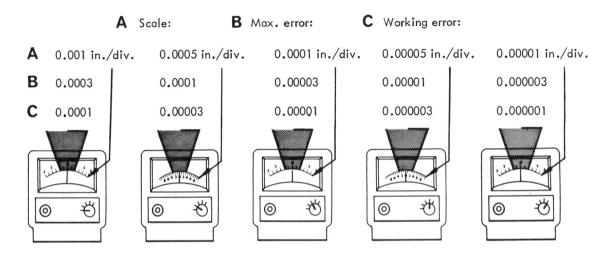

	A Scale:	B Max. error:	C Working error:		
A	0.001 in./div.	0.0005 in./div.	0.0001 in./div.	0.00005 in./div.	0.00001 in./div.
B	0.0003	0.0001	0.00003	0.00001	0.000003
C	0.0001	0.00003	0.00001	0.000003	0.000001

Fig. 13-22   In a multiple-scale instrument the percentage of error will be the same for all scales but the magnitudes will vary in proportion to the amplification. Note that if measurement is restricted to one third of the total scale range (shaded portion), instrument capability exceeds general measuring capability in all scales except the lowest amplification.

The scales of multiple-scale electronic comparators must be calibrated separately. At a glance it might appear that if the highest-amplification scale is calibrated all the lower scales should also be within calibration. Referring back to the generalized measurement system, Fig. 11-12 shows why this is not true. The electronic comparator consists of several distinctly separate elements that work together. Each of the elements is subject to its own errors, dependent, independent and correlated. The total error for each element may also be dependent, independent or correlated in respect to the other elements.

To illustrate this, suppose that the detector-transducer stage (the gage head) has a massive error when its probe is between 5° and 5 1/2° with the horizontal. If the intermediate-modifying stage (the amplifier) and the indicator-recorder-controller stage (the meter) were perfectly linear, the gage head error would react differently according to the amplification of the scale selected. At the lowest scale the error would appear as uncertainty in one portion of the scale only, A in Fig. 13-23. That would be a random error from the standpoint of the instrument because it occurs only in one area with no pattern or predictability. It would be a systematic error from the standpoint of measurement with the instrument, because it could always be predicted that a measurement would or would not contain this error. If either the amplifier or the meter also contained errors, the gage head error would be independent and compensate.

At a higher-amplification scale the gage head error may encompass the entire scale range, B in Fig. 13-23. The error of the head is then independent of the amplifier and meter errors. Both occur one upon the other. These then are independent errors, in that they compensate. But if the measurement takes place on either side of the gage head error, it does not affect the amplifier or the meter errors. Then the errors are correlated.

C in Fig. 13-23 shows the result of the gage head error at the high-amplification scales. Now a dependent error condition exists. For the major portion of the range of the gage head the error is avoided. If measurements are taken in the error-producing portion of the range, the error of the gage head may be so enormous in respect to the total range of the meter that for all practical purposes it is linear and does not affect the results.

Moral — always use compatible equipment. The gage head, and the amplifier-meter should match in sensitivity and accuracy. And these elements should be supported by stands or holders that also match.

Generally it is advisable to calibrate a multiscale comparator starting at the low scales and proceeding to the high ones. If excessive error is found in any scale, the gage head should be checked with another amplifier-meter, and the amplifier-meter should be checked with another gage head. Reversible gage heads should be calibrated in both directions.

ERROR — TYPE VS. AMPLIFICATION

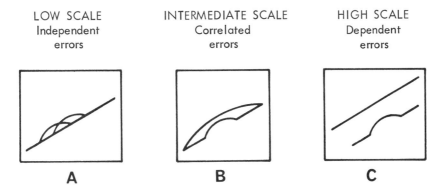

LOW SCALE Independent errors	INTERMEDIATE SCALE Correlated errors	HIGH SCALE Dependent errors
A	B	C

Fig. 13-23 This example assumes an error in one portion of the gage head range, a linear amplifier and a meter with an error in one portion of its range. The relationship of these errors changes with the amplification of the system as shown.

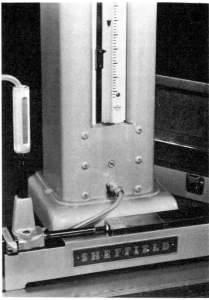

(The Sheffield Corp.)

Fig. 13-24 These are fixtures for setting pneumatic gages. The two on the left are themselves cali-brated to gage blocks. The one on the right uses gage blocks directly to make up various openings.

CALIBRATION OF SETTING STANDARDS

Setting standards are gages used to set or calibrate other gages. While fixed gages are not used for extremely precise work, setting standards are. They are often used to set in-dicator gages with discrimination to *one-tenth mil* (0.0001 inch). The rule of ten to one thus requires that the setting standards must be held to within 10 *mike* (0.000010 inch). This is seldom achieved. A practical interpreta-tion is that the setting standards are cali-brated with the master standards (gage blocks) as closely as the available instrumen-tation, personnel and environment will permit.

All of the precautions required for the cali-bration of the high-amplification comparators are brought into play, *not once, but twice.* The instrument must be zero set to the gage blocks and the setting standard measured to the same degree of precision.

Perhaps the most practical setting stan-dards are gage blocks. This accounts for their increasing frequency of direct use as in Fig. 13-24. Gage blocks can be calibrated to a sur-prising degree of reliability using readily available equipment. Because of the impor-tance of this special class of setting standards calibration, the next chapter is devoted to the general subject.

CALIBRATION OF MASTER STANDARDS AND TRACEABILITY

This degree of precision is beyond all ex-cept the best-equipped metrology labora-tories. Interferometers are used to compare master gage blocks with grand masters. In some cases the blocks are measured directly in terms of light waves, Fig. 13-25.

Instructions on the procedure would be of little value. However, it is appropriate to drive home the importance of calibration of the master standards. All accuracy arises at the international standard, Fig. 13-2, and is only transmitted to the measurement you make next by means of successive calibra-tions. In view of this, failure to keep the mas-ter standards (almost always a set of gage blocks) in current calibration may be an ex-tremely costly mistake.

Careful organizations maintain two com-plete sets of master standards so that one may be at the National Bureau of Standards or at a metrology laboratory for calibration at all times while another is ready to check lesser standards at all times.

The importance of calibration is attested by the fact that it is now a matter of law. Con-tractors who are working on government con-

Fig. 13-25   This instrument measures by counting the light waves that equal the length of the spindle displacement. An electronic counter is used to provide readings directly in inches. *(Continental Machine Co.)*

tracts must be able to document traceability. That means they must be able to prove a series of calibrations that lead back to the National Bureau of Standards. In view of the number of subcontractors involved on large government projects, the traceability requirements affect much of the industry in the United States.

RECORD AND CORRECT CALIBRATION READINGS

Except for the lowest grades of gage blocks, calibration data will be available for each individual block. This states the amount that the block varied from its basic size at its most recent calibration. It is stated as plus or minus and is an independent systematic error. If only one or two blocks are involved and the rule of ten is followed, this may be ignored. If, as is often the case in calibration, the rule cannot be followed, then the block error should be taken into consideration. Be warned, however, that it may sneak up on you even in cases where the rule is being carefully followed. If a large number of blocks are involved, their combined errors may alter the direction that the final result is rounded off. In some cases, it could even change a significant figure which was not rounded off.

For this reason, plus some others, it is good practice to systematically record and correct all readings. You should not use any old scrap of paper that might be handy. Keep a pocket-size notebook for this purpose. You should not mentally correct the readings and jot down answers. Too risky, and in case of mistake there is no way to recheck. You should record all of the readings at the maximum number of significant figures and eliminate the unnecessary or inappropriate places at the end of the computation. That brings up our old friend, *rounding off*, that you first met in Chapter 3 and presumably have been using ever since.

ROUNDING OFF CALIBRATION RESULTS

In Chapter 3 a means was given for rounding off unnecessary decimals that is statistically sound. Two other methods are used in calibration. These are not statistically sound at all, and actually are exact opposites. The author has never heard these methods called by any specific names and will simply call them by their predominant characters, conservative and liberal.

In politics these are the good guys and the bad guys. Which is which usually depends, not upon facts, but upon whom our parents are, what our friends believe in, or how much we want to impress someone. Actually, a liberal is one who believes that government itself should be used for social progress. A conservative is one who believes that government should only provide the framework in which society can take care of its own advancement. So it is in rounding off.

If the rounding off is used to affect the conclusion, it is liberal in nature. If it is used oppositely, to assure even greater separation from observational error, it is conservative. In both of these cases it is the wear variable that is manipulated.

WHERE OH WEAR?

Wear is omnipresent in measurement - and just about omnipotent when it comes to causing trouble. Every single time any gage or measurement instrument is used, it wears. Every time a part is measured it wears. This wear is miniscule, but is definitely measurable, and must be taken into consideration. This, unfortunately, is misunderstood. For example, literally tens of thousands of very expensive class XX plug gages are purchased every year for applications where they wear into class X tolerance within the first few minutes of use. The moderately priced class X gages could have been bought in the first place. Don't laugh — it's your money that is being wasted. You see it whenever you look at an inflated price tag or a deflated dividend or profit-sharing check.

Wear is so important that it is recognized in the specifications of gages that have been standardized. It is called the wear allowance. Those who go into industrial measurement will become well acquainted with it. Because we are concerned now only with the principle involved, we will just consider one elementary but typical case.

In Fig. 13-26 an inside diameter is shown with its maximum and minimum metal limits. The difference between these limits is the tolerance zone. The function of the gage is to assure that no bad parts pass. That is as anachronistic as the mustache cup in this day of statistical quality control, but so it is. Suppose that the diameter of the go plug is ex-

actly at the smallest size permitted by the tolerance. Each time it is used to check a part it wears slightly. Perhaps the first few times the wear is imperceptible, but sooner or later it is passing undersized parts. The solution is a wear allowance. The allowance permits the gage to have a satisfactory life, wearing away gradually. Eventually, calibration shows that it is at the tolerance limit and it is replaced, or ought to be.

The not-go end does not present the same problem. Presumably it does not go into very many parts. Therefore the wear is slight. Furthermore no amount of wear on the not-go end will ever cause out-of-tolerance parts to be passed.

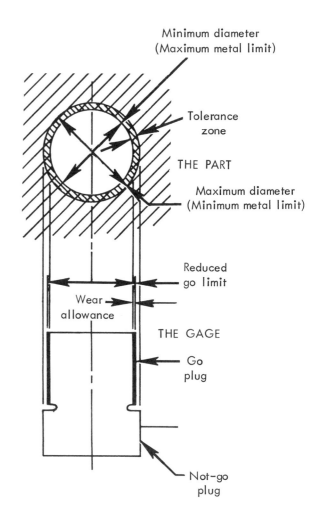

Fig. 13-26 The go plug should enter the hole in the part. The not-go plug should not. The wear allowance on the go plug reduces the tolerance specified for the hole.

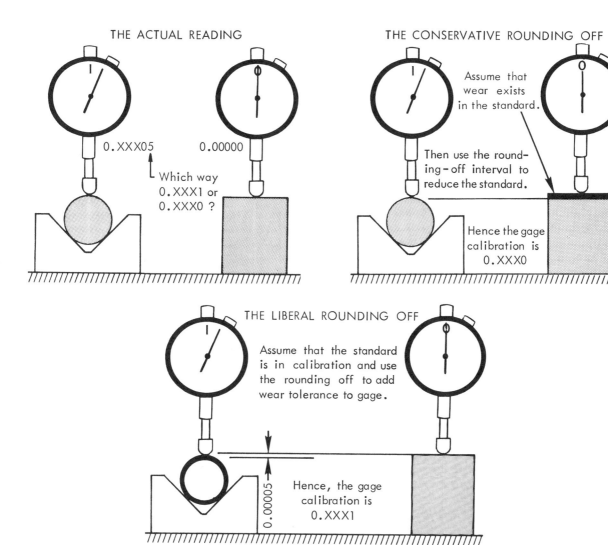

Fig. 13-27 Take your pick. Both methods for rounding off have merit.

That is the first step, the calibration step comes next. Periodically the go-not-go gage is rechecked to determine its condition. Only when the last decimal place is a 5 do you have a decision to make. Figure 13-27 does not show the consequences of your decision. It does give you arguments to support your decision, whichever it may be.

The conservative approach is to assume that the standard may have worn since its last use. Therefore, the zero setting really represents something less than the presumed basic dimension. In that case the rounding-off interval is used to bring the standard back to size. It is added to the standard. This is the same as subtracting it from the gage being calibrated. So, the rounding off drops the 5 from the readings and leaves the digit before as it was.

In the liberal approach, the optimistic attitude is taken and the standard is assumed to be in calibration. Therefore, we have the rounding-off interval to play with exclusively for its effect on the gage itself. With this license, it is logical to round up, adding it to the gage. This is equivalent to adding wear tolerance to the gage.

Of course, the sobering aspect of this is the realization that in either case, you have become a Don Quixote of the gage room. He jousted with windmills, thinking them to be giants. In these examples, we jousted with numbers, mistaking them for metal.

Summary — Measurement for Measurement's Sake

● Calibration is the measurement of measurement instruments and standards. Its purpose is to establish their relation to the international standard, their accuracy. It also establishes their other metrological features, such as precision and linearity. Because it relates to outside considerations, calibration requires the consideration of the line of authority in measurement, usually called traceability. This follows the same principles in measurement organization as it does in business or the military. A confused line of command results in error or inefficiency.

● Error is the enemy in measurement. The various errors can be distinguished in several ways. Random errors affect precision. Systematic errors affect accuracy. There are also errors that should not happen at all, the illegitimate errors. In calibration we attempt to separate the various errors, determine their magnitude and occurrence, and then determine their interaction. This may be either dependent or independent.

● Gage blocks are the practical standard for most calibration. They are used in comparison with the instrument being calibrated. Efficient calibration is that which uses the fewest comparisons to determine the required metrological data. Efficiency is obtained by selecting calibration points that check more than one variable at the same time. Generally, this is expedited by bridging the area being investigated, and then by successive halving, zeroing-in on the specific area.

● Because the metrological characteristics of instruments are very different, no general procedure can be given for calibration. It is necessary to study the instrument carefully before beginning. Fortunately, the general procedure has been formalized for most instruments.

● Instrument calibration cannot be separated from instrument use. This insinuates human considerations into all precision calibration. The user of the instrument must be trained to the calibrated instrument, and vice versa. Furthermore, all of the elements of the measurement system enter into the calibration. This becomes particularly important for instruments such as comparators. Calibrating the final stage alone is not enough.

● Calibration accuracy depends upon the more subtle aspects of measurement. Cleanliness is absolutely essential, and temperature control is only slightly less important. When setting standards are calibrated, extreme attention to these details is tantamount to accuracy.

QUESTIONS

1. What does instrument calibration determine?
2. How is instrument calibration accomplished?
3. What is meant by the uncertainty of a measuring instrument?
4. Of what does calibration consist?
5. Of what value are gage blocks for measuring the instrument capability?
6. Why is it necessary to distinguish between the several possible errors in an instrument before you can check them?
7. How can systematic errors be detected?
8. Which errors are particularly difficult to detect?
9. What is definition error?
10. How is definition error prevented?
11. What is meant by interaction of errors?
12. What is the proper method for determining linearity errors?
13. How does pitch error differ from lead error in a micrometer thread?

14. What is usually the first step when calibrating a micrometer or vernier instrument?
15. What is the most reliable method for checking lead error?
16. Of what importance is variation of feel in calibration?
17. How do changes in feel have the same effect as zero setting?
18. What can be done to assure minimum feel differences in the total observational variable?
19. What is the most serious observational variable?
20. How can interpolation errors be minimized?
21. Why are internal instruments more difficult to calibrate than external measuring instruments?
22. What determines the sensitivity variable in dial indicators?
23. What determines the contact method used between standard and part?
24. What is the most common form of contact used with comparators?
25. Why is this type of contact used?
26. Why should sensitivity be calibrated before accuracy?
27. For calibration of many instruments, the measurements do not have to be made at a specific temperature. Why?
28. What is meant by the term "compatible equipment" in reference to electronic comparators?

DISCUSSION TOPICS

1. Explain the classification of measurement errors.
2. Discuss the necessary steps required for precalibration.
3. Explain the types of errors that must be considered when determining interaction of errors.
4. Discuss the proper procedure for calibrating the variables of a vernier caliper.
5. Explain how observational errors can be minimized through use of high-amplification comparators?
6. Discuss the various types of contacts between the reference standard and part.
7. Explain the type of errors prevalent in the various comparator amplifications.

> ... let it not be feared that erroneous deductions may be made from recorded facts: the errors which arise from the absence of facts are far more numerous and durable than those which result from unsound reasoning respect-true data.
>
> Charles Babbage
> (19th century English inventor who conceived the computer as an instrument for research)

Measurement with Optical Flats
Surface With the Fringe On Top

This book has progressed in order of amplification, starting with rules which have one power. Now we arrive at the end of the road. Using light waves for measurement, amplification as high as one million power is practical and in daily use.

Among the other instruments, each move to a higher amplification meant more expensive, less familiar and more delicate instruments. This jump reverses all of that. The equipment for measuring with light waves is inexpensive, quite familiar and very rugged, Fig. 14-16A is an example.

The principal instrument is the optical flat. This is simply a transparent plate with one face finished to nearly perfect flatness. When this face is placed on another nearly flat surface, light bands, or fringes, are seen. This phenomenon is known as interference. The fringes correspond to separation distances between two surfaces. Because the wave length of light is known, the separation is measurable by counting the fringes.

IMPORTANCE OF STANDARDS

The fact that there is agreement between the industrial nations as to the relationship of their standards of length would seem sufficient for all general purposes. It is sufficient as long as the smallest division of any of the standards is greater than the error to which the standards are defined. It must be remembered that no matter how precisely a standard is defined, the absolute is never reached. *Whenever a measurement is made requiring a subdivision of the inch (or meter) that is smaller than the error known to exist in the standard, then that measurement is not in conformation to the standard, Fig. 14-2.*

Obviously the smaller the subdivision that the standard permits, the greater the discrimination that is possible in measurement. As mentioned in Chapter 11, the need for fine discrimination increases as (1) the speed of

Fig. 14-1   A remarkable phenomenon of light provides fringe patterns -- the contour maps for micromeasurements. *(Carl Zeiss, Inc.)*

the mechanism increases, (2) the number of component parts increases, and (3) the importance (or cost) of the end result increases. There is no way that manufacturing can economically supply the demands of society without constantly improving means of measurement. It was these pressures that forced the adoption of light wave standards.

LIGHT WAVES AS STANDARDS

The value of light waves for length standards was recognized long before the practical needs forced the adoption. The final breakthrough was a by-product of another quest for knowledge. Albert A. Michelson and W. L. Morley wanted to test the theory

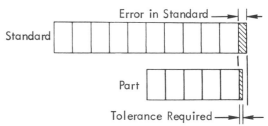

Fig. 14-2 Whenever a measurement is made requiring a subdivision of the standard that is smaller than the known error in the standard, a measurement out of conformation to the standard results.

of ether drift. This theory was that outer space behaved much as a transparent liquid instead of as complete emptiness. In order to test this, they used an interferometer to measure the path difference of light that had passed a tremendous distance through space as compared to a portion of the same beam when it traveled only a short distance.

To improve the precision of this experiment, it was found necessary to measure the wave length of a particular color of light, the red cadmium spectral line being chosen. For this, an interferometer was used. In the experiment *the wave length was measured in terms of the meter. Therefore, the reverse was also true — the meter was expressed in terms of the wave length of that particular color of light.*

Until Michelson's work the unit of length and the standard of length were always separate. The meter bar was never really the meter. It simply represented the meter as closely as it could be constructed. As long as these two were separate there would always be a discrepancy between them. *Michelson's work for the first time created a standard of length which was also a unit of length.*

The advantages were immediately recognized. The wave length was stable beyond any material that had hitherto been used for the standard. The length of the waves was so short that, at that time at least, it did not appear that they would ever have to be divided further for any conceivable discrimination requirement. Moreover, the light was relatively easy to reproduce anywhere.

Much work has been performed since the turn of the century when Michelson and co-workers performed their experiments. The light emitted by an isotope (number 86) of krypton is now the accepted standard. Using it, an inch is defined as 41,929,399 wave lengths. The difference between the various international inches was not reconciled until 1959. The United States, Canadian and British inch are now all equal to 25.4 mm. At last!

FROM WAVE FRONTS TO RAYS

It is not necessary to know anything about the phenomenon of interference to use optical flats for measurement – but it helps and it is

interesting. By now you have undoubtedly caught the idea that light is a wave motion. That conflicts with our usual notions about light rays. Actually these concepts are not contradictory.

Light is energy emitted by an electron moving from an outer orbit around the nucleus to an inner orbit. It is a pulse that expands as a wave from the point of origin. Any point along the expanding wave acts as another point of origin from which a new wave can be considered to emerge. Because every point along the wave is acting in this manner, the advancing pulse can be considered to be a wave front, Fig. 14-3.

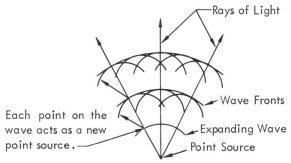

Fig. 14-3   In 1678 a Dutch physicist, Christian Huygens, developed this concept of wave motion.

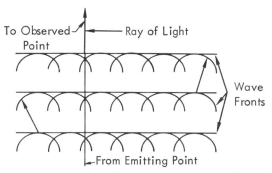

Fig. 14-4   When the wave fronts are far from the emitting point, thay may be considered as planes and the ray may be considered to be any line perpendicular to them.

While the *wave fronts are real, rays are a convenience* man invented to explain the behavior of light. By definition, a ray is a line from the emitting point to the observed point. The radius of the wave front becomes very great as its distance from the source is increased. It becomes so great, in fact, that the wave front can be considered to be a plane, Fig. 14-4. Then the rays are perpendicular to the wave fronts and parallel to their line of travel.

FROM RAYS TO SINE WAVES

Rays were invented because they provide a means to trace the passage of light through an optical system, whether complex or a simple lens. This is known as *ray tracing* and it is a technique of *geometric optics*.

Going back to wave fronts, remember that they represent energy pulses. Sound waves are pressure pulses in air. Light waves are similar except that they can move through any transparent material, are incredibly shorter and faster than sound waves, and are electromagnetic pulses. As the pulses pass any given point along the ray, they will vary from low energy to high energy repeating this cycle as long as the light continues. Similarly, if the action could be stopped and examined, the energy levels along the ray would cycle back and forth between maximum and minimum amplitudes as shown in Fig. 14-5.

Using standard graphical methods the amplitudes could be drawn along the ray. The result is the sine curve shown in the figure. This sine wave is a familiar device that is used universally to explain interference. It is used so universally, in fact, that you might get the idea that it is real, that something actually looks like that if it could be magnified sufficiently. Forget it. The ray and the sine wave are phonies. We made them up in order to demonstrate a phenomenon about light. *The wave fronts are real but the rest is purely a matter of universally accepted shorthand notation.*

Figure 14-6 shows some of the conventions that you need to know about waves. At the top are the terms used. The other two drawings show the interaction of waves of the same frequency traveling in the same direction. Although theoretically possible, the odds against waves from opposite directions merging just the right way to interact are so enormous that this does not need to be considered. As shown, waves that are in phase add their amplitudes, and waves that are 180 degrees out of phase cancel. For any phase combination in between the waves, add algebraically.

FROM SINE WAVES TO FRINGE BANDS

To see how fringes are formed consider what takes place when waves from one point

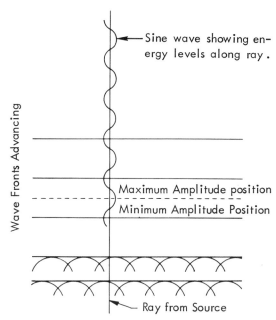

Fig. 14-5   If the energy levels of the fronts are graphed along a ray, a sine wave results.

RULES FOR WAVES

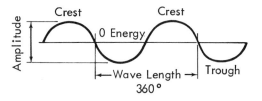

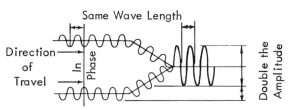

When in-phase waves meet, their amplitudes add but their length remains the same.

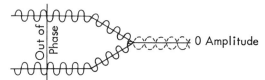

When out-of-phase waves meet, their amplitudes cancel, leaving zero amplitude. If any part remains, it will have the same wave length as the out-of-phase waves.

Fig. 14-6   Waves of the same frequency interact when they meet traveling in the same direction.

FRINGE FORMATION

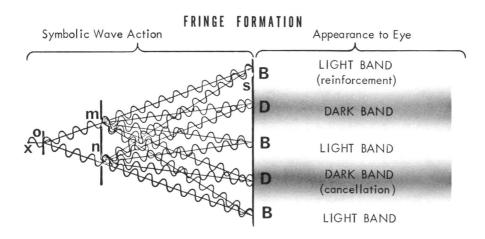

Fig. 14-7   Fringes are formed by alternate reinforcement and cancellation of converging waves.

are caused to converge at a surface. In Fig. 14-7, the emitting point is X. Passing through the slit 0 divides the ray into many rays, two of which are shown. (Remember your pin-hole camera?) The slits M and N repeat the action and again act as separate emitting sources. The rays from M and N meet at the surface S. At each point along the surface a ray from M meets a ray from N. If these rays have in-phase waves, the surface is illuminated. If they are out of phase, the surface is not illuminated, hence dark. These are the alternate light and dark bands.

It should not be thought that this happens just here and there. The interaction is continuous across the surface. This is shown by the enlargement of one dark wave in Fig. 14-8.

So far so good, but what does this have to do with measurement, you might ask? By rearranging the light paths we can use the fringe bands to measure distances. The basic method is shown in Fig. 14-9. Light from the source passes through the optical flat to the reflecting surface and from there to the eye. The working surface of the flat also reflects part of the light which also travels to the eye.

In the drawing all that would be seen would be two reflections of the source X. Consider what happens as you reduce the air gap. The ray *ac* approaches the position of ray *bc*. Remember that the light in both rays is the same wave length and traveling the same direction. When these rays very nearly occupy the same path, their energy waves interact. If the waves are in phase a bright band will be seen. If they cancel, a band of much less light is seen.

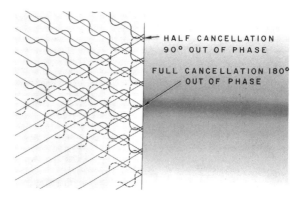

Fig. 14-8   This blowup of one band shows gradients from cancellation to reinforcement.

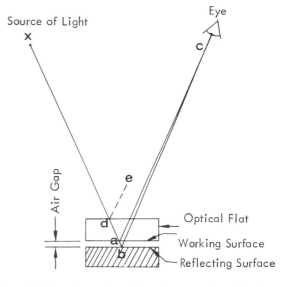

Fig. 14-9   Ignoring refraction and all reflections except the ones we are interested in, the air gap causes two rays to converge at the eye.

Those who have studied physics will immediately recognize that Fig. 14-9 has been simplified. No refraction is shown. Reflections that play no part in fringe formation are also eliminated, such as *de* from the top surface of the flat. There would also be internal reflections. All that the reflections do is diminish the amount of light energy that remains in the fringe pattern. Refractions cancel out.

THE PHASE SHIFT MATTER

Failure to explain the other oversimplified factor is one of the nastiest tricks ever played on the serious student. Because measurement with optical flats is easy and useful, much has been written about it. Too many writers of technical books and articles simply "cut and paste" the work of earlier writers. Some place along the line the matter of phase shift got lost. And pity the poor student who, not knowing about it, tries to work out the principles of measurement by interference on his own. (The author literally wasted days on this when he first encountered it.)

When light is reflected some of the light penetrates into the reflecting surface. This causes a *phase shift*, which simply means that the phase of the reflected light is slightly retarded in comparison to the incident light. The amount of retardation may vary from nearly 0 degrees to 180 degrees. It varies with the angle of reflection, surface finish of the material and, surprisingly enough, with the material itself. Because it is subject to so many variables, you cannot account for it except by laborious computation. Furthermore, in the end it does not change the results. The latter is probably the reason that someone, long ago, failed to mention it. The reason that writers since that time have failed to mention it is moot.

FROM FRINGES TO MEASUREMENT

The reason that the air gap can be measured by fringe bands is because of the difference in the lengths of the two paths. Obviously, in Fig. 14-10, the path from X to C by way of surface R is always longer than when it is reflected at S. But a change in the air gap distance results in nearly twice as much change in the path difference.

This change in path difference is important because it results in a change in the phase relationship. In Fig. 14-11 the paths are shown as wave fronts, and sine waves are used to show the energy in these fronts. A swing to the left is minimum energy, one to the right is maximum. Dark shading thus accompanies the left part of the cycles to show absence of light.

In the illustration the two reflected paths of light are in phase. The maximum phases are opposite as are the minimum. Therefore an observer would see them combined as a bright band. This resulted from an air gap separation of 1 1/2-wave lengths. How much is that in inches? It depends upon the wave length of the particular light being used.

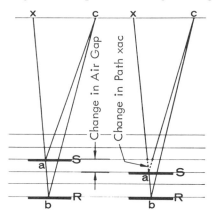

Fig. 14-10 The difference in length of the two paths xac and xabc changes nearly doubles the rate that the space between S and R changes.

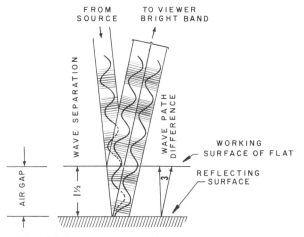

Fig. 14-11 Ignoring phase shifts from reflections, these are the interaction of the two reflected rays. Each ray is simply a line to show the travel of wave fronts. The energy pulses of the wave fronts are shown by shading.

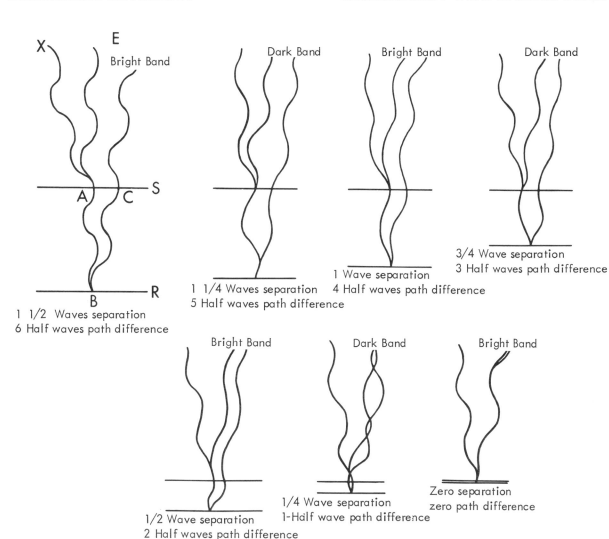

Fig. 14-12   The bands change by half-wave path differences.

Figure 14-12 shows what happens when the air gap is reduced. At 1 1/4-wave separation the path difference has changed one-half wave. This causes the two reflected paths to oppose each other. A maximum energy point on one is opposite a minimum energy point on the other. Hence they cancel. Because they cancel all along, at no time does the observer see much light. This is a dark band. Close the gap another one-quarter of a wave and the path difference becomes four. Again both paths reinforce and create a bright band.

It does not take much study of Fig. 14-12, to show that even-numbered half waves of path difference produce bright bands and odd-numbered half waves produce dark bands. Even more useful is the observation that the separation from one dark band to the next dark band represents 1/2-wave length. This

is true of bright bands also. Dark bands are easier to use, however, and discussions usually concern them.

If the flat is raised so slowly that the fringes can be counted, every time a dark band passes, the flat will have been raised 1/2-wave length – but 1/2-wave length from where? Common sense tells us that we should begin counting at zero, or no separation. If you do, the first band occurs at 1/4-wave length. They then re-occur at 3/4-, 1 1/4-, 1 3/4-wave lengths, and so on. Yet every book or article you are likely to consult, will state that you measure separation by counting the dark bands and multiplying by 1/2-wave length of the light being used. These two statements differ by 1/4-wave length. Both cannot be correct. Which is?

Neither one is entirely correct or entirely incorrect. If the air gap could really begin at zero and if the net results of all phase shifts cancel out, we would have the situation in the top of Fig. 14-13. Unfortunately, there almost always is some phase shift. Furthermore, it is very difficult to force two nearly perfectly flat surfaces into intimate contact. That sounds almost unbelievable but you can demonstrate it very neatly.

Clean scrupulously a gage block as described in Chapter 12. Similarly clean an optical flat. Place the gage block on a solid, smooth surface. Place the flat on it with the working surface down. (The working surface is marked as you will see in the photos.) Without wringing simply push the flat down on the gage block until fringes are seen. Ordinary light will do, but monochromatic light will give more vivid results. These are described later in this chapter. Note that as you push the flat against the gage block the fringes diminish. Now try to force the flat so tightly against one edge of the gage block that you get intimate contact. If you succeed, it will look like the top illustration in Fig. 14-13. More likely, no matter how hard you try, all you will do is broaden the first dark fringe in the lower illustration. The moment you release the pressure the air will rush back in and many fringes will again be seen.

Because of this, many people using optical flats think that contact begins with the dark fringe. There are actually many published illustrations showing this. You will certainly get into arguments with the "experts" if you state otherwise. Here is the way you win the argument.

Take a thin gage block and, after the usual thorough cleaning, wring it to a flat (always to the working surface). If you push on the center of the block you will bend it. This you can see by the fringes rapidly moving away from the point of pressure. No matter how hard you push by hand all you will get is a wide dark fringe. That seems to prove the point for the opposition. To disprove it, lay the wrung block and flat on a small smooth rod (a 1/4-in. dowel pin will do) resting on a solid surface, Fig. 14-14. Now apply pressure on the flat. The fringes will rapidly depart from both sides. The remaining dark fringe will widen; and, with continued downward force, the dark fringe will split, moving

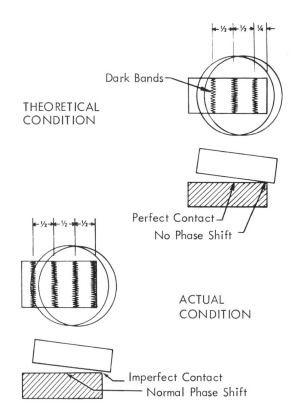

Fig. 14-13   Theory thus far calls for the top condition, but most books and articles show the bottom condition.

almost exactly one-half of a normal band width to either side of a bright area. If examined with a magnifying glass, this bright area will be seen to be different than an ordinary bright band. That is because the two surfaces are in intimate contact at the center.

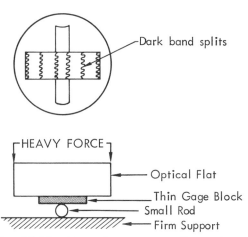

Fig. 14-14   Heavy force on a small area is required to achieve intimate contact between a gage block and the flat.

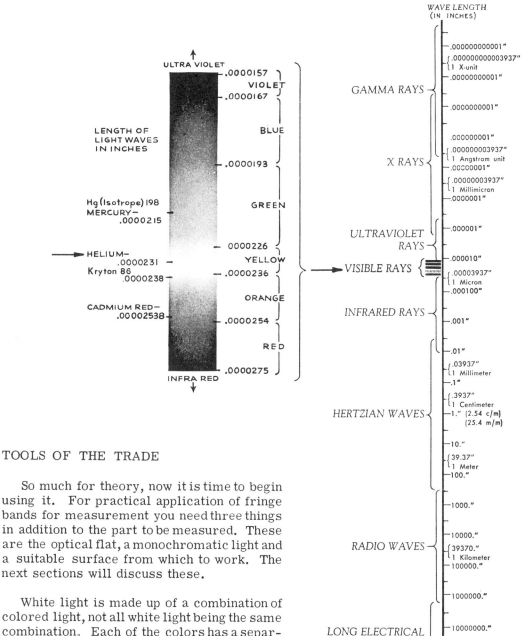

Fig. 14-15   Light waves represent only a very small part of the electromagnetic spectrum.

TOOLS OF THE TRADE

So much for theory, now it is time to begin using it. For practical application of fringe bands for measurement you need three things in addition to the part to be measured. These are the optical flat, a monochromatic light and a suitable surface from which to work. The next sections will discuss these.

White light is made up of a combination of colored light, not all white light being the same combination. Each of the colors has a separate and distinct wave length. For the international standard great care had to be used in the selection of the best particular color. It had to be very definite in wave length. That is, each time it was reproduced anywhere in the world it had to be exactly the same. Krypton 86 gas excited electrically emitted a light that met that requirement.

For practical use other factors must also be considered. One of these is the definition of the fringe bands, how easily they are seen. Other considerations are cost and conveni-

ence. The light emitted by helium gas has proved most practical when all considerations are balanced. The relationship of helium light to the balance of the spectrum is shown in Fig. 14-15. Several light sources are available, the three most popular types being shown in Fig. 14-16.

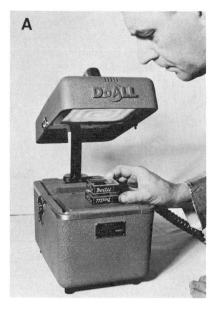

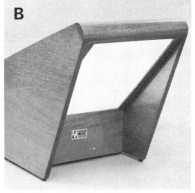

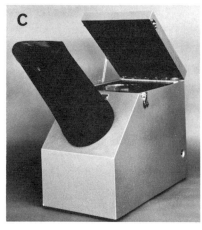

Fig. 14-16 Many monochromatic light sources are available, these being the most popular types. A and B direct the light onto the flat from above. C is a reflex type in which the light is directed from below.

Optical flats are made in a range of sizes, shapes and materials. Although they are available in sizes as large as 10 inches in diameter, the ones you are most likely to encounter are from one to three inches in diameter. Materials range from inexpensive glass to very expensive sapphire. The majority are high-quality optical quartz.

The measuring surface generally has one of three degrees of flatness. *Working flats* are *four mike* (0.000004 in.). That is a unilateral tolerance, in that no point will deviate in height from any other point by more than that amount. *Master flats* are *two mike* (0.000002 in.) and *reference flats one mike* (0.000001 in.). Some manufacturers also furnish a commercial grade which is *eight mike.*

One or both surfaces may be finished for measurement. A finished surface is usually shown by an arrow on the edge of the flat. The second surface adds very little additional cost to the purchase price. This love of a bargain, real or imagined, is the only reason the author can find to explain double-surfaced flats. Generally when one surface is worn out of tolerance it is scratched and no longer clear. In order to use the other surface you must look through the worn out surface. Therefore, expectancy of double life is foolish.

The flats are available at extra charge with *coated surfaces.* This refers to a thin film, usually titanium oxide, on the surface to reduce the light lost by reflection. The

less light that is lost by unwanted reflection, the clearer the fringe bands, Fig. 14-17. This photograph shows also that the coating is so thin that it does not affect the position of the fringe bands. That is the good side. The bad side is that a coated flat requires even greater care than an uncoated one. There are some relatively rare cases in which uncoated flats provide better fringe bands. Carbon seals are an example.

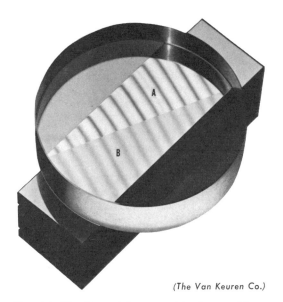

(The Van Keuren Co.)

Fig. 14-17 Coated flats provide sharper fringe bands in most cases. In this photograph the flat has been half coated. A is coated and shows the fringes much better than B.

The supporting surface on which optical flat measurements are made is relatively unimportant in one case and very important in another. In every case it must afford a clean, rigid platform. If the measurement consists of the changes on one surface, little more is needed. If, however, the measurement involves the comparison of two surfaces, the supporting surface becomes the limiting factor. The precision of the measurements can be no better than that of the supporting surface.

For the latter reason, optical flats are often used as support for the part. Steel flats, known as *toolmaker's flats* or *gagemaker's flats*, are also available. These are simply steel optical flats to which parts and gage blocks may be wrung for measurement. Of course, other precision-finished surfaces, such as comparator anvils, may also be used. It must be remembered, however, that the errors contributed by the supporting surface are of the independent type. They may combine with the measurement errors, thereby increasing the uncertainty of the overall system.

SETUP FOR MEASUREMENT

The main points under this heading have been noted over and over again, but bear repeating. Cleanliness is of tremendous importance. Even a stray particle of dust that might alight on the part before the flat is placed over it can completely destroy any chance for reliable measurement. In fact, one particle of dust can prevent the formation of fringe bands in borderline cases. It is advisable to have a camel's-hair brush at hand. At the last moment before placing the flat on the part and/or gage block, whisk it off. Why camel's hair? Because it does not shed – or at least, so it is said.

Temperature changes are more apparent when using optical flats than with any other kind of measurement. Fortunately, most of the measurement involves relatively small parts that normalize quickly. Most optical flats have a lower coefficient of conduction than the metal parts with which they are used. This is a mixed blessing. It means that they are not heated as rapidly by handling as they might be. It also means that once heated or cooled, it takes the flat longer to regain the ambient temperature (surrounding temperature, usually referring to air).

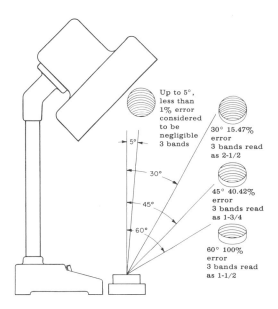

Fig. 14-18 The smaller the angle, the more accurate the measurement will be.

This can easily be demonstrated by wringing a gage block to an optical flat until three or four bands appear. Place the gage block and flat on a surface under a monochromatic light. Then drip four or five drops of solvent on the flat near one edge. Blow on the solvent to cause it to evaporate rapidly, cooling one part of the flat. You will see no immediate change in the fringe pattern. However, return in a few minutes, perhaps five, and you will find a changed fringe pattern. The flat will have bent because of the contraction of the cool portion. Permit it to stand for an hour and the fringe pattern will change again as the flat normalizes, but will not return to exactly the same configuration as at the beginning of the experiment.

Figure 14-18 shows the proper way to view an object through an optical flat. The more nearly perpendicular the line of sight is to the surface, the more accurate the measurement will be. This led to the development of the reflex type of monochromatic light such as C in Fig. 14-16. In this type a beam-splitter mirror is used to permit both the line of sight and the monochromatic illumination to be perpendicular to the measurement surface.

To achieve maximum clarity the measurement surface should be as close to the light source as convenient. That is the purpose of the adjustable light support in type A of Fig. 14-16. It does no good unless adjusted.

Wring a gage block or part to a flat much the same as you would to another block. It will begin to wring almost immediately unless one of three things prevent it. These are (1) insufficiently flat surface, (2) insufficiently fine surface finish, and (3) improperly cleaned surfaces. Excessive wringing wears the flat.

There is difference of opinion about the degree of surface finish necessary for wringing. This results from the difficulty of measuring very fine surface finishes. Some say that a four microinch AA surface is required. Other published figures are as fine as 1.2 microinch. As a basis for comparison, the best gage blocks have less than 1.0 microinch AA surface finish.

As soon as the block has begun to wring you will see fringe bands. As you continue to wring, you will find that you can cause the bands to run any direction across the part. With practice you can maneuver them at will. The reason for this will be clear in the next section, but first a warning.

Never leave a gage block, part or another flat wrung to an optical flat beyond the time required for measurement. If left overnight or longer you may have to break the flat to separate them. If they must be forced apart use a wood block, never metal. First try soaking in solvent. Now on to measurement.

PARALLEL SEPARATION PLANES CONCEPT

Actually the fringe bands form in the air between the observer and the measurement surface.* Therefore, to make fringe bands a practical measurement tool we must adopt a convention, the parallel separation planes concept. While completely nonexistent from the theoretical standpoint, it is a great aid in actual measurement.

The convention consists of a set of imaginary planes all parallel to the working surface of the flat and one-half wave length apart, Fig. 14-19. The intersections of these planes and the part are the dark fringe lines. The number of fringes thus represents separation between the surfaces in units of half-wave lengths.

* Candler, "Location of Interference Fringes," Modern Interferometers, London, Hilger Division, Hilger & Watts Ltd., 1951, p. 69.

Whenever this is discussed at least one person cries "foul" because of cosine error. True, it does exist but in a real case, unlike the exaggerated drawing, the two surfaces are so nearly parallel (1 to 10 seconds of arc) that the error is in *billionths of an inch*.

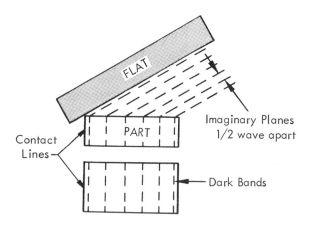

Fig. 14-19 The parallel separation planes concept envisions planes parallel to the working surface of the flat and one-half wave length apart. Dark fringes occur at their intersection with the part.

THE BASIC AIR-WEDGE CONFIGURATION

Gage blocks are convenient objects on which to practice surface inspection with optical flats, but any other sufficiently flat parts will do. Other familiar and suitable parts are refrigeration compressor seals and automatic transmission seals. Because of their availability this discussion will be based on gage blocks, but will be equally valid for any other suitable parts.

After thoroughly cleaning the gage block and optical flat, wring them together. As described earlier, work with the wringing until the fringe pattern crosses the block as shown in Fig. 14-20. The five dark bands or fringes tell us that we have an air wedge of five, half-wave lengths height separating the flat from the gage block. We do not know which way the wedge is facing. Either the right or left end may be the open one. This is found by applying force to the ends. If there is no change you are pushing along the line of contact, Fig. 14-21. When the force causes the fringes to spread out we know that we are squeezing the wedge closed, Fig. 14-22. Therefore we are pushing against the open end. The situation greatly exaggerated is shown in Fig. 14-23.

Fig. 14-20 The five fringes represent an air wedge with a height of five half-wave lengths. They do not show which end is the open one.

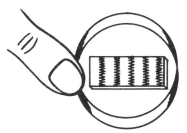

Fig. 14-21 If force at one end does not change the fringe pattern, that end is in contact with the gage block or part.

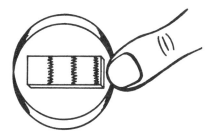

Fig. 14-22 When force does widen the fringe bands, you know that you are squeezing the air wedge closed. Only slight force is required.

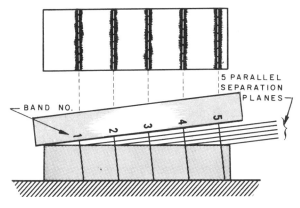

Fig. 14-23 The situation in Fig. 14-20 (greatly exaggerated) shows that the height of the wedge is represented by 5 parallel separation planes, each equal to 1/2 wave length.

The air wedge forms the basic configuration for fringe patterns. It would have applied equally well if we had oriented the fringes lengthwise along the block, Fig. 14-24. The contact can always be found by applied force, and the height at the widest point can be determined by multiplying the number of fringes by one half the wave length of the light used. This will be 11.6 *mike* (0.0000116 in.) for helium light. The general relationships are summed up in Fig. 14-25. They are: *the fewer the bands, the narrower the angle; and the more numerous the bands, the greater the angle.* This always applies.

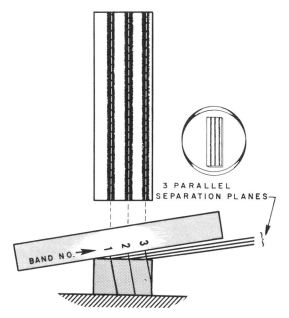

Fig. 14-24 If the fringe bands had run lengthwise, the air wedge would still exist and the contact could be found by applied force.

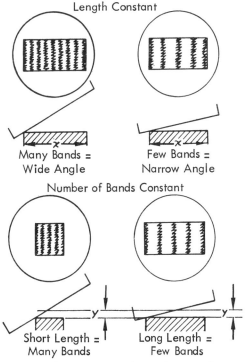

Fig. 14-25 These relationships apply to air wedge measurements.

More important, is the understanding that *the number of bands is a measure of height difference, not of absolute height.* As in all measurement there must be a reference from which every length is expressed.

SURFACE INSPECTION

In surface inspection the matter of reference is easy. Some part of the surface is arbitrarily chosen as the reference from which the other parts are expressed. Going back to Fig. 14-24, we can tell from the fringe pattern that the surface is flat. That is known because the fringe bands are straight and uniformly spaced.

If there had been a sharp drop off, that would have shown up by a change in the pattern as in Fig. 14-26. In this case the closely spaced bands show that the angle is larger along one edge than along the other. As in the previous case, no measurements can be made unless a reference is known. In this instance it is the left edge of the block.

This principle is extended in Fig. 14-27. In A the reference is a line, the lower edge of the block. All we know about the block is that it is nearly flat along its length. In B the bands curve *toward* the line of contact showing that the surface is convex and high in the center. The reverse is shown in C. This surface is concave and low in the center. D indicates a surface that drops at the outer edges. Note the importance of the reference line. If

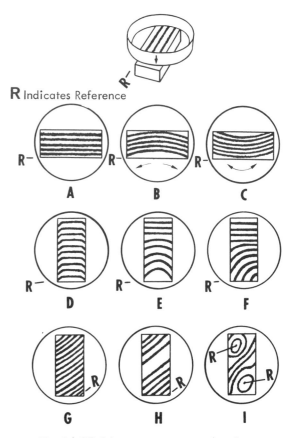

R Indicates Reference

Fig.14-27 Fringe patterns reveal surface conditions like contour lines on a map.

it had been at the top, the same fringe pattern would indicate a surface that rises at the outer edges.

A surface that is flat at one end but becomes increasingly convex is shown in E. In F is shown a surface that is progressively lower toward the bottom left-hand corner. The bands turn toward the line of contact and get progressively farther apart. From lower right, in G to upper left, this surface is flat, but the slight curvature of the bands away from the line of contact indicates it is slightly concave. In H the surface is flat in the direction that the bands run. But diagonally across the center, where the bands are widely spaced, it is higher than at the ends. The two points of contact in I are frequently found. They show two high points surrounded by lower areas.

Once you recognize surface configurations from their fringe patterns, it is an easy step to measurement of the configurations. However, the fundamentals in Fig. 14-28 must be understood.

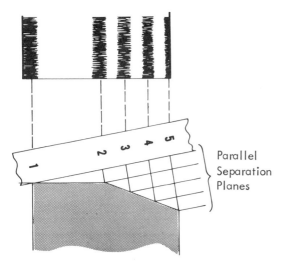

Parallel Separation Planes

Fig. 14-26 The sharp dropoff is clearly shown by the close bands on the right.

FRINGE READING FUNDAMENTALS

Wedge Method

1. In order to interpret the extent of band curvature from absolute straightness it is necessary to know the reference line or point from which the measurements can be expressed. Imaginary lines parallel to the line of contact are expressed. They should pass through the center of the bands.

2. Even though the distance between bands varies, the height difference from band to band is always 11.6 mike and is always counted from the line of contact.

3. The extent of curvature is always measured against the distance between bands.

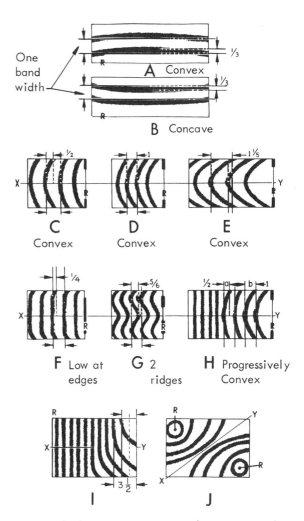

Fig. 14-28 These basic considerations apply to all optical flat measurement.

Fig. 14-29 Measurements are the deviation of bands compared to the distance between bands.

In A of Fig. 14-29 the surface is convex by one-third of a band. One-third of 11.6 equals 3.866 *mike* or 3.9 *mike* rounded off. The reverse is true in B. The surface is concave or low in the center by the same amount. C, D, and E are all convex but by different amounts. C is one-half wave high in the center, D is one wave and E is one and one-half waves. In millionths of an inch these are 5.8, 11.6 and 13.9 respectively.

A very common type of surface conformation is shown in F. This surface is flat except at the edges. The edges drop off one-quarter wave length or 2.9 *mike*. The surface in G has two low troughs with center and edges the same height. The ridges are five-sixths of 11.6 or 9.7 *mike* (0.0000097 in.) high.

Few surfaces are uniform as the above examples have been. Most are changing from end to end. Typical is H. It starts with a flat surface that becomes progressively more convex towards the right. At a, the surface is 5.8 *mike* convex. Further along, at b, it is 11.6 *mike* convex. In I is a surface that is flat near the reference line but rises at the right edge. The top right edge is about one wave length high while the lower right edge is about three and one-half wave lengths.

In J there are two high points with a low trough between them. The bottom of this trough is approximated by the line XY. Note that there are four convex bands on each side of the high points. That indicates that the trough is four and one-half wave lengths or 52.3 *mike* (0.0000523 in.) low.

WHICH WAY SHOULD THE BANDS RUN?

Note in each case the fringe pattern could have been perpendicular to the ones chosen. These fringe patterns would have been just as useful as contour maps to show the surface configuration but would have made measurement more difficult. The reason, of course, is the point made earlier. The number of bands is a measure of height difference, not of absolute height.

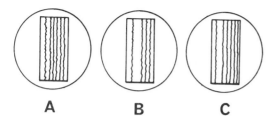

Fig. 14-30 Spacing of the bands depends upon the angle of the air wedge. A shows the same pattern as Fig. 14-26. B is the same surface but with a smaller angle. C is a larger angle.

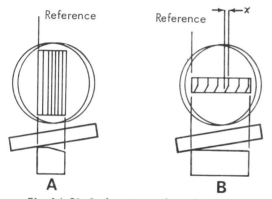

Fig. 14-31 Both patterns show the surface conformation but B permits direct measurement of the dropoff.

This point is shown in Fig. 14-30. All three fringe patterns show that the surface is straight along its length but drops off along one edge. The reason that the band spacing varies is that the angle of the air wedge in B is smaller than in A. In C it is greater than A. From these patterns the amount of drop off could be calculated. However, it is much easier to reorient the pattern. This is done in Fig. 14-31. The exaggerated drawing shows the relationship of the two fringe patterns. A in Fig. 14-31 corresponds to A in Fig. 14-30. In B it is only necessary to determine what fraction of one band is represented by x in order to determine the amount of drop off.

In summary, any fringe pattern shows the surface conformation — if the contact point is known. If a change in elevation is desired, it is most easily measured by bands that cross the area in which the change takes place. Then the change can be measured by the amount that the band deviates from straightness.

MEASUREMENTS OF PARALLELISM

The parallelism of planes can be measured to great accuracy with optical flats. There is a severe limitation to the measurement of parallelism. Fringes are only visible with the unaided eye when the separation distance is very short, seldom over ten-fringe bands or 100 *mike* (0.0001 in.). Interferometers make greater separation distances possible, but their use is not yet as familiar as simple optical flats.

Again gage blocks come to the rescue. Because a standard set of gage blocks will make up over 200,000 dimensions to *tenth mil* (or 100 *mike*), it is never necessary to have separations beyond the resolution of optical flats. (Note that sensitivity of optical instruments is usually termed resolution.)

Assume that the part whose parallelism we wish to check is a gage block of questionable condition. If this unknown block is wrung to a flat along with a known block we then know that the wrung surface of each is parallel to the other. They are parallel to the limit of flatness of the optical flat. Assuming that the known block is in calibration, we now have constructed a plane of known parallelism to the reference plane, Fig. 14-32. Not only that, it is almost parallel to the top surface of the unknown block. The amount that it misses parallelism is what we want to determine.

The lack of parallelism may be along two of the three coordinates, Fig. 14-33. To measure parallelism the rotation about both of these axes must be determined. With optical flats, they are determined simultaneously. Thanks to gage blocks the substitute reference plane can be constructed very near to the top plane of the unknown part. Therefore fringe patterns can be seen. The comparison of those on the known part with those on the unknown is the measure of parallelism.

Figure 14-34 is an example. It reveals three facts. First, the unknown surface is parallel longitudinally to the known because it produces the same number of bands. Second, it is not parallel across the width because the bands on the unknown are at an angle to those on the known surface. Third, the amount that the unknown is out of parallel to the known is one-half band in one width.

PARALLELISM MEASUREMENT

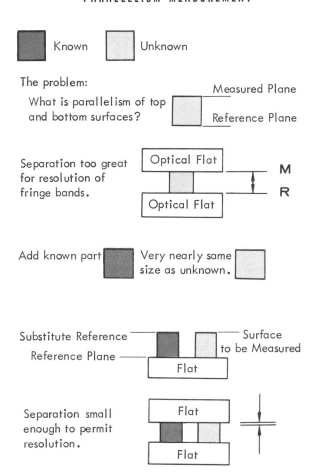

■ Known   □ Unknown

The problem:

What is parallelism of top and bottom surfaces?

— Measured Plane

— Reference Plane

Separation too great for resolution of fringe bands.

Optical Flat

M

R

Optical Flat

Add known part   Very nearly same size as unknown.

Substitute Reference —

Reference Plane —

— Surface to be Measured

Flat

Separation small enough to permit resolution.

Flat

Flat

Fig. 14-32   With gage blocks a surface can always be constructed within the resolution range of optical flats.

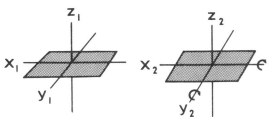

Fig. 14-33   Parallelism of two planes is a measure of their relative rotation about the x and y axes. Rotation about the z axis and translation (movement) along any axis does not disturb parallelism.

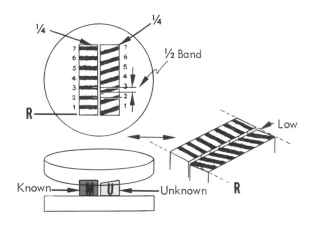

½ Band

R

Low

Known   M   U   Unknown

R

Fig. 14-34   In this example both surfaces have the same number of bands, but those on the unknown, slant.

The example in Fig. 14-35 also answers three questions. What is the relationship of the planes in the x axis? What is the relationship in the y axis? How much are the planes out of parallel? In this case the fringe bands are parallel on both surfaces. This shows us that the surfaces are parallel across their widths, or their x axes. But because there are more bands on the unknown than on the known surface, we know that the air wedge between the unknown and the flat must be greater than between the known and the flat. There are two more which indicate that the out-of-parallel condition must be two bands in the length of the block. Another way of considering it is that the unknown is 23.1 *mike* (0.0000231 in.) lower at its far end than the known. See Fig. 14-46 for values.

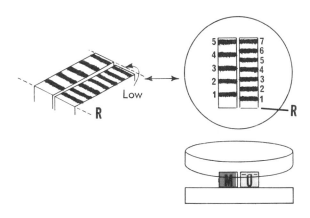

Low

R

R

M   U

Fig. 14-35   More bands, in this example, are seen on the unknown than on the known surface.

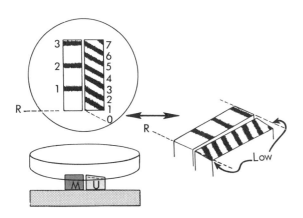

Fig. 14-36 Surfaces can be out of parallelism in two axes simultaneously. This example shows the resulting fringe pattern for one case.

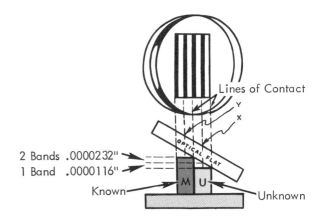

Fig. 14-37 Counting the bands on the unknown part provides the measurement of the height difference. x and y are points to apply force to find the lower part.

In Fig. 14-33 the out-of-parallel condition was in the y axis. In Fig. 14-35 it was in the x axis. Figure 14-36 shows that the out-of-parallelism can exist in both the x and y axis at the same time. The unknown surface in this example is out of parallel by one band in the x axis and three bands in the y axis.

COMPARATIVE MEASUREMENT WITH OPTICAL FLATS

In all of the parallelism examples a convenient assumption was made. We assumed that the unknown part had the same basic size as the known part, or gage block. That rarely happens, but it permitted us to develop some points about parallelism.

In most cases there would be a small difference in height. The effect of this is shown, greatly exaggerated, in Fig. 14-37. Note the basic difference from the parallelism examples. In them the point or line of contact of both the known and the unknown part always have the same height above the reference surface to which they were wrung. In the height comparison examples, the contacts are at different heights.

Obviously before you would compare the heights, you would first inspect the surfaces of both the known and the unknown. To keep the following simple, we will assume that these surfaces are very nearly flat. Note also, in the examples that follow, the parts are in close contact with each other.

After the known and the unknown have been wrung tightly to a flat and are positioned alongside each other, it is necessary to determine which one is the higher. This is done by placing the second flat over them, orienting the fringe pattern to run lengthwise, and then applying force to a point above the center of each part. If force applied at x in Fig. 14-37 does not spread out the pattern, but force applied at y does, then we know that the known part is the higher. The reverse would, of course, apply if it were lower.

In the example the air-wedge triangles formed over each part are identical. Therefore the fringe patterns are identical. Counting the fringes across the unknown shows that it is two bands lower than the known. Now you can appreciate the parallel separation concept. Although we see the fringes along the hypotenuse of the triangle, they provide a direct measurement of the height of the triangle. To avoid confusion, it is suggested that you do not mention this to your geometry teacher unless you have the explanation readily at hand.

This method is not restricted to crosswise comparison. Figure 14-38 shows the height difference of two parts compared along their lengths.

MEASUREMENT OF SMALL HEIGHT DIFFERENCES

The above examples purposely chose cases in which the differences were considerable.

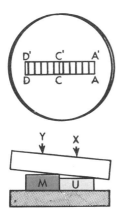

Fig. 14-38 In this example the unknown is 5 1/2 bands lower than the known, or 63.8 microinches lower.

Does the same method apply when the differences are very small, say less than two bands? The principle applies but the method must be altered. When the differences are very small, pressure at either x or y causes the bands to spread out.

When this happens, manipulate the top flat until the band pattern runs diagonally, as in Fig. 14-40. Then if the bands on the known exactly correspond to those on the unknown, as at A, the parts are the same height. If they do not correspond in number or do not line up, the blocks are not the same height. Then to determine the difference in height between the two blocks follow the steps in Fig. 14-39.

In B of Fig. 14-40 the reference line intersects band one on the known block and the band two on the unknown block. Therefore, 2 bands minus 1 band multiplied by 11.6 equals

---

STEPS FOR MEASURING HEIGHT DIFFERENCE

1. Run an imaginary reference line through a line on the known block and continue it across the unknown block. The reference line may or may not coincide with an interference band on the unknown block.

2. Count the number of bands on the known block from the reference line back to the point of contact. Count the number of bands and estimated fraction that occur between the reference line and point of contact on the unknown block.

3. Multiply the difference by 11.6 to obtain the result in mike (millionths of an inch).

4. Because, in these examples, the master and unknown have flat and parallel surfaces, it makes no difference where the reference line is established.

Fig. 14-39 The simplicity of these steps is deceiving. Reliable results require practice.

11.6 *mike*. The unknown block is then 0.0000116 in. lower than the known. The reference line could have been anywhere else along the fringe pattern. The difference would have been the same. In C of Fig. 14-40 the reference line intersects the first band on the known block and then crosses the unknown half between the second and third bands. The difference is one and one-half mike (0.0000174 in.).

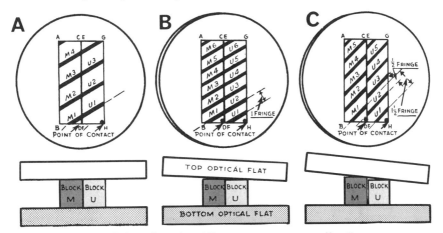

Fig. 14-40 When the differences are very small, diagonal patterns are used to measure height differences.

"CHEESE CUTTERS"

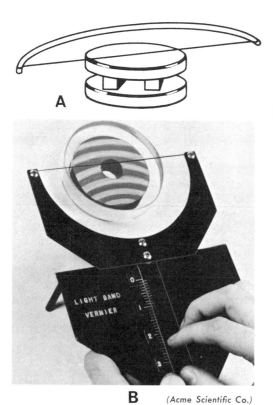

A

B   *(Acme Scientific Co.)*

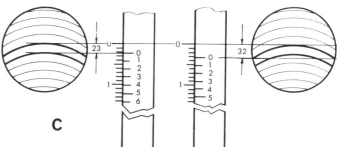

C

Fig. 14-41 "Cheese cutters" (A) are usually improvised taut-wire holders that provide a datum line for
counting fringes and estimating their separations.
The light-band vernier (B) eliminates much of the
guesswork. First a band is measured, then the curvature. These are arbitrary units. A monograph table
(C) supplied with the instrument, converts these
readings to <u>mikes</u> (millionths of an inch).

Measurement from an arbitrary line is fa-
cilitated by a device called a "cheese cutter".
This is simply a fine taut wire held by hand
or in a bow-like frame, Fig. 14-41. No doubt
a cheese cutter would do very well. One make
of monochromatic light has cross hairs on its
diffuser glass. These lines are projected onto
the work area and serve the same purpose.
Their use is limited, unfortunately, because
the lines are fixed and rarely occur where
wanted. A unique instrument called a light
wave vernier, Fig. 14-46, eliminates this
problem, and provides direct numerical
readings.

These same techniques and principles
apply when the surfaces being compared are
not nearly flat as they have been in these examples. It becomes vastly more complex.
Only practice and patience will develop any
degree of reliability. It is suggested that the
beginner sketch the surface configuration of
each block before making height comparisons.
Reference to the sketches will often clarify
confusing patterns formed by the two blocks
together.

An extension of the height comparison
method is shown in Fig. 14-42. Cardboard
templates are used to hold parts in fixed relationships to gage blocks of the same basic
heights. The change in the patterns on the
gage blocks gives the size deviations. No attempt is made here to discuss the mathematics of this method. Those wishing to pursue
it should remember that the distance between
the unknown part and the gage block must be
taken into consideration.*

APPLICATIONS OF OPTICAL FLAT
MEASUREMENT

There are a number of laboratory and industrial uses for which optical flats are frequently preferred. They are unsurpassed for
close examination of surface configurations.
Figure 14-43 shows two examples of ring seal
inspection.

* <u>Precision Measurement in the Metalworking
Industry</u>, Dept. of Education, Syracuse, New
York, International Business Machines Corp.,
Syracuse University Press, 1952, p. 252.

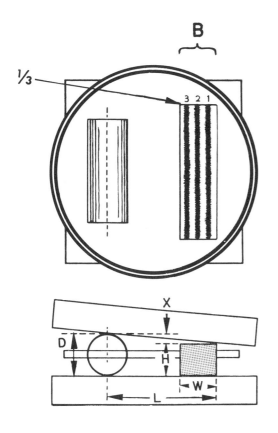

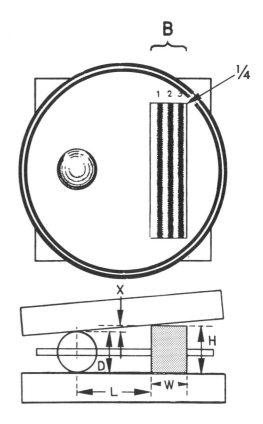

If gage block is 0.30000 inch
and L = 1.125 inch, cylinder
is 0.300116 inch.

If gage block is 0.50000 inch
and L = 0.750 inch, ball is
0.4999246 inch.

The Formula: $\dfrac{X \ 11.6 \ X \ B \ X \ L}{W}$

X   =   the difference in height between the ob-
            ject and the block.

11.6   =   .0000116 inch *

B   =   the number of interference bands or frac-
            tions thereof.

L   =   the distance between the perpendicular
            axis of the object and the edge, or line
            of contact.

W   =   the width of the gage block used.

H   =   the height of the gage block used.

D   =   the perpendicular diameter of the object.

* Note: This constant is only suitable for
rough work. It is obtained by rounding
off a more accurate value 0.00001157.
In large multiples, the difference could
be considerable. More accurate values
are given in Fig. 14-46.

Fig. 14-42   Use of cardboard templets enables cylinders, balls
and similar parts, to be measured with optical flats and gage blocks.

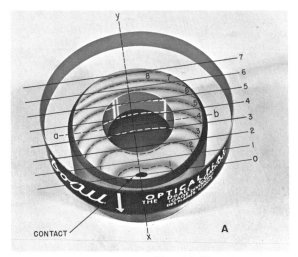
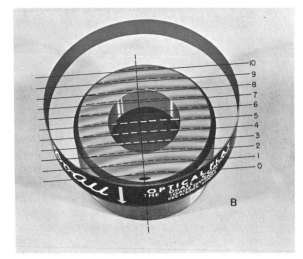

Fig. 14-43 Fringe patterns are used to inspect precision seals for refrigeration compressors, among other parts.

Example A shows the band pattern produced by a rotating seal that is not flat, and which has the greatest variation from flatness approximately across the center.

The spot at the bottom is the point of contact. From this point the scale is established as shown. Seven scale lines are involved. Across the center, line a-b is constructed, terminating at the ends of band number 5. The line x-y is constructed at right angles to the scale lines.

Against this scale we can now interpret the flatness of the rotating seal at various points, and measure variations in flatness. All bands are numbered, and, for easy reference, white dashes have been drawn on the photo to direct the eye to the continuation of each band crossing the bore.

First we will determine the degree of flatness across the center. Observe that the outer ends of band 5 join reference line a-b, and that, across the center, line a-b is about halfway between scale lines 3 and 4 (or, is at scale 3.5). Observe also that at scale line x-y, band 5 coincides with scale line 5. The difference between the outer ends of the band and the center, therefore, is one and one-half scale lines or 17.4 *mike*.

If we follow along line x-y from the *x* end, and compare the distance between bands with the distance between scale lines, we see that the bands formed are closer together toward the y end. This means that the surface is falling away from the reference plane established by the scale lines. Note that band 8 and scale line 7 coincide. Hence, the surface at this point is low by one band or 11.6 *mike*.

Observe that bands 2, 3, 4 and 5 have a relatively sharper curve as they approach the outer edges and the edge of the bore. This condition indicates that the edges have been lapped down more rapidly than the main part of the surface, due to the piling up of loose abrasives on the laps. The amount of wear is determined by constructing profile lines as previously described. Note that band 2 follows a course around the bore and curves toward the point of contact. This indicates that the edge of the bore is rounded off and is low about 7 *mike*.

Example B in Fig. 14-43, shows the surface of a rotating seal that is not rounded at the edges by piled-up or loose abrasives on the lap. This is indicated by the relative straightness of the bands. The bands do curve slightly, however, indicating that the outer edges are slightly lower. Along scale line 5 this deviation from flatness would be read as about one-fifth of a band, or 2.3 *mike*.

This method is so helpful for the examination of seals that it is often considered practical to make up special optical flats where necessary. Figure 14-44 shows three examples. At A an extra thick flat is used to inspect a recessed seal. The additional thickness is needed for mechanical rather than optical reasons. The protruding portion is used to manipulate it.

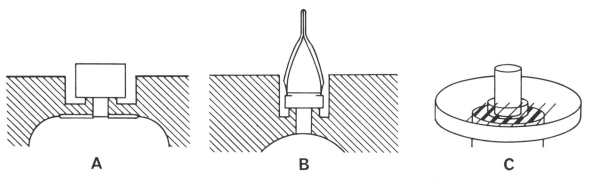

Fig. 14-44 Special optical flats are used to examine surfaces inaccessible with standard ones.

When the recess is too deep for a thick flat, flats with gripping surfaces may be used. This is illustrated at B, a tweezer being used to handle it. In many instances the seal is on a shaft and some part of the shaft must pass through the flat. An example of this is shown at C.

Fig. 14-45 The flatness of this plate can be checked by successive moves of the optical flat.

Figure 14-45 shows that even a large surface, such as a high precision bench plate, can be checked for flatness with a relatively small optical flat. If a 2-in. flat is used and the band pattern reveals an error in flatness of, say four *mike* (0.000004 in.) across its two-inch width, it may be assumed that this rate of error continues in the direction in which the bands run. In this case the error will be 24 *mike* (0.000024 in.) across twelve inches of the surface. However, this assumption should be checked. It is done by taking successive readings as the flat is moved two inches at a time in the direction in which bands appear. In like manner, the entire surface area can be checked and its degree of flatness made known. This is a practical, and very precise, method for small areas. It is generally too time-consuming to be used for surface plates, however.

CONVERSION TABLE

Number of Bands	Microinches (Millionths of an inch)	Inches	Millimeters
0.1	1.2	0.0000012	0.000029
.2	2.3	.0000023	.000059
.3	3.5	.0000035	.000088
.4	4.6	.0000046	.000118
.5	5.8	.0000058	.000147
.6	6.9	.0000069	.000176
.7	8.1	.0000081	.000206
.8	9.3	.0000093	.000235
.9	10.4	.0000104	.000264
1.	11.6	.0000116	.000294
2.	23.1	.0000231	.000588
3.	34.7	.0000347	.000881
4.	46.3	.0000463	.001175
5.	57.8	.0000578	.001469
6.	69.4	.0000694	.001763
7.	81.0	.0000810	.002056
8.	92.5	.0000925	.002350
9.	104.1	.0001041	.002644
10.	115.7	.0001157	.002938
11.	127.2	.0001272	.003232
12.	138.8	.0001388	.003525
13.	150.4	.0001504	.003819
14.	161.9	.0001619	.004113
15.	173.5	.0001735	.004407
16.	185.1	.0001851	.004700
17.	196.6	.0001966	.004994
18.	208.2	.0002082	.005288
19.	219.8	.0002198	.005582
20.	231.3	.0002313	.005876

Fig. 14-46 When approximations are sufficient, one band is considered to be ten mike (0.000001 in.) or 0.0003 mm.

METROLOGICAL DATA MEASUREMENT WITH OPTICAL FLATS

INSTRUMENT	TYPE OF MEASUREMENT	NORMAL RANGE	DESIGNATED PRECISION	DISCRIMINATION	SENSITIVITY	LINEARITY	RELIABILITY	
							PRACTICAL TOLERANCE FOR SKILLED MEASUREMENT	PRACTICAL MANUFAC- TURING TOLERANCE
Optical Flats Flatness Measurement	direct	to 0.0001 in.	one to ten mike	one to ten mike	one to ten mike	one to ten mike	one to three mike	ten mike
Surface Finish Measurement	direct	to 0.0001 in.	one to ten mike	one to ten mike	one to ten mike	one to ten mike	one to three mike	ten mike
Length Com- parison Meas.	direct	to 0.0001 in.	one to ten mike	one to ten mike	one to ten mike	one to ten mike	three to five mike	twenty mike

Fig. 14-47   These values are highly arbitrary because of the difficulty in making a science of the skill of measuring with optical flats. For example, a two-band pattern usually permits one band to be divided into ten parts; but a ten-band pattern on the same part permits virtually no interpolation.

Summary — Limitations of Optical Flat Methods

● Direct measurement with optical flats is restricted to an extremely short range. Beyond this range, the fringe pattern corresponds to the transducer stage only of the generalized measurement system, Fig. 11-12.

● Optical flat methods are unexcelled for the examination of surfaces. They are not as important for comparison measurement as they once were because of the development of the high-amplification, electronic comparator instruments. These instruments are more rapid and require relatively less skill than optical flats.

● The use of optical flats requires considerable manipulation. This results in a serious heat transfer problem. The flats themselves do not heat rapidly. Unfortunately, this also means they normalize very slowly. When measuring to a discrimination of one band or 11.6 *mike* the heat transfer to steel parts can quickly render the measurements nearly valueless.

● On the other hand, there are fewer variables to cause errors with the optical flat method than with any other method of comparable precision and accuracy. If care is taken to assure uniform wringing intervals, and if ample time is available for normalizing, this method provides greater reliability than comparable methods.

● In an attempt to have the best of both worlds, interferometers are rapidly coming into industrial use. They have long been a mainstay of dimensional measurement in the physical sciences. Their principle is the same as optical flat measurement. The mechanical design minimizes the time-consuming manipulation. And because of their greater stability, magnifying optical systems can be used which permit resolution over longer distances and discrimination to fractions of bands. Now the use of lasers has greatly extended the potential range and discrimination of the interferometers.

TERMINOLOGY

Light	The electromagnetic energy radiation in frequencies to which the human eye is sensitive as well as those of slightly longer and slightly shorter wave lengths.
Light wave	The pulsation in space that transmits light energy.
Wave front	An advancing pulse of energy.
Ray	An imaginary line perpendicular to a wave front.
Ray tracing, geometric optics	The study of optical systems by means of rays.
Interference	The interaction of energy pulses when two rays are brought together.
Fringe bands	The alternate light and dark stripes resulting from interference.
Fringe pattern	The general appearance of a group of fringe bands.
Sine wave	The graphical form of the energy pulses along a light ray.
Phase	The relation of a sine wave to time.
White light	Light made up of a combination of colored light wave lengths. Various sources differ in proportion.
Monochromatic light	Light consisting of one wave length (or color) only.
Krypton 86	The gas which when electrically excited emits a very stable wave length of light as the basis for the international standard.
Helium light	The light emitted by excited helium. A practical source of mono-chromatic light.
Flats	Reference surfaces, usually small, with very nearly flat surfaces.
Optical flats	Flats of quartz, glass, sapphire or other transparent materials.
Toolmaker's flats	Flats of steel or other opaque material.
Working surface	The finished surface or surfaces of a flat.
Reference flats	Flats with working surfaces of highest precision. Used for calibration of master flats only.
Master flats	Flats with sufficiently accurate surfaces to calibrate working flats.
Working flats	Flats used for most routine inspection.
Coated flats	Flats with working surfaces treated to be less reflective than untreated surfaces. Provide clearer fringe patterns for most work.
Parallel separation planes	A concept that simplifies measurement with optical flats. It consists of a set of planes parallel to the working surface and one-half wave length apart. Their intersection with the part is seen as fringe bands.
Air wedge configuration	Fringe pattern that generally results when an optical flat rests on a part. Tested by applying force.
Hole	A low area usually about as long as it is wide.
Trough	A low area usually longer than it is wide.
Peak	A high area usually about as long as it is wide.
Crest, ridge	A high area usually longer than it is wide.

Fig. 14-48   The terminology for measurement with optical flats sounds like physics. It is applied physics and finds equal application in both science and industry.

QUESTIONS

1. Why are light waves used for measurement?
2. How can light be a standard?
3. What is an optical flat?
4. Why are light wave measurements becoming increasingly important?
5. When does the need for fine discrimination increase?
6. Who was responsible for creating a standard of length (light wave) that was also a unit of length?
7. What is light?
8. What is the present accepted light wave standard?
9. What is interferometry?
10. What is monochromatic light?
11. Why is monochromatic light desired for measurement purposes?
12. What light source is most commonly used for practical shop work?
13. What general patterns of fringe bands are usually found?
14. What do straight parallel bands indicate?
15. What do curved bands show?
16. Are concentric circles high or low areas?
17. Can gage block lengths be checked using fringe bands?
18. Why are temperature changes and cleanliness more apparent when using optical flats than with most other measuring equipment?

DISCUSSION TOPICS

1. Explain the use and application of the optical flat.
2. Explain the reason why fringe patterns are obtained.
3. Explain why light is actually a wave motion.
4. Explain how light wave motion takes the form of a sine curve.
5. Discuss phase shift and its implication concerning measurement with light.

> *The language common to all that ties a product together is the language of the metrologist. Not too long ago this was relatively simple: a thousandth of an inch, minute of arc, polished or ground surface, degrees of temperature, mass, hardness. All these were easily and completely expressed by the draftsman on his print, and understood by the machine operator and inspection department, maybe not as the designer meant them to be understood, but with perfect agreement arrived at among the shop personnel. . . . a little juggling to meet the print specifications, and the products were built.*
>
> Arnold W. Young
> Engis Equipment Company

Reference Planes
The Point of Known Return

From the beginning it was emphasized that every linear measurement starts at a reference point and ends at a measured point. Many of the instruments, met thus far, contained their own reference points. These were usually small planes in the form of anvils, bases or jaws. When an auxiliary reference was needed, we simply called it a reference surface and let it go at that.

Whether large or small, these surfaces are often, and quite properly called "datum planes". Now we will examine the most important one, the surface plate. That will lead us right into right angles. We will then have our standards for the most essential angles, 0, 90 and 180 degrees. They give our linear and angular measurements, points of known return. This is as essential for positional accuracy as the length standard is for linear accuracy, Fig. 15-1.

Flatness appears so basic that you might assume that we always had it. Unlike the other geometric forms such as the circle and regular polygons, *flatness is a stranger to nature.*

Its invention (or was it discovery?) marks one of the epochal breakthroughs that prepared the world for the Industrial Revolution. Before man gained command of flatness the world was very different. Mechanical devices were dependent upon swinging and turning movements. Without sliding movements even simple reciprocating devices were a maze of linkages. It appears that James Watt devoted more time to his "parallel motion" patent than he did to his initial steam engine patent.*

It was not until 1817 that Richard Roberts invented the planer, Fig. 15-2. For the first time parallel surfaces could be duplicated readily and the world of sliding motions and flat surfaces was born. These were extremely inaccurate surfaces. But who could define accuracy when there was no reference on which to base it? Attention was drawn to this critical matter in 1840 by Joseph Whitworth – the most important "drop out" of all times.

---

*Roe, Joseph Wickham, English and American Tool Builders, New Haven, Yale University Press, 1916, p. 3

Fig. 15-1 For positional accuracy we must have a reliable reference plane and a means for establishing perpendiculars to it. Shown here are the surface plate and cylindrical square.

Fig. 15 - 2 The planer invented by Richard Roberts in 1817 was the first machine tool that developed flat surfaces.

SIR JOSEPH WHITWORTH

It is impossible to estimate how much we owe to this unpleasant, crotchety denizen of the machine shops, or "engineering works" as they were known in his day. At 14 his father placed him in the office of a cotton spinner to learn the business. He was more interested in the textile machines, which he concluded were poorly designed. He ran away to London and began his career as a mechanic.

His second job was in the leading shop of the time, Maudslay & Field. That is the same Maudslay who made the "Lord Chancellor" micrometer shown in Fig. 7-2. He was "placed next to" the shop's best workman. He then decided that the surface plates of the shop were not sufficiently accurate. While the "old-timers" were still roaring with laughter, he proved it. His method was not new. His technique was. Far more important, however, was his understanding of the importance of the task.

The romanticists like to wrap up history in neat little bundles. They make it sound as if the international length standard evolved directly from the Egyptian cubit, that Eli Whitney single-handedly invented mass production, that Johansson invented gage blocks, and that Whitworth invented the three-plate method for generating flatness. In our preoccupation with invention, far greater contributions are overlooked. Whitworth's case demonstrates this.

The working of three plates against each other was already the established practice when Whitworth became a mechanic.* It appears that this was done by lapping with abrasive particles, and probably evolved by trial and error. It appears that Whitworth substituted scraping for lapping.** That is the removal of minute layers of metal by drawing a hardened cutting edge along the surface. This permitted Whitworth to perfect the small areas of the plates that were missed by the overall lapping. Was this invention? Was it of any great importance?

Whitworth knew its importance. He perceived quite correctly that there could be no precision manufacture without precision measurement, and that there could be no precision measurement without true reference planes. This remained a consuming interest of his. At the age of 30 he opened his own shop. His eventual partnership with Sir William Armstrong founded the firm with the dubious distinction of being one of the world's greatest arms

Much of Whitworth's success can be attributed to the perfection of his reference surfaces. The superiority of his machines was called to the attention of the world at London Exhibition of 1851. Then in 1856 he delivered a paper to the Institute of Mechanical Engineers emphasizing the importance of flatness and systemizing its attainment. He stated, "a true surface instead of being in common use, is almost unknown".

Before discussing the method, Whitworth's other enormous contributions should be mentioned. It was Whitworth who recognized the advantages of end standards over line standards. He recognized that the sense of feel when magnified could be applied far more easily to most practical work than eyesight. He found that he could attain sensitivity in millionths of an inch by the falling of a tumbler held by friction between parallel planes. He built a measuring machine and made this comment about it: "We have in this mode of measurement all the accuracy we can desire; and we find in practice in the workshop that it is easier to work to the ten-thousandth of an inch from standards of end measurement, than to one-hundredth of an inch from lines on a two-foot rule". And that was over 100 years ago!

It was Whitworth who recognized that huge masses of metal did not assure either strength or stability. He revolutionized machinery construction by the use of hollow frame members, rounded corners, smooth surfaces and even by the addition of paint where appropriate. He is best known for his standardization of the screw thread. The standard thread system in Great Britain is still known by his name. He is credited also with the origination of plug and ring gages. The author has been unable to trace this reference, however.

His apologists excuse his bad disposition on the years of battling with the bureaucrats in the British government. He had developed clearly superior methods for mass slaughter that nearly of all the civilized (sic) world had adopted except his own lethargic government.

*Roe, English and American Tool Builders, p. 44
**Roe, English and American Tool Builders, p. 99

If you have ever battled a bureaucrat in or out of government you will understand. Whitworth was the last of the great British machine innovators. With his passing, the leadership moved to a small remote area known as New England. However, if the available technical writing is any indication, our English brothers got the message about flatness and measurement better than we have. So let's look at the matter closer ourselves.

DEGRADATION OF WORKMANSHIP

This begins with what appears to be a contradiction. The plane surface can be developed to a high order of accuracy because we have no standard for it. The reason this is true is known as "degradation of workmanship".* This applies to all machine work. The product is not as precise as the machines. The machines are not as precise as the gages. The gages are not . . . ., etc.

This can be shown by an example, albeit, a greatly oversimplified one.** In the next chapter you will meet dividing heads, which are used to divide a circle into spaces. The principal parts of the dividing head are a spindle, a worm wheel, the worm and an index plate, Fig. 17-51. The indexing error will depend upon the interaction of separate errors in these parts. Suppose it is set up for cutting a gear. At some places the errors will cancel out, but at others they will add and be maximum. Suppose the resulting gear is used as a change gear on a lathe and that a thread is cut. Now the errors will have been extended from the dividing head to the thread picking up some additional lathe errors. The thread will have greater errors than any of the previous ones.

---

*Halsey, Frederick A., Methods of Machine Shop Work

**Pennington, E. Willard, "The Accuracy and Geometry of Precision," Know-How Library Series, Vol. 2, Cleveland, Huebner Publications, Inc., 1959, p. 2

Now we make the full circle. Suppose this thread is a worm for another dividing head. Think of the errors that this head will have. Isn't that degrading? The more times we repeat this cycle the worse it gets. Conversely the fewer repetitions, the better. Hence, the generalization that the less work that is performed the better. This should come as a profound relief to many readers who assumed it all along. Carried to its extreme, the greatest accuracy is obtained when there is no work. Work in this sense means the reliance on previously manufactured articles. Thus a method to make plane surfaces that does not require any previously manufactured articles should be capable of theoretical perfection.

THE THREE-PLATE METHOD

In Whitworth's time, surface plates were chiefly cast iron. Granite has now replaced cast iron, and various methods that you will meet later have replaced the three-plate method for generating flatness. The technique is of fundamental importance. Much of it actually can be used in practical problems in toolmaking and machine construction. Furthermore, it provides an example of the *reversal technique* or *reversal process* of which you will hear much more as we go on.

First, three plates are cast using a ribbed construction to provide rigidity under load without excessive weight. They are then rough machined along their edges and top surfaces.

Next comes normalizing. This usually consists of being stacked outside for a year or more so that temperature changes relieve the stresses from the casting and rough machining. They are then finish machined. The flatness at this point depends upon the machine tool used for the finishing. The plates are marked, #1, #2, and #3; and we are ready to begin scraping, Fig. 15-3.

PRIOR TO SCRAPING

Fig. 15-3 Greatly exaggerated, this shows the possible surface conformation of three plates after finish machining.

Step One – Plates #1 and #2 are scraped together. First, #1 is blued by wiping on a thin coat of Prussian blue (or special colors made for this purpose). The plates are placed together and the bluing comes off of #1 onto #2 showing where they contact. These spots are scraped. Then #2 is blued, #1 is wiped clean and the procedure is repeated. *Note that both plates are scraped.* This is continued only until there is reasonable agreement; because as yet we have no way of knowing if we are scraping the plates more flat or less flat. As shown in Fig. 15-4, one could very well be convex and the other concave.

Step Two – Plates #1 and #3 are now scraped together. This step requires an important change. Plate #1 is made the *control* plate. Bluing is applied to #1 only and *metal is scraped only from #3*, Fig. 15-5. We now have some agreement but still none is known to be flat. Plates #2 and #3 now have the same

amount of error as #1. The errors in #2 and #3 are in the opposite direction as #1. Here is the beauty of the *reversal process.* When #3 is turned over to place it on #2 the relative direction of its error is reversed. Thus the error is doubled. Plates #2 and #3 are out of agreement by twice the amount that the control plate #1 is concave or convex. Remember that in actual practice you do not know up from down.

Step Three – Clearly the next move is to get plates #2 and #3 paired off, and reduce the doubled error. To do this both of them are scraped. It is impossible to know that we are removing metal uniformly from each. Therefore there is no point in carrying this past general agreement. The score is now as shown in Fig. 15-6. Still nothing is flat but #2 and #3 are known to be flatter than #1. Therefore, we will now make #2 the control plate.

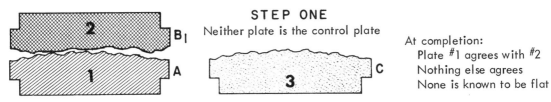

STEP ONE

Neither plate is the control plate

At completion:
Plate #1 agrees with #2
Nothing else agrees
None is known to be flat

Fig. 15-4   This step is carried only far enough to get general agreement between #1 and #2.

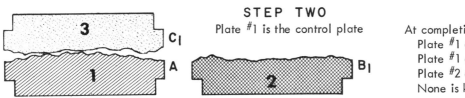

STEP TWO

Plate #1 is the control plate

At completion:
Plate #1 agrees with #2
Plate #1 agrees with #3
Plate #2 does not agree with #3
None is known to be flat

Fig. 15-5   At the completion of this step both #2 and #3 will have picked up #1's error.

STEP THREE

Neither plate is the control plate

At completion:
Plate #1 does not agree with #2
Plate #1 does not agree with #3
Plate #2 agrees with #3
None is known to be flat
#2 and #3 are known to be
flatter than #1

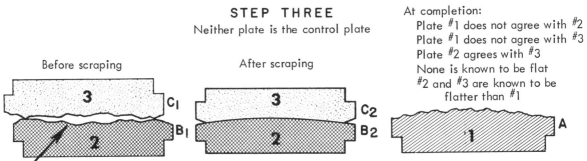

Before scraping

After scraping

Twice the error
from plate #1

Fig. 15-6   By scraping some of #1's error off of #2 and some off of #3 we get closer to flatness for these two plates.

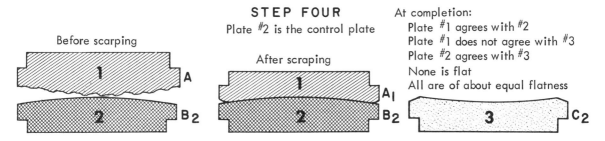

STEP FOUR
Plate #2 is the control plate

Before scarping

After scraping

At completion:
Plate #1 agrees with #2
Plate #1 does not agree with #3
Plate #2 agrees with #3
None is flat
All are of about equal flatness

Fig. 15-7   At the completion of this step the plates may be of approximately the same flatness but one is convex while two are concave.

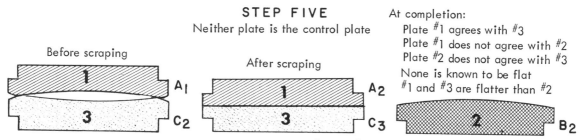

STEP FIVE
Neither plate is the control plate

Before scraping

After scraping

At completion:
Plate #1 agrees with #3
Plate #1 does not agree with #2
Plate #2 does not agree with #3
None is known to be flat
#1 and #3 are flatter than #2

Fig. 15-8   The first time plates #1 and #3 were brought together, only #3 was scraped. This time both are.

Step Four – Removing metal only from #1 now makes it approximately the same as #3 and with much less flatness error than it had before, Fig. 15-7.

Step Five – In this step both #1 and #3 are scraped to remove the doubled error, Fig. 15-8. Each time any two plates are brought together the opposite procedure is used from the previous time. In step two, #1 and #3 meet. That time #1 was the control plate and only #3 was scraped.

Step Six – Plates #2 and #3 meet up again in Fig. 15-9. The previous time in step three neither was the control plate, and both were scraped. This time only #2 gets the treatment, #3 already being in agreement with #1.

This procedure is then continued until the desired degree of agreement is reached. As the work proceeds, dry red lead is substituted for the greasy blue stuff. At the final stages only the shine of rubbed high spots are scraped. When the three plates are fully interchangeable perfect flatness has been achieved. This must be so because of a basic axiom: *Two things that are equal to the same thing are equal to each other.*

Perfect flatness is never achieved, however. Philosophically it is impossible to reach perfection. On a practical level, it is not practical. In the first place the cost of further improvement soars as we approach perfection. Is a Cadillac *worth* twice the price of a Chevrolet? In the second place, nearly

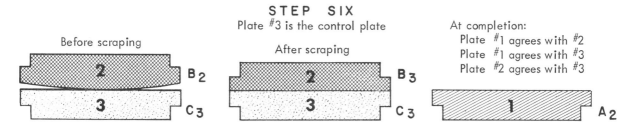

STEP SIX
Plate #3 is the control plate

Before scraping

After scraping

At completion:
Plate #1 agrees with #2
Plate #1 agrees with #3
Plate #2 agrees with #3

Fig. 15-9   Although this step brings the three plates into approximate agreement, it does not mean that they are flat. The entire procedure is repeated until the desired flatness is obtained.

perfect surfaces wring together which makes handling difficult. At each successive scraping more and more spots "drop in". The plates usually are considered satisfactory when all agree with approximately 45 per cent of their surfaces in contact,* Fig. 15-10.

Exactly the same procedure can be applied to straightedges; because, however narrow, they must have some width. Thus they are actually elongated surface plates. Surface plates and straightedges provide us with references and the standards for 0 and 180-deggree angles.

PERPENDICULARITY

Next only to flatness, the most important reference is perpendicularity. In a broader sense this is simply the right angle. Like flatness it can be generated by the interchange of three unknowns. However, it is just slightly degenerate. While flatness can be generated without reference to any previously performed operations, generation of the right angle requires a previous 180-degree angle or *straight angle*. Therefore, the right angle will always contain some error from the straight angle.

Right angles in their most familiar form are known as *squares*. In the next chapter you will meet quite a number of these that are used as standards. In other words, they are used to judge the squareness of other things. Here we are concerned with squares that are primarily reference surfaces. The most ap-

Fig. 15-10 A well scraped surface is a thing of beauty. Scraping represents a very high skill. Its chief use today is for machine surfaces. The minute irregularities retain lubrication and for that reason are used for ways and sliding surfaces. Even in this role it is being replaced by hardened and ground surfaces. *(The Challenge Machinery Co.)*

parent difference is that they have greater areas on their contact surfaces.

It will not come as a surprise to find that Whitworth had his scraper in this, too. If he did not originate the method for generating squares, he at least formalized it. The making of master squares begins about the same way: casting, rough machining, normalizing and finish machining. The machined squares are then scraped in pairs in the same order as used for plane surfaces, Fig. 15-11. As before, it is continued to the desired degree of precision. Unlike the previous case, the generation of squares is dependent upon previously performed work, the plane surface.

*Judge, A.W., Engineering Precision Measurement, London, Chapman & Hall Ltd., 1957, p. 298

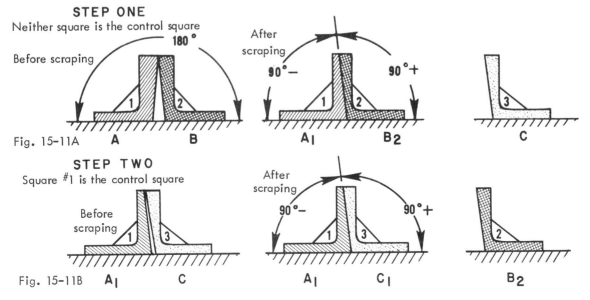

STEP ONE
Neither square is the control square

Fig. 15-11A

STEP TWO
Square #1 is the control square

Fig. 15-11B

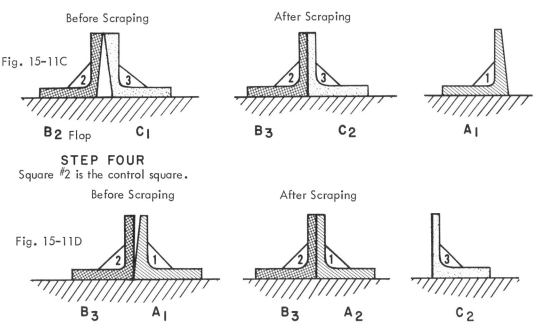

STEP THREE
Neither square is the control square.

Fig. 15-11C

Before Scraping

After Scraping

B₂ Flop   C₁   B₃   C₂   A₁

STEP FOUR
Square #2 is the control square.

Before Scraping

After Scraping

Fig. 15-11D

B₃   A₁   B₃   A₂   C₂

Fig. 15-11   Master squares can be generated by much the same technique as was used for generating plane surfaces.   However, the accuracy of the resulting squares is dependent upon the accuracy of the reference surface upon which the generation took place.

Figure 15-12 shows that the generation could be performed on a convex surface. In this case, everything would match up fine — but the results would not be squares.

In all of these examples we have been solving simultaneous equations. The three surfaces in each case are x, y and z. The equations are solved when x = y = z. This results when their coefficients are all equal. The coefficients were subtracted by the scraping away of thicknesses of metal. As far as the "math" is involved, it matters not whether we are dealing with acres of terra firma, cast iron, or hardened steel. In fact, we could turn the whole thing inside out and the theory would still apply.

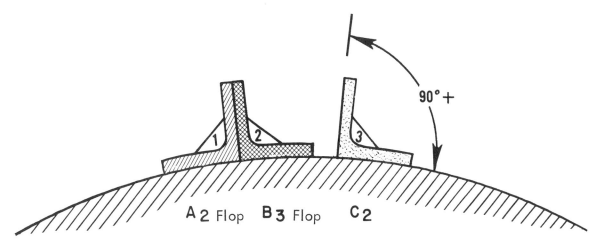

90° +

A₂ Flop   B₃ Flop   C₂

Fig. 15-12   The generation of squares picks up the error of the reference surface as this example shows.

That has been done in Fig. 15-13 in which an inside square is calibrated against a master square. The surprising thing is that the *reversal process* permits us to develop the master square at the same time that it is being used for calibration.

In A of the figure, the rough male square has been placed in the inside square. Its errors merge with the error of the inside square but metal is removed until it fits. It is turned 90 degrees at B. Of course, it does not fit until again adjusted. This is repeated at C. Then on the next 90 degree turn, at D, the divergence x represents four times the combined errors. Now both the magnitude and direction of the error is known so it can be corrected in the inside square. Following the correction the four steps are repeated. The remaining error is again revealed by its fourfold magnification. This is repeated until the desired accuracy is reached. Note that both the male and female squares are corrected simultaneously.

Suppose that a second inside square is desired, would the entire procedure be repeated? Not even in Whitworth's day. The finished male square would be used to correct directly the new inside square. Only when wear or doubt enters in would the scraping begin from scratch.

All of this is somewhat outmoded today. Commercial surface plates and the various right-angle plate accessories are generally granite instead of cast iron. Autocollimators are used to establish the geometry and lapping processes are used in place of handscraping.

GENERATING AN INSIDE SQUARE

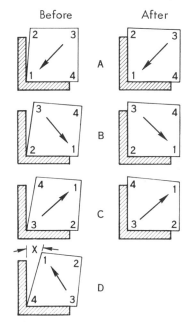

Fig. 15-13 The same general technique can be used to simultaneously generate an inside square and a master square. x is the error magnified four times.

Cast-iron surface plates are still available. According to the "old-timers" they are supposed to have some superiority. Offhand, the author cannot think of any. Various, scraped cast-iron references are shown in Fig. 15-14. The technique, however, is invaluable and finds many other applications. In the next chapter, for example, the method for generating inside squares will make a much modified appearance for the generation of lapped, precision squares.

Fig. 15-14 These are typical references made of hand scraped cast iron. They function as supports as much as reference surfaces and, therefore, are heavy and rugged. *(The Challenge Machinery Co.)*

GRANITE SURFACE PLATES

Proof of the superiority of granite surface plates seems to be their general acceptance. It is difficult to find a manufacturing plant that is not using them. In the plants devoted to the newer technologies, such as missile hardware, they are used exclusively in preference to cast iron. Metal is rarely used for separate reference surfaces except those 6 in. in diameter and smaller, such as toolmaker's flats. References that are built into instruments and equipment are usually metal; but here, too, granite is rapidly coming into use. But, if history teaches anything, it is that general acceptance does not prove merit. Therefore, let's examine granite more closely.

There are only two things certain about granite surface plates. They are an American contribution to metrology and they are overwhelmingly superior to cast-iron plates. Beyond that, it is difficult to say anything that will not provoke an argument from one or another of the granite partisans. Most manufacturers of measurement instruments are so conservative that it is difficult to get them to admit an important feature, even when asked. Well, this does not apply to the granite surface plate crowd. In their scramble to "debunk" all competitors, real facts are hard to find. It is not even possible to state with any conviction who first used "stone" surface plates. Commercial use was started about 1942.

Three major controversies flourish. There are several kinds of granite. There are several approaches to flatness. And there are various functional features, such as ledges and inserts. Which are best? The author will present the conflicting claims as honestly as he can. However, never in human activity can bias be overlooked. You should know that the author spent many years with a manufacturer of granite surface plates and has patents in this field.

The most important advantages of granite surface plates are closer tolerances and lower prices. The next most important feature is their noncorroding and nonrusting property. They are not subject to contact interference. They are nonmagnetic, hard, stable, long wearing, easy on the eyes, easy to clean, and have exceptional thermal stability. Now let's consider what these features mean to reliable measurement.

Fig. 15-15 Granite boulders are sliced into slabs by wire saws. The slabs are then inspected and suitable ones become surface plates. *(Continental Granite, Inc.)*

It is a small marvel of modern technology that granite surface plates can be produced to flatness in *mikes* (millionths of an inch) at such remarkably low prices as compared to cast-iron and steel reference surfaces. Boulders are quarried in the usual fashion. They are then cut into slabs with wire saws. These are giant, ganged band saws using twisted steel cable charged for cutting with abrasive slurry, Fig. 15-15.

Only a small portion of the slabs are suitable for surface plates. These are separated by diamond-edged circular saws. The slabs then go to honing beds where they are roughly surfaced to within a few *mil* (0.001 in.). The next operations are performed in temperature-controlled rooms, Fig. 15-16. The procedures vary. Usually the rough plates are

(Continental Granite, Inc.)

Fig. 15-16 Surface plates are carefully lapped in constant temperature rooms using autocollimators to map contours of the plate profiles.

surfaced on rotary lapping machines that bring them very close to the desired flatness. Autocollimators then are used to draw a profile of their surfaces. Area lapping and rechecking with autocollimators successively eliminates high spots until the plates are within tolerance. Note that although Whitworth's method played no part in this, it laid the foundation upon which a flatness standard could exist.

Two advantages of granite surface plates come from their tolerance to water. They can take it or leave it alone. It does not hurt them. They do not rust or corrode. Therefore, they can be scrubbed scrupulously clean. Later we will discuss care and maintenance. The denser granites absorb so little moisture that they do not cause fine surfaces to rust, even after prolonged contact.

In contrast, cast-iron plates not only rust but they induce rusting on iron or steel surfaces placed upon them. The finer the surfaces, the greater the problem. Overnight exposure can be too long. This is remedied by oil -- and eternal vigilance. Oil collects dirt. Dirt is the major enemy to reliable measurement. It adds measurable error; and, being abrasive, it accelerates wear. While it is a good practice to keep all precision surfaces covered, cast-iron ones must be.

The thermal stability is a little discussed feature, but important. Granite plates have great thermal stability. In contrast, cast-iron plates respond quickly to drafts, bright lights, and warm torsos. The reverse of this, of course, is that they require much longer to normalize than cast-iron plates. In some circumstances this can nearly undo the nonrusting advantage. A cold granite surface plate requires a long time to reach ambient temperature. If the humidity is high, moisture will condense on its surfaces until it warms up. This will rust fine surfaces upon the plate, of course. Because of the thermal stability, this may go on for days. There is no advantage that does not have a disadvantage. For reliable measurement this must never be forgotten.

The term *contact interference* refers to surface conditions that are damaging to measurement. They may be metrologically or functionally damaging, often both. A spot of paint or patch of protective coating will add measurement errors. In a cast-iron or steel plate, a scratch or a dent displaces material above the datum plane. This material adds errors; and if in the form of a burr, it tears up mating surfaces. A much touted advantage of granite plates is that this cannot happen. A scratch or impact knocks out crystaline particles which can then be dusted off, Fig. 15-17. If an advantage of granite surface plates is superior hardness, then broken off particles must also have superior hardness and be murderous to mating surfaces. Furthermore, the remaining grooves and cavities collect abrasive dust and take their toll as well. *Thus, use of granite plates is no substitute for care and cleanliness.*

It is frequently stated that one of the disadvantages of granite plates is that you cannot wring to them. That simply is not so. They do not wring nearly as strongly as metals and many other materials. But, a granite plate that will not show some signs of wringing is poorly finished.

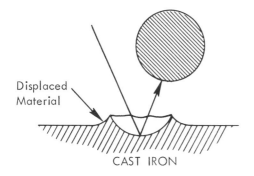
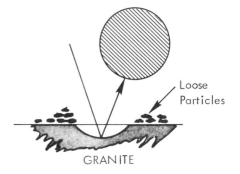

Fig. 15-17 An impact on a cast-iron surface plate raises the displaced material around the edge of the crater, thus disturbing the reference plane. A similar impact on a granite plate loosens particles which are wiped off. The reference plane is not disturbed.

This is a mixed blessing. Wringing does not assure a predictable contact error for very precise measurements. Such measurements, however, are usually made on comparators which incorporate their own reference that has been lapped to a light band or less. Most surface plate measurements require the shifting about of the work and instruments. Unwanted wringing can be a nuisance in this case and it is rarely required because of contact error considerations. Granite plates provide *casual contact*, Fig. 13-18, which is precisely what is wanted for surface plate work.

Fig. 15-18 Ledges were quite logical on cast-iron plates. They were carried over to granite plates even though there is no good reason for them.

FUNCTIONAL CONSIDERATIONS

Ledges on surface plates are one of the controversial subjects. The use of ledges on the edges of plates is so ingrained that 4 ledges have come to denote a laboratory grade plate, 2 ledges an inspection grade, and no ledges a toolroom grade. To trust a plate because of its ledges is shamanistic, and has as much place in metrology as voodoo drums. Why, then, are there ledges?

As we all know the reason the engine is in the front of an automobile is because once upon a time the horse was up there; the reason for ledges on surface plates is similar. The logical design for a ribbed cast-iron plate provides a ledge around the outer edges, Fig. 15-10. The makers of the first granite plates slavishly copied this, and so we are stuck with the custom today. The old-timer has his usual glib explanation for this -- clamping. This is not good enough. Edge clamping limits wear to the edges of the plate and invites clamping error, Fig. 15-19.

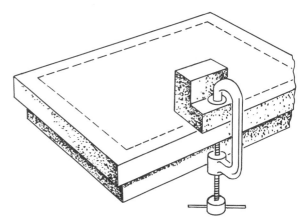

Fig. 15-19 Edge clamping localizes wear, while the reverse is desired.

Most important, edge clamping is not necessary and not economical. The ledges are cut with diamond saws. This is an expensive operation. More practical and much less costly is the use of threaded inserts imbedded in the surface. These can be replaced anywhere that best serves the intended use of the plate. Quick acting clamps that attach to the inserts are available, Fig. 15-20. The T-slot seen in this illustration is also a means to eliminate ledges. Whether a T-slot or inserts should be chosen depends entirely upon the application.

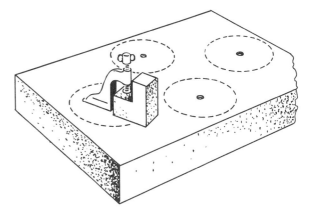

Fig. 15-20 Threaded inserts in the plate permit quick-acting clamps to be used. 360 degrees of the plate around the insert are used, thereby spreading wear over a sizable area.

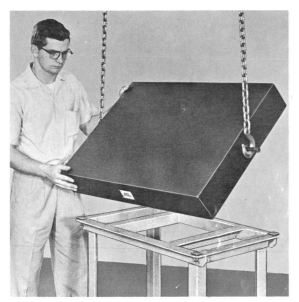

Fig. 15-21 Double-faced plates have twice the life of single-faced plates. The faces may be of different accuracy grades.

Fig. 15-22 A large variety of granite accessories is available. (DoALL Co.)

One of the important modern advances is the double-surfaced plate, Fig. 15-21. This permits a plate to have two usable sides. At each end of the long axis of the plate a large threaded insert is inserted. Whenever it is desired to turn the plate, standard bolts are inserted for lifting trunnions.

First of all, the main reason given for double-surfaced plates is the very real saving. For a small company this permits a high tolerance side to be on reserve for inspection while the lower tolerance side is in use for toolroom and other work. For larger organizations, it reduces the frequency that the plate must be removed from use for refinishing.

Then, there is a metrological advantage to this that is not mentioned in the ads. The usual rough-finished lower surface has more surface area than the top. Therefore, it is more susceptible to temperature change, causing a tendency to distort. The finished lower surface minimizes this.

Most accessories that were developed for cast-iron surface plates - plus a few space-age innovations, are now available in granite. In most cases they furnish both support and reference, Fig. 15-22. The straightedges are an exception in that they are generally used for reference only.

HOW FLAT IS FLAT?

One of the raging controversies concerns flatness. To eliminate any sophomoric thinking note Fig. 15-23, and then try to come up with your own definition. Webster is no help at all. As a result of this confusion, each manufacturer claims the particular configuration that results from his method is really true flatness, and everybody else is either stupid or lying. Unfortunately, each has amassed impressive documentation to support his view.

IMPORTANCE OF FLATNESS

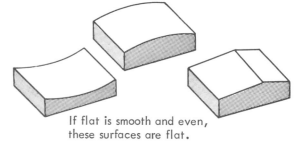

If flat is smooth and even, these surfaces are flat.

This surface is not smooth, but is it flat?

Fig. 15-23 These exaggerated drawings show that a lot of geometry ordinarily gets confused with flatness.

DEFINING A PLANE

Any 3 Points

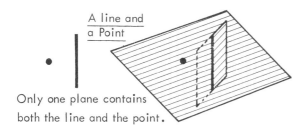

Only one plane contains
all three points.

A line and
a Point

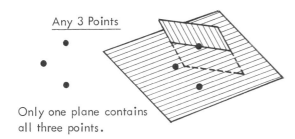

Only one plane contains
both the line and the point.

Two Intersecting
Lines

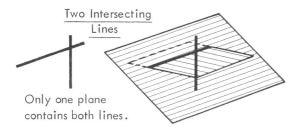

Only one plane
contains both lines.

Fig.15-24 The small planes cutting the large planes
show that only one in each case can contain all of
the elements.

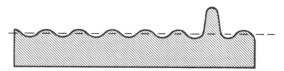

If the highs and lows are averaged, this plate has
perfect flatness.

If the maximum deviation from the mean is used,
the same plate has a flatness of x.

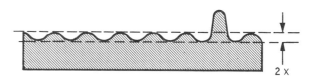

If total deviation is used, the flatness becomes 2x.

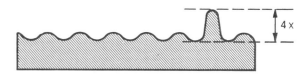

If one point varies, does that make the expression
of flatness double?

Fig. 15-25 This illustration shows that the statement
of accuracy must specify precisely how the accuracy
is defined.

First of all, consider the plane all by itself. Visualize three points in space -- any
three points. Only one plane can cut through
that space and contain those three points. A
line is an infinite series of points. Therefore
the three points could define three lines.
Each line could have any length greater than
the distance between the points that defined
it. Thus, to define a plane you must have
three points, or one point and one line, or two
lines that intersect, Fig. 15-24.

In defining the plane we have dealt entirely
with concepts that have no thickness, - points
and lines. The plane itself has no thickness.
Therefore, it has no up or down, top or bottom, front or back. This theoretical plane
would be ideal as a reference plane. Unfortunately, in the real world we must add sub-
stance to the theoretical plane in order to use it.

Because we cannot make anything perfectly, we must have a tolerance on the surface. This tolerance is thickness, the amount
that points on the surface lie outside of the
theoretical plane. The size of the tolerance
is limited by our skill in producing flat surfaces -- and by the way we choose to express
it. The latter problem is great because the
reference plane itself is imaginary. No measurements can actually be taken from it. Our
knowledge of it is only through our measurements of points, none of which we know are
exactly in the plane. Therefore, in every
statement of surface flatness there is a
purely arbitrary selection of reference plane.
Figure 15-25 shows that one surface plate
may be read as zero deviation, x deviation,
2x or even 4x.

SIZE AND ACCURACY

Size				Work surface accuracy	
Work surface		Grade A thickness (minimum)	Grade B thickness (minimum)	Grade A tolerance plus or minus	Grade B tolerance plus or minus
Width	Length				
Inches	*Inches*	*Inches*	*Inches*	*Inch*	*Inch*
3½	4	1	1	0.000025	0.0001
8	12	3	3	.000025	.0001
12	12	3	3	.000025	.0001
12	18	4	4	.000025	.0001
18	18	4	4	.000025	.0001
24	24	4	4	.000025	.0001
24	36	6	5	.000025	.0001
24	48	8	6	.000050	.00015
36	48	8	6	.000075	.00015
36	72	12	10	.00015	.0003
48	96	14	12	.0002	.0004
48	144	24	20	.0004	.001

Fig. 15-26   This table is from Federal Specification GGG-P463b.
Note that the tolerance widens as the size increases.

The Federal Specification GGG-P-463b applies to granite surface plates. Table 1 from the specification is shown in Fig. 15-26. Grade A is usually called *inspection grade.* Grade A plates are spoken of as "50-millionths" plates although the tolerance widens for large sizes. Grade B is *toolroom grade* and spoken of as "100 millionths". Reflecting the upgrading of measurement, grade B is more and more being referred to as *shop grade*, and most manufacturers supply grade AA for the highest accuracy work. These plates are called *laboratory grade* or "25-millionths" plates.

The federal specification is clear but not precise enough to quell the conflicting claims. It states that no point may deviate more than the specified tolerance from a mean plane. Clearly you must find the mean plane before you can measure deviation from it. This is no mean trick and rates the major portion of the later Chapter devoted to surface plate calibration.

Figure 15-27 shows that two plates can lie within tolerance and yet have very different configurations. The one on the right is a *crowned* plate, a type advocated by some leading manufacturers. Both the crowned-plate and the anticrowned-plate people make lavish claims for their own views and spare no effort in running down the opposition. The author will try to present each side as fairly as possible and then throw in his own views.

It is claimed that crowned plates have better repeatability, accuracy under load and wear life than wavy plates. They undoubtedly do -- but what does this mean? Are these important considerations?

First of all consider repeatability. As pointed out much earlier in this text, repeatability is an indication of precision not of accuracy. It is a useful test but valueless, in fact harmful, if erroneous measurements are being repeated. Even with that proviso the crowned plate shows up much better than a wavy one, Fig. 15-28.

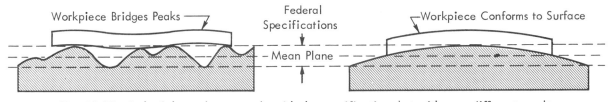

Fig. 15-27   Both of these plates comply with the specifications but with very different results.

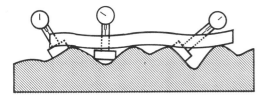 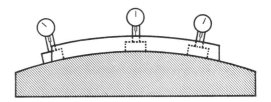

Fig. 15-28   The wavy plate certainly results in poor repeatability as compared to the crowned one.

Accuracy under load refers primarily to deformation of the part, not of the plate. In the figure the part conforms closely to the plate.   This is probably quite true for large parts reclining directly on the plate. Bear in mind that although we speak of the plate as crowned the radius of curvature is very large. It is, in fact, 45.5 miles for a 24x24-in. plate with *50-mike* (0.000050-in.) tolerance.*

Wear life is determined by the amount of material that can be worn away before the surface passes through the lower limit of the tolerance.   The crowned plate has a major portion of its surface near the high side of the tolerance. Clearly this is a plate that can take considerable wear before being under the limit. But how does a plate wear? Is this additional material in the area that will wear? What happens if the edges receive most use?

*Rahn, R. J., "Check Surface Plate Flatness by Applying the Geometry of the Sphere," Tooling & Production, Cleveland, Huebner Publications, Inc., August 1960, p. 42

The anticrowned-plate partisans come up with some cogent arguments of their own. They begin by yelling "foul", that the wrong comparison has been made. Just because a plate is not crowned does not make it wavy, and that it is crowned does not mean it is free of waves. They state that the comparison in Fig. 15-29 is more nearly like it. Some waviness remains in either case, and they suggest that it is better to have this waviness without the added error of the crown. They go on to claim that the reason for the crown is not that anyone actually wants it, but that it results from the use of large circular laps instead of more time-consuming area lapping.

The geometric portion of the argument is that curvature does not alter measurement as long as Abbe's law is obeyed, as it appeared to be in Fig. 15-28. The other side asks us to take a side look, Fig. 15-30. This shows overhang. The overhang magnifies any error contributed by the curvature. Then as height above the plate is increased this error is accentuated.

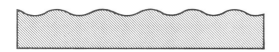 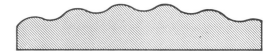

Fig. 15-29   This is the comparison that the anticrowned-plate people feel should be made.

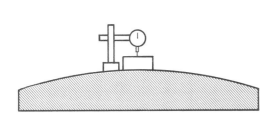 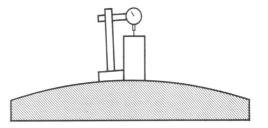

Fig. 15-30   The effect that a crown has on measurement is increased when either overhang or height are increased.

The crowned-plate advocates have an answer for this. They state that all deviations from absolute flatness will be concave or convex curvature. It is better to have one uniform convex curve for a surface than a haphazard assortment of curves.

The author's personal opinion is that for all practical purposes the matter is academic. If the highest measurement practices are adhered to, the errors contributed by the surface plate will be of secondary concern. Both sides must be approximately equal because the leading proponents of each make the same guarantee. A measurement may be repeated anywhere on the plate with repeatability within the plate tolerance. If measurements of the highest reliability are to be made, the most nearly flat surface available should be preferred. In measurement there should be no rationalization. Manufactured items, including surface plates, must be produced at prices that preclude the greatest perfection of their surfaces. This fact is only permissible when it is not concealed by confusing claims.

Similar controversies rage over the granite used. Surface flatness claims used sufficiently vague terms to permit legitimate differences of opinion, but the claims for the various granites are outright contradictory. Either black granite is harder than pink or it is not. If the Federal Specification GGG-P-463b can reliably be used as a guide, the

Fig. 15-31 Measurements taken from surface plate setups can be no more accurate than the setups themselves. The positional requirements for accuracy are one of the most important considerations. *(Northrop Aircraft)*

black granites are clearly superior to the other granites for surface plates. That still permits some artful dodging because nowhere is it specified at what degree of blackness gray granite becomes black granite. Furthermore, an oil treatment can greatly darken the gray granites. Does this make them black granite?

Summary — We Don't Scrape for Accuracy Anymore

● Many instruments do not contain their own reference points. The most important auxiliary reference used with these instruments is the surface plate. This is nothing more than a reference plane (which is imaginary) with a material backing. The measure of its perfection is its flatness.

● Flatness is a stranger to nature and did not evolve until the Industrial Revolution, which could not have reached fruition without a formal recognition of flatness. For this we owe much to Sir Joseph Whitworth.

● The less dependent work is on previously performed work, the more accurate it can be. Plane surfaces can be generated completely independent of previous work. Therefore, they are potentially among the most accurate standards known. They provide the essential 0 and 180-degree angles for use in measurement.

● The basic method used to generate flatness consists of the interchange of three surfaces. The reversal process is utilized to compound errors for their detection and elimination. The technique requires scraping and rarely is used today. The principle, however, has many applications throughout measurement.

● This same method can be extended to generate the other basic reference, 90 degrees. In this case it is not quite so accurate because it relies upon the previously created flat surface. Both inside and outside squares can be produced.

● Although they still have their loyal defenders, cast-iron surface plates are rapidly being replaced by granite ones. The granite plates have many functional advantages in addition to the fact that they are less expensive than the cast-iron ones.

● The lavish claims made by producers of granite surface plates becloud all real facts about them. One thing is certain, however. The use of ledges is anachronistic. Much better means for holding parts to plates is

provided by threaded inserts. In some instances, T-slots in the plates may be preferred.

● The definition of flatness is a highly partisan matter. Several opposing schools have developed. Their specious claims may be ignored if high-quality plates are purchased from reliable manufacturers -- and if they are recalibrated at regular intervals.

● Although government specifications only require two grades of surface plate, industrial standards require three, grades AA (laboratory), A (inspection) and B (shop). Of the types of granite available, black appears superior.

TERMINOLOGY

Reference plane Datum plane }	Plane in which reference points lie. Perpendicular to line of measurement.
Flatness	The measure of deviation from a reference plane.
Surface plate	A horizontal reference plane of sufficient strength and rigidity that measurement operations may be supported on it.
Flats	Small surface plates, usually of high accuracy.
Plate work	General term for measurements made from a surface plate.
Degradation of workmanship	Generalization that no part feature is as accurate as the process or machine that produced it. Thus, the less dependent a part feature is on previous operations, the greater the probability of accuracy.
Reversal technique Reversal process }	Method for detecting or canceling of small changes by comparing a variable with itself but with reversed algebraic sign.
Three-plate method	Technique for generating flat surfaces by the systematic working together of three surfaces.
Perpendicularity	Condition of a line in which all angles to a reference plane are right angles. In plate work, perpendicular is generally synonymous to vertical.
Square	Condition of being at a right angle to a line or plane. Also, a right angle in material form.
Master square	Square used as a reference standard to check other squares.
Scrape	Use of hardened edged hand tool to remove minute layers of metal.
Granite	A hard, igneous natural stone used for industrial purposes.
Black granite	A particular granite of exceptional density and hardness.
Ledges	Cut back lower edges of surface plate, serving no useful purpose.
Inserts	Internally threaded metal plugs cemented into surface plate for clamping.
Lapping	Abrasive stock removal process using loose abrasive particles.
Area lapping	Selective lapping of specific portions of a surface.
Ambient temperature	Surrounding temperature.
Contact interference	Condition preventing intimate contact between two plane surfaces without damage.
Crowned plate	A surface plate whose reference surface is a portion of a sphere of very large radius.

Fig. 15-32 Many terms pertaining to surface plates are familiar words that have taken on specialized meanings.

QUESTIONS

1. What is an auxiliary reference?
2. What are the two most common auxiliary references?
3. What is the usual form for the reference plane?
4. What is the geometrical characteristic of the reference plane?
5. In what important respect does this characteristic differ from other familiar geometric forms such as the circle and the regular polygons?
6. With what important advances do we associate Sir Joseph Whitworth?
7. What is the minimum and maximum number of surfaces?
8. What is the control surface?
9. Can squares be generated as accurately as surface plates?
10. What are the principal benefits of granite surface plates?
11. Does this method have other applications?
12. What are the three controversial matters concerning granite surface plates?
13. How is a granite surface plate worked to final flatness?
14. How is this process different from grinding?
15. How is the flatness checked?
16. What is meant by contact interference?
17. What happens if a cast-iron and a granite surface plate both receive a damaging blow?
18. What remedial action is required for each?
19. What type of contact is provided by surface plates?
20. What is the metrological function of a surface plate?
21. What are the functional features of a surface plate?
22. An infinite number of planes can pass through a point in space. How can all planes except one be eliminated?
23. What is the exception to this?
24. How is flatness defined?
25. What principal claim is made for crowned plates?
26. What principal claim is made by the anticrowned plate faction?
27. What is the most practical defense to the arguments surrounding flatness?

DISCUSSION TOPICS

1. Relate degradation of workmanship theory to problems in communications. Show equivalents to channel, noise, units of information. Show communication equivalents to all of the errors discussed in Chapter 11.
2. If the three-plate method is not in general use today, what is its significance?
3. Explain the difficulty in expressing surface flatness.
4. When opposing arguments are encountered in life the two expedient steps are to find out in what respect (if any) they relate to our own real needs, and to determine what safeguards are available to protect us from injury. Explain how we do this for surface plates.
5. Set forth some guidelines for selecting a reliable manufacturer of any product and show their reference to the selection of surface plates.

We are in greater danger from the blunders of our friends than from the stratagems of our enemies.

Pericles

Angle Measurement I
A Standard in Nature

All length standards are arbitrary inventions of man. This is true even of the light wave standard. Although light is a natural phenomenon, it took a bit of doing to make a length standard of it. Now we come to a standard that is not an arbitrary act of man but one that actually exists in nature – the circle.

Whether the circle is the path of an electron around its nucleus or the circumference of Aurigae, its geometry is always the same. Its parts always bear the same relationships. Here we have a standard that can be recreated anywhere, anytime. It is the standard by which angles are measured. Its universality is no wonder. It is inescapable in engineering and the sciences. Angular measurement is used in every phase of life from billiards to botany. Squares, in all their diverse forms, are the most basic of the angle-measurement instruments, Fig. 16-1.

AT THE STARTING LINE

The most basic element of geometry is the point. Give it a shove and its path is a line. Like a point, a line has no width or thickness. It does have length. Philosophers dispute this, but metrologists don't. Give a line a shove, and the area it sweeps is a plane. This, too, is highly imaginary serving only to contain lines and providing a place from which to reach.

Visualize a line with one end fixed to a plane and the other end free to move, Fig. 16-2. That is the *initial line*. Now move the line in the only way it can move, by revolving it. Any new position will result in a *terminal line*. The difference in position is an angle. Obviously, the two angles in the lower part of Fig. 16-2 are different. Describing their difference requires more than the statement of linear dimensions. The easiest way, quite obviously, is to state the portions of circles through which they sweep. This is true whether the lines are long, short, up, down, in, or out. Any angle can be expressed as

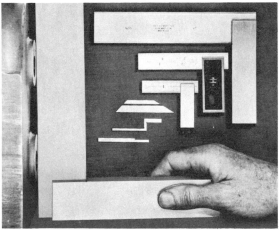

Fig. 16-1   Squares, in all their diverse forms, are the most basic of the angle measurement instruments.
(The L. S. Starrett Co.)

BIRTH OF AN ANGLE

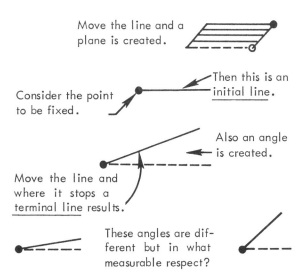

Fig. 16-2   Some elementary concepts are worth review.

353

uniform divisions of the circle. Even if the angle itself is not an exact fractional division, it is formed by combinations of uniform divisions. Thus, the circle is the standard of measurement for angles. As such, it deserves our attention.

THE CIRCLE

A curve consisting of points in a plane all equally distant from a center point is a circle. *It is distinguished among curves by being alike at all points.* If the circle is turned about its center (and in its plane) all new positions are exactly like the original position. This is the characteristic property of circles called *roundness.*

A circle is formed by the continuous motion of a fixed length about a point. The perfection of the circle is independent of the instrument used to scribe it. This is in marked contrast to the creation of a straight line. When a line is drawn along a straightedge, it duplicates the errors of the straightedge.

Romanticists love to create the idea that measurement has been a continuous evolution from the Egyptian cubit to modern light wave standards. Unfortunately, nothing could be further from the truth as far as linear measurements are concerned. When cultures died so did their measurements, not to be known again until dug up by archaeologists centuries later. Only one connection can be deduced. It is that wherever there has been a culture, a system of measurement has resulted. Not so, however, with angle measurement. The circle as a standard extends deep into antiquity, probably into Biblical Babylon. It has been passed on to us for angle measurement in an unbroken history.

It is known that the Babylonians used the *sexagesimal* system of notation for their weights, time and currency. What we call the decimal system is more properly termed the "decimal-fraction" system because any number may be written as a multiplier of ten (decimo) and/or a fraction of ten. For convenience, we have substituted a decimal point for the denominator of the fraction. The sexagesimal system used a fraction with sixty as the denominator instead of ten.

Recently a more convenient method for specifying these systems has come into use. The decimal system is the *base-ten system,* the binary system for computers is the *base-two system.* In this system the jawbreaker *sexagesimal system* becomes simply *base-sixty system.*

It appears that Greek astronomers borrowed the base-sixty scale from the Babylonians or their successors and developed it. They chose 120 units to represent the diameter of a circle, probably because the numerous possible factors (2, 3, 4, 5, 6, 10, 12, 15, 20, 30, 40, and 60) made fractions easy. At this time, π (Pi) was considered to be three. It was thought that the circumference of a circle was three times the diameter. Therefore, the circle became 360 units. It is often assumed that angular notations of minutes and seconds were derived from time or at least had something to do with time. This is quite similar to the popular pianist in each era who makes a feature of playing *Chopin's Minute Waltz* in less than one minute. They take advantage of the public's ignorance of the word "minute." It is from the Latin word *minutus* which means "small". Therefore, *both* time measurement and angular measurement received their minor divisions for the same reason. Each major unit was divided by 60 to become the "first *minute* part" or "first small part." The next division was " *second* minute part" or "second small part." These with use became simply minutes and seconds.

These divisions of the circle were preserved through the middle ages by astronomers and have reached us intact. The notation is familiar to all: The degree sign is °; the minute sign, '; the second sign, ". The latter two are unfortunate and confusion with feet and inches is inevitable. To give an idea of the potential discrimination of this system of angle notation, *one second of arc subtends a distance equal to a basketball 29 miles away, or 1.15 miles on the moon.*

Today some other angle measurements are used. The radian is one. This is the angle subtended by an arc of circle equal to the radius of the circle. This is used chiefly in artillery measurements. Decimal divisions of the degree are even more common, being used both in industry and science.

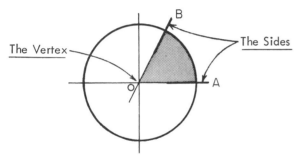

Fig. 16-3 The angle is defined as AOB. It refers to the directions of the sides, not to the space between them.

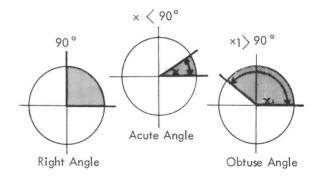

Fig. 16-4 A right angle is 90 degrees. If smaller, it is acute; if larger, obtuse.

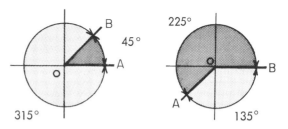

Fig. 16-5 What is the measure of ∠ AOB in each case?

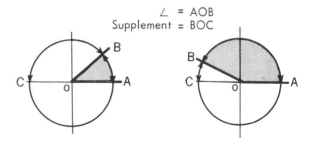

Fig. 16-6 Supplementary angles total 180 degrees.

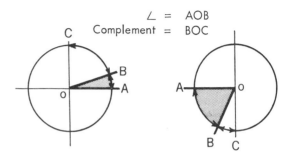

Fig. 16-7 Complementary angles total 90 degrees.

SHORT COURSE ON ANGLES

As a refresher for those who were not paying close enough attention to their geometry teacher, a few basic considerations about angles will be mentioned before going into angle measurement.

In the first place, angles deal with directions, not spaces – whether Webster agrees or not. In Fig. 16-3 the angle is the relationship of the two lines. This relationship is measurable because the lines intersect if extended. The intersection is known as the *vertex*. The lines are *sides*. Angles are conveniently designated by three letters, AOB in this example. The symbol ∠ is frequently used to designate an angle, such as ∠ ABC. Unfortunately, it can be confused with the mathematical notation for *larger than* ($>$) and *smaller than* ($<$).

A *right angle* is simply one-fourth of a circle, or one *quadrant*, Fig. 16-4. If an angle is less than the 90 degrees of a right angle, it is called *acute*. If more, it is *obtuse*. An angle can be measured from either direction. In Fig. 16-5 the left angle could be 315 degrees, or 45 degrees. The other one could be 225 degrees or 135 degrees. To minimize ambiguity, the terms supplementary and complementary are used.

Two angles that total 180 degrees are *supplementary angles*. Figure 16-6 shows that this is true regardless of whether the angle is acute or obtuse. Similarly, angles that total 90 degrees are known as complimentary angles, Fig. 16-7.

How do you know what is the angle and what is the supplement? Unless otherwise speci-

fied, the angle is the smaller of the two. Frequently the application is the deciding factor, such as in the addition or subtraction of angles.

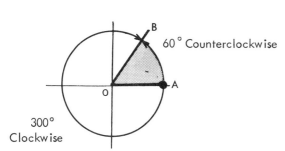

Fig. 16-8   The clock rotation analogy is used to describe angles.

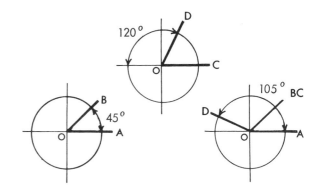

WRONG   45° + 120° = 165°
RIGHT   45° + (180°- 120°) = 105°

Fig. 16 - 9   To add, angle must be expressed as rotation in one direction.

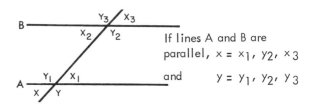

If lines A and B are parallel, $x = x_1, y_2, x_3$

and $y = y_1, y_2, y_3$

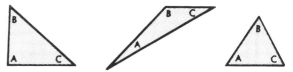

The sum of $\angle a$, $\angle b$ and $\angle c$ equals 180°

Fig. 16 - 10   These relationships find extensive use in practical measurement.

Two precautions must be observed. The first is that the angles are expressions of rotation in the same direction. Figure 16-8 shows that every angle can be expressed as either a clockwise or a counterclockwise rotation. What this means to addition is shown in Fig. 16-9. To add angles, the second angle must be considered to be a continuation of the opening that created the first angle. The reverse is true in subtraction. Some other points you all know but which bear repeating are shown in Fig. 16-10.

Many measurement operations involve right triangles. So many, in fact, that the ancients attributed mystical properties to this triangle. It is known that the Egyptians discovered that a string knotted at 3-, 4- and 5-unit intervals could be used to lay out a right triangle. Probably the first uses of the right triangle were in surveying. We are all familiar with the Pythagorean theorem which states that the square of sum of the sides is equal to the square of the hypotenuse. However useful, it is difficult to understand what it had to do with eating beans -- an awful thing to Pythagoras.

Right triangles have one 90-degree angle. It can be any one. The sum of the angles of a triangle is always 180 degrees or two right angles. If one angle is 90 degrees the sum of the other two must also be 90 degrees. They are therefore complementary angles. If one is known, the other can always be determined by subtracting the known angle from 90 degrees.

Triangles provide one of the most useful metrology tools, because they permit us to apply the basic tenet of the mathematical method to practical measurement situations. All mathematical advancement rests on a foundation of *abstraction* and *proof*. Abstrac-

tion is the ability to recognize a characteristic that is common to different objects or situations. Axe handles, violins and rolling pins are all made of wood, for example. Proof is an argument starting at a *premise* and reaching a *conclusion* in which no flaws can be found in any step.* Consider a 30-60-90-degree triangle. The proportions are exactly the same if it is part of a minute printed circuit or the relationship between two galaxies and the earth, Fig. 16-11. Triangles provide the readiest of abstractions; and, because their angles may always be resolved in relation to circles, they furnish hard-to-refute proof.

*Berganini, David, Mathematics, Life Science Library, New York, Time, Inc., 1963, p.39

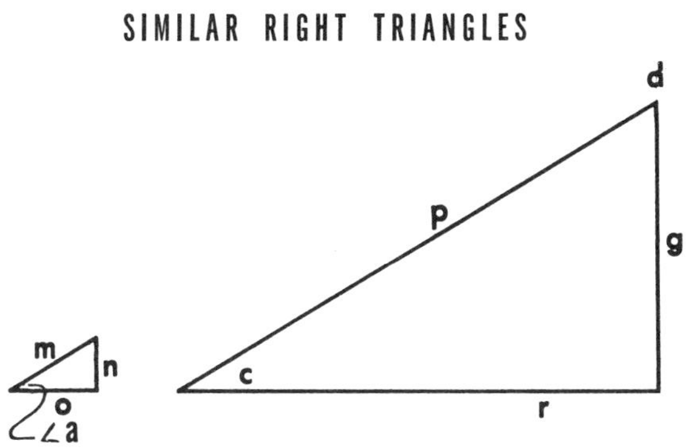

SIMILAR RIGHT TRIANGLES

Fig. 16-11   If ∠ a of the smaller triangle equals ∠ c of the large one, side m has the same relation to side o as side p has to side r.

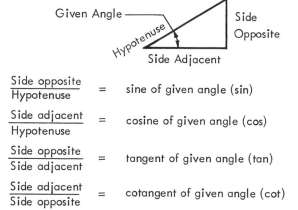

$$\frac{\text{Side opposite}}{\text{Hypotenuse}} = \text{sine of given angle (sin)}$$

$$\frac{\text{Side adjacent}}{\text{Hypotenuse}} = \text{cosine of given angle (cos)}$$

$$\frac{\text{Side opposite}}{\text{Side adjacent}} = \text{tangent of given angle (tan)}$$

$$\frac{\text{Side adjacent}}{\text{Side opposite}} = \text{cotangent of given angle (cot)}$$

Fig. 16-12   These are the most common of the trigonometric functions.

For example, if the hypotenuse of the small triangle is 10 and the *side opposite* the given angle is 5, then in the larger triangle if the hypotenuse measures 100 we know the *side opposite* the given angle must be 50.

Having introduced the term *side opposite* it is only necessary to add *side adjacent* to your triangle lexicon and we are ready for trigonometric functions. These are shown in Fig. 16-12 along with their usual abbreviations. They are important because they provide a method for designating angles by the ratios of the sides of triangles. These ratios have been worked out to long decimals. As you will soon see, they are among the useful tools of measurement.

ANGLE MEASUREMENT

The instruments for angle measurement are equivalent to those for linear measurement. They also range from simple scaled instruments to highly sophisticated types using interferometry. The general relationships are shown in Fig. 16-13.

The most familiar angle measurement instruments are variations of the *simple protractor*. In the *vernier protractors* the discrimination is improved by means of a vernier scale. The various *dividing heads* and rotary table devices are protractors in which additional mechanical support has been added for greater range and/or reliability. Some of these are very elaborate, but this does not alter the basic principles.

EQUIVALENT INSTRUMENTS

Linear Measurement	Type	Angular Measurement
Steel Rule	scaled	Plain Protractor
Combination Square	scaled	Protractor Head of combination set
Vernier Caliper	vernier	Vernier Protractor
Micrometer	mechanical	Index Heads
Gage Blocks	standards	Angle Blocks
Comparators	comparison	Sine Devices with comparators
Measuring Microscopes	optical	Autocollimators

Fig. 16-13   The angular measurement instruments closely approximate their linear measurement equivalents.

The angle measurement equivalent of the measuring microscope is the family of instruments referred to as *optical tooling*. The most familiar are the *collimator* and the *autocollimator*. They have the advantage of measurement over large distances. Perhaps the most refined of all angle measurement instruments is the *pointing interferometer*.

The objective of angle measurement is not always to measure angles – as strange as that may sound. The extremely sensitive instruments, such as autocollimators, are usually used to measure alignments, straightness and flatness. The angle readings which they record are measurements of error.

All of the above are direct-measurement instruments. Similar to linear measurement, there is also a group of indirect-measurement instruments for measuring angles. These are based on the sine of angles, on the geometry of regular polygons and on blocks similar to gage blocks. The most familiar are those based on the *sine* of an angle. This is the relationship of the height of a right triangle to its hypotenuse. The instruments for this are appropriately named: *sine bar, sine plate, sine table*, etc. They are all used in conjunction with comparison-measurement instruments such as dial indicators or electronic comparators.

THE IMPORTANCE OF SQUARES

When we bisect a circle we automatically get two 180-degree angles. Any point on a reference surface may be considered the vertex of a 180-degree angle. Much of the work in measurement involves distances from reference planes. When the reference plane is a horizontal one, as it usually is in practical work, the distance is a height. *Measurement reliability depends upon the degree to which that height is perpendicular to the plane.*

This, again, is a measurement of angle. That is why the right angle, or 90-degree angle, is the most important one in measurement. Fortunately, it is self-checking. You were told, in the beginning, that to have reliable measurement you must have a standard and the measurement must be provable. In the last chapter you were shown that the right angle can be generated from nothing more than a plane surface. This is the most irrefutable proof possible. The right angle is one-fourth of our standard for angle measurement, the circle; and it is provable. Add four of them and you are back where you started. Subtract two and you have no angle at all. It is the logical place to begin a study of applied angle measurement.

Squares are hardened-steel right angles. They are available in a wide range of sizes and shapes, Fig. 16-14. The difference between the squares in this chapter and the ones in the previous chapter is one of use. The previous ones were large and rugged. They are intended primarily to hold parts at right angles to a reference surface. The ones you are now meeting are measuring instruments intended primarily to determine whether or not angles are right angles.

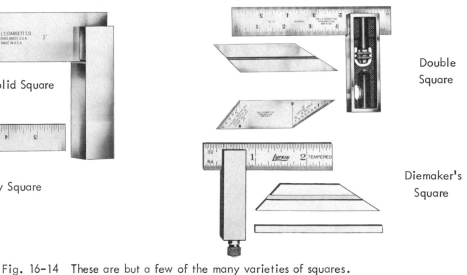

Solid Square

Try Square

Double Square

Diemaker's Square

Fig. 16-14   These are but a few of the many varieties of squares.

FUNCTIONAL FEATURES   **PRECISION SQUARES**   METROLOGICAL FEATURES

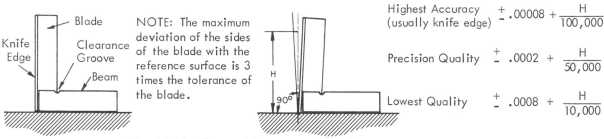

Highest Accuracy (usually knife edge) $\pm\,.00008 + \dfrac{H}{100,000}$

Precision Quality $\pm\,.0002 + \dfrac{H}{50,000}$

Lowest Quality $\pm\,.0008 + \dfrac{H}{10,000}$

NOTE: The maximum deviation of the sides of the blade with the reference surface is 3 times the tolerance of the blade.

Fig. 16-15 The precision square differs from many instruments in that its precision is built in and not determined by adjustments.

The most familiar of these, the combination square, you met in Chapter 5. Nomenclature of the most familiar types is shown in Fig. 16-15. These squares are comparison instruments. Either the inside or the outside may be used. The most common method is simply by observation, using the outside surfaces of the square. Both the square and the part being checked are placed on a reference surface. Little attention usually is given to the bases of the part and the square. Both should be inspected and free of burrs, or dirt particles. The reference surface also should be looked at, felt and cleaned.

The square is then slid into contact with the part feature to be checked. Whether the surfaces permit wringing or not, the general motions should be gone through. The squareness is usually determined by observation. If a white sheet of paper is placed behind the square and the part, a crack between them of *one-tenth mil* (0.0001 in.) may be detected. If greater precision is required, a magnifying glass should be used.

For best results a light box or other good source of light should be placed behind the part and the square. A gap of *one-tenth mil* (0.0001 in.) is readily seen. Between 50 and 70 *mike* (0.000050 in. to 0.000070 in.) the light appears red.* At about 30 *mike* it changes to blue. This is due to diffraction caused by the narrow slit, the same principle used in spectroscopy.

Another method is to insert narrow strips of paper between the part and the square at the tip and bottom of the feature. If when the square is pushed into the part both strips are

held in place, the part is square within the thickness of the paper. If the papers are pulled gently and the same amount of tension is required for each, the squareness is less than the paper thickness. Cigarette paper or cellophane is about *one-mil* (0.001-in.) thick. Therefore, this method is considered accurate to *five-tenth mil* (0.0005 in.) — but not by this author.

There is a natural tendency to turn the square at a slight angle to the surface being checked in order to get a sharp edge to use for comparison, Fig. 16-16. Unfortunately, only the narrow edges of the blade are accurately perpendicular to the beam. The blade is unstable in its thin direction and can flex freely. If the square is turned, a portion of the inaccurate side is added to the accurate edge. The error in squareness you see may be in instrument and not the part. To eliminate this tendency, squares are available with

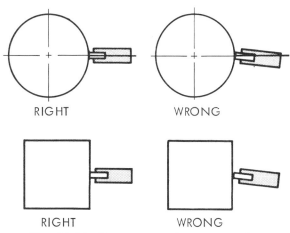

RIGHT   WRONG

RIGHT   WRONG

Fig. 16-16 Squares are only accurate when used with the narrow edge of their blades along a diameter of round parts or perpendicular to the surface of flat parts.

---

* Hume, K. J., <u>Engineering Metrology</u>, London, Macdonald & Co., Ltd., 1951, p. 172.

knife edge. It should be remembered that the less contact area, the faster the instrument will wear out of square. These squares should be reserved for the highly precise work in which a crack of light will be the measure of squareness. Another highly precise square is shown in Fig. 16-17.

THE CYLINDRICAL SQUARE

How is a square calibrated? Usually by comparison to a cylindrical square, Fig. 16-18. The cylindrical square is a heavy-wall, steel cylinder that has been precisely ground and lapped so that it is nearly a true cylinder with end faces square with the axis. The ends are therefore square to any line along the outer surface that is parallel to the axis. Notches on the ends provide spaces for particles of dirt, but are no substitute for cleanliness. Cylindrical squares are made in 4-, 6- and 12-inch heights.

Because it is a cylinder this square provides line contact with the surface being checked no matter how it is turned. This is equivalent to a beam with an infinite number of blades spreading out in 360 degrees.

Placing the 12-inch cylindrical square on a surface plate is very similar to holding onto a greased pig. It is almost certain to get away, and unless you are very careful much damage will be done and you may get hurt. The weight, over 60 pounds, coupled with the exceedingly smooth surface causes the trouble.

Fig. 16-17 The master square provides a stable base and four reference edges.

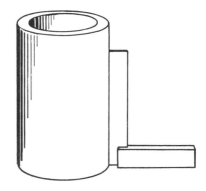

Fig. 16-18 The cylindrical square is used to calibrate squares.

If you try to lower it vertically onto the plate, Fig. 16-19, there is real danger of it slipping. A better method is to place its storage box on the plate. The square rests hori-

HANDLING CYLINDRICAL SQUARE

WRONG

RIGHT

Go this route — ruined square, nicked surface plate and smashed fingers result.

Roll the horizontal square up over edge onto plate.

Then carefully raise one end into vertical position.

Fig. 16-19   Handling a 12-inch cylindrical square requires care.

zontally in the box. The square is lifted out and placed horizontally on the plate. It is then raised from one end, the lower end being steadied to keep it from sliding. If the square is moved on and off of a plate, it is easiest to roll it up over the edge of the plate and then proceed as before.

An interesting variation of the cylindrical square is the direct-reading model, Fig. 16-20. This 6-inch, chrome-plated square provides a numerical reading of the amount that the part feature is out of square. This is achieved by lapping one end slightly out of square and calibrating the sides. The sides are laid out in elliptical curves, marked by small punch marks. These are numbered at the top and from 0 to 12 in steps of two. The numbers represent *tenth-mil* increments (0.0002, 0.0004 in.). The zero deviation position is a line of vertical punch marks.

The square is placed in contact with the part feature and turned until the light is shut out, Fig. 16-20. The highest curve is followed to the top where its number shows the out-of-squareness in *two-tenth mil* increments (0.0002 in.), per 6 inches of height. The reading may be made from two places on the circumference. This makes the direct-reading cylindrical square self-checking. It may also be used in the conventional manner by using the unnumbered end as the base.

HOW SQUARE IS SQUARE?

For erection of machinery and much scientific work, squareness error is measured in seconds of arc. In machine shop work, however, it is more convenient to measure the linear amount of out of squareness, for example, the number of *mil* in a specified dis-

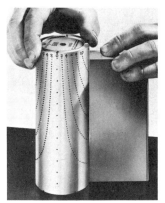

Fig. 16-20 One end of this cylindrical square is deliberately out-of-square. Therefore, lines on its surface are used to measure small angular deviations from squareness.

tance. The direct-reading cylindrical square is typical but provides readings to *two-tenth mil* increments.

When no light passes between the square and the part it is quite easy to determine squareness. You may safely assume that the part is square to within 10 times the squareness tolerance of the square. It may be even more accurately square but it is difficult to be sure because of the effect of the reference surface. This you will hear more about later.

The problem arises when the part is not square. How much out of square does a crack of light of a given size designate? We have color to go by for very narrow cracks but how do we distinguish between, say, *one mil* (0.001 in.) and *two mil* (0.002 in.)? The only reason that the experts in the past usually did not disagree, is that they had no way of knowing. Fortunately, we do. It is not difficult to make a setup that places a dial indicator or electronic comparator so that its travel is in the horizontal plane at some fixed distance above the reference plane. By comparing horizontal readings against an unknown square with a known one, a numerical measure of squareness is obtained, Fig. 16-21. This is such a

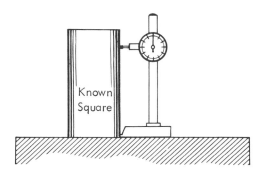
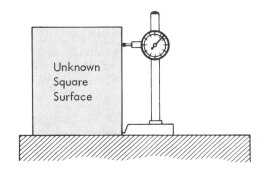

Fig. 16-21 A crack of light is open to argument. So is an indicator reading — but not nearly as much.

useful technique that devices are available specially for it. The one shown in Fig. 16-22 uses a multirange electronic comparator for maximum versatility and sensitivity.

FROM PARALLELISM TO SQUARENESS

This is a good time to demonstrate one of the axioms with which the study of angular metrology started -- the standard for angle measurement exists in nature. In order to create a square, all we need is a flat reference plane and a means for constructing lines parallel to that reference. The method, shown in Fig. 16-23, can be used with height gage and file, or milling machine, or surface grinder. Its accuracy is only limited by the flatness of the reference surface and the precision with which we can make parallel surfaces.

It is easy to make parallel surfaces. Assume that the last grind has just been taken resulting in the rectangular parallelepiped in A of Fig. 16-23. It is turned on its side in B and a cut is made across it. This time a narrow land is left along one edge. The part is now turned over in C. Because of the land, one edge is raised. The next cut thereby produces a surface at a *slightly smaller angle to the vertical edges* than the previous one. In D the part is turned over and that side ground parallel to the previous one. The broken lines in E show the original shape. The new shape clearly is more nearly square. This is repeated in F, G and H for still greater squareness. Of course, this technique is used only after ordinary machining methods have produced a close approximation of squareness. Even at the first time around the step in the first cut should be less than *one mil* (0.001 in.). Area lapping can then be used to complete the job.

When a square of sufficient accuracy has been developed it can be used to calibrate other squares. A cylindrical master square could also be used. Simple comparison provides the results. Thanks to the reversal technique even these masters are not required for the calibration of squareness.

THE REVERSAL TECHNIQUE

Reversal technique or *reversal process* refers to methods for doubling apparent error. This halves the uncertainty in solving

Fig. 16-22 The electronic comparator makes measurement of squareness to a very high level of precision.

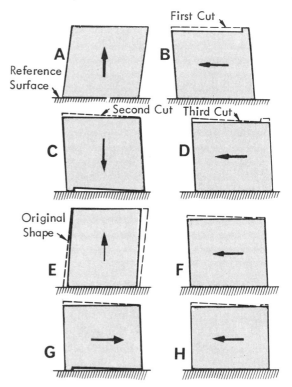

Fig. 16-23 Method for attaining squareness, devised by the National Physical Laboratory in England shows that only a flat reference surface is needed to create a right angle.

for the error. For example, consider two blocks that are identical in height. What is meant by "identical"? Mr. Webster was not sufficiently precise for our purposes. As normally used in measurement, it means that two things are so nearly alike that *available instrumentation* does not show any measurable difference. Figure 16-24 shows that if the two blocks are measured together and their combined length halved, each one will contain only half of the measured error. This is generally true even when the added error of the extra step is included. Now, here is the way it can be used for square calibration.

Only a parallel, no master, is needed. The parallel does have to be accurate, and any flatness error in the reference surface will cause calibration errors.* You will learn later that with a few other tricks the parallel does not have to be particularly accurate. There is no substitution for reference surface accuracy, however.

First, set up the parallel in a vertical position. It can be clamped to any stable support. It is set approximately by sighting along the square to be calibrated. In Fig. 16-25 the square is placed on the right of the parallel. The stock of the square is held firmly to the surface plate and gently slid towards the parallel until it engages a 0.1005-in. gage block. Apply only enough force to keep the block from falling under its own weight. It must not rest on the plate.

With the lower block in place, try to insert a 0.1004-in. block in the top position. If loose, try a 0.1006-in. block. Continue this until one is found that is too large. This will cause the lower block to fall. Using the bridging technique select a block halfway between the too large block and the one tried before it. If the 0.1004-in. block is too small and the 0.1006-in. block is too large, then the next choice is 0.1005 in.

But if, you have only one set of blocks you have already used the 0.1005 in. one and must try something else. One solution is to extend the range. You could use 0.220 in. for the lower position, made up from 0.108 and 0.112-in. blocks. This would leave the *tenth-mil* (0.0001-in.) series blocks available for use in the top position. For example, they could be combined with 0.119, 0.120 and 0.121-in. blocks. Why was this not suggested in the first place? Generally, it is not necessary and single gage blocks are handled more reliably than stacks of blocks. Take care not to apply more force than is needed to hold the square firmly on the plate and against the lower block.

* Hume, K. J., Sharp, G. H., Practical Metrology, London, Macdonald & Co., Ltd., 1958, p. 57.

DOUBLING TECHNIQUE

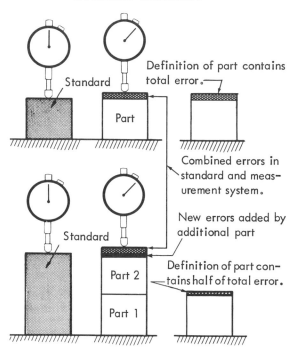

Fig. 16-24 Measuring the additional part adds error. The resulting error in the measurement of the part is less than the error from the measurement of a single part.

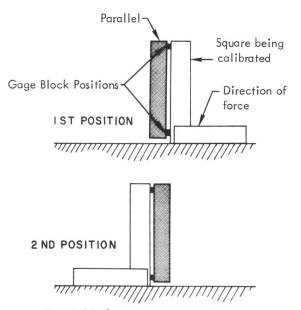

Fig. 16-25 These two steps provide a measurement of double the squareness error.

The setup is now reversed. Care must be taken not to disturb the parallel. The entire procedure is repeated. There should be less

REVERSAL PROCESS CALIBRATION

Right side of parallel:	Bottom gage blocks	0.1005 in.
	Top gage blocks	0.1008 in.
	Out of squareness	$-$ 0.0003 in.
Left side of parallel:	Bottom gage blocks (0.108 + 0.112)	0.2200 in.
	Top gage blocks (0.118 + 0.1001)	0.2181 in.
	Out of squareness	$+$ 0.0019 in.

Amount of error: $\dfrac{\text{Algebraic sum of errors}}{2} = \dfrac{-0.0003 + 0.0019}{2}$

$= +0.0008 \text{ in.}$

Direction of error: excess metal at top (shown by plus sign).

Fig. 16-26   The only precaution needed is to keep track of the plus and minus signs.

trial and error, because you now know the direction of the error. Typical calculations are shown in Fig. 16-26.

ON THE LEVEL

One of the most useful measurement instruments is the level. It is widely used in engineering metrology, but not in the United States. Some school boards might find this line of thought un-American. Therefore, we will immediately examine the instrument and its use. Of course, you are already familiar with the restless bubble in a glass tube that enables you to get things level. You may not have realized it, but this is angle measurement. This will only introduce you to levels or "bubble instruments". You will meet them again many times as parts of other instruments.

In metrology we are concerned mostly with *precision levels, clinometers,* and *theodolites. Bench levels* and *mechanic's levels* are shop instruments. They, too, are capable of high precision when properly used. All of these use bubbles in fluid-filled tubes, and let gravity do the work. To keep them from freezing the tubes were once filled with "spirits of wine", hence the general term "spirit level". The general design of *block levels* is quite similar, Fig. 16-27. Clinometers, theodolites and some of the sophisticated bubble-splitting levels bear no resemblance to their simple brothers.

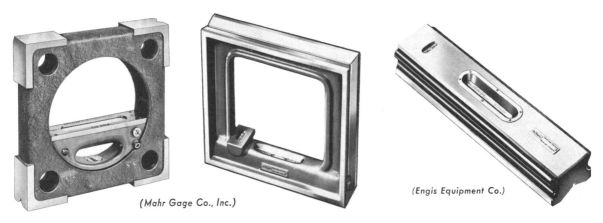

(Mahr Gage Co., Inc.)

(Engis Equipment Co.)

Fig. 16-27   The block level is one of the most useful angular measurement instruments. They are available in several forms and in a range of sensitivities.

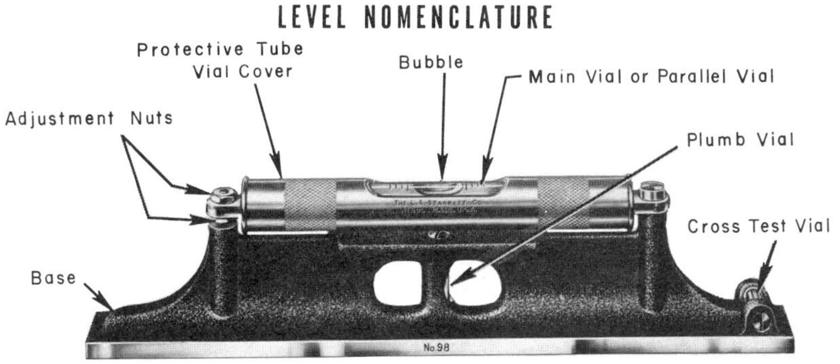

LEVEL NOMENCLATURE

Protective Tube
Vial Cover

Bubble

Main Vial or Parallel Vial

Adjustment Nuts

Plumb Vial

Cross Test Vial

Base

No. 98

Fig. 16-28 Most block levels are of this general type. The least precise ones do not have the test vials and have shorter bases.

The precision depends upon the curvature of the glass tube, Fig. 16-29. Cheap levels actually have a bent tube. The precision types have straight tubes that have been ground internally for the desired radius. The larger the radius, the greater the sensitivity. We have now used two of the basic metrology terms, *precision* and *sensitivity*. For a level they are the same, and are designated by the *discrimination* of the instrument. That leaves *accuracy*. All too often it is stated incorrectly as the precision. Accuracy refers to calibration only. It is the amount that the reading of an angle varies from the true angle. Of these terms, sensitivity is the one most often encountered. When you use one on an unstable surface you will understand why.

The metrological features of a level, Fig. 16-30, consist entirely of the geometrical relationship between the bubble and two references. The first reference is the effect of gravity acting at the center of curvature. The second is the scale against which the bubble

SPIRIT LEVELS

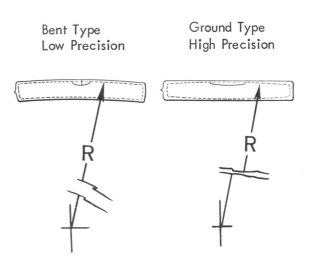

Bent Type
Low Precision

Ground Type
High Precision

R

R

Fig. 16-29 The longer the radius of curvature, the more precise the level will be.

PRECISION LEVEL

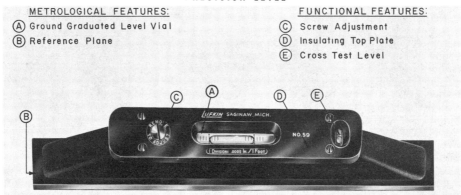

METROLOGICAL FEATURES:
(A) Ground Graduated Level Vial
(B) Reference Plane

FUNCTIONAL FEATURES:
(C) Screw Adjustment
(D) Insulating Top Plate
(E) Cross Test Level

LUFKIN SAGINAW, MICH.

NO. 59

1 Divisions .0005 in. / 1 Foot

Fig. 16-30 This typical 10-second level has graduations that represent five-tenth mil (0.0005 in.) per foot.

LEVEL PRINCIPLE

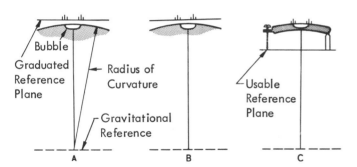

Fig. 16-31   Nature provides the reference from which a level expresses its values. In the center it is seen that the graduated plane moves, the bubble remains stationary.

position is read ("fiducial plane" if you want a name.) These are shown at A in Fig. 16-31. When the reference planes are parallel the bubble is centered. When the planes are not parallel the bubble appears to move. B of the figure shows that the bubble actually "stays put." A few alterations change theory to practice. At C the graduations have been placed on the vial to eliminate parallax, among other things. Then the vial is connected to a usable reference plane, which is the base of the instrument. Now any changes in the angle between the base plane and the reference nature provided show up as bubble displacement.

Each division on most precision levels is approximately one-tenth of an inch long. Sensitivity is designated as seconds per division. The most useful sensitivity for most precision measurement is 10 seconds per division.* The table in Fig. 16-32 shows that this requires a radius of 343 feet.

The level shown in Fig. 16-28 is a 60-second instrument. In keeping with the previously mentioned disinterest in levels in this country, the manufacturer does not bother to

---

* Hume, K.J., Engineering Metrology, London, Macdonald & Co., Ltd., 1951, p. 133.

state its sensitivity in the catalog -- the ultimate insult to a fine instrument. The one shown in Fig. 16-30 is nearly the same in construction. It appears quite different, however, because its 10-second sensitivity requires greater thermal precautions. For example, the top plate that nearly conceals the vial is of a low-conduction material. The 15-inch-long body is of a specially selected cast iron for thermal reasons. It is important not to select a higher sensitivity than is actually required. A five-second sensitivity, for example, requires great care in use, is very sensitive to temperature change and takes a long time to *settle down*.

The *square level* in the center of Fig. 16-27 is particularly useful for machine-tool inspection. Its body is an 8-inch square whose sides are square and parallel to half a bubble division. Those with 10-second sensitivity thus have *two-tenth mil* (0.0002 in.) squareness tolerance.

THE BUBBLE SHRINKER

Again and again in the study of metrology we find ourselves face to face with the problem of temperature change. With levels heat be-

RELATIONSHIPS OF CURVATURE TO ANGLE TO RISE							
Radius in feet	1718	687	343	171	114	57	0.93
Angle per division							
Seconds	2	5	10	20	30	60	3600
Minutes						1	60
Degrees							1
Rise, inch per foot per division	0.0001	0.00025	0.0005	0.001	0.0015	0.003	0.180

Fig. 16-32   The precision of a level is usually expressed as seconds per division. The actual rise per division will vary among instruments of various design, depending upon width of graduations.

comes a bubble shrinker. Like other high-amplification instruments that we have discussed the high-sensitivity level is calibrated at the standard temperature (68°F. or 20°C.). When the level is warmed, the liquid expands, thereby reducing the size of the bubble. This reduces its length.*

If the warming takes place over the same period of time that a series of measurements are made, the readings towards the end tend to be smaller in respect to true values than those at the beginning.

This could be a serious source of error but is partially compensated by proper reading. If only one end of the bubble is read, the full temperature change error is included in the reading. If both ends are read and averaged, the error will cancel out. Of course, the bubble will fall short of the calibration markings and some residual observational error will remain.

Sensitivity depends upon the viscosity of the fluid in the vial. Those with low viscosity also have high vapor pressure. As might be expected this results in some highly erratic readings when the vial is heated unevenly.

This is easily demonstrated by placing a finger on one end of the vial for a few minutes and watching the results. A precision level that has been fully normalized and in a dark room, will actually respond to a flashlight shining on one end of the vial.

---

* Judge, A. W., Engineering Precision Measurements, London, Chapman and Hall, Ltd., 1957, 3rd Edition, p. 181.

READING LEVELS

Levels are read as shown in Fig. 16-33. The centered position is shown by long graduations that bridge the bubble as exactly as can be read with the unaided eye. These are sometimes marked with a dot at each end. The graduations at the end of the bubble duplicate each other. They automatically enable you to corroborate your readings.

In the discussion of the rule of ten to one it was mentioned that there are numerous exceptions. We meet two of them in the reading of levels. First, is the exceptional ability of the eye to resolve displacements of one line in relation to another. The second is the increased precision possible with a repeatability of five seconds.* This is many times greater than predicted by the rule of ten to one. However, do not abandon the rule.

The last example (E) in Fig. 16-33 is not meant to be funny. If the surface is far out of the range of the level, the bubble will not be seen. Caution: Be sure you know at which end of the vial the bubble is hiding before you begin to adjust. You can waste a lot of time adjusting before you discover that you are going the wrong way.

In Fig. 16-33 note the convention for reading. Left bubble is negative, right bubble positive. *These are in relation to the observer, not in relation to the level.* They remain the same even though the level may be reversed.

---

* Hume, K. J., Engineering Metrology, London, Macdonald & Co., Ltd., 1951, p. 135.

READING A LEVEL

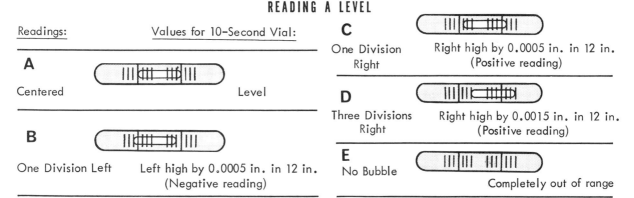

Fig. 16-33 The readings are the number of divisions that the bubble moves. Having two ends, the reading is visually corroborated as read.

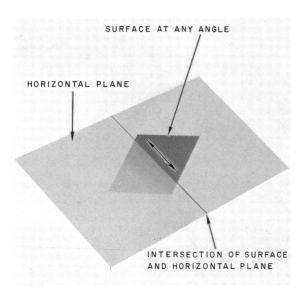

Fig. 16-34 No matter at what angle a surface may be, one direction across it is horizontal.

LEVEL ADJUSTMENT

All levels of 20-seconds sensitivity or greater have an adjustment screw. In Fig. 16-31, it is seen that this adjustment aligns the vial to the base. Adjustment is very simple when the high precision is considered. Only a stable flat surface is required. Surprisingly enough at first thought, the surface does not need to be accurately level. *Regardless of the position of a flat surface, one direction across it is horizontal.* Please see Fig. 16-34 before arguing this point.

First orient the level on the surface until the bubble is centered. Clamp a straightedge along one edge as shown in the top drawing of Fig. 16-35. The reading is shown at A. A straightedge is clamped to the surface along one side of the level. When the level is reversed, the error is read as at B. It is then repeated, orienting to a new location on the surface, D. This is repeated until the desired *accuracy* is achieved. Note that the factor being adjusted is calibration accuracy, not precision or sensitivity.

This is another example of the extremely useful *reversal process* that was mentioned in conjunction with squares. Figure 16-36 shows its workings for steps A, B and C of Fig. 16-35. How far should this be carried for ordinary requirements? Generally one-fifth of a division is considered satisfactory.

STEPS FOR LEVEL ADJUSTMENT

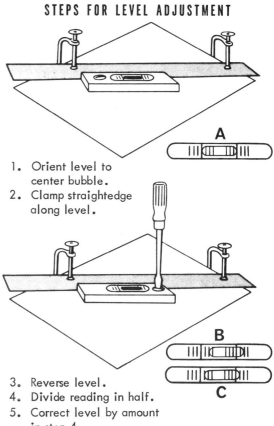

1. Orient level to center bubble.
2. Clamp straightedge along level.

3. Reverse level.
4. Divide reading in half.
5. Correct level by amount in step 4.

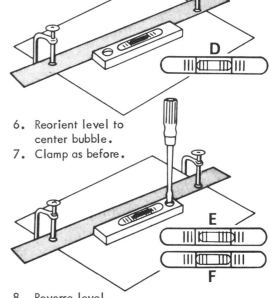

6. Reorient level to center bubble.
7. Clamp as before.

8. Reverse level.
9. Divide remaining reading.
10. Correct as before.

Fig. 16-35 These steps are repeated until desired accuracy setting is achieved.

REVERSAL PROCESS

Bubble centered, errors subtracted

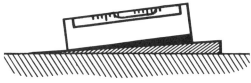

Level reversed, errors added

Error is divided by adjustment

Fig. 16-36   The reason the surface did not have to be level to adjust the level is figuratively shown here.

If the level is so far out of calibration that the bubble is off of the scale, adjustment may be long and tedious.   Figure 16-37 shows a trick that will get the job done quickly.   Simply select two lines across the surface that provide the same reading.   Bisect the angle they form.   This will be the approximate level line needed to *"bring in"* the instrument.

By means of a tilting table the sensitivity can be calibrated.   It is not within the scope of this book.   Let it suffice to add that the sensitivity tolerance should not be greater than plus or minus 10 per cent.   This is two-tenths of a division.

REVERSAL PROCESS

The reversal process not only helps adjust levels, it also provides their exceptional reliability in use.   Whenever precise measurements are made with a level, readings should be taken at both ends of the bubble, the level should be reversed and the readings repeated, Fig. 16-38.   Averaging the four readings provides a possible accuracy of one-tenth of a division.   This is possible accuracy.   No amount of reversing will make sloppy work reliable.

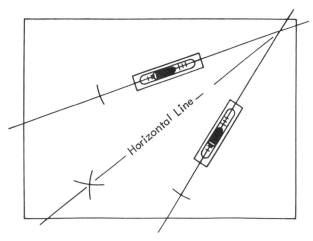

Fig. 16-37   If the angle formed by two lines of the same level reading is bisected, a line approximately horizontal will be found.

RELIABILITY WITH LEVELS

For Precise Measurement:

1. Take readings from both ends of the vial.

2. Reverse level.

3. Repeat readings from both ends.

4. Average the four readings.

5. Repeat all steps for critical cases.

Fig. 16-38   These steps enable level measurements to be relied upon to one-tenth division.

In most practical work very small angles are more conveniently dealt with in terms of linear displacement than in seconds. It is important to keep in mind that a level measures angles even though we may translate them into linear measurements.

The linear displacement is expressed in terms of a *measurement base*. Measurement base has nothing to do with the length of the base of the instrument. It is the distance in which the linear rise per division is expressed. It is generally 12 in. in the United States. That is the base upon which Fig. 16-32 is based. It is very likely that you will use British levels. These, Fig. 16-27, have a measurement of 10 in.

LINEAR SENSITIVITY

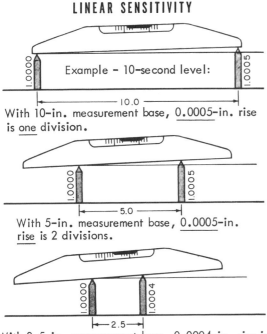

Example – 10-second level:

With 10-in. measurement base, 0.0005-in. rise is one division.

With 5-in. measurement base, 0.0005-in. rise is 2 divisions.

With 2.5-in. measurement base, 0.0004-in. rise is 3 divisions. And with 1.0-in. measurement base, 0.000015 in. is 3 divisions.

Fig. 16-39 The shorter the measurement base, the greater the linear sensitivity.

The linear sensitivity is inversely proportional to the length of the measurement base. Figure 16-39 gives meaning to this statement. In the top example a *five-tenths mil* (0.0005 in.) rise acting on a 10-in. measurement base moves the bubble one division. In the center example the measurement base is reduced to 5 in. With no change in the rise, this registers as two divisions, in the bottom example, a 2.5-in. measurement base requires only *four-tenth mil* (0.0004 in.) rises for three divisions, or full scale displacement. If the measurement base is reduced further to 1.0 in. a rise of only 150 *mike* (0.00015 in.) similarly uses the full scale. Taken another way, a 1-in. measurement base increases the linear sensitivity to 50 *mike* (0.00005 in.) from the *five-tenth mil* (0.0005-in.) linear sensitivity of the same instrument with a 10-in. measurement base.

PUTTING LEVELS TO WORK

The tremendous linear sensitivity of the precision level coupled with the reliability of the reversal process has not gone unnoticed.

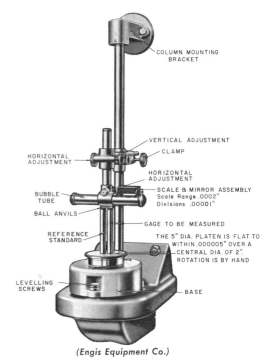

(Engis Equipment Co.)

Fig. 16-40 The linear sensitivity of the precision level, together with the reversal process, give the level comparator the unusual combination of high amplification, high reliability and simplicity.

An example of its versatility is the *level comparator*. The instrument shown in Fig. 16-40 has two cylindrical gage blocks ready for comparison on its platen. The head is simply a precision level whose base consists of two precision balls. One end of the very long bubble is read by means of a mirror and scale assembly. The scale is graduated to read linear measurement directly, one division being equal to 10 *mike* (0.00001 in.). To use the reversal process, the platen turns.

Figure 16-41 shows an example of measurement with the level comparator. In A the master and the unknown have been wrung to the platen and the head lowered onto them. The scale shows the unknown to be higher than the master and reads 10.8 divisions. Note: It does *not* read 10.8 divisions *higher*. The head is raised, the platen turned 180 degrees, and the head is lowered again. Now the height difference is reversed, and the reading at B is 10.2 divisions. The difference in readings is 0.6 divisions. Thanks to the reversal process this is twice the height difference. The height difference thus is known to be 0.3 divisions or three *mike* (0.000003 in.). C and D are other examples. The level comparator

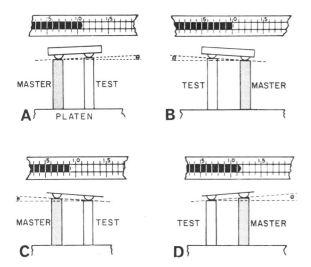

Fig. 16-41 These two examples show that the level comparator does not have to be completely level to produce reliable results.

is generally used at a lower level of precision. When measuring to *mike* increments (millionths of an inch) it must be housed away from drafts and preferably operated by remote control.

The level comparator has been detailed because it shows that levels can be used for highly precise length comparison and to show that the surface upon which they are used does not have to be exactly level itself. A more familiar application of this is the calibration of surface plates. A precision level can compete quite well with an autocollimator except at the highest orders of precision. This will be met later in metrology. The uses in machine alignment and setup are so straightforward that they hardly need discussion. The important point is that they by no means need to be restricted to work which is level.

OTHER LEVELS

An interesting level for the alignment of very large devices is the *micrometer water level*, Fig. 16-42. It consists of two glass cylinders interconnected by tubing. The fine reading is made by lowering the micrometer dipper into the meniscus of the water column. It measures differences in level up to 1 in. to an accuracy of *one mil* (0.001 in.) or 0.02 mm. The instrument can be read directly to height differences of *point one in.* (0.1 in.).

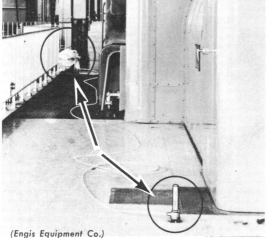

(Engis Equipment Co.)
Fig. 16-42 The micrometer water level provides a convenient means for leveling and aligning very large devices.

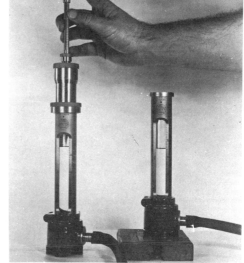

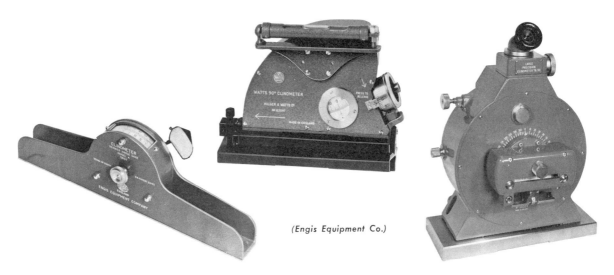

(Engis Equipment Co.)

Fig. 16-43 A few of the many types of clinometers are shown. The one on the left uses a pendulum and reads to one minute. The center one also reads to one minute but uses a spirit level for greater sensitivity.   The right clinometer is very precise, using coincidence reading to one second.

There is virtually no practical limit to the distance over which the instrument can be used. Sensitivity reduces, however, over 25 feet. For greater discrimination the micrometer dipper is used. As you already know, in order to increase precision it is necessary to increase the amplification of the measurement system. Try to lower a needle until it just contacts a solid surface. Notice how difficult it is to perceive of the last few *mil* movement. Now repeat using a brim-full glass of water. When contact is made, the water almost snaps from its flat surface to a ringed surface that surrounds the point. This surface clings to the point as it is withdrawn, for a way, and then snaps back flat again. This phenomenon is caused by surface tension and is used to amplify our senses in the micrometer water level.

The micrometer dipper is very similar to an ordinary micrometer except that its thimble divisions are *half mil* (0.0005 in.). The dipper telescopes for rough positioning. The micrometer is used for the final reading. The metric instrument discriminates to 0.01 mm. Interpolation is not recommended.

The block level is restricted to relatively small angles. This restriction is removed in the *clinometer*. It is a level mounted in a frame so that the frame may be turned at any desired angle to the horizontal reference. Three instruments ranging from simple to sophisticated are shown in Fig. 16-43.

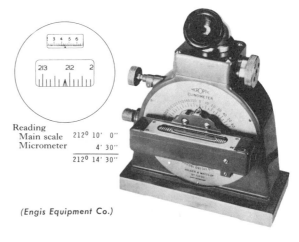

Reading
Main scale   212° 10' 0"
Micrometer   4' 30"
            212° 14' 30"

(Engis Equipment Co.)

Fig. 16-44 This clinometer reads to 5 seconds. A typical view through the eyepiece is shown. (The lower scale is divided in 10-minute intervals and numbered in degrees.) The top scale is divided in 5-second intervals and numbered in minutes.

To measure with a clinometer its base is placed against the surface with the circle clamp in the free position. The level is rotated until approximately level. The external scale seen in Fig. 16-44 may help make the approximate setting. The circle is then clamped and the bubble is centered with the fine adjustment. The scale is then viewed through the eyepiece. First the micrometer knob is turned until a reading on the degree scale is in line with the fiducial arrowhead. Then the reading on top scale is added to the degree scale.

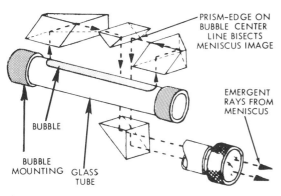

Fig. 16-45   The coincidence reading system splits the bubble and brings both ends together.

COINCIDENCE READING

The vial has no graduations.

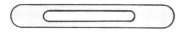

The optical system splits the bubble.

Two ends of the bubble are brought together.

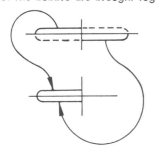

View when in coincidence.

View when separated.

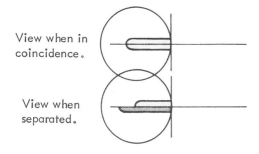

Fig. 16-46   The coincidence reading system is another example of the leveling process.

In clinometers the bubble is read only to provide the reference plane. The angle must be read from another scale. It is, of course, the relationship between the vial and the frame of the instrument. The measurement can be no better than the combined observations. It is no wonder that the reading arrangements have become highly developed.

One of the most precise means is the co-incidence reading. This also is found on many other precise instruments. Its use on the clinometer is shown in Fig. 16-45. In its most developed form it splits the bubble in half and swings one end around to match up with the other one, Fig. 16-46. Note that this is another application of the reversal process. The separation of the two ends of the split bubble is double the actual bubble movement. Co-incidence reading provides such obvious advantages both in ease of reading and improved reliability that it has been extended even to the simpler levels, Fig. 16-47.

There are four principal disadvantages of levels. By far the worst is the general ignorance about them. The second is the length of time required to *settle down*. The third is their single sensitivity characteristic. And fourth is the fact that they do not produce an output that can be used as the input *loading* of a measurement system, Fig. 11-12.

Fig. 16-47   Although small and compact, this level employs coincidence reading for reliability at high sensitivity.

All three of these are eliminated by a relatively recent adaptation of electronics to levels. This in principle is a pendulum whose displacement is converted into electrical signals by an LVDT, Chapter 11. The example in Fig. 16-48, has a sensitivity of one second, yet settles down in only one second. Its sensitivity is adjustable; and because its output is electrical, it can be used to lead recorders or to monitor continuous operations.

Levels, with and without coincidence reading systems, are widely used in scientific instruments. They have long been an important part of surveying instruments. Since World War II they have found a permanent place in the theodolites of metrology. These are level and telescope combinations. It is nearly certain that they will find increased use in dimensional metrology as tolerances continue to tighten down.

Fig. 16-48   *(Engis Equipment Co.)*

This is a good place to start another chapter, if for no reason than to keep the figure numbers from becoming unwieldy. It is as smart in measurement as it is in mountain climbing to look back now and then to find out where we have been before climbing on.

METROLOGY DATA FOR SQUARES

INSTRUMENT	TYPE OF MEASUREMENT	NORMAL RANGE	DESIGNATED PRECISION	DISCRIMINATION	SENSITIVITY	LINEARITY	RELIABILITY	
							PRACTICAL TOLERANCE FOR SKILLED MEASUREMENT	PRACTICAL MANUFACTURING TOLERANCE
Combination square	comparison	none	none	not applicable	beyond accuracy	not applicable	30'	1°
Precision square	comparison	none	none	not applicable	beyond accuracy	not applicable	30"	1'
Surface plate square	comparison	none	none	not applicable	beyond accuracy	not applicable	10"	30"
Cylindrical square	comparison	none	none	not applicable	beyond accuracy	not applicable	5"	30"
Graduated cylindrical square	comparison	0. to 0.0012"	0.0001" in 6"	0.0002" in 6"	beyond accuracy	50 mike within 6"	0.0002" in 6"	0.0004" in 6"
Square and transfer stand	All factors limited by metrological data of transfer instrument							
Mechanic's level	direct	6°	1°	1°	30'	30'	1°	2°
Precision level	direct	1'20"	10"	10"	5"	5"	10"	30"
Clinometer (average)	direct	0° to 360°	10"	10"	2"	2"	5"	15"

Fig. 16-49   It is possible to use a slit of light to compare squareness to such a high degree of sensitivity that this factor is of relatively little importance when evaluating squares.

TERMINOLOGY

Complimentary angles	Angles which when added equal 90 degrees.
Triangle	Figure formed by three lines intersecting in pairs in three points. Three sided closed figure with three angles totalling 180 degrees.
Right triangle	Triangle in which one angle is a right angle.
Protractor	Direct measurement instrument for angles, consisting of a circumferential divided scale.
Solid square	A fixed right angle standard in a standardized design but in many sizes and tolerances.
Cylindrical square	A master right angle reference in form of a heavy cylinder.
Parallelogram	A four-sided closed figure whose opposite sides are parallel.
Rectangle	A parallelogram whose internal angles are right angles.
Rectangular parallelepiped	A rectangular solid whose sides are rectangular.
Reversal process Reversal technique	Method of comparing some part of the measurement problem with itself to eliminate variables.
Identical	As nearly alike as can be determined by available instrumentation.
Level	Instrument that establishes a horizontal reference by the action of gravity.
Bench level Mechanic's level	Low sensitivity levels.
Precision level	High sensitivity level.
Block level	Most common functional form of level. Block with mounted vial.
Clinometer	Level and protractor combined.
Theodolite	Level, protractor and telescope combined.
Square level	Level and square combined.
Fiducial	Reference usually applied to lines or points.
Circle	A closed plane curve, all of whose points are equidistant from one enclosed point, the center. The standard for angular measurement.
Angle	The rotation necessary to bring one line into coincidence with (or parallel to) another line in the same plane.
Initial line	The reference line when two intersecting lines form an angle.

Terminal line	The measured line when two intersecting lines form an angle.
Roundness	The characteristic that all parts of a circle are identical.
Sexagesimal system	Measurement system based on the number 60. Used for angle measurement.
Circumference	The perimeter of a circle. Usually means the length of the perimeter.
Chord	A line cutting circumference at two places.
Diameter	A chord passing through center.
Radius	A line from the center to the circumference. Thus, one half of diameter.
Pi (π)	Ratio of circumference to diameter. (3.1415926536)
Degree	In angle measurement, the 360th part of the circumference.
Minute	One 60th part of a degree. One 21,600th part of a circle.
Second	One 60th part of a minute. One 3600th part of a degree. One 1,296,000th part of a circle.
Vertex	The intersection of lines forming an angle.
Right angle	One-fourth part of a circle, or 90 degrees.
Acute angle	An angle smaller than a right angle.
Obtuse angle	An angle larger than a right angle.
Supplementary angle	Angles which when added equal 180 degrees.
Level comparator	A length comparator using a level as the transducer converting length changes into linear values.
Water level	Two connected vessels that permit a liquid level to be used as a datum plane for height measurement.
Coincidence reading	Comparison of two images of one fiducial moving in opposite directions.

Fig. 16-50 Many of these terms have slightly different meanings in ordinary use. It is important to know which way they are intended.

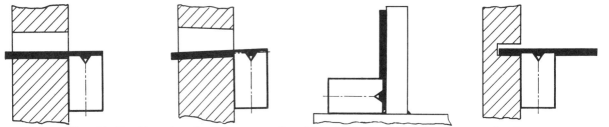

Fig. 16-51 Now that the importance of positional relationships is being recognized, specialized squares are being developed. This design is an example. It combines a V-block with a straight pin to form a highly versatile square. *(Mahr Gage Company, Inc.)*

Summary — Completing the Circle

● In this chapter you heard about the unique status of the circle -- a standard in nature. From division of the circle all of our angular measurement is derived and provable. The ready convertability back and forth from angular measurement and linear measurement permits us to measure angles linearly and vice versa. To systematize this there is geometry and trigonometry.

● Like linear measurement the lowest amplification angular measurement is by a scaled instrument, the protractor. Following the same pattern, the vernier protractor adds amplification. A host of other devices extends the amplification up to and beyond those found in most linear-measurement instruments. These instruments are reserved for later chapters because two special angles deserve attention first. These are the right angle (90°) and the flat angle (180°).

● In measurement at least, squares are indispensable. They provide the axis for height measurement. They are self checking. For most work they can be visually compared to an unknown part feature. This can be done with great precision, less than *one-tenth mil* (0.0001 in.). This means that they must be calibrated to even greater precision. Several methods can be devised that require no master. A faster method is comparison with a cylindrical square, a real "knuckle buster".

● Calibration without a master uses the reversal process, one of the most helpful measurement techniques. It is used to cancel out unwanted variables, and to use the geometry of the part as a reference in the measurement of that part.

● The horizontal is every bit as important as the vertical. The level is to it what the square is to the vertical. If there were more carpenters working in toolrooms and inspection departments than butchers, the usefulness of this instrument would be better known. It is to angular measurement what the optical flat is to linear measurement. It provides high precision with almost unbelievable simplicity. Unfortunately, it is slow.

● The important lesson about levels is that they are not restricted to leveling. They are comparators. Any two surfaces with the same bubble reading are parallel along that axis. Bubble readings are angular measurements, usually in seconds. They are readily converted into linear measurements. The shorter the measurement base the higher the precision of these linear measurements. Like squares, precision levels can be calibrated without a standard. To put it another way, the standard is always available, being gravity; and every flat surface has one line across it which is horizontal.

● Because of its versatility, many variations of levels are available. Square levels, for example, provide both horizontal and vertical references. In large-scale metrology water levels are used to compare surfaces separated by large distances and with obstructions in between. The nerve center of the clinometer is the level. Therefore they were included in this chapter. However, unlike the other instruments discussed they measure any angle. They would have been quite at home in the next chapter. And that lets you know what to expect.

QUESTIONS

1. What is the fundamental difference between length standards and angular standards?
2. Is an angle the two intersecting lines, or the space between them, or the distance between them?
3. If an angle cannot be defined by a single linear measurement, how is it defined?
4. Are all circles curves and vice versa?
5. When is a curve a circle?
6. What is this characteristic called?
7. In what field was angular measurement first developed?
8. What is meant by the base-sixty system?

9. In addition to our base-ten system (the decimal system), what other base is in extensive use?
10. What do minute and second mean in angular measurement?
11. What are the two most important angles?
12. What are two angles called if they total 90°? 180°?
13. What precaution must be observed when adding or subtracting angles?
14. To what linear measurement instrument does the simple protractor correspond?
15. Name a situation in which angular measurement instruments are used, but in which you are not concerned with angles in the usual sense.
16. What is an example of indirect measurement, or comparison measurement, of angles?
17. Which angle is of vital importance to surface plate work?
18. What is the most common form for the right angle?
19. Why is the square particularly important to measurement?
20. How is a square read?
21. How small a gap can light be seen through?
22. How is a square calibrated?
23. What variable must constantly be kept in mind when calibrating squares?
24. What is a direct reading cylindrical square?
25. What is a more reliable method to compare squareness than using the light gap?
26. What is the prerequisite to the generation of an accurate square?
27. What is meant by the reversal process?
28. The reversal process may be used to cancel errors. Can it be used in any other way?
29. What is an example of the use of the reversal process?
30. What is a level?
31. What is the most common type?
32. In what units do levels discriminate?
33. What is the greatest sensitivity usually found in precision levels?
34. What are frequent modifications of the level?
35. Which end of a level is read?
36. Does the ten-to-one rule apply to levels?
37. Must the reference surface be accurately level in order to use a level upon it?
38. What is the quickest way to find a level line?
39. What conditions are needed for reliable use of a precision level?
40. What is a clinometer?
41. What is a water level?
42. What other kinds of levels are there besides spirit levels?

DISCUSSION TOPICS

1. Evaluate the importance of the base for the measurement system by considering a base-eight system.
2. Explain the importance of the reversal process.
3. Explain the importance of levels in metrology.

Angle Measurement II
From Zero to Ninety to Naught

The previous chapter dealt with simple instruments. True, in some cases they were capable of very great precision. They gained their precision at the expense of range, but without them we would have no references for the instruments to come. This chapter discusses the instruments for measuring angles that are not 0, 90 and 180 degrees.

The instruments are discussed more or less in the order in which you are likely to meet them. Each of them is available in several different types. Some have been refined to the point that they are unrecognizable. Only the principles will be developed. It would take an entire book to do any justice to some of them, the dividing heads, for example.

THE PROTRACTOR

The simple protractor for measuring angles is equivalent to the rule for measuring lengths. Its discrimination and sensitivity are its finest division. This is one degree except for very large protractors.

Like a steel rule, the simple protractor has limited use. But mechanical additions to the

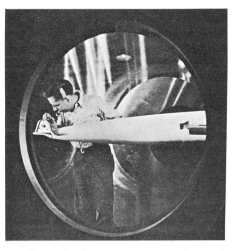

Fig. 17-1 Angle measurement combines the techniques of both science and engineering. A clinometer is being used here to check a critical angle on a part that may end up in orbit. (*"Missiles and Rockets" Magazine*)

rule made it the extremely versatile combination square, and a vernier made it into the highly precise vernier caliper or height gage. Similar modifications of the simple protractor result in the *universal bevel protractor*, Fig. 17-2.

UNIVERSAL BEVEL PROTRACTOR

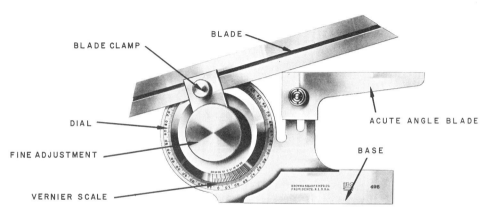

BLADE

BLADE CLAMP

DIAL

FINE ADJUSTMENT

VERNIER SCALE

ACUTE ANGLE BLADE

BASE

BROWN & SHARPE MFG.CO. PROVIDENCE, R.I., U.S.A.

496

(Brown & Sharpe Mfg. Co.)

Fig. 17-2 The universal bevel protractor is to angle measurement what the combination square is to linear measurement. On some instruments, the blade clamp and the fine adjustment are all located at the center.

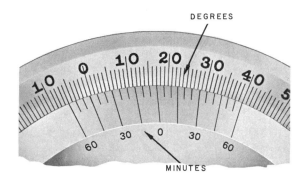

Fig. 17-4 Care must be used to read the minutes from the correct pair of lined-up graduations.

The heart of this instrument is the dial, which is graduated in degrees. These are divided into four 90-degree quadrants. The degrees are numbered to read either way from zero to 90 and then back to the zero opposite the starting zero.

The vernier scale is divided into 24 spaces, 12 each side of zero. These are numbered 60-0-60. The 24 spaces equal 46 spaces on the dial. Thus one vernier division equals 1/12 of 23 degrees or 1 11/12 degrees. The difference between one vernier division and two dial divisions is 1/12 degree or 5 minutes. When the angle is an exact degree the zero graduations of the vernier and the two 60-minute graduations line up with dial divisions, Fig. 17-3.

Sometimes there is confusion about which direction to read the vernier. For example, in Fig. 17-4 is the reading 12° plus 50' or plus 25', or minus 25'? This confusion is quickly eliminated by a simple rule. *Always read the vernier in the same direction from zero that you read the dial* and add the minutes to the dial degrees, Fig. 17-5. Thus the reading in Fig. 17-4 is 12° 50'.

APPLICATIONS FOR VERNIER PROTRACTORS

Angles of any degree of arc can be measured with the universal bevel protractor -- but you have to keep your wits about you. For every position of the blade four angles are formed. Two are read on the dial and vernier.

RULE FOR READING VERNIER PROTRACTORS

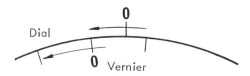

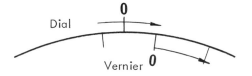

The other two are supplements. Of course there are times when the blade is not extended through the base. At those times one angle and one supplement are not usable.

In Fig. 17-6 the center protractor is set at 90°. Then all four angles are as read. If the blade is turned *counterclockwise* as at A, the angle read will represent only the angles formed from the blade to the base *counterclockwise*. This happens in two positions as shown. If the blade is turned *clockwise* as at C the angle read will be formed only in two places again. These places are always from the blade to the base rotating clockwise.

This cannot be used as a general rule because the reverse applies when reading from

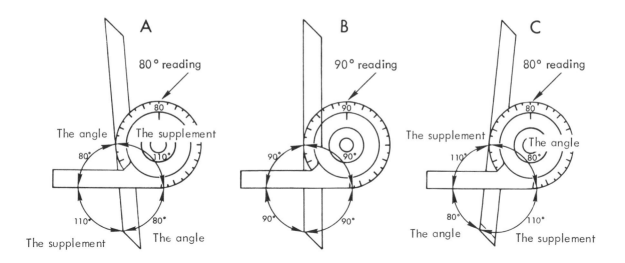

Fig. 17-6   When reading from 90°, these are the positions where the angle and its supplement are found.

a straight angle (180° or 0°). As shown in Fig. 17-7, this is not a problem. It would be difficult to confuse a 10° angle with 170° supplement. No rule is needed because usually you can tell the desired angle from its supplement. One is acute, the other obtuse. The problem only arises near 45°. It may be difficult to tell at a glance if a given angle is 43° or 47°. Caution must be used.

The way this affects an actual measurement situation is shown in Fig. 17-8. Two of the four measurements required to check a hexagon are read directly. The other two must be converted from supplements.

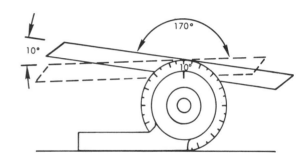

Fig. 17-7   When reading from 0°, there is little danger of confusing the angle and its supplement.

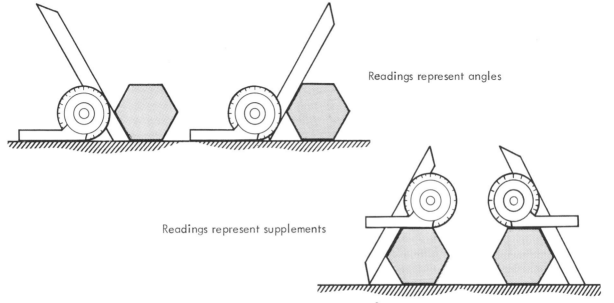

Readings represent angles

Readings represent supplements

Fig. 17-8   In every case, the above readings would be 60° when measuring a hexagon.

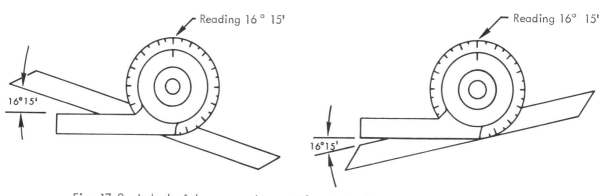

Fig. 17-9   In both of these examples, a 15° angle has been set. Unfortunately, so little length of the blade and base enclose the angle that it is not usable for most measurement.

MEASURING ACUTE ANGLES

Although the dial of the universal bevel protractor rotates through 360°, not all of that represents usable range. The reason is shown in Fig. 17-9. So much of the blade is covered by the base that remaining edges that define the angle are too short for most measurement.

This situation is remedied by a simple attachment, Fig. 17-10. It simply forms a reference edge at 90° to the base. The long edges thus define an angle. Because it is at a right angle to the base the reading is the complement of the desired angle. This, of course, is adjusted by subtracting the reading from 90°.

Another type of acute-angle attachment is shown in Fig. 17-11. This is a parallel extension of the base. It enables the blade to be used at the top of the dial, providing direct readings of angles. As you remember, any time an operation, manipulative, visual, computational or any other kind, can be eliminated, reliability is enhanced, because a chance for error is eliminated. The disadvantage of this attachment is that in some cases it is too bulky to be placed in the area of the angle.

PRECAUTIONS FOR USING THE UNIVERSAL BEVEL PROTRACTOR

Remember that the instrument does not measure the angle *on the part*. It measures the angle *between its own parts*. It provides knowledge of the part under observation (or being inspected) only under two conditions.

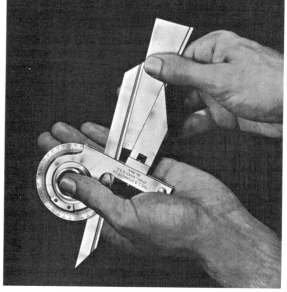

(The L. S. Starrett Co.)

Fig. 17-10   The acute angle attachment is being used here to measure a 10° angle.

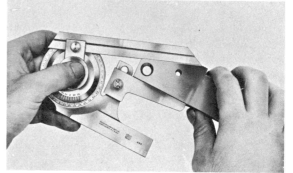

(Brown & Sharpe Mfg. Co.)

Fig. 17-11   In this form of acute angle attachment, the readings are direct and need not be converted.

First, the physical parts of the protractor must be in intimate contact with features of the part. The closer this can be done, the more accurate the readings. Second, the protractor must be in the same plane as the angle being measured, or in a plane parallel to it.

Figure 17-12 shows the effect of contact error. The desired situation is in A. Blade contact error is at B and base contact error at C. These errors usually result from outright carelessness, surface obstructions, and overconstraint. Repetition is the safeguard for the first. True, the error may be repeated, However, sufficient attentiveness to recheck the measurement will usually disclose an illegitimate error.

Cleanliness and surface preparation have been emphasized over and over. They return for protractors with renewed vigor. The effect of shortening the measurement base was shown for levels. It applies here, too. The shorter the part features, the more seriously a burr or grain of dust will disrupt reliability.

In an effort to achieve reliable results it is easy to overconstrain the instrument and induce an error. This is shown at C of Fig. 17-12. In order to force the blade firmly into contact with the bevel on the part, the entire instrument has "climbed" up the bevel. Apply only sufficient force to assure close contact.

Generally contact between the base and the part is assured by applied force. Contact between the blade and the part must rely on more subtle means. Sometimes the dial is left free to rotate and the force of the part against the blade, above or below center, rotates the blade

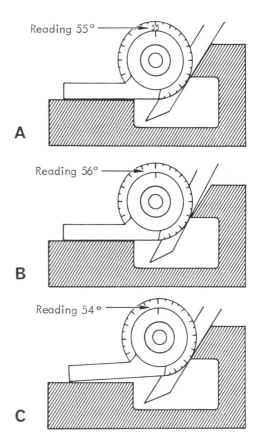

Fig. 17-12   The instrument must duplicate the angle before it can measure the angle.

and dial into position, Fig. 17-13. This is only reliable when the part feature is relatively long and extends from above to below the center of rotation of the dial. Figure 17-14 shows what happens when the part feature is short and/or entirely on one side of the center of the dial.

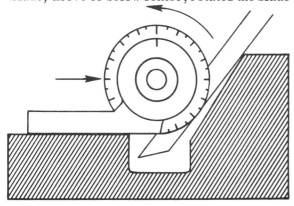

Fig. 17-13   When the movement to the right halts, the blade will be mated to the bevel surface.

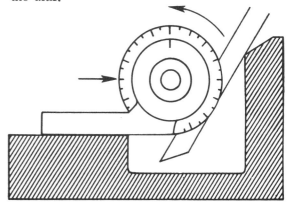

Fig. 17-14   The blade has already passed the mating position and further movement causes still greater error.

One method to determine if the blade is in contact with the part is to place a light behind it and adjust. If the surface finish of the part is good, all light can be shut out. If poor, adjust so that the light that "leaks" through is uniform. For this method the dial is locked when approximately positioned and the fine adjustment, Fig. 17-2, used. When it is not possible to place a light behind the blade, the paper strip method may be used. This was described in conjunction with squares.

Figure 17-15 shows that the true angle is measured only when the instrument is in a plane parallel to the plane of the angle. When the protractor is resting on the part as at A it is constrained from translation in the z axis (up and down). It is free to move in the y axis away from the part feature, and to slide back and forth along the part in the x axis. This is particularly bad because its rotation is constrained only in one axis, x. It is the rotational position in that axis that is the measure of the angle. B of the figure shows what happens.

The effect of rotation around the z axis is demonstrated in Fig. 17-16. In this example the reading is 38° when the instrument is parallel to the plane of the angle. If it is rotated, the upper end of the blade will move away from the part feature even though the bottom remains in contact. If the dial is rotated so that the blade can be brought into contact again, the reading will be reduced.

Exactly the same errors would be introduced if the instrument was slanted. That would be rotation about the y axis. These are *independent errors*. When they exist you never know whether they are adding or canceling. You do know that reliability has been lost.

RELIABILITY WITH THE UNIVERSAL BEVEL PROTRACTOR

A general check list is given in Fig. 17-17. In general, the same precautions that are recommended for any instrument in the vernier class apply to the universal bevel protractor. It is important not to expect greater precision than the instrument discrimination (5 min.). The ten-to-one rule applies, but as for linear measurement it must be tempered by practical considerations. If your only instrument

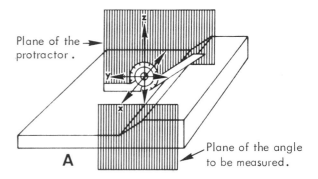

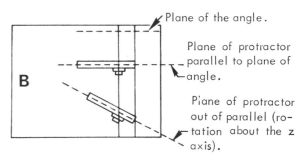

Fig. 17-15 The true angle is measured only when the protractor is in a plane parallel to that of the angle.

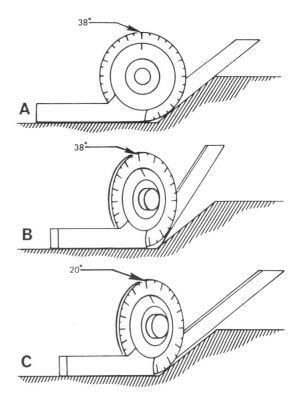

Fig. 17-16 Importance of positional relationship in angle measurement is shown here.

is the universal bevel protractor, or vernier protractor as it is usually called, and you are required to measure to 5 min. what would you do? Only a coward would give up! Go ahead and do it, but exercise every precaution you know. Then, if possible, have someone else check you.

A note of caution about care of the instrument is required. They are far more delicate by nature than anything mentioned thus far, including the high-amplification instruments. Figure 17-18 shows the kind of treatment vernier protractors expect.

PUTTING NATURAL FUNCTIONS TO WORK

The trigonometric functions are among the most useful tools in practical angle measurement. They are formed by the sides of triangles. These are always straight lines and can be measured by linear measuring instruments. Once we know two sides, we can easily express the fraction formed by them in decimal form.

RELIABILITY WITH PROTRACTORS

Mechanical considerations:
    1. Can the base and the blade both reach their respective surfaces unobstructed?
    2. Is overconstraint causing erronious contact?
    3. Do burrs, dirt or excessive roughness interfere with intimate contact?

Positional considerations:
  (Consider angle in yz plane.)
    1. Is vertical axis of instrument parallel to the plane of the angle?
    2. Is horizontal axis of instrument parallel to the plane of the angle?

Observational considerations:
    1. Is the reading the complement of the angle being measured?
    2. Is the reading the supplement of the angle being measured?
    3. Does parallax error exist?
    4. Are you conscious of bias?

Fig. 17-17   This check list will disclose chances for error in angle measurement.

CARE OF THE UNIVERSAL BEVEL PROTRACTOR

Before use:
    1. Wipe off dust and oil.
    2. Examine for visual signs of damage or abuse.
    3. Run fingers along base and blade to detect burrs.
    4. Check mechanical movement for freedom.
    5. Check clamps for security.
    6. Allow instrument to normalize.
    7. Determine that the instrument has been recently calibrated.

During use:
    1. Keep case nearby so that instrument may be placed in case rather than on hard surface when not being used.
    2. Avoid excessive handling to minimize heat transfer.
    3. Do not slide along abrasive surfaces.
    4. Do not overtighten clamps.
    5. Do not spring or bend by overconstraint.
    6. Take precautions to avoid dropping instrument and to avoid dropping objects on it.
    7. Avoid work near heat sources.

After use:
    1. Clean thoroughly. Do not use compressed air which could drive particles into instrument. Dip in solvent and shake dry if exposed to cutting fluids.
    2. Lubricate moving parts.
    3. Apply thin rust-preventative lubricant.
    4. Replace in case.

Fig. 17-18   These precautions require only minutes but they assure minutes accuracy.

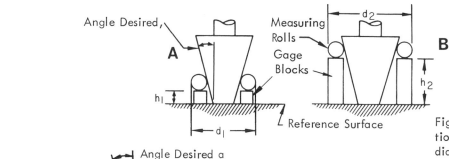

Fig. 17-19 Solving the fraction formed by the heights and diameters differences results in a decimal value for the tangent. From this the angle can be looked up in "trig tables".

$$\text{Tan } a = \frac{\text{side opposite}}{\text{side adjacent}} = \frac{h_2 - h_1}{\dfrac{d_2 - d_1}{2}}$$

Their usefulness was discovered early, particularly by astronomers. Rheticus, a student of Copernicus, published a 10-decimal place trigonometric table in Germany in 1596. By 1613 he had expanded the table to *15 places*. The following year Napier invented logarithms. They greatly simplified computation and stand as one of the man's major triumphs. From then until modern electronic computers took over, further development of trigonometric tables consisted of *logarithmic functions*. From that day to this there has been little change, other than error correction.

For elementary angle computation natural functions are adequate. With a "trig table," all you have to know is the function in order to look up the angle. Whenever the multiplication and division becomes laborious remember the logarithmic functions.

This general procedure is used both directly and to sire an entire family of instruments, the sine bars, sine plates, sine tables, etc. The direct method is the most difficult, so we will dispense with it first.

GETTING A TAN

One of the trigonometric functions used in practical angle measurement is the *tangent*. This is abbreviated *tan* and pronounced just that way. A typical application is the example chosen, measuring the angle that the side of an external taper makes with its axis.

Consider the setup shown in Fig. 17-19. A tapered part is set on a reference surface and two cylinders are supported one on either side. First the cylinders are on low stacks of gage blocks as at A, then on high stacks as at B. A micrometer or other instrument is used to measure the separation of the cylinder at both positions. The cylinders used for this purpose in machine shop practice are called *measuring rolls or measuring rods*.

The difference in height of the gage block stacks and the difference in the measurements over the cylinders provide the triangle which we must find in order to obtain the necessary sine, C in Fig. 17-19. In that triangle we do not know the hypotenuse but we do know the side opposite and the side adjacent to the angle. Making a fraction of these with the side opposite as the numerator gives us the tangent of the angle. Note that we divided the horizontal measurement in half because we are only concerned with the angle on one side of center.

The balance is easy. We simply divide out the fraction to obtain a decimal value for the tangent. Then this is looked up in a "trig table," and the angle is found. The table will read to minutes with provisions for interpolating to seconds. Does this mean that you have measured the angle to seconds? To minutes? Certainly not. The measurement can be to no greater precision than the least precise step in making it. And the *uncertainty* (lack of accuracy) will be the total accumulated effect of the separate errors. This point is so important that it will be discussed in detail later in the chapter. For now, *en garde!*

SIN IN METROLOGY

The sine of the angle is one of the most useful in measurement. I am not referring to the profusion of poor puns made possible by its abbreviation, *sin*, (pronounced *sine*, fortunately). It finds applications all the way from submicroscopic research to manufacturing inspection and on to astrophysics. Why is it so important? It provides the angle by which one path deviates from another, if their distances apart are known. Like all generalities you can poke this one full of holes but the basic fact remains.

Consider the part in Fig. 17-20. This is a bracket that supports another part at a fixed angle to the horizontal reference plane. The problem is to measure this angle. I have purposely selected awkward dimensions for this example, because in the next section I want you to welcome the simplicity of the sine bar.

To measure the angle we first require two *measuring rolls*. These are clamped to the part where shown. Then with any suitable instrument the distance between the cylinders is measured. If they are the same diameter, one diameter is added to the measurement to obtain the center distance. This is the hypotenuse of the triangle. Subtracting the height of each cylinder from the reference plane yields the side opposite the angle in question. The sine is then formed as a fraction, divided out, and looked up in the trig table. The result is the needed angle measurement.

The computations for this involve considerable work. Bear in mind that the computational steps are variables each containing an element of uncertainty, the same as the manipulative steps. First, the hypotenuse must be determined. If measurement between the rolls with a vernier caliper reads 7.106 in., then one radius should be added to each end to get the center-to-center measurement, 7.106 plus 0.375 or 7.856 in. The side opposite is found by measuring the heights of the rolls above the reference surface and subtracting, 6.125 minus 3.375 or 2.750 in.

This now gives us the ingredients to work out the sine of angle A. It is 2.750 in. divided by 7.856 in. The resulting decimal is 0.35005 or 0.3501. Note that this is just plain 0.3501, not 0.3501 <u>inch</u>. It is a ratio and would be the

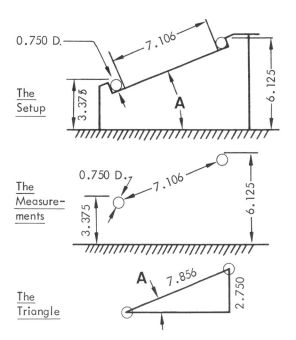

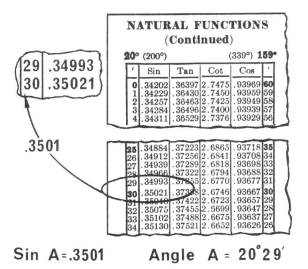

Fig. 17-20 The setup must be viewed as a problem in trigonometry to find the angle.

Sin A = .3501   Angle A = 20°29′

Fig. 17-21 The sine from Fig. 17-20 falls between the sine 20° 29′ and 20° 30′. Why was 29′ chosen instead of interpolating to seconds?

same if the original distances had been in centimeters, rods or Soviet archines.

The last step is to consult the table of natural functions, or trig tables, and locate the angle, Fig. 17-21. In order to be consistent with the precision of the measurements (vernier caliper), the angle was rounded off to the nearest minute instead of being interpolated to seconds.

THE SINE BAR

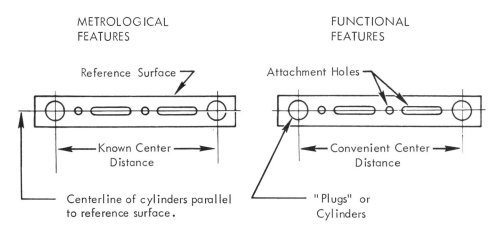

Fig. 17-22   The sine bar is a hypotenuse of a triangle frozen in steel with a length selected to minimize computations.

THE EASY WAY TO MEASURE ANGLES

A sine bar, Fig. 17-22, (or similar sine instrument) provides knowledge of angles by the same method that we have been using with one important difference -- part of the problem is built into the instrument and resolved.

The sine bar is a hypotenuse formed by a steel bar terminating in a cylinder near each end, Fig. 17-23. The distance between the cylinders is selected to make the computation easy. It is usually 5 or 10 in. The working surface is as parallel to the center line of the cylinders as it is practical to make it. Sine bars must be very carefully constructed because by their nature they introduce several elements of uncertainty, Fig. 17-24.

When one of the cylinders is resting on a surface, the bar can be set at any desired angle by simply raising the second cylinder. The desired angle is obtained when the height difference between the cylinders is equal to the sine of the angle multiplied by the distance between the centers of the cylinders.

Referring back to Fig. 17-20, it can be seen that a sine bar allows us to solve the same type of problem. The two important exceptions are that the hypotenuse is always the same, and its length is an easy-to-use constant. Consider the same measurement, but this time using a sine bar instead of two loose measuring rods.

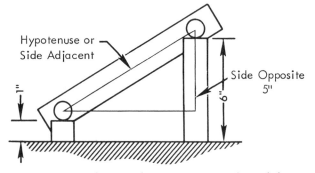

Fig. 17-23   The sine bar forms a triangle with hypotenuse either 5 in. or 6 in. The side opposite is the difference in the height of the cylinder supports.

SINE BAR MEASUREMENT VARIABLES

Geometric:
  1. Parallelism of the working surface to the centerline of the cylinders.
  2. Squareness of the axes of the cylinders to the instrument.
  3. Roundness of the cylinders.

Mechanical:
  1. Error in center-to-center distance.
  2. Differences in cylinder diameters.
  3. Surface imperfections such as insufficient flatness of working surface.

Setup:
  1. Error in two sets of height supports.
  2. Imperfect reference surface.

Fig. 17-24   These three sources of error should be recognized when using the sine bar.

In the first place, the setup is easier because the sine bar holds the cylinders. Separate clamping is not needed. The cylinders rest on the surface in question and the entire sine bar is lightly clamped, Fig. 17-25. In the example, an instrument such as a height gage is then used to measure the distance between the cylinders and the reference surface. No measurement is required between the cylinders, of course. Furthermore, the diameters of the cylinders do not enter into the problem as long as they are equal. Subtracting the heights provides the triangle needed.

Now, however, we do not need to divide out the sine. We simply refer to a table of constants for the 5-in. sine bar, Fig. 17-26. Reading across the degrees at the top we see that 1.750 is greater than 20° (1.7101) and less than 21° (1.7918). Therefore, we go down the 20° column and find that 1.750 is closer to 1.7496 than to 1.7510. The corresponding minutes are shown in the left column. The angle accordingly is 20° 29'.

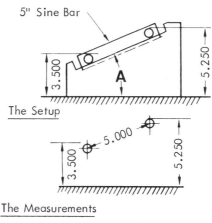

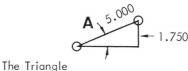

Fig. 17-25 The sine bar greatly simplifies the angle measurement in Fig. 17-20.

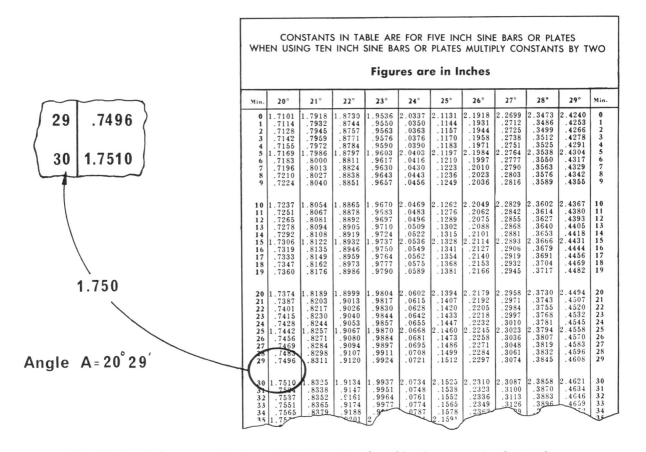

	29	.7496
	30	1.7510

1.750

Angle A = 20° 29'

CONSTANTS IN TABLE ARE FOR FIVE INCH SINE BARS OR PLATES
WHEN USING TEN INCH SINE BARS OR PLATES MULTIPLY CONSTANTS BY TWO

Figures are in Inches

Min.	20°	21°	22°	23°	24°	25°	26°	27°	28°	29°	Min.
0	1.7101	1.7918	1.8730	1.9536	2.0337	2.1131	2.1918	2.2699	2.3473	2.4240	0
1	.7114	.7932	.8744	.9550	.0350	.1144	.1931	.2712	.3486	.4253	1
2	.7128	.7945	.8757	.9563	.0363	.1157	.1944	.2725	.3499	.4266	2
3	.7142	.7959	.8771	.9576	.0376	.1170	.1958	.2738	.3512	.4278	3
4	.7155	.7972	.8784	.9590	.0390	.1183	.1971	.2751	.3525	.4291	4
5	1.7169	1.7986	1.8797	1.9603	2.0403	2.1197	2.1984	2.2764	2.3538	2.4304	5
6	.7183	.8000	.8811	.9617	.0416	.1210	.1997	.2777	.3550	.4317	6
7	.7196	.8013	.8824	.9630	.0430	.1223	.2010	.2790	.3563	.4329	7
8	.7210	.8027	.8838	.9643	.0443	.1236	.2023	.2803	.3576	.4342	8
9	.7224	.8040	.8851	.9657	.0456	.1249	.2036	.2816	.3589	.4355	9
10	1.7237	1.8054	1.8865	1.9670	2.0469	2.1262	2.2049	2.2829	2.3602	2.4367	10
11	.7251	.8067	.8878	.9683	.0483	.1276	.2062	.2842	.3614	.4380	11
12	.7265	.8081	.8892	.9697	.0496	.1289	.2075	.2855	.3627	.4393	12
13	.7278	.8094	.8905	.9710	.0509	.1302	.2088	.2868	.3640	.4405	13
14	.7292	.8108	.8919	.9724	.0522	.1315	.2101	.2881	.3653	.4418	14
15	1.7306	1.8122	1.8932	1.9737	2.0536	2.1328	2.2114	2.2893	2.3666	2.4431	15
16	.7319	.8135	.8946	.9750	.0549	.1341	.2127	.2906	.3679	.4444	16
17	.7333	.8149	.8959	.9764	.0562	.1354	.2140	.2919	.3691	.4456	17
18	.7347	.8162	.8973	.9777	.0575	.1368	.2153	.2932	.3704	.4469	18
19	.7360	.8176	.8986	.9790	.0589	.1381	.2166	.2945	.3717	.4482	19
20	1.7374	1.8189	1.8999	1.9804	2.0602	2.1394	2.2179	2.2958	2.3730	2.4494	20
21	.7387	.8203	.9013	.9817	.0615	.1407	.2192	.2971	.3743	.4507	21
22	.7401	.8217	.9026	.9830	.0628	.1420	.2205	.2984	.3755	.4520	22
23	.7415	.8230	.9040	.9844	.0642	.1433	.2218	.2997	.3768	.4532	23
24	.7428	.8244	.9053	.9857	.0655	.1447	.2232	.3010	.3781	.4545	24
25	1.7442	1.8257	1.9067	1.9870	2.0668	2.1460	2.2245	2.3023	2.3794	2.4558	25
26	.7456	.8271	.9080	.9884	.0681	.1473	.2258	.3036	.3807	.4570	26
27	.7469	.8284	.9094	.9897	.0695	.1486	.2271	.3048	.3819	.4583	27
28	.7483	.8298	.9107	.9911	.0708	.1499	.2284	.3061	.3832	.4596	28
29	.7496	.8311	.9120	.9924	.0721	.1512	.2297	.3074	.3845	.4608	29
30	1.7510	.8325	1.9134	1.9937	2.0734	2.1525	2.2310	2.3087	2.3858	2.4621	30
31	.7524	.8338	.9147	.9951	.0748	.1538	.2323	.3100	.3870	.4634	31
32	.7537	.8352	.9161	.9964	.0761	.1552	.2336	.3113	.3883	.4646	32
33	.7551	.8365	.9174	.9977	.0774	.1565	.2349	.3126	.3896	.4659	33
34	.7565	.8379	.9188	.9	.0787	.1578	.2362				34
35	1.757		.9201	2.		2.1591					35

Fig. 17-26 Little computation is required when using the table of constants for the sine bar.

SHORT-CUT CALCULATIONS FOR FIVE-INCH SINE BARS

There is a good reason that five inches was chosen as the standard center distance for most sine instruments. Since multiplying by five is the same as multiplying by ten and dividing by two, computation is very easy even when a table of constants is not available and the table of natural functions must be used.

If you want to set a five-inch sine bar to an angle, simply look up the sine in a table of natural functions, move the decimal point one place to the right and divide by two. The result is the height that one cylinder must be elevated above the other one. *To find the angle when you know the height difference move the decimal point one place to the left, multiply by two and look up the result in the table.*

To see if this works, go back to Fig. 17-25. The height difference was 1.750. Moving the decimal point one place to the left results in 0.1750. Multiply by two and this becomes 0.350. That is close enough to the 0.3501 that we originally had when we did it the hard way, Fig. 17-21, to prove the point.

For a 10-inch sine instrument it is only necessary to move the decimal point right or left. For the rather uncommon 20-inch instrument the decimal point is moved but instead of dividing by two when locating the sine of a given angle, you multiply by two. And, of course, you divide by two when looking up the angle from a known height difference.

Suppose that a table of constants for a five-inch sine bar is available but that the instrument is some other length. In that case, the figures in the table can be converted by multiplying by constants. These are given in Fig. 17-27 for five sizes.

COMPARISON MEASUREMENT WITH SINE BARS

The above method, while useful, has serious limitations. What is the limit of *precision* in this method? It is the *sensitivity* of the measuring instrument. In the example chosen, as in most direct measurement with the sine bar, the instrument was the vernier height gage. This automatically limited the precision to *one mil* (0.001 in.). What is the *accuracy*? It can never be better than the precision and may be poorer because of the combination of independent errors. The accuracy of the vernier height gage is of such a low order, compared to the workmanship of the sine bar or the flatness of the surface plate, that it is probably the determining factor.

In order to improve the accuracy, it is obviously necessary to replace the height gage with an instrument of greater sensitivity. When studying surface plate work, you will meet a group of instruments that measure heights with micrometer accuracy or better. Suppose, however, that this is not good enough and you desire to measure the angle with high-amplification instrumentation. Easy, you use gage blocks and measure by comparison.

In this case, the sine bar is used to construct an angle equal to the angle being measured, *but in the opposite direction*. The part is then supported by the sine bar. The two angles compensate and cancel out if exactly equal. Any deviation can be detected by measuring the parallelism between the part feature and the reference surface, Fig. 17-28.

SINE BAR CONSTANT FACTORS

Based on 5-inch sine bar constants	
When using:	Multiply constant by:
20" sine bar	4
10" sine bar	2
2 1/2" sine bar	0.5
3" sine bar	0.6
4" sine bar	0.8

Fig. 17-27 Many sine bars use center distances other than 5 in. Obviously an error in setting a 10-in. sine bar would cause half the inaccuracy in measurement that would be caused by the same error with a 5-in. sine bar.

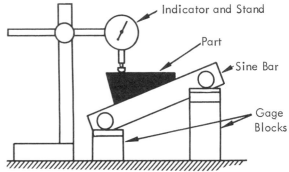

Fig. 17-28 For comparison measurement, the sine bar is used to cancel out the angle being measured.

If instead of a dial indicator as shown in the example, a high-amplification comparator is used, the precision of the measurement may approach the accuracy of the gage blocks. This is seldom practical, because the uncertainty in the setup would then be much greater than that of the instrumentation.

SINE BAR EVOLUTION

Not self-supporting. Interference at vertex end.

A

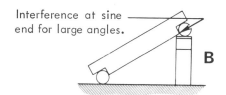

Interference at sine end for large angles.

B

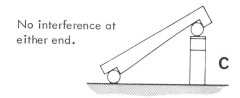

No interference at either end.

C

Fig. 17-29 All three types of sine bars are in extensive use, although the more refined types clearly evolved after the simple type at A.

Figure 17-29 shows the evolution of the sine bar. The one at A was an obvious first. It is still extensively used because of low price and convenience for small parts. It is not self-supporting and must always be clamped for vertical use. The chief disadvantage is that the lower plug cannot rest directly on the reference surface. The type in B is more difficult to make but it is self-supporting. Unfortunately it, too, has interference in a portion of its range. That interference is removed in the design at C. While an apprentice, the author discovered for himself still another reason for the C design. After scrapping two bars of the B-type by lapping them undersized, it became clear that with the C-type you can always correct an undersize condition by lapping the opposite plug-locator shoulder. This more than compensates for the extra machining.

SINE BLOCKS AND SINE PLATES

Sine blocks are wide sine bars. *Sine plates* are wider sine blocks. *Sine tables* are still wider. Where the dividing lines are depends upon whose catalog you use. Not all manufacturers even make the same distinctions between sine instruments. MIL-STD-120 skips all terms except sine bar and sine plate. In it all sine instruments over one-inch wide are sine plates. If we must derive our own distinction, the following is practical: *A sine instrument wide enough to stand unsupported is a sine block, Fig. 17-30. If it rests on an integral base it becomes a sine plate, Fig. 17-31. If it is an integral part of another device, such as a machine tool, it is a sine table.*

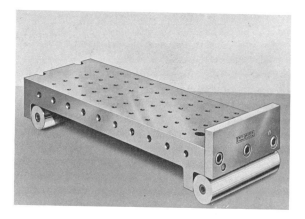

Fig. 17-30 A sine block is a wide sine bar. They usually have tapped holes for the attachment of parts and a stop to prevent parts from sliding off. *(Taft & Pierce Co.)*

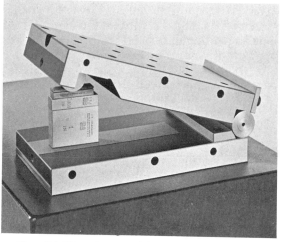

Fig. 17-31 A sine table is a sine block with an attached base. *(Johansson Gage Co.)*

All of these sine instruments differ from sine bars in one important respect. The part rests on them. Rarely can a setup be made in which they rest on the part as in Fig. 17-25. This feature restricts them on one hand and extends them on the other. The only practical way they can be used to measure angles is by comparison. But they can also be used to hold parts so that angles may be produced on them. Some, in fact, are made specifically for the latter purpose, Fig. 17-32, and can hardly be considered in the measurement instrument class. Tolerances for these instruments vary. Those of one high-grade manufacturer are shown in Fig. 17-33.

Fig. 17-32   This heavy-duty sine plate is rugged enough to hold parts for the machining of angles as well as for the inspection of angles. *(Taft & Pierce Co.)*

TOLERANCE OF SINE INSTRUMENTS

	BAR	CYLINDERS		
Size	Working Surface to be Flat, Square with Sides and Parallel Within	Cylinders to be Alike, Round and Straight, Within	Cylinders to be Parallel with Each Other and With Working Surface of Bar Within	Cylinders to be at Nominal Center Distance (±)
IN.	COMMERCIAL CLASS			
5	0.00010	0.00005	0.00010	0.0002
10	.00015	.00005	.00015	.0003
20	.00020	.00006	.00020	.0004
IN.	LABORATORY CLASS			
5	0.000050	0.00003	0.000050	0.00010
10	.000075	.00003	.000075	.00015
20	.000100	.00004	.000100	.00020

Fig. 17-33   These are typical tolerances for quality sine bars, sine blocks and sine plates.

WHEN A COMPLEMENT MEANS A LOT

The precision of sine measurement decreases as the angle increases. The reason is shown in Fig. 17-34. Remember that the sine is a fraction formed by dividing the side opposite the angle by the hypotenuse, and that the hypotenuse is always the same with any particular sine instrument.

At 1° the side opposite is extremely minute as compared with the hypotenuse. This makes the sine very small, 0.01745 in this case. An increase of only one degree increases the size opposite tremendously compared to its original size. At 2° the sine is almost double the sine of one degree, of 0.03490. The jump to 3° also makes a large change, but not nearly as much as the previous one. Each successive increase sees the amount of change in the sine reduced.

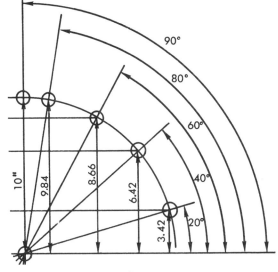

Fig. 17-34   The sine changes rapidly near 0° and very slowly as it approaches 90°.

If these changes seem large consider what happens for really small angles. The sine of 0° 0' 10" is 0.000,048,481,4.* Ten *seconds* larger and the sine is 0.000,096,962,7. It misses doubling by only *one-tenth of a billionth.* This might seem obscure when measuring in the mass production of refrigerators, but an error of a like amount could easily cause a landing on the moon to end up as an unwanted satellite.

If we consider what happens at the opposite extreme, near 90°, the difference is striking. The sine of 90° is unity to whatever number of places you carry it because both the hypotenuse and the side opposite are equal. At 89° the side opposite has decreased only slightly. It is nearly as large as the hypotenuse and the sine is reduced only from 1.00000 to 0.99985. Between 89° and 88° the change in the sine is only 0.00046. At this end of the 0 to 90 degrees quadrant, a large change in angle results in a very small change in the sine. This is exactly opposite from the other end of the quadrant.

This has an important effect upon measurement. The reason is shown in Fig. 17-35. At A the part is being checked to determine whether or not the angle is 65°. In the example it is not exactly 65 degrees because the indicator reads *five mil* (0.005 in.) lack of parallelism in 2 inches. At B the setup has been made to measure the complement of the angle or 25°. Again the lack of parallelism is *five mil* (0.005 in.) in 2 inches. Does the *five mil* (0.005 in.) in a 2-inch indicator reading in each case show that the angles have been read with precison? Not at all. That parallelism change at the steep angle at A does not represent nearly the same angle change as it does for the complement shown at B.

The larger the angle, the more important this becomes. At 80° a 10-inch sine bar requires only a *five-tenth mil* (0.0005 in.) height change to change the angle one minute. At 10 degrees a one-minute angle change is caused by a *two point eight mil* (0.0028 in.) height change. This is over five times the amount required at the complement. Obviously anything that disturbs the measurement at 10 de-

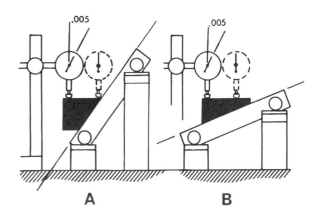

Fig. 17-35   In A, the indicator shows that the out-of-parallel condition is 0.005 in 2 inches. In B, it reads the same. Does this mean that both angles are being read to the same precision?

grees will have only one-fifth the effect that it would have at 80 degrees. *When an angle greater than 60 degrees is measured with sine instruments, the complement of the angle should be used.* Some authorities do not recommend using sine instruments at angles over 45 degrees.**

USE OF GAGE BLOCKS

This sounds easy but a question logically arises. How do you use a sine instrument to measure an angle whose sine is smaller than the thinnest gage block? Although gage blocks are made as thin as *ten mil* (0.010 in.) they are uncommon. Most sets start at *one hundred mil* (0.100 in.). To make it worse, rarely will you require a sine that is the exact size of any one gage block. In most cases you must combine two or more blocks which raises the height of the stack even more.

No problem results when using sine bars of the type shown in Fig. 17-29, and Fig. 17-35. The lower cylinder must rest on a gage block or stack of gage blocks so the sine will always be the difference in heights of the stacks. The same principle applies when using sine blocks, Fig. 17-30.

* Sines, Cosines and Tangents 0°-6°, Coast and Geodetic Survey, Washington, D. C., U. S. Government Printing Office, Spec. Publication No. 246, p. 1, Price $0.30.

* * Military Standard Gage Inspection, MIL-STD-120, Washington, D. C., U.S. Government Printing Office, December 12, 1950, 8.5.3.4, p. 93.

Sine tables complicate this because of their hinge construction. You cannot place gage blocks under the hinge cylinder, of course. To permit blocks to form the sines of very small angles a step is provided in the base of these instruments, Fig. 17-36.

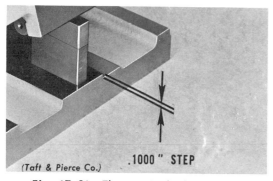

(Taft & Pierce Co.)   .1000 " STEP

Fig. 17-36 The step in the base permits angles to be set whose sines are smaller than the thinnest gage blocks.

Should the gage blocks be placed with their long dimension in line with the cylinders or across the cylinders? All of the illustrations show them across the cylinders, yet from the metrological standpoint the other way would be best. In the discussion of comparators the high unit pressures caused by a spherical surface was mentioned. Similarly, a line contact such as that of the cylinders of a sine instrument also create high pressures. This slightly deforms the surfaces on which they rest and the cylinders themselves. Obviously, the longer the line of contact, the less deformation. From this we assume that all of the illustrations have been wrong. This is a case

in which practical considerations have over-ruled the metrological ones. The setups shown are more stable and less subject to toppling over, and in these cases they won out.

A far better compromise is to *use square gage blocks* whenever they are available. When heavy setups are made double stacks of gage blocks are recommended. These should be as widely spaced as the cylinders and finished reference surfaces permit.

The use of gage blocks between the lower cylinder and the reference surface is recommended even when not required mathematically to make up the sine of the angle. The line contact of the cylinder may fall into low portions of the reference surface. Furthermore, its high unit pressure may deform the surface, as has been mentioned. The very hard and dense surface of gage blocks is preferable for the cylinder to rest upon.

The two cases are shown in Fig. 17-37. How do you know which will be best? Uncommon "common sense" will give the answer. In lieu of that, analyze the errors. A large number of errors are common to each method. The minute compression of the cylinders, when under weight load, is an example. We are concerned, in cases such as this, with the errors that are inherent only with each specific case under consideration. In A, the two important errors are the surface imperfections and compressibility of the reference surface. In B, these have been minimized by the insertion of a gage block between the lower cylinder and the reference surface. But this

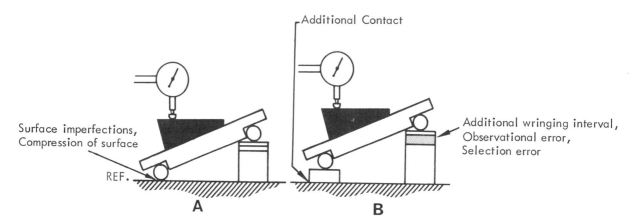

Additional Contact

Surface imperfections, Compression of surface

REF.

Additional wringing interval, Observational error, Selection error

A   B

Fig. 17-37 The comparator does not know the difference between A and B except for the errors caused by the two setups.

requires an equivalent addition to the other stack of gage blocks. With this comes a host of errors: imperfect contact between the added gage block and the reference surface, selection error in combining the higher stack, added observational error in reading the stack and added wringing error as more blocks are required.

Suppose the reference surface was a toolmaker's flat. Then nothing possibly could be gained by adding the block under the lower cylinder. On the other hand, if the reference surface is only a rough machined portion of a large part, then all of the added chances of error are minor compared to the one being minimized. Note, as a preview of coming attractions, that a rough machined surface can be very flat.

COMPOUND ANGLES

It is beyond the scope of this text to present a full treatment of compound angles. This subject usually is covered in mathematics courses. It is impossible to do much practical measurement either in the shop or the laboratory without encountering them. And that is the only excuse the author required to introduce this interesting phase of measurement.

Simple angles lie in one plane. Compound angles are the angles formed by the edges of triangles which lie in different planes. If these triangles lie in different planes they define the boundaries of solids, real or imaginary. Their edges are the intersections of the planes.

FACE ANGLES

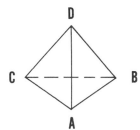

Fig. 17-38 In this pyramid there are four sets of face angles, three around each point A, B, C and D.

In a solid formed by the intersections of planes (we will conveniently avoid curved figures), the angles on the surface planes are called *face angles*. This is shown for a pyramid in Fig. 17-38. There are four sets, one for each point. Around each point there are three face angles. Around A for example, there are ABC, BAC and CAD.

DIHEDRAL ANGLE

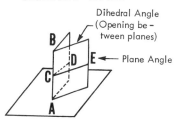

Fig. 17-39 The plane angle lies in a plane perpendicular to the intersection of the two planes.

A *dihedral angle* is the opening between two intersecting planes. The *plane angle* of a dihedral angle is an angle whose sides are in two intersecting planes and perpendicular to the line of intersection at the same point. If this is difficult to visualize, Fig. 17-39 will help. Here we have two planes intersecting at line AB. The dihedral angle is the opening between the planes. To express the size of this angle we construct the plane angle, CDE. This angle lies in a plane perpendicular to the line AB.

A dihedral angle is formed between every pair of faces in any solid figure. Only in the case of cubes and rectangular parallelepipeds are the face angles equal to the plane angles. Whenever one or more sides are at any angle other than a multiple of 90 degrees, the plane angles will differ from the face angles.

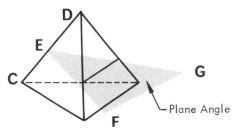

Fig. 17-40 Each pair of sides of the pyramid has a plane angle. These plane angles differ from the face angles and determine the shape of the pyramid.

This is shown in Fig. 17-40 for the pyramid we have used as an example. Of the 12 face angles, 10 could be altered without changing the plane angle FEG. There are as many plane angles as there are edges. In this example there are six plane angles. This is half as many as there are face angles. From all of this it is clear that the plane angles are a more efficient means for defining the shape of a solid than the face angles. Therefore, it is the plane angles that must be dealt with when either machining solid figures or in measuring them.

Remembering your plane trigonometry you will recall that no matter how odd-looking a triangle might be, it can always be divided into right triangles. So it is for compound angles. Some combination of two angles will form every compound angle. These combinations fall into five types, Fig. 17-41.

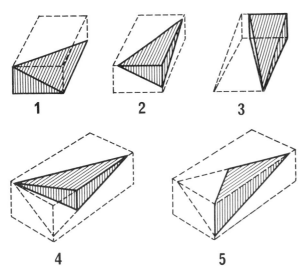

1 **2** **3**

4 **5**

Fig. 17-41 In order to find the compound angle, most solid figures reduce to one of these five types.

Each consists of two planes perpendicular to a third. These planes provide the references without which there could be no measurement. The fourth plane that closes the figure is the one that creates the compound angles. Because each pair of planes has a dihedral angle between them, the closing plane creates three compound angles. In each case one side of the dihedral angle is one of the reference planes. This does not imply that you will encounter compound angle measurement problems that resemble these five types. It means that the ones you do encounter can be divided into two or more parts, each of which will fit one of the five general types.

In all except type 3 the figure can be visualized as circumscribed by a rectangle. In type 3 a wedge circumscribes the figure. This is important because it shows the reason why sine plates can be used to measure compound angles. First, however, a reminder will do no harm. Remember that sine instruments do not measure angles directly. They are set up to cancel out the angle, so that a comparator can verify the angle by checking parallelism to the reference surface.

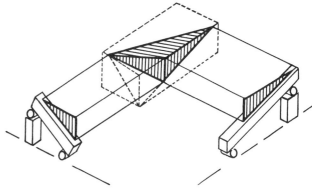

Fig. 17-42 All compound angles may be reduced to simple angles, providing that adequate reference surfaces are located.

Figure 17-42 shows that sine instruments can be set to the face angles of the figure by projecting them to the sides of the imaginary rectangular parallelepiped that circumscribes them. Type 4 is used in the example. To do this the pivoting axes of the sine instruments are placed at right angles to each other. The next step, logically, is to place them one on top of the other. And that precisely is the principle of the *compound sine angle plate.*

THE COMPOUND SINE ANGLE PLATE

The result of two superimposed sine plates is shown schematically in Fig. 17-43. The base creates one plane and the top is the second plane. Any two planes that are not parallel intersect. In practical cases this intersection will be an imaginary line in space. The line along which these planes intersect passes through the intersection of the hinge lines. The plane angle, which is the compound angle we desire, lies in a plane perpendicular to the line of intersection of the two planes. Note that it is different from either of the angles set in the upper or lower parts of the sine plate, and that the plane in which it lies is *not* perpendicular to any of the planes of the instrument.

A typical compound sine angle plate is shown in Fig. 17-44. These are used to form compound angles as extensively as to measure them. Thus they are generally massive and can support heavy loads. The rough forming of angles, simple and compound, generally depends upon the mechanical movements built into milling machines, boring mills and other

COMPOUND SINE ANGLE PLATE

METROLOGICAL FEATURES   FUNCTIONAL FEATURES

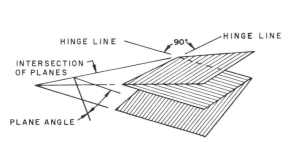

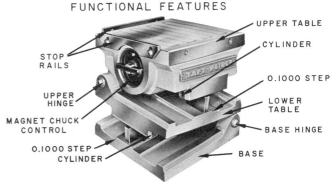

Fig. 17-43 The plane angle is formed by the inter-action of the two sine plates. It is neither of the angles set in the separate sections.

Fig. 17-44 The compound sine angle plate is two simple sine plates with hinge lines at right angles to each orner. *(Taft & Pierce Co.)*

machine tools. The sine plates are not required until the finishing operation which is usually performed on grinding machines. For this purpose, both simple and compound sine plates are available with built-in magnetic chucks, Fig. 17-44 and 17-45.

A typical inspection application of the compound sine angle plate is shown in Fig. 17-46.

Frequently, it is not necessary to know the plane angle in order to set up a compound sine angle plate. Often the values of the angle in each of its two component planes are known. In these cases the instrument is used as if it were two simple sine plates. When the plane angle must be set, it is necessary to compute it. The computation is not difficult but is too lengthy for discussion here. It is included in all thorough texts on machine shop mathematics.*

Before leaving compound angles, a little trick should be mentioned that may save you early gray hairs trying to visualize compound angles. Modeling clay is an excellent short cut for visualization. A part can be easily modeled, turned about and looked at until the desired angle is clearly determined. As a quick check, you can slice through the model perpendicular to any edge and find the plane angle. Although your tolerances may have to be plus or minus 5 degrees or more, you will at least know whether or not you are on the right track.

*State University of New York, <u>Machine Shop Mathematics</u>, Albany, N.Y., Delmar Publishers, Inc., 1941

Fig. 17-45 Sine plates with magnetic chucks are used to finish work accurately to the desired angles, either simple as shown here or compound as shown in Fig. 17-44. *(Taft & Pierce Co.)*

Fig. 17-46 For inspection work or measurement, the compound sine angle plate and comparator are both placed on an accurate reference surface.

MECHANICAL ANGLE MEASUREMENT

It was pointed out that the compound sine angle plate finds as much use in machining angles as it does in measuring angles. Another method, called *mechanical indexing,* was developed specifically for machining rather than measurement. These are the *dividing heads, indexing heads or index heads.* All three names are correct. This book will use the latter. They are more machines than instruments. However, they are very valuable for angle measurement, and are included here. The mention will be brief because they are covered extensively in texts on machine tool operation.*

As the name implies, these devices were originally developed to divide circles into equal divisions. Milling teeth in cutters and flutes in reamers, and spacing bolt circles are familiar operations. Because the circle is the standard for angles, any division of a circle is angle measurement. This is just as true when we speak of the number of divisions as when we speak of degrees and minutes.

There are three principal classes of index heads: dial, plain and universal. Each of these can be divided into still further subclasses, particularly the latter which can be extended to very sophisticated devices. Like the other instruments discussed in this book, the mechanical angle measuring devices range from low amplification to relatively high amplification.

---

*State University of New York, <u>Milling Machine Work,</u> Albany, N. Y., Delmar Publishers, Inc., 1953, p. 220.

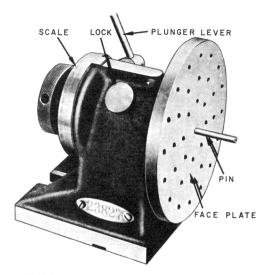

Fig. 17-47   The dial index head quickly divides rotation into 24 divisions 15 degrees apart.

THE DIAL INDEX HEAD

The lowest amplification is the dial index head. This instrument with one power (1X) is now relatively unimportant. An understanding of it, however, will aid your grasp of the higher power variations. It consists of a horizontal spindle mounted on a base that rests on a reference surface, Fig. 17-47. On one end of the spindle is a plate for clamping parts. On the other end is a knob or wheel for turning the spindle and usually a ring calibrated in degrees. The plate has another function. It is also an index plate. As such it contains holes which a plunger in the stationary housing engages. Usually there are 24 holes. This provides 360 degrees of rotation in 15-degree increments.

DIAL INDEX HEAD CAPABILITY

Method of Operation	Discrimination	Sensitivity	Accuracy
Graduations	1°	30'	30'
Index Plate	15°	30"	1'

Fig. 17-48   The widely divergent precision of the two methods of operation shows the contrast between human muscle and eye coordination and the precision that can be built into mechanical devices. The above values are typical but vary among manufacturers of dial index heads.

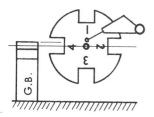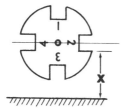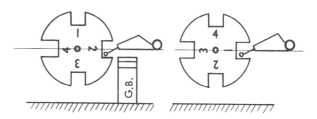

1st Step
Set head on zero. Clamp part loosely. Zero in — dicator on pin. Using gage blocks or height gage, determine height of pin.

2nd Step
Subtract half of pin diameter from pin height. Subtract half of slot width in order to find height x.

3rd Step
Zero indicator on dimension x. Indicate slot. Now note deviation from zero.

4th Step
Index to next slot. Indicate slot. Note deviation from zero. Continue for remaining slots.

Fig. 17-49 For clarity the dial index head is not shown in this sequence of operation steps. The deviation read by the indicator will contain both the index error and slot width error.

There are two types of sensitivity and accuracy for this instrument and the contrast is interesting, Fig. 17-48. It drives home a point made earlier in the book. *You cannot judge the accuracy of an instrument by its graduations.* By using the graduations you may set the head to 1 degree. If reliability is important, better not plan to repeat the setting closer than 1/2 degree. Hence the 30-minute sensitivity shown in Fig. 17-48. In this case sensitivity uncertainty dwarfs instrument uncertainty. This also limits the accuracy to 30 minutes.

Then consider settings using the index plate. These have a discrimination of 15 degrees but their sensitivity and accuracy are limited only by the care used in the design and manufacture of the head. The ones shown in Fig. 17-48 are typical, but vary considerably among manufacturers and in amount of wear.

Parts using the index head for machining or measurement have an axis of rotation and usually an open center. This enables them to be mounted simply to the plate. At the center of the plate is a hardened pin. Bushings are slipped into the pin to adapt it to the inside diameter of the part. This centers the part to the limit of accuracy of the pin, bushings and part inside diameter. It is then held to the plate with clamps. The pin is also used as a reference for measurement.

Use of the dial index head is covered in books on inspection and machining.* Figure 17-49 shows the general principles for a typical problem. Note that this measurement is influenced by two independent variables, the index uncertainty and slot width uncertainty.

Does this mean that you cannot measure the angular positions of the slots without being concerned with slot widths? No, either can be determined. If only the angle measurement is required, you would simply repeat steps two through four from the left side of the part. The resulting set of readings can be used in conjunction with the first set to determine slot width error or angle error — another example of the reversal process.

The shortcomings of the dial index head are the lack of discrimination when using the index plate and the lack of sensitivity when using the graduations. Improvement in either requires additional amplification. First, consider added amplification for the graduated ring method of angle measurement. You already are familiar with vernier scales and micrometer screws for increasing amplification. The microscope is also an important method. All have been applied to indexing heads in various combinations.

* "Dial Index Head," Precision Measurement in the Metalworking Industry, Syracuse, N. Y., Dept. of Education of International Business Machines Corp., Syracuse University Press, 1939, 1952 Revision, p. 204.

Fig. 17-50 This index head uses vernier micrometer and microscope principles to increase the sensitivity.

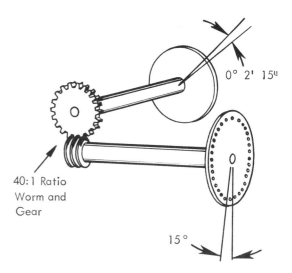

Fig. 17-51 If 40 to 1 gearing is placed between the dial index head, index plate and the face plate, the rotation of the face plate per index is reduced from 15° to 0° 2' 15".

One highly sophisticated head that incorporates all of these methods is shown in Fig. 17-50. This head has a discrimination of one second and an inherent accuracy of nearly the same amount. In expert hands the sensitivity may be considered one second. This is equivalent to 1,800X when compared to the 30-minute sensitivity of the dial index head.

PLAIN INDEX HEAD

The plain index head evolved from the dial index head by applying amplification mechanically. Note that when using the index plate of the dial index head, we have relatively high sensitivity and a low error variable. Unfortunately the discrimination limited us to 15-degree increments. What is to stop us from adding more index holes in the plate? Space. The holes have to have some size and that forever prevents high discrimination by that means. Why not increase the size of the plate? Size again stops us. A plate with holes of a quarter-inch diameter would have to be at least 24 *feet* in diameter to have a discrimination of one minute.

The solution came from gearing. Suppose the index plate is connected to the spindle and face plate in such a way that it makes two revolutions to one of the spindle. An index plate spacing of 24 holes (15 degrees) would have to revolve twice to turn the face plate once. It would have made 48 stops along the

way, thereby dividing the face plate rotation into increments of 7 1/2 degrees. Thus the discrimination is doubled.

The plain index head does not stop at 2X amplification. A worm and a gear are utilized to provide a ratio of 40 to 1, or 40X, Fig. 17-51. One turn of the index plate turns the spindle only 9 degrees. If the same 24-position index plate that was used in the dial index head is considered, the discrimination leaps from 15° to 0°2' 15".

In order to make full use of the increased amplification, the index plates are fixed and a crank is turned. The crank carries a pin which can be located in the index plate holes. The index plates are interchangeable. A great variety of construction is available from the various manufacturers. The plain index head in Fig. 17-52 is typical.

(Cincinnati Milling Machine Co.)
Fig. 17-52 The interchangeable index plates enable the plain index head to divide a circle in a wide range of spacings.

An index plate contains several circles of equally spaced holes. Each circle, of course, contains a different number of holes. As mentioned, a complete revolution of the crank turns the spindle 9 degrees. This can begin from any *initial position*, that is, from any hole in which the pin is inserted. If angles in multiples of 9 degrees are desired, the crank is turned the correct number of turns ending at the *terminal position*. If the desired angle is less than 9 degrees, or between multiples of 9 degrees, a terminal position is selected different from the initial position.

Consider, for example, turning the spindle 30 degrees when the index plates has six rows of holes, 18, 24, 29, 30, 34, and 37. Three turns of the crank result in 27 degrees. Three more degrees are needed. That is one third of a revolution of the crank. Any row of holes equally divisible by 3 could be selected. If the 24-hole circle is selected, the crank would be advanced 8 spaces beyond the initial hole. Note that to advance 8 spaces requires that the pin is *placed in the 9th hole*, counting the initial hole as one.

If 30°12' had been required it could have been figured just as easily. Form a fraction whose denominator is the total minutes of spindle movement for one turn of the crank. That is 9 times 60 or 540 minutes. The numerator is the number of additional minutes required after the last complete turn of 9 degrees. For 30°12', that would be 3°12' or 192 minutes. The fraction is 192 over 540. Reduce the fraction until the denominator is the number of holes in one row on an index plate. Dividing by 8 reduces the fraction to 24 over 30. Therefore, the number of additional spaces required on the 30-hole row is 24.

In many practical problems, several identical rotations are required. This is simplified by an adjustable segment that rotates around the crank shaft, Fig. 17-53. The selector arms of the segment can be spaced to span the added number of holes required past the last full turn of the crank. The bevel edge of one arm is alongside of the initial position hole. The other is on the opposite side of the terminal position hole. The crank is rotated the required number of full turns and then stopped at the terminal position hole. The pin is inserted and the segment rotated to bring the edge of the other arm against the pin. The head is then set for the next indexing.

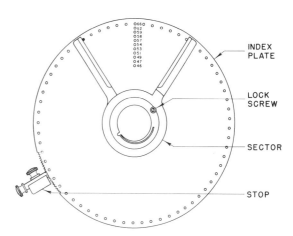

Fig. 17-53 These movable segments permit an angle setting to be repeated without counting holes.

UNIVERSAL INDEX HEAD

This head looks quite similar to the plain index head except somewhat bulkier, Fig. 17-54. Internally it is quite different. The head is in a trunnion so that it can be tilted in respect to the base. Even more important, its construction permits *differential indexing*. This is required when the angle desired cannot be made up of any combination of full turns and hole spacing provided with the plain index head.

The index plate on the universal head is fixed to the housing for plain indexing. When this will not produce the desired angle, it is connected by gearing to the crank. For example, assume that the crank must be moved 1/9 turn to produce the desired angle but that the index plate has 8 holes.

(Cincinnati Milling Machine Co.)

Fig. 17-54 The universal index head differs from the plain index head in that it is mounted in a trunnion and that it permits differential indexing.

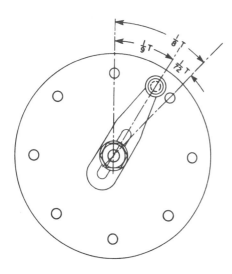

Fig. 17-55 Differential indexing enables the index plate to advance or recede with crank rotation, thereby greatly increasing the number of places the crank can stop and have its pin enter an index hole.

If the crank were turned 1/9 turn from any initial hole, it would fall short of the next hole by 1/8 times 1/9 or 1/72 of a turn. In order to use that hole, the plate would have to be advanced by the same amount, Fig. 17-55. Obviously this simple case would not have required differential indexing because several rows of holes are divisible by 9. It is essential, however, for many divisions such as 57, 271 or 319.

METROLOGY OF MECHANICAL INDEXING

Mechanical index heads developed from the needs of the shop. They are rugged and have functional features unlike any of the previously discussed instruments. They are made not only to support heavy parts but to withstand the forces of machining. On a high quality head the spindle base error will be less than 0.00025 in. The spindle is aligned with the reference plane of the base so that there will not be more than one mil (0.001 in.) runout at the end of an 18-in.-long test bar.

We are concerned primarily, in this book, with their metrological characteristics, their accuracy, sensitivity and precision. The accuracy of index heads is usually stated as total accumulative error. Indexing is analagous to the stepping off of a circle with dividers back in Fig. 5-21. For a high quality dividing head the accuracy is stated as follows: "*The accumulated error, in indexing from one hole*

to the next through a complete circle, must be within 0.0015 in. on a 12 in. dia." This moot statement means that the angular error is in proportion to the size of the angle, not in proportion to the number of divisions. Note that in practical work, angles do not necessarily stop and begin all over again at 360 degrees. In many problems it is easier to consider angles of multiples of 360 degrees. The accuracy as stated is not particularly astonishing, but look at it another way. This is equivalent to *eighteen mil* (0.018 in.) in 12 feet.

You probably have noticed that precision and discrimination are the same under certain circumstances. They are different when the instrument reads in response to something done to it, like shoving a gage block under the spindle. They are the same when the instrument is the doer; for example, when an angle is created by a sine bar. Dividing heads are of the latter case. They create angles to which an unknown angle may be compared.

The precision of an index head is the fineness to which it can be set to angles, or the extent to which it can divide the circle. This is dependent upon the gear ratio, Fig. 17-51, and number of holes in the index plate. Standard index plates have eleven rows of holes on each side. If all combinations of full turns and hole spaces are considered, many angles will be duplicated. Even after these have been eliminated a tremendous number of separate angles remain. The more important ones are listed in Fig. 17-56. Many others can be formed. For example, with the 54-hole circle all degrees may be obtained. This discrimination is extended even further by the high-number index plates, Fig. 17-57.

The amount of indexing discrimination that can be used is limited by the error. The maximum of 0.0015 in. in a 12-in. diameter is one part in 25,133. It generally would be unreliable to attempt a setting closer than ten times the uncertainty. That reduces the usable discrimination to one part in 2,513, or about 8 minutes.

The mechanical indexing principle has been extended to heads that provide very high precision. The example in Fig. 17-58 is a universal head with a *wide-range divider*. It provides divisions from 2 to 400,000 with angular divisions in intervals of 6 seconds.

DIVISIONS WITH STANDARD INDEX PLATE

No. of Divisions	Circle	Turns	Holes	No. of Divisions	Circle	Turns	Holes	No. of Divisions	Circle	Holes	No. of Divisions	Circle	Holes	No. of Divisions	Circle	Holes	No. of Divisions	Circle	Holes	No. of Divisions	Circle	Holes	No. of Divisions	Circle	Holes
2	any	20	....	29	58	1	22	56	28	20	96	24	10	152	38	10	224	28	5	340	34	4	510	51	4
3	24	13	8	30	24	1	8	57	57	40	98	49	20	155	62	16	228	57	10	344	43	5	520	39	3
4	any	10	....	31	62	1	18	58	58	40	100	25	10	156	39	10	230	46	8	360	54	6	528	66	5
5	any	8	....	32	28	1	7	59	59	40	102	51	20	160	28	7	232	58	10	368	46	5	530	53	4
6	24	6	16	33	66	1	14	60	42	28	104	39	15	164	41	10	235	47	8	370	37	4	540	54	4
7	28	5	20	34	34	1	6	62	62	40	105	42	16	165	56	16	236	59	10	376	47	5	560	28	2
8	any	5	....	35	28	1	4	64	24	15	106	53	20	168	42	10	240	66	11	380	38	4	570	57	4
9	54	4	24	36	54	1	6	65	39	24	108	54	20	170	34	8	245	49	8	390	39	4	580	58	4
10	any	4	....	37	37	1	3	66	66	40	110	66	24	172	43	10	248	62	10	392	49	5	590	59	4
11	66	3	42	38	38	1	2	68	34	20	112	28	10	176	66	15	250	25	4	400	30	3	600	30	2
12	24	3	8	39	39	1	1	70	28	16	114	57	20	180	54	12	255	51	8	408	51	5	620	62	4
13	39	3	3	40	any	1	....	72	54	30	115	46	16	184	46	10	260	39	6	410	41	4	660	66	4
14	49	2	42	41	41	....	40	74	37	20	116	58	20	185	37	8	264	66	10	420	42	4	680	34	2
15	24	2	16	42	42	....	40	75	30	16	118	59	20	188	47	10	270	54	8	424	53	5	720	54	3
16	24	2	12	43	43	....	40	76	38	20	120	66	22	190	38	8	272	34	5	430	43	5	740	37	2
17	34	2	12	44	66	....	60	78	39	20	124	62	20	192	24	5	280	28	4	432	54	5	760	38	2
18	54	2	12	45	54	....	48	80	34	17	125	25	8	195	39	8	290	58	8	440	66	6	780	39	2
19	38	2	4	46	46	....	40	82	41	20	130	39	12	196	49	10	296	37	5	456	57	5	820	41	2
20	any	2	....	47	47	....	40	84	42	20	132	66	20	200	30	6	300	30	4	460	46	4	840	42	2
21	42	1	38	48	24	....	20	85	34	16	135	54	16	204	51	10	304	38	5	464	58	5	860	43	2
22	66	1	54	49	49	....	40	86	43	20	136	34	10	205	41	8	310	62	8	470	47	4	880	66	3
23	46	1	34	50	25	....	20	88	66	30	140	28	8	210	42	8	312	39	5	472	59	5	920	46	2
24	24	1	16	51	51	....	40	90	54	24	144	54	15	212	53	10	320	24	3	480	24	2	940	47	2
25	25	1	15	52	39	....	30	92	46	20	145	58	16	215	43	8	328	41	5	490	49	4	960	24	1
26	39	1	21	53	53	....	40	94	47	20	148	37	10	216	54	10	330	66	8	496	62	5	980	49	2
27	54	1	6	54	54	....	40	95	38	16	150	30	8	220	66	12	336	42	5	500	25	2	1000	25	1
28	42	1	18	55	66	....	48																		

First Side: 24,   25,   28,   30,   34,   37,   38,   39,   41,   42, and 43.

Second Side: 46,   47,   49,   51,   53,   54,   57,   58,   59,   62, and 66.

Fig. 17-56   Each of the divisions of the circle is, of course, an angle.

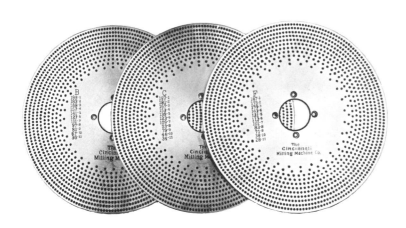

Fig. 17-57   The high-number index plates add 66 hole circles to those available on the standard plate.

(Cincinnati Milling Machine Co.)

Fig. 17-58   This universal index head has a range from 2 to 400,000 divisions and is appropriately known as a wide-range divider.

(Cincinnati Milling Machine Co.)

TERMINOLOGY

Protractor	A calibrated arc for use in angle measurement by the displacement method.
Universal bevel protractor	Precision protractor with a slidable reference surface and vernier, discrimination to fifteen minutes of arc.
Trigonometry	The branch of mathematics dealing with the relationships of sides of triangles to the angles formed by them.
Trigonometric functions	The ratios of pairs of triangle sides. Useful to designate angles.
Natural functions	Tables of trigonometric functions divided out into decimals.
Logarithmic functions	Tables of natural functions expressed as logarithms of the decimal values.
Logarithms	Generally used to mean "common logarithms." These are the exponents to which ten must be raised to equal the number. For example, the logarithm of 100 is 2 because $10^2 = 100$. The logarithm of 23 is 1.36173 because $10^{1.36173} = 23$.
Measuring rolls Measuring rods }	Precisely finished cylinders used in the measurement of tapers, etc.
Sine bar	A bar to which two identical cylinders are attached with a known separation and a known relation to the reference surface of the bar. The bar becomes the hypotenuse of a triangle for angle measurement.
Sine bar constants	Table of sine and angle relationships for angle settings with sine devices.
Sine blocks	Wide sine bars.
Sine plates	Still wider sine bars.
Sine tables	Sine devices incorporated into machines or mechanisms.
Compound angles	Angles formed by the edges of triangles that lie in different planes.
Face angles	The angles formed by the intersection of edges of a solid.
Dihedral angles	The openings between intersecting planes.
Mechanical indexing	A mechanical device that steps off intervals either linear or angular.
Dividing heads Indexing heads } Index heads	Simple mechanical devices for dividing the circle into spaces.
Dial index head	Index head consisting of an index plate integral with a rotating spindle and a means for stopping against the plate.
Index head	A plate with circles of equally spaced holes into which a stop pin may be inserted.
Plain index head	An index head in which the index plate rotation is connected to the spindle through worm gearing. Note that although the index plate establishes the rotation, it is usually stationary while the pin stop rotates about it.
Universal index head	An index head with differential gearing and generally with a trunion to tilt the axis of the spindle.
Wide-range divider	A means of compounding index plates for very high amplification.

Fig. 17-59   The terminology for angular measurement draws heavily on geometry and trigonometry.

Summary — From Minutes to Seconds

● Before going on it might be wise to sum up what we have covered about angles. Chapter 16 took up our grand master standard for angles, the circle; and introduced its useful secondary standards, the 90 and 180 degree angles. In Chapter 17 the angles in between have been considered, protractors being the most convenient.

● The universal bevel protractor is to angle measurement what the combination square is to linear measurement. It is adaptable to many diverse measurements and has a discrimination, or least count, of 5 minutes. Although simple in construction and use, it requires careful attention to the compliments and supplements of angles. The same considerations emphasized for reliable linear measurement apply to angular measurement: cleanliness, bias, temperature change, "cramping", and positional relationships.

● The next most frequently encountered angle measurement devices are the various sine instruments, of which the sine bar is the most common. These devices are based on a triangle of which the hypotenuse is constant. Changes in the side opposite the angle are converted into angles by means of tables. They can be used to set up an angle or to cancel out an angle. Thus they only provide standards and depend upon comparators for actual measurement. They are capable of discrimination in seconds but are affected by too many factors to be considered reliable finer than about 30 seconds. The huskier sine devices are known as sine plates and sine tables. Double sine plates with their hinges at right angles to each other form compound angles. The measure of a compound angle is its plane angle. All compound angles may be reduced to one or more of five types.

● An entirely different family of angle devices evolved from mechanically rotating spindles. One branch of this family achieves precision and accuracy by high amplification read-out devices, usually microscopes reading calibrated scales. The other branch of the family places little confidence in the ability of the observer but high reliance in the craftsmanship of the builder. It uses index plates with holes for positive stops with index pins. The lowest precision of these has no amplification, the index plate turning with the spindle. The higher precision ones have the index plate connected to the spindle through a worm gear so that many plate revolutions are equivalent to one spindle revolution. Still higher amplifications are obtained by differential gearing. With it the index plate is rotated along the worm but at a slightly different rate. The highest precision is achieved by compound index heads which use index plates to control the rotation of index plates. For circular division, an important form of angular measurement, dividing heads have few peers. Keep this in mind next time you need to divide a circle into a quarter of a million spaces or so.

● Although seconds threaded their way in and out of this and the previous chapter, a different class of devices is used to measure them when reliability is paramount. These are mentioned in Chapter 18. You now have at your command both linear and angular measurement with precision and accuracy enough for most practical measurement situations.

QUESTIONS

1. What is a universal bevel protractor?
2. What is the discrimination of a vernier protractor?
3. How many angles are read at each position of the vernier protractor?
4. The vernier scale extends both ways from zero. Which direction is it read?
5. When does a problem arise about which angles are represented by the reading?

6. What is the most important consideration for reliable measurement with the vernier protractor?
7. What are the two problems to be guarded against?
8. What positional precaution must be observed?
9. In which direction from zero is the vernier scale read?
10. What is the function of the acute angle attachment?
11. Which trigonometric function finds the most use in ordinary angular measurement?
12. How is this function defined?
13. What is the class of instrument called that utilizes this function?
14. What are the names of the most familiar forms of these instruments?
15. What is a sine bar?
16. What is the most common length for sine bars?
17. What is general reliable limit for sine instrument measurement?
18. How does a sine block differ from a sine bar? From a sine plate? From a sine table?
19. When should the setup be made for the complement of the angle?
20. How is the sine usually formed with these instruments?
21. What is a compound angle?
22. How many basic types of compound angles are there?
23. How is a compound angle usually formed in shop work?
24. What are the instruments called that divide the circle into angles mechanically?
25. What is the standard in the simplest mechanical indexing devices?
26. What is the limitation of this type of dividing?
27. What two functions do the mechanical indexing devices perform?
28. How is amplification achieved in mechanical indexing devices?
29. Basically what is the function of the index plate?
30. How does a universal index head differ from a plain index head?

DISCUSSION TOPICS

1. It was emphasized that a vernier protractor measures angles between its own parts, not angles on the part being checked. What was the significance of this?
2. What are principal positional considerations for the vernier protractor?
3. Discuss importance of triangles for angle measurement.
4. Explain the basic principle by which sine instruments are used for angular measurement.
5. Explain the relationship between angle and precision with a sine instrument.
6. Explain the fundamental difference between mechanical indexing and sine instruments, and its importance.

> *No one before the non-Euclidians perceived that arithmetic and geometry stand on quite different footings, the former being continuous with pure logic and independent of experience, the latter being continuous with physics and dependent upon physical data.*
>
> Bertrand Russell
> The Analysis of Matter

Role of Metrology
Do You Measure Up?

You now know the basic principles for the measurement of part features. They make rather dry reading, but come to life whenever applied to making things or studying physical phenomena.

Knowing these principles does not qualify you to do much more than poke criticism at the practical people who are actually using measurement: the machinists, toolmakers, inspectors, laboratory technicians and miscellaneous scientists. For the sophomore (of any age or position) few thrills compare with watching the Ph. D. use a micrometer like a C-clamp or the director of Q. C. heat up the gage blocks in his hot little fist. Their ability to gather data falls far short of their ability to use data. That is where you come in. The value of your service is directly proportional to the value of your data – accurate measurements. Before previewing the specific roles to which your measurements must be hitched to have salable value, let us first review what makes your measurements worth considering.

THE DISCIPLINE OF MEASUREMENT

Measurement is for communication. It utilizes completely arbitrary units and terms. And it is a social activity. Unless there is need to communicate there is no need for measurement. The only legitimacy that measurement units have, whether inches, meters or archines*, lies in their acceptance and usefulness. This acceptance requires the cooperation of other people. Hence, the social aspect.

Communication requires language. Thus we must first determine that all parties understand each other. Two terms are central to an understanding of measurement – *precision* and *accuracy*. Roughly, precision pertains to the degree of fineness while accuracy pertains to conformity with an accepted standard.

*One archine equals 28 inches in the U.S.S.R.

Communications are subject to many serious distortions. In measurement we know these as errors. For our data to be useful we must recognize the errors – the *random errors* that destroy precision, the *systematic errors* that destroy accuracy, and the *illegitimate errors* that destroy intent. We must be able to detect the various ways these errors creep into our measurements. Errors of calibration, parallax, bad technique, temperature and cramping are examples of systemic errors; fluctuating conditions, interpolation and definition cause random errors; and deliberate deception as well as inexcusable carelessness are the destroyers of intent.

A study of measurement errors is beneficial beyond the gathering of data for production or research, because we are bombarded with similar errors in all areas of life. As in measurement, some are easily recognized but others are so accepted that it takes a strong will to admit their possibility. The bald-faced lie is expected from the used car, encyclopedia or bible salesman; but how hard it is to question the clergyman, doctor or statesman – particularly on points with which we agree. Similarly we know that visible dirt, parallax or loose clamps will

Fig. 18-1 Whenever you make a measurement, its accuracy, or lack of it, may be traced to the International Bureau of Weights and Measures, Sèvres, France.

impair measurement, but it requires conscious effort to keep in mind the equally damaging effect of temperature fluctuations, cosine error and hysteresis. *Error always exists. The only questions are: How much? Where? To what effect? And, what can I do about it?*

The precision of measurements often may be checked by repeatability. The accuracy can only be checked by comparison with a higher standard. Because of the importance of the convertability and interchangeability of measurement data, the methods for traceability to the international standard have been formalized by governments and by trade groups, Fig. 18-1. The major considerations are summarized in Fig. 18-2.

PUTTING MEASUREMENT TO WORK

A course in strength of materials will not make you a bridge designer. But full grasp of the course may make you an exceptional bridge designer. So it is in measurement. The principles in this text are, at their best, generalizations. In order to apply them they must be related to specific measurement situations. These are as diversified as the needs of man.

It is rare that anyone can achieve experience in all of the special roles of measurement. This is most unfortunate. The resulting parochialism has held back progress. Science struggles along with antiquated reliance on vernier calipers while becoming superbly proficient with optical instruments. Industry, on the other hand, long ago obsoleted the vernier instruments for critical work, but largely ignores the possibilities of the optical instruments.

It is the function of Part II of this text to provide an overall view of the entire field, and to correct a serious omission in this first part. Among the basic principles of metrology is optical magnification. Because of the great importance of this subject, space required that it be advanced to Part II. For those who do not continue in a formal course of study, some idea of the scope of actively applied metrology will be given. There are four general areas of interest to which the basic principles in this text may be applied, Fig. 18-3.

DISCIPLINE OF MEASUREMENT

1. The accuracy of a measurement can never be as great as that of the standard.

2. The accuracy is diminished by errors.

3. Calibration, comparison to a higher standard, is the test for accuracy.

4. Repeatability is the usual test for precision.

5. Every measurement alters the object being measured and the measurement system.

6. The potential errors are reduced by:

   Elimination of separate measurement acts.
   Elimination of separate parts of the measurement system.
   Elimination of separate fluctuating conditions.
   Elimination of separate positional variables.

7. The least positional error results when the line of measurement, standard and axis of the comparison instrument are all in line.

Fig. 18-2 These considerations are of extreme importance and must be both understood and practiced.

SCOPE OF APPLIED METROLOGY

Group A. Applied Metrology

    1. Plate work — measurements performed on and from surface plates.
    2. Coordinate measurement — technique for use of x, y, z axes in measurement.
    3. Roundness — problems of defining, attaining and measuring.
    4. Thread measurement — geometrical aspects and formalized systems.
    5. Gear measurement — geometrical and functional considerations.
    6. Cam measurement — calculation, layout and verification of curves.

Group B. Rules and Regulations

    1. Tolerances and dimensioning — language of applied measurement.
    2. Attribute gaging — role and metrological considerations of industrial gaging.
    3. Inspection — administrative considerations correlated with metrology.
    4. Introduction to quality control — statistics applied to measurement data.
    5. Reliability — consideration of the ultimate consequences.
    6. Government compliance — statutory guides and considerations.

Group C. Optical Measurement

    1. Optical magnification — advantages and geometry of light in measurement.
    2. Projector measurement — the most familiar optical instrument, the contour projector.
    3. Micromeasurement — microscopes measure the minute.
    4. Macromeasurement — telescopes bring distant measurements up close.
    5. Autocollimation — parallel light simplifies alignment and angular measurement.
    6. Interferometry — phase differences in light provide accuracy at highest precision.

Group D. Derivative Metrology

    1. Angle measurement — use of angle gage blocks, optical polygons and auto-collimators.
    2. Reference metrology — precise investigation of surface flatness.
    3. Measuring machines — from toolmaker's microscopes to jig borers.
    4. Coordinate measuring machines — moire fringe and other methods for rapid layout and inspection.
    5. Large-scale metrology — tapes, chains, lines and levels for extremely large measurements.
    6. Optical metrology — optics extend large-scale metrology.
    7. Ultraprecise measurement — considerations of measurement at the metrology laboratory level.
    8. Thin films — special problems of measuring thicknesses of less than five mil. (0.005 in.)
    9. Dynamics metrology — measurement of continuously changing values.
    10. Surface texture and hardness measurement — measurement applied to metallurgical features.
    11. Magnetic particle inspection — detection of metallurgical flaws.
    12. Physical testing — measurement of physical values.

Fig. 18-3 Although measurement extends into nearly everything we do, in these areas it is so important that special techniques and terminologies have been developed.

APPLIED METROLOGY

The first one, applied metrology, will have immediate value for those going into industry. *Plate work* is the denouement of measurement principles where you will have your greatest opportunity to apply this text – and, alas, your greatest temptation for sinning. For adequate support or need of a reference plane, much measurement is performed on surface plates, Fig. 18-4. In recent years this has ushered in a new genus of instruments, the *height gage size units* and *transfer stands*. With these units the *zero transfer* technique provides speed plus reliability. Unfortunately it is little understood. The new instruments are often a step backward instead of forward.

Coordinate measurement is a technique for speeding up layout, machining and inspection by having all part features dimensioned from the three rectangular coordinates, x, y, and z. A bolt circle, for example, would have each hole located by two dimensions, one from the vertical and the other from the horizontal axis. No angles and radii would be shown. It is important to know both when and how to use this technique.

One of the most illusive things in measurement is *roundness*. When tolerances fall below *one mil* (0.001 in.) it becomes very difficult to make anything round and even more difficult to prove it. One geometric quirk called *lobing* actually results in parts that "mike" round but are actually far from it, Fig. 18-5. The study of roundness takes up *V-blocks, centers* and assorted sophisticated electronic instruments.

Fig. 18-4 Plate work refers to measurements taken from setups on surface plates. These are so useful that a new family of instruments and techniques has been developed for this purpose. *(Cadillac Gage Co.)*

The only aspects of measurement that come immediately to mind in which the inch system cannot be converted readily with the metric system are screw threads and gears. These involve some of the "hairiest" problems in measurement. The problems are usually solved "by rule and by rote". This is fine until an error is made. Unlike solving problems by deductive reasoning, the error does not call attention to itself and escapes unnoticed. In industry, at least, an understanding of the fundamentals of thread-and-gear measurement is invaluable.

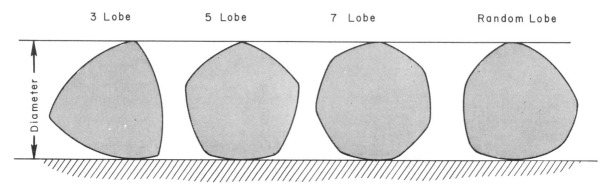

Fig. 18-5 All of these figures have the same measurement across any diameter, but are obviously, far from round.

Then comes *cam measurement*. You either need it or you don't. The chances are that any technician in any of the branches of mechanical engineering will. The constantly changing curves bring to mind haunting fears of the calculus. In actual practice, you will find short cuts that reliably reduce a cam to a series of measurable points.

RULES AND REGULATIONS

When things derive from natural phenomena we need little in the way of laws, statutes and agreements. No country has ever passed a law of gravitation. But when arbitrary inventions of man are involved, we surround them with bastions that would make the fortifications of Vauban look puny. Virtually the entire foundation of measurement is arbitrary — boy, do we have rules! There are international agreements, national standards, various other government standards such as the military standards for our defense (sic), trade association standards, commercial standards, plant standards and department standards.

It is generally agreed, outside of the military, that people are more efficient when they understand what they are going. Therefore, the basic language of measurement is expanded. New terms are needed to communicate about *dimensions* and *tolerances;* and shamanistic symbols are required to convey messages on drawings, Fig. 18-6.

When many parts must be made to close tolerances and be inspected by persons who would often prefer unemployment compensation, gages are substituted for measuring instruments. Gages speak the language of computers, yes and no. Therefore, great skill is required so that the fine line between these opposites will be distinct, and located where it will do the most good and the least harm.

Inspection recognizes the social element. It is the administration of measurement activity. Its role varies, as does its instrumentation; but certain principles are common among orderly run operations. If not strictly controlled, the chief product of the inspection function, data, will contain excessive error. The statisticians have all too often been in an ivory tower busily extracting conclusions to the 27th decimal place from data accurate to the third decimal place. The resulting ridicule has beclouded the tremendous benefits they can provide to any program in which sufficient data are involved. In industry this activity belongs to quality control. *Its results can be no better than your measurements.*

The question always exists, "How much measurement is required?" This cannot be answered without considering costs, production methods, and results. When the number of parts, the speed of operation, and the value of the results increase, the importance of inspection as well as its cost increases logarithmically. It is the function of the *reliability* people to balance results against the other factors.

GEOMETRIC CHARACTERISTIC SYMBOLS

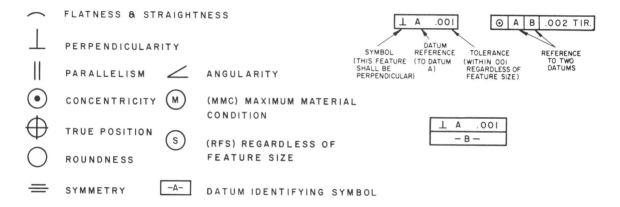

Fig. 18-6   These are but a few of the symbols used in dimensional metrology.

In recognition of the fact that each of these groups might get so wrapped up in its own problems that it overlooks the needs of the others, to say nothing of profits, we have the formal government requirements. These apply not only to companies dealing directly with the various government agencies, but are passed on down the scalar succession until they reach nearly everyone. Hence the importance of understanding *government compliance*.

OPTICAL MEASUREMENT

Optics were omitted from this first part for one reason only, lack of space. No other basic measurement method can boast equivalent versatility, range, precision and accuracy. The chief advantages are summarized in Fig. 18-7. Of the basic methods discussed nearly all have had their greatest development in industry and have later been adapted to science. The reverse is true for optics. In the United States, it has only been in relatively recent years that industry has discovered the great advantages of optical methods. Fortunately, we are moving rapidly to regain lost ground.

The ignorance of optical measurement is difficult to understand in view of the popularity of one isolated type of optical measurement, the *contour projector*, Fig. 18-8. This instrument is relied upon for the inspection of parts with difficult-to-measure features such as form tools. It embodies many of the features of the *toolmakers' microscope*, yet that mainstay of European metrology is still a rarity here, Fig. 18-9.

The microscopes and projectors are examples of *micrometrology*. They increase the apparent size of the part feature, or object as it is known. Because the standard with which this object is compared is not increased proportionally (if at all), the object not the errors is magnified. The opposite extreme is found in *macrometrology* in which the basic instrument is the telescope. It decreases the apparent distance between the viewer and the object. Theoretically these instruments are the same. In practical metrology they are quite different. The telescope, on one hand, becomes the cathetometer of science, Fig. 18-10; and on the other, becomes the alignment scope of *optical metrology*.

TEN REASONS

why measuring microscopes are often preferred over other types of measuring instruments.

1. Microscopes do not disturb or distort the object because there is no physical contact.
2. Microscopes can measure otherwise inaccessible dimensions.
3. With proper precautions, specimens at high temperature do not affect microscopes and are not affected by them.
4. An object can be inspected under magnification and measured simultaneously.
5. Angular, as well as linear measurements are possible.
6. Microscopes are easily set up and can be attached to other machines.
7. Comparatively low power illumination is required, thereby eliminating possible effects of heat on the object.
8. The image is recognizable, not a shadow on a ground glass.
9. Long life and comparatively low cost of microscopes makes depreciation small.
10. Microscopes are easy to understand and operate.

Fig. 18-7 Most of these advantages of measuring microscopes also apply to the other optical measuring instruments.

(Bausch & Lomb, Inc.)

Fig. 18-8 The contour projector is widely used throughout the metalworking industries.

Fig. 18-9 The toolmaker's microscope consists essentially of a high-powered microscope provided with a long travel calibrated mechanical stage and adequate illumination. *(Carl Zeiss, Inc.)*

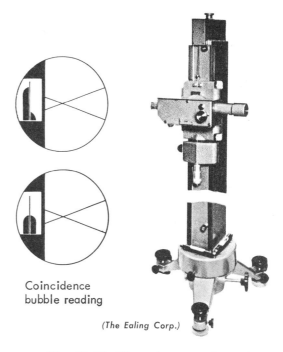

Coincidence
bubble reading

(The Ealing Corp.)

Fig. 18-10 The cathetometer is nearly unknown outside of scientific research, yet could be invaluable in industry.

Collimation means to render parallel, grammar notwithstanding. When this is done to light rays we gain a measurement tool of almost unbelievable sensitivity. In its most familiar form, the *autocollimator*, parallel rays from crosshairs are projected out of the instrument. A remote mirror reflects them back into the instrument. The difference in position between the outgoing and incoming crosshairs is proportional to the angle of the mirror. So great is the sensitivity that there are commercial instruments that will discriminate to 0.02 seconds of arc – sixty-five millionths of a circle.

Optical metrology reaches its ultimate precision in *interferometry*. These instruments were mentioned in Chapter 14. In that chapter it was shown that optical flats could be used to recombine a separated light path; and by means of fringe bands, show the difference in the distance each path traveled. Optical flats are very limited in application. Interferometers are those instruments that have been developed to extend this principle to many other measurement tasks. They rank among the most precise of all measuring instruments in use today.

DERIVATIVE METROLOGY

Measurement like everything else in life refuses to be divided into categories that can all be stored in neat little boxes. Some of life's greatest absurdities come from the general insistence that this be done. These range from little irritations (such as the index that under "water softeners" states "see softeners, water" when it would have taken far less space to go ahead and give the page reference) to fundamental errors, such as the belief that everyone who is not against our enemies must, therefore, be against us. To avoid such anomalies the author uses the term *derivative metrology*. It includes all of those instruments that require two or more of the basic principles; and the determination of attributes which are not properly dimensional measurements, but require dimensional measurements for their expression. Derivative or not, this includes some of the most fascinating and useful aspects of measurement.

Only one of the high-amplification methods for angular measurement has been touched upon, levels. They use the direct-reading method, and are limited to very short ranges.

Fig. 18-11   An optical polygon is being calibrated with an autocollimator. It uses the reversal process to actually calibrate against itself. J. C. Moody and J. M. Bunch, "Making Precision Measurements", Tool Engineer, Detroit, American Society of Tool and Manufacturing Engineers, February, 1958, p. 6.

Angular measurement with *angle gage blocks* and *optical polygons* employs the principle of comparison measurement, and provides 360-degree range. They use autocollimators in the role that you now know comparator or indicators. Best of all, highly precise measurement of angles permit us to use the often mentioned *reversal process* at its best. It works so well in fact that we appear to lift ourselves by our bootstraps. It is actually possible to use a part feature as the standard by which it is itself calibrated. Impossible? Not at all, but there are some "ifs" and "buts" that you will learn when you get to precision angular measurement, Fig. 18-11.

Another important aspect of measurement that also relies on the autocollimator as a comparator is *reference metrology*. This is the study of reference surfaces and is particularly involved in *surface plate calibration*. An electronic computer comes in handy in this field, but there are tricks of the trade which should be known to everyone involved with surface plate calibration.

Measuring machines, Fig. 18-12, are the aristocrats of mechanical measurement. They may incorporate any combination of the basic measurement principles. They usually rate a room of their own — a very fancy one at that with special lighting, ventilation, and

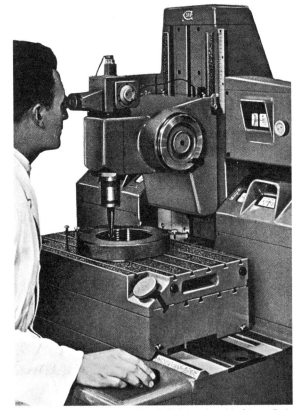

(Marshall & Huschart Machinery Co.)

Fig. 18-12   Measuring machines are available in many ranges and styles. All represent modern technology at its highest level.

temperature control. All too often they are bought as status symbols, "taken off the top" profitwise and intended to dazzle the not-too-bright officials who award important contracts on the basis of facilities lists. Properly applied they represent the greatest integrity in applied metrology. Some future archeologist, upon finding one in the atomic rubble, may wonder what could have gone wrong with a society that produced such a wonder. Far more responsibility lies in the hands of the operator of a measuring machine than in the most prestigious justices of the peace.

Coordinate measuring machine, Fig. 18-13, sounds like the previous instrument. If you want to be literal about it, the previous one answers this description. In general use, however, this is a group of machines that are noted for their great speed and convenience. They are used at a highly precise level but not at the metrology laboratory level of the previous instruments. They automate the coordinate measurement technique mentioned earlier.

The most popular coordinate measurement machines today use *moire fringes*. At one time or another you have probably noticed that when two screens are looked through, a curious pattern of light and dark is formed. These are known as *moire fringes*. They are a relatively new, and highly practical means to achieve magnification in measurement by optical means.

Large-scale metrology is the branch of measurement that treats the very long measurements involved in the construction, erection, inspection and maintenance of very large equipment. This includes paper mills, ship building, radar installations, power generating plants, airframe and missile construction, and a multitude of uses in science such as nuclear accelerators.

This work formerly relied heavily on tightly stretched strings, tapes, chains, levels (especially the water level), and surveying instruments. It was from the surveying instruments that the specialized instruments for large-scale metrology developed. These

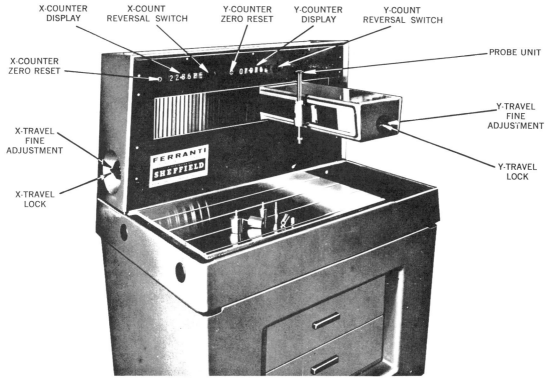

(Sheffield Ferranti — Coordinate Measuring Machines pg. 3)

Fig. 18-13 This coordinate measuring machine electronically counts moire fringes for linear measurement. The large digital read-out eliminates all interpolation errors.

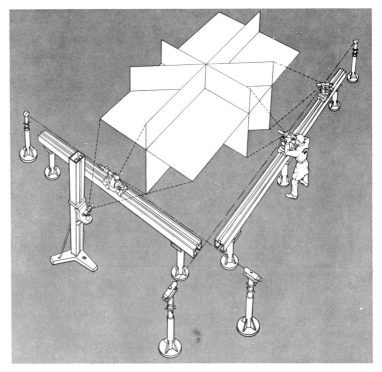

Fig. 18-14 Optical metrology permits both linear and angular measurements of very large devices — and without physical contact. *(Keuffel & Esser Co.)*

are optical instruments and their use is known as *optical metrology*, Fig. 18-14.

Bearing very little resemblance to this are the special skills and facilities required for *ultraprecise measurement*. How precise is ultraprecise? That depends largely upon the length. Any measurement to *ten mike* (0.000010 in.) or less is ultraprecise. When the part length is over 5 or 6 or 7 inches (there are no absolutes in this rarefied atmosphere), *50 mike* (0.000050 in.) is ultraprecise. The would-be sophisticates consider this hilarious. They are quick to tell you how they measure to less than that everyday on their "left-hand astridulator" line. True they do — but you and I know that *measuring to 50 millionths precision and to 50 millionths accuracy are two different things*. Ultraprecise measurement requires a lot more than "white rooms", Fig. 18-15, and expensive measuring machines.

Thin films sound like "much ado about nothing". They are mighty close to nothing, but the little that is left is important. That is why we must measure them both precisely and accurately. The measurement of these films is important throughout both industry and science. Paint is a good practical example. Any physics student knows the complications that the thin film of a soap bubble can provide.

Suppose that you have carefully developed the fastest aluminum-foil mill, what would your greatest concern be? Selling the stuff, obviously. After that, however, you would begin to think about quality control. If it is too thin, it will not handle as it appears on the TV commercials. If too thick — boy, are you losing money! You cannot stop the mill to measure the in-process sheets and after they come off the rolls, it is too late. To achieve the best product at the lowest price you must measure the thickness en route. That requires measuring a continuously changing variable. It is called *dynamic metrology*, and is encountered in most process industries.

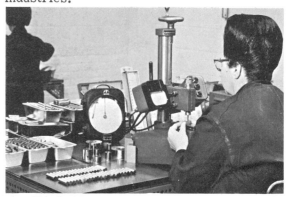

Fig. 18-15 Very precise measurement, such as the calibration of gages, requires "controlled environment" rooms in which temperature and other variables are kept very nearly constant.

Surface texture (we used to call it simply "surface finish") and hardness both are expressed in dimensional terms even though they more or less apply to physical or metallurgical attributes. There is very little one can do in industry without meeting both of these attributes. It is helpful to know what the measurements really mean.

Magnetic particle inspection is of a similar nature. It is an inspection that discloses metallurgical defects, but these are expressed in dimensional terms. Any of you who kick up the dust on the dirt tracks know that this is one area in which your life may literally depend upon your accuracy.

Physical testing may be somewhat more remote than detecting a flaw in the front axle before the main Sunday afternoon "heat" but the principle is the same. Somewhere along the line, someone must determine that ma-terials will behave according to the theories of the "slip-stick" boys. This is done in the physical test laboratory, and again the raw data are largely dimensional measurements. Their value depends upon the measurer.

This by no means completes the activities in which you might encounter dimensional metrology. It does include those most frequently met. In each of them your success will depend mostly upon luck. But in a dynamic economy which has a severe shortage of technicians who get a charge out of the solution of a problem, luck is all on your side. If measurement bores you stiff, forget all about becoming an inspector, a technician, an engineer or a scientist. Try dentistry, if your folks have the resources, it pays better. But if you are genuinely interested in the standard of living or knowledge of the universe let your measurements be as accurate as they are precise.

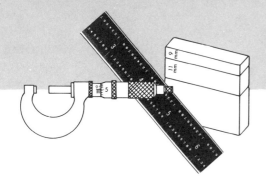

Index

INDEX INSTRUCTIONS

The nine most frequently sought topics are listed first. If, for example, you are looking for the basic considerations of measurement, you will find them grouped under *Principles*. An asterisk (*) after any item means that you will find additional related items in the General Index.

Page references are listed in *order of importance,* not numerically as in most indices. They may refer to either text or illustrations, usually both.

The items under *Reminders* in the Topics Index are boxes or tabular material that condense sections of the text. They are intended for quick reference after you are familiar with the complete text.

TOPICS INDEX

GENERAL INDEX

A

7/85 (4C1644F)